With our b

Adrian Bejan

Sylvie Lorente.

Design with Constructal Theory

Design with Constructal Theory

Adrian Bejan
Duke University, Durham, North Carolina

Sylvie Lorente
University of Toulouse, INSA, LMDC Toulouse, France

John Wiley & Sons, Inc.

This book is printed on acid-free paper. ♾

Published by John Wiley & Sons, Inc., Hoboken, New Jersey
Published simultaneously in Canada

Library of Congress Cataloging-in-Publication Data

Bejan, Adrian.
Design with constructal theory / Adrian Bejan, Sylvie Lorente.
p. cm.
Includes index.
ISBN 978-0-471-99816-7 (cloth)
1. Design, Industrial. I. Lorente, Sylvie. II. Title.
TS171.4.B43 2008
745.2–dc22

2008003739

Printed in the United States of America

10 9 8 7 6 5 4 3 2 1

Contents

About the Authors

Adrian Bejan received all of his degrees (BS, 1971; MS, 1972; PhD, 1975) in mechanical engineering from the Massachusetts Institute of Technology. He has held the J. A. Jones distinguished professorship at Duke University since 1989. His work covers thermodynamics, convective heat transfer, porous media, and constructal theory of design in nature. He developed the methods of entropy generation minimization, scale analysis, heatlines and masslines, intersection of asymptotes, and the constructal law. Professor Bejan is ranked among the 100 most cited authors in engineering, all disciplines all countries. He received the Max Jakob Memorial Award and 15 honorary doctorates from universities in ten countries.

Sylvie Lorente received all her degrees in civil engineering (BS, 1992; MS, 1992; PhD, 1996) from the National Institute of Applied Sciences (INSA), Toulouse. She is Professor of Civil Engineering at the University of Toulouse, INSA, and is affiliated with the Laboratory of Durability and Construction of Materials, LMDC. Her work covers several fields, including heat transfer in building structures, fluid mechanics, and transport mechanisms in cement based materials. She is the author of 70 peer-referred articles and three books. Sylvie Lorente received the 2004 Edward F. Obert Award and the 2005 Bergles-Rohsenow Young Investigator Award in Heat Transfer from the American Society of Mechanical Engineers, and the 2007 James P. Hartnett Award from the International Center of Heat and Mass Transfer.

Constructal theory advances are posted at www.constructal.org.

Preface

This book is the new design course that we have developed on several campuses during the past five years. The approach is new because it is based on constructal theory—the view that flow configuration (geometry, design) can be reasoned on the basis of a principle of configuration generation and evolution in time toward greater global flow access in systems that are free to morph. The generation of flow configuration is viewed as a physics phenomenon, and the principle that sums up its universal occurrence in nature (the *constructal law*, p. 2) is deterministic.

Constructal theory provides a broad coverage of "designedness" everywhere, from engineering to geophysics and biology. To see the generality of the method, consider the following metaphor, which we use in the introductory segment of the course. Imagine the formation of a river drainage basin, which has the function of providing flow access from an area (the plain) to one point (the river mouth). The constructal law calls for configurations with successively smaller global flow resistances in time. The invocation of this law leads to a balancing of all the internal flow resistance, from the seepage along the hill slopes to the flow along all the channels. Resistances (imperfection) cannot be eliminated. They can be matched neighbor to neighbor, and distributed so that their global effect is minimal, and the whole basin is the least imperfect that it can be. The river basin morphs and tends toward an equilibrium flow-access configuration.

The visible and valuable product of this way of thinking is the *configuration*: the river basin, the lung, the tree of cooling channels in an electronics package, and so on. The configuration is the big unknown in design: the constructal law draws attention to it as the unknown and guides our thoughts in the direction of discovering it.

In the river basin example, the configuration that the constructal law uncovers is a tree-shaped flow, with balances between highly dissimilar flow resistances such as seepage (Darcy flow) and river channel flow. The tree-shaped flow is the theoretical way of providing effective flow access between one point (source, sink) and an infinity of points (area, volume). The tree is a complex flow structure, which has multiple-length scales that are distributed nonuniformly over the available area or volume.

All these features, the tree shape and the multiple scales, are found in any other flow system whose purpose is to provide access between one point and an area or volume. Think of the trees of electronics, vascularized tissues, and city traffic, and you will get a sense of the universality of the principle that was used to generate and to discover the tree configuration.

Vascularized is a good name for the complex energy systems that the new thermal sciences is covering. Vascularized is everything, the animate, the inanimate, and the engineered, from the muscle and the river basin to the cooling of high-density electronics. The tissues of energy flows, like the fabric of society and all the tissues of biology, are designed (patterned, purposeful) architectures. The climbing to this high level of performance is the transdisciplinary effort: the balance between seemingly unrelated flows, territories, and disciplines. This balancing act—the optimal distribution of imperfection—generates the very design of the process, power plant, city, geography, and economics.

Trees are not the only class of configurations that result from invoking the constructal law. Straight tubes with round cross-sections are discovered when one favors the access for fluid flow between two points. Round tubes are found in many natural and engineered flow systems (blood vessels, subterranean rivers, volcanic shafts, piping, etc.). Optimal spacings between solid components are discovered by invoking the constructal law. Examples are the spacings between fins in heat exchangers and the spacings between heat-generating electronics in a package. Optimal intermittence (rhythm) is discovered in the same way, and, once again, the examples unite nature with engineering, from human respiration (in- and outflow), to the periodic shutdown and cleaning of heat exchangers in power plants.

Loops and grid-shaped flow patterns are useful because they add resilience and robustness to the tree-shaped flow configurations that they serve. Robustness and redundancy are precious properties in design, and our course teaches how to endow designs with such properties.

In summary, this design course provides the student with strategy for how to pursue and discover design (configuration, pattern) in both space and time. Constructal theory pushes design thinking closer to science and away from art. It tears down the walls between engineering and natural sciences. Because the design (the configuration generation phenomenon) has scientific principles that are now becoming known, it is possible to learn where to expect *opportunities* for discovering new configurations that are stepwise more effective. How to pursue the discovery with less effort and time (i.e., with strategy) is another merit of learning design generation as a scientific subject.

At the end of the day, this new *design as science* paradigm makes a solid contribution to physics, to predicting nature. The drawings made in this book are qualitatively the same as those of natural porous and vascularized materials. Most valuable are the similarities that emerge between natural structures and the ones derived here based on principle. They shed light on the natural tendency that *generates* multiple scales, hierarchy, complexity, and heterogeneity in flow systems such as hill slope drainage, forests, and living tissues. The fact that natural flow structures—the champions of flow perfection—have features similar to those discovered in constructal design lends confidence in the pursuit of better engineering design with constructal theory.

* * *

This book and solutions manual are based on an original fourth-year undergraduate and first-year graduate design course developed by the two of us at Duke University—course ME166 Constructal Theory and Design. We also taught constructal theory and design in short-course format at the University of Évora, Portugal; University of Lausanne, Switzerland; Yildiz University, Turkey; Memorial University, Canada; Shanghai Jiaotong University, People's Republic of China; and the University of Pretoria, South Africa.

We thank the students, who stoked the fire of our inquiry with questions and new ideas. In particular, we acknowledge the graphic contributions of our doctoral students: Sunwoo Kim, Kuan-Min Wang, Jaedal Lee, Yong Sung Kim, Luiz Rocha, Tunde Bello-Ochende, Wishsanuruk Wechsatol, Louis Gosselin, and Alexandre da Silva.

Our deepest gratitude goes to Deborah Fraze, who put the whole book together in spite of the meanness of the times.

During the writing of this book we benefited from research support for constructal theory from the Air Force Office of Scientific Research and the National Science Foundation. We thank Drs. Victor Giurgiutiu, Les Lee, and Hugh Delong of AFOSR; Drs. Rita Teutonico and Sandra Schneider of NSF; Dr. David Moorhouse of the Air Force Research Laboratory; and Professor Scott White and his colleagues at the University of Illinois.

Constructal theory and vascularization is a new paradigm and a worldwide activity that continues to grow (see www.constructal.org). We thank our friends and partners in the questioning of authority, in particular Heitor Reis, Antonio Miguel, Houlei Zhang, Stephen Périn, Gil Merkx, Ed Tiryakian, and Ken Land.

Adrian Bejan
Durham, North Carolina
Sylvie Lorente
Toulouse, France
January 2008

List of Symbols

a	acceleration, m/s^2
a, b	aspect ratios, Eq. (10.118)
a, b	lengths, m
A	area, m^2
AR	aspect ratio, Table 1.1
b	length, m
B	dimensionless group, Eq. (9.26)
B_1, B_2	global thermal resistances, Eqs. (5.22) and (5.50)
$\dot{B}$	brake power, W, Eq. (2.50)
Be	Bejan number, Eqs. (3.35) and (8.35)
Bi	Biot number, Eq. (1.45)
c	concentration of one species per unit volume of solution, mol/m^3
c	specific heat, J/kg · K
c_P	specific heat at constant pressure, J/kg · K
c, C	constants
C	pump work requirement, J, Eq. (1.33)
C	thermal conductance, Eq. (6.16)
C_D	drag coefficient, Eq. (1.37)
C_f	skin friction coefficient
COP	coefficient of performance, Eqs. (2.18)–(2.19)
d	depth, smallest dimension, m
D	diameter, spacing, m
D	effective diffusion coefficient, m^2/s
D_h	hydraulic diameter, m, Eq. (1.21)
E	energy, J
E	modulus of elasticity, Pa
f	flow resistance, dimensionless, Eq. (4.57)
f	friction factor, Eq. (1.16)
f	ratio, Eq. (6.59)
f, F	function
f_c'	strength of concrete
f_s	strength of steel
F	Faraday constant, 9 648 J/V mol
F	force, N
g	gravitational acceleration, m/s^2
h	heat transfer coefficient, W/mK, Eq. (1.56)
h	specific enthalpy, J/kg

h_{sf}	latent heat of melting, J/kg
h, H, H_m	height, m
I	area moment of inertia, m^4
I	current, A
I	integral
j	current density, A/m^2
J	diffusive flux, mol/m^2s
k	thermal conductivity, $W/m \cdot K$
k_s	roughness height, m
K	local-loss coefficient, Eq. (1.31)
K	permeability, m^2
l	length, m
l	mean free path, m
L	length, thickness, m
m, M	mass, kg
m	number
$\dot{m}$	mass flow rate, kg/s
M	dimensionless mass flow rate, Eqs. (7.14) and (7.38)
M	moment, Nm
n, N	number
N	number of heat loss units, Eq. (7.100)
Nu	Nusselt number, Eq. (1.60)
p	number of pairing (or bifurcation) levels
p	porosity
p	wetted perimeter, m
P	force, N
P	pressure, Pa
$\hat{P}$	dimensionless pressure drop, Eq. (7.49); see also Be, Eq. (3.35)
Po	Poiseuille constant, Eq. (1.23)
Pr	Prandtl number, Eq. (1.60)
q	heat current, W
q', Q'	heat current per unit length W/m
q''	heat flux, W/m^2
q'''	volumetric heat generation rate, W/m^3
Q	heat transfer, J
Q	volumetric fluid flow rate, m^3/s
Q'	heat source per unit length, J/m
Q''	heat source per unit area, J/m^2
$\dot{Q}$	heat transfer rate, W
r	radial position, m
r	ratio
r_0	pipe radius, m
R	radial distance, radius, m

R	ideal gas constant, J/kg · K
R	resistance
R	universal gas constant, 8.314 J/K mol
Ra_y	Rayleigh number based on y, Eq. (1.76)
Re_D	Reynolds number based on D, Eq. (1.14)
R_t	thermal resistance, K/W, Eq. (1.40)
s	specific entropy, J/kg · K
s	stress, Pa
S	entropy, J/K
S	spacing, m
S	sum
S	surface, m
Sc	Schmidt number, ν/D
S_{gen}	entropy generation, J/K
St	Stanton number, Eq. (1.70)
Sv	Svelteness number, Eq. (1.1)
t	thickness, m
t	time, s
T	temperature, K
u, v	velocity components, m/s
U	average longitudinal velocity, m/s
U	overall heat transfer coefficient, $\mathrm{W/m^2K}$
U	potential, V
v	specific volume, $\mathrm{m^3/kg}$
V	velocity, m/s
V	volume, $\mathrm{m^3}$
W	width, m
W	work, J
$\dot{W}$	power, W
$\dot{W}'$	power per unit length, W/m
x, y, z	Cartesian coordinates, m
X	flow entrance length, m
X_T	thermal entrance length, m
z	charge number
Z	thickness, m

Greek Letters

α, β	angles, rad
α	thermal diffusivity, $\mathrm{m^2/s}$
β	coefficient of volumetric thermal expansion, $\mathrm{K^{-1}}$
γ	ratio, Eq. (8.41)
δ	deflection, m

δ thickness, m
ΔP pressure difference, Pa
ΔT temperature difference, K
ε effectiveness, Eq. (7.75)
ε small quantity
η fin efficiency, Eq. (1.44)
η_{I} first-law efficiency, Eq. (2.15)
η_{II} second-law efficiency, Eq. (2.16)
θ angle, rad
θ dimensionless temperature difference, Eq. (7.123)
θ temperature difference, K
λ critical length scale, m
λ Lagrange multiplier
λ thickness, m
μ viscosity, kg/s m
ν kinematic viscosity, m^2/s
ρ radius of curvature, m
ρ density, kg/m^3
α stress, Pa
τ shear stress, Pa
ξ aspect ratio, Eq. (4.44)
ξ pressure loss, Eq. (1.32)
ϕ volume fraction, porosity; see also p
φ electrical potential, V
ψ dimensionless global flow resistance, Eq. (8.51)

Subscripts

a air
avg average
b base
b body
b brick
b bulk
B branch
c canopy
c central
c channels
C compressor
C conduction
D diffuser, drag
E east

exp	exposed
f	fluid, frontal
FC	forced convection
g	ground
H	high
i	inner, species, rank
in	inlet
L	low
lm	log-mean
m	maximum
m	mean
m	melting
m	minimized
ma	maximum allowable
mm	minimized twice
mmm	minimized three times
max	maximum
min	minimum
N	north
N	nozzle
NC	natural convection
o	optimized
oo	optimized twice
ooo	optimized three times
o	outer
opt	optimum
out	outlet
p	path
p	pipes
p	pump
r	radial
ref	reference
rev	reversible
s	sector, solid, steel
S	south
t	thermal
t	trunk
T	turbine
W	west
w	wall
z	longitudinal
$\sum$	total, summed
∞	free stream, far field

Superscripts

b	bulk
n	nano-size
$(\ )^*$	optimized
$(\bar{\ })$	averaged
$(\tilde{\ })$, $(\hat{\ })$	dimensionless
$(\ )'$	per unit length
$(\ )''$	per unit area
$(\ )'''$	per unit volume
$(\dot{\ })$	rate, per unit time
P	power plant
R	refrigeration plant

1

FLOW SYSTEMS

1.1 CONSTRUCTAL LAW, VASCULARIZATION, AND SVELTENESS

A flow represents the movement of one entity relative to another (the background). To describe a flow, we speak of *what* the flow carries (fluid, heat, or mass), *how much* it carries (mass flow rate, heat current, etc.), and *where* the stream is located. The *where* is the focus of this new course. A flow system has configuration, drawing, that is, *design*.

In science, the origin (genesis) of the configuration of flow systems has been overlooked. Design has been taken for granted—at best, it has been attributed to chance, inspiration, talent, and art. Our own education in the sciences is based on sketches of streams into and out of boxes, sketches that bear no relation to reality, to the position that the stream occupies in space and in time. This book is our attempt to change this attitude.

The benefits from thinking of *design as science* are great. The march toward smaller dimensions (micro, nano)—the miniaturization revolution—is not about making smaller and smaller components that are to be dumped like sand into a sack. This revolution is about the "living sack," in which every single component is kept alive with flows that connect it to all the other components. Each component is put in the right place, like the neurons in the brain, or the alveoli in the lung. It is the configuration of these extremely numerous components that makes the sack perform at impressively high levels.

Because natural flow systems have configuration, in this book we treat the emergence of flow configuration as a physics *phenomenon* that is based on a scientific principle. Constructal theory is the mental viewing that the generation of the flow structures that we see everywhere in nature (river basins, lungs, atmospheric circulation, vascularized tissues, etc.) can be reasoned based on an evolutionary principle of increase of flow access in time, i.e. the time arrow of the animated movie of successive configurations. That principle is the *constructal law* [1–4]:

> For a finite-size flow system to persist in time (to live), its configuration must change in time such that it provides easier and easier access to its currents (fluid, energy, species, etc.).

Geometry or drawing is not a figure that always existed and now is available to look at, or worse, to look through and take for granted. The figure is the persistent movement, struggle, contortion, and mechanism by which the flow system achieves global objective under global constraints. When the flow stops, the figure becomes the flow fossil (e.g., dry river bed, snowflake, animal skeleton, abandoned technology, and pyramids of Egypt).

What is the flow system, and what flows through it? These are the questions that must be formulated and answered at the start of every search for architectures that provide progressively greater access to their currents. In this book, we illustrate this thinking as a *design method*, mainly with examples from engineering. The method, however, is universally applicable and has been used in a *predictive* sense to predict and explain many features of *design in nature* [1–4, 10].

Constructal theory has brought many researchers and educators together, on several campuses (Duke; Toulouse; Lausanne; Évora, Portugal; Istanbul; St. John's, Newfoundland; Pretoria; Shanghai) and in a new direction: to use the constructal law for better engineering and for better organization of the movement and connecting of people, goods, and information [2–4]. We call this direction *constructal design*, and with it we seek not only better configurations but also better (faster, cheaper, direct, reliable) *strategies* for generating the geometry that is missing.

For example, the best configurations that connect one component with very many components are tree-shaped, and for this reason *dendritic* flow architectures occupy a central position in this book. Trees are flows that make connections between points and continua, that is, infinities of points, namely, between a volume and one point, an area and one point, and a curve and one point. The flow may proceed in either direction, for example, volume-to-point and point-to-volume.

Trees are not the only class of multiscale designs to be discovered and used. We also teach how to develop multiscale spacings that are distributed nonuniformly through a flow package, flow structures with more than one objective, and, especially, structures that must perform both flow and mechanical support functions. Along this route, we unveil designs that have more and more in common with animal design. We do all this by invoking a single principle (the constructal law), not by copying from nature.

With "animal design" as an icon of ideality in nature, the better name for the miniaturization trends that we see emerging is *vascularization*. Every multiscale solid structure that is to be cooled, heated, or serviced by our fluid streams must be and will be vascularized. This means trees and spacings and solid walls, with every geometric detail sized and positioned in the right place in the available space. These will be solid-fluid structures with multiple scales that are distributed nonuniformly through the volume—so nonuniformly that the "design" may be mistaken as random (chance) by those who do not quite grasp the generating principle, just like in the prevailing view of animal design, where diversity is mistaken for randomness, when in fact it is the fingerprint of the constructal law [1–4].

We see two reasons why the future of engineering belongs to the vascularized. The first is geometric. Our "hands" (streams, inlets, outlets) are few, but they must reach the infinity of points of the volume of material that serves us (the devices, the artifacts, i.e. the engineered extensions of the human body). Point-volume and point-area flows call for the use of tree-shaped configurations. The second reason is that the time to do such work is now. To design highly complex architectures one needs strategy (theory) and computational power, which now we possess.

The comparison with the vascularization of animal tissue (or urban design, at larger scales) is another way to see that the design philosophy of this book is the philosophy of the future. Our machines are moving toward animal-design configurations: distributed power generation on the landscape and on vehicles, distributed drives, distributed refrigeration, distributed computing, and so on. All these distributed schemes mean trees mating with trees, that is, vascularization.

A flow system (or "nonequilibrium system" in thermodynamics; see Chapter 2) has new *properties* that are complementary to those recognized in thermodynamics until now. A flow system has *configuration* (layout, drawing, architecture), which is characterized by *external size* (e.g., external length scale L), and *internal size* (e.g., total volume of ducts V, or internal length scale $V^{1/3}$). This means that a flow system has *svelteness*, Sv, which is the global geometric property defined as [5]

$$\text{Sv} = \frac{\text{external flow length scale}}{\text{internal flow length scale}} \tag{1.1}$$

This novel concept is important because it is a property of the global flow architecture, not flow kinematics and dynamics. In duct flow, this property describes the relative importance of friction pressure losses distributed along the ducts and local pressure losses concentrated at junctions, bends, contractions, and expansions. It describes the "thinness" of all the lines of the drawing of the flow architecture (cf. Fig. 1.1).

To illustrate the use of the concept of svelteness, consider the flow through two co-linear pipes with different diameters, $D_1 < D_2$ (Fig. 1.2). The pipe lengths are L_1 and L_2. The sudden enlargement of the flow cross-section leads to recirculation and dissipation (imperfection, Chapter 2) immediately downstream of the expansion. This effect is measured as a local pressure drop, which in Example 1.1 (p. 16) is derived by invoking the momentum theorem:

$$\Delta P_{local} = \left[1 - \left(\frac{D_1}{D_2}\right)^2\right]^2 \frac{1}{2}\rho V_1^2 \tag{1.2}$$

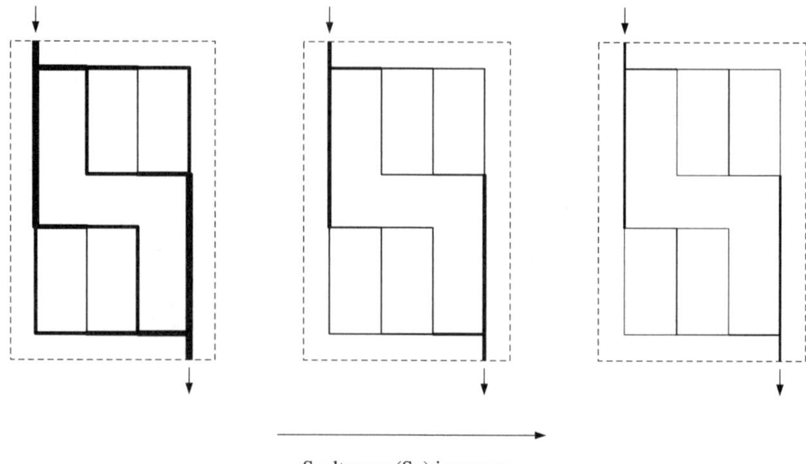

Figure 1.1 The svelteness property Sv of a complex flow architecture: Sv increases from left to right, the line thicknesses decrease, and the drawing becomes sharper and lighter, that is more svelte. The drawing does not change, but its "weight" changes.

Equation (1.2) is known as the Borda formula. In the calculation of total pressure losses in a complex flow network, it is often convenient to neglect the local pressure losses. But is it *correct* to neglect the local losses?

The calculation of the svelteness of the network helps answer this question. The svelteness of the flow geometry of Fig. 1.2 is

$$\text{Sv} = \frac{L_1 + L_2}{V^{1/3}} \tag{1.3}$$

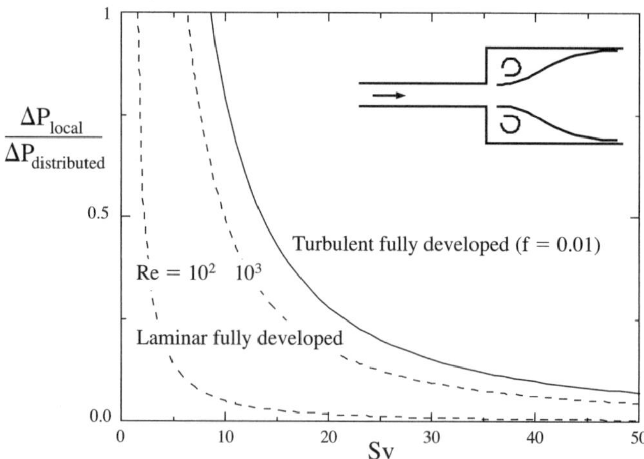

Figure 1.2 The effect of svelteness (Sv) on the importance of local pressure losses relative to distributed friction losses in a pipe with sudden enlargement in cross-section (cf. Example 1.1).

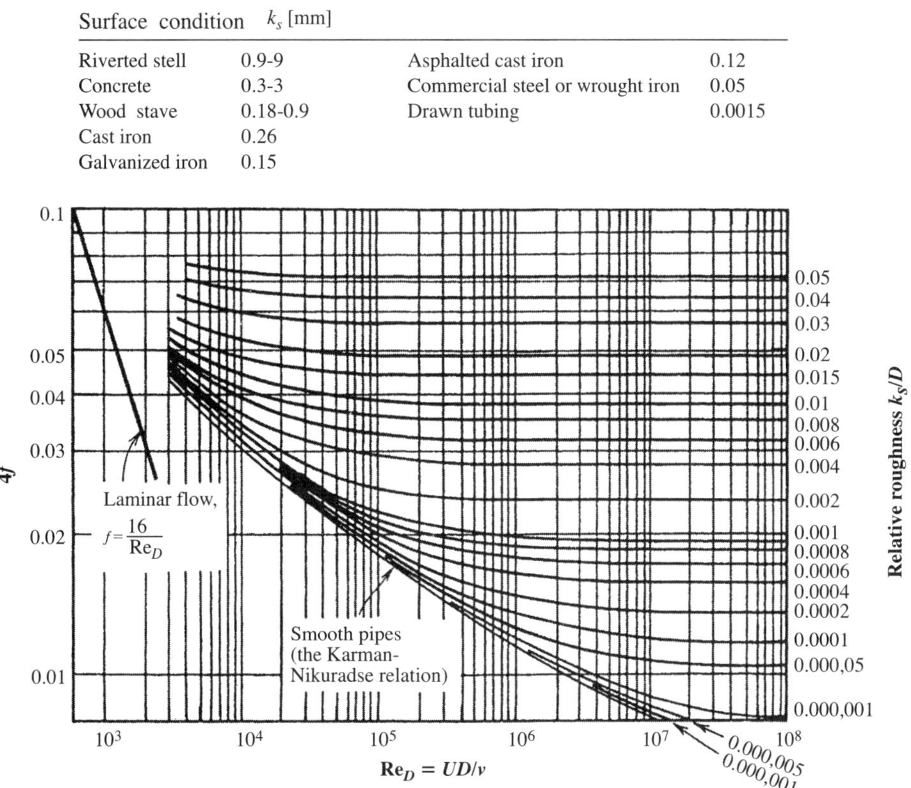

Surface condition	k_s [mm]		
Riverted stell	0.9-9	Asphalted cast iron	0.12
Concrete	0.3-3	Commercial steel or wrought iron	0.05
Wood stave	0.18-0.9	Drawn tubing	0.0015
Cast iron	0.26		
Galvanized iron	0.15		

Figure 1.3 The Moody chart for friction factor for duct flow [6].

where V is the total flow volume, namely $V = (\pi/4)(D_1^2 L_1 + D_2^2 L_2)$. The distributed friction losses ($\Delta P_{\text{distributed}}$) are associated with fully developed (laminar or turbulent) flow along L_1 and L_2, and have the form given later in Eq. (1.22), for which the friction factors are furnished by the Moody chart (Fig. 1.3).

The derivation of the curves plotted in Fig. 1.2 is detailed in Example 1.1. The ratio $\Delta P_{\text{local}}/\Delta P_{\text{distributed}}$ decreases sharply as Sv increases. When Sv exceeds the order of 10, local losses become negligible in comparison with distributed losses. We retain from this simple example that Sv is a global property of the flow space "inventory." This property guides the engineer in the evaluation of the performance of the flow design.

A flow system is also characterized by *performance* (function, objective, direction of morphing). Unlike in the black boxes of classical thermodynamics, a flow system has a *drawing*. The drawing is not only the configuration (the collection of black lines, all in their right places on the white background), but also the thicknesses of the black lines. The latter gives each line its slenderness, and, in ensemble, the slendernesses of all the lines account for the svelteness of the drawing.

The constructal design method guides the designer (in time) toward flow architectures that have greater and greater *global performance* for the *specified flow access conditions* (fluid flow, heat flow, flow of stresses). The architecture discovered in this manner for the set of conditions "1" is the "constructal configuration 1". For another set of conditions, called "2", the method guides the designer to the "constructal configuration 2". In other words, a configuration developed for one set of conditions is not necessarily the recommended configuration for another set of conditions. The constructal configuration "1" is not universal—it is not the solution to other design problems. Universal is the constructal law, not one of its designs.

For example, we will learn in Fig. 4.1 that the best way to size the diameters of the tubes that make a Y-shaped junction with laminar flow is such that the ratio of the mother/daughter tube diameters is $2^{1/3}$. The geometric result is good for many flow architectures that resemble Fig. 4.1, but it is not good for all situations in which channels with two diameters are present. For example, the $2^{1/3}$ ratio is not necessarily the best ratio for the large/small diameters of the parallel channels illustrated in Problem 3.7. For the latter, a different search for the constructal configuration must be performed, and the right diameter ratio will emerge at the end of that search.

In this chapter and the next, we review the milestones of heat and fluid flow sciences. We accomplish this with a run through the main concepts and results, such as the Poiseuille and Fourier formulas. We do not *derive* these formulas from the first principles, for example, the Navier-Stokes equations ($F = ma$) and first law of thermodynamics (the energy conservation equation). Their derivation is the object of the *disciplines* on which the all-encompassing and new science of design rests.

In this course we do even better, because in the analysis of various configurations we derive the formulas for *how* heat, fluid, and mass should flow. In other words, we teach the disciplines with purpose, on a case-by-case basis, not in the abstract. We teach the disciplines for a second time.

In this review the emphasis is placed on the relationships between flow rates and the "forces" that drive the streams. These are the relations that speak of the *flow resistances* that the streams overcome as they flow. In Chapter 2 we review the principles of thermodynamics in order to explain why streams that flow against resistances represent *imperfections* in the greater flow architectures to which they belong. The body of this course and book teaches how to distribute these imperfections so that the global system is the least imperfect that it can be. The constructal design method is about the *optimal distribution of imperfection*.

1.2 FLUID FLOW

In this brief review of fluid mechanics we assume that the flow is steady and the fluid is newtonian with constant properties. Each flow example is simple enough so that its derivation may be pursued as an exercise, in the classroom and as homework. Indeed, some of the problems proposed at the end of chapters are of this type. Our presentation, however, focuses on the formulas to use in design, and on the commonality of these flows, which in the case of duct flow is the relation between pressure difference and flow rate. The presentation proceeds from the simple toward more complex flow configurations.

1.2.1 Internal Flow: Distributed Friction Losses

We start with the Bernoulli equation, written for perfect (frictionless) fluid flow along a streamline,

$$P + \rho g z + \frac{1}{2}\rho V^2 = \text{constant} \tag{1.4}$$

where P, z, and V are the local pressure, elevation, and speed. For *real* fluid flow from cross-section 1 to cross-section 2 of a stream tube, we write

$$P_1 + \rho g z_1 + \frac{1}{2}\rho V_1^2 = P_2 + \rho g z_2 + \frac{1}{2}\rho V_2^2 + \Delta P \tag{1.5}$$

where ΔP represents the sum of the pressure losses (the fluid flow imperfection) that occurs between cross-sections 1 and 2. The losses ΔP may be due to distributed friction losses, local losses (junctions, bends, sudden changes in cross-section), or combinations of distributed and local losses (cf. Example 1.1).

Fully developed laminar flow occurs inside a straight duct when the duct is sufficiently slender and the Reynolds number sufficiently small. For example, if the duct is a round tube of inner diameter D, the velocity profile in the duct cross-section is parabolic:

$$u = 2U\left[1 - \left(\frac{r}{r_0}\right)^2\right] \tag{1.6}$$

In this expression, u is the longitudinal fluid velocity, U is the mean fluid velocity (i.e., u averaged over the tube cross-section), and r_0 is the tube radius, $r_0 = D/2$. In hydraulic engineering, more common is the use of the volumetric flowrate $Q[\mathrm{m}^3/\mathrm{s}]$, which for a round pipe is defined as

$$Q = U\pi r_0^2 \tag{1.7}$$

The radial position r is measured from the centerline ($r = 0$) to the tube wall ($r = r_0$). This flow regime is known as Hagen-Poiseuille flow, or Poiseuille flow for short. The derivation of Eq. (1.6) can be found in Ref. [6], for example.

How the pressure along the tube drives the flow is determined from a global balance of forces in the longitudinal direction. If the tube length is L and the pressure difference between entrance and exit is ΔP, then the longitudinal force balance is

$$\Delta P \pi r_0^2 = \tau 2\pi r_0 L \tag{1.8}$$

The fluid shear stress at the wall is

$$\tau = \mu \left(-\frac{\partial u}{\partial r} \right)_{r=r_0} \tag{1.9}$$

Equations (1.8) and (1.9) are valid for laminar and turbulent flow. By combining Eqs. (1.6) through (1.9), we conclude that for laminar flow the mean velocity is proportional to the longitudinal pressure gradient

$$U = \frac{r_0^2}{8\mu} \frac{\Delta P}{L} \tag{1.10}$$

Alternatively, we use the mass flow rate

$$\dot{m} = \rho U \pi r_0^2 \tag{1.11}$$

to rewrite Eq. (1.10) as a proportionality between the "across" variable (ΔP) and the "through" variable ($\dot{m}$):

$$\Delta P = \dot{m} \frac{L}{r_0^4} \frac{8}{\pi} \nu \tag{1.12}$$

The tandem of "across" and "through" variables has analogues in other flow systems, for example, voltage and electric current, temperature difference and heat current, and concentration difference and flow rate of chemical species (see Chapter 2, Fig. 2.4). In Eq. (1.12), the ratio $\Delta P/\dot{m}$ is the flow resistance of the tube length L in the Poiseuille regime. This resistance is proportional to the geometric group L/r_0^4, or L/D^4:

$$\frac{\Delta P}{\dot{m}} = \frac{L}{D^4} \frac{128}{\pi} \nu \tag{1.13}$$

The flow is laminar provided that the Reynolds number

$$\mathrm{Re}_D = \frac{UD}{\nu} \tag{1.14}$$

is less than approximately 2000. The tube length L is occupied mainly by Poiseuille flow if L is greater (in order of magnitude sense) than the laminar entrance length of the flow, which is $X \simeq D\mathrm{Re}_D$ [6]. The condition for a negligible entrance length is therefore

$$\frac{L}{D} \gg \mathrm{Re}_D \tag{1.15}$$

The flow resistance solution (1.12) has been recorded alternatively in terms of a friction factor, which is defined as

$$f = \frac{\tau}{\frac{1}{2}\rho U^2} \tag{1.16}$$

After using Eqs. (1.6) and (1.9), the friction factor for Poiseuille flow through a round tube becomes

$$f = \frac{16}{\mathrm{Re}_D} \tag{1.17}$$

The Poiseuille flow results for straight ducts that have cross-sections other than round are recorded in a notation that parallels what we have just reviewed for round tubes. Let A and p be the area and wetted perimeter of the arbitrary duct cross-section. The average fluid velocity U is defined as

$$\rho A U = \iint \rho u \, dA = \rho Q \tag{1.18}$$

Instead of Eq. (1.8), we have the force balance

$$\Delta P \cdot A = \tau p L \tag{1.19}$$

Instead of Re_D, we use the Reynolds number based on the hydraulic diameter of the general cross-section,

$$\mathrm{Re}_{D_h} = \frac{U D_h}{\nu} \tag{1.20}$$

$$D_h = 4\frac{A}{p} \tag{1.21}$$

Laminar fully developed flow prevails if $\mathrm{Re}_{D_h} \leq 2000$. The duct length L is swept almost entirely by Poiseuille flow if $L/D_h \gg \mathrm{Re}_{D_h}$, cf. Eq. (1.15).

The friction factor definition (1.16) continues to hold. Now, if we combine Eqs. (1.16) and (1.19) through (1.21), we arrive at the general pressure drop formula

$$\Delta P = f\frac{4L}{D_h}\frac{1}{2}\rho U^2 \tag{1.22}$$

For Poiseuille flow, the friction factor assumes the general form

$$f = \frac{\mathrm{Po}}{\mathrm{Re}_{D_h}} \tag{1.23}$$

where Po is the Poiseuille constant, for example Po = 16 for round tubes, and Po = 24 for channels formed between parallel plates. We see that every cross-sectional shape has its own Poiseuille constant, and that this constant is not much different than 16. A channel with square cross-section has Po = 14.2, and one with equilateral triangular cross-section has Po = 13.3. We return to this subject in Table 1.2.

A word of caution about the D_h and f definitions is in order. The factor 4 is used in Eq. (1.21) so that in the case of a round pipe of diameter D the hydraulic diameter D_h is the same as D [substitute $A = (\pi/4)D^2$ and $p = \pi D$ in Eq. (1.21), and obtain $D_h = D$]. A segment of the older literature defines D_h without the factor 4,

$$D'_h = \frac{A}{p} \tag{1.24}$$

and this convention leads to a different version of Eq. (1.22):

$$\Delta P = f' \frac{L}{D'_h} \frac{1}{2} \rho U^2 \tag{1.25}$$

The alternate friction factor f' obeys the definition (1.16). We also note that $D_h = 4\,D'_h$ and $f' = 4f$, such that for a round pipe with diameter D and Poiseuille flow the formulas are $f = 16/\mathrm{Re}_D$ and $f' = 64/\mathrm{Re}_D$. The numerators (16 vs. 64) are the first clues to remind the user which D_h definition was used, Eq. (1.21) or Eq. (1.24). The $4f$ plotted on the ordinate of Fig. 1.3 suggests that this chart was originally drawn with f' on the ordinate.

To summarize, by combining Eqs. (1.22) and (1.23) we conclude that all the laminar fully developed (Poiseuille) flows are characterized by a proportionality between ΔP and U, or ΔP and $\dot{m}$:

$$\frac{\Delta P}{\dot{m}} = 2\mathrm{Po}\nu \frac{L}{D_h^2 A} \tag{1.26}$$

Verify that for a round tube ($\mathrm{Po} = 16$, $D_h = D$, $A = \pi D^2/4$), Eq. (1.26) leads back to Eq. (1.23). The general proportionality (1.26) is a straight line drawn at $\mathrm{Re}_{D_h} < 2000$ in Fig. 1.3 – one line for each tube cross-sectional shape. In Fig. 1.3, only the line for the round tube is shown. The Moody chart is highly useful because it is a compilation of many empirical formulas (f vs. Re_{D_h}) in a single place–a bird's-eye view of distributed friction losses in all ducts with fully developed laminar or turbulent flow.

For turbulent flow through the same ducts, the ΔP vs. $\dot{m}$ relation is different. One relation that is independent of the flow regime is the pressure drop formula (1.22), because this formula is a rewriting of the force balance (1.19) in combination with the friction factor definition (1.16). Different for turbulent flow is the manner in which f depends on the flow rate, or Re_{D_h}.

The Moody chart (Fig. 1.3) shows that in turbulent flow f is a function of Re_{D_h} and the roughness of the wall inner surface. Plotted on the abscissa is Re_D (round tube), however, the curves for turbulent flow hold for all cross-sectional shapes provided that Re_D is replaced by Re_{D_h}. To assist computer-aided design and analysis, the friction factor function $f(\mathrm{Re}_D, k_s/D)$ is available in several alternative mathematical forms. The Colebook relation is often used for turbulent flow:

$$\frac{1}{f^{1/2}} = -4 \log \left(\frac{k_s/D}{3.7} + \frac{1.256}{f^{1/2}\text{Re}_D} \right) \tag{1.27}$$

Because this expression is implicit in f, iteration is required to obtain the friction factor for a specified Re_D and k_s/D. Several explicit approximations are available for smooth ducts:

$$f \cong 0.079\text{Re}_D^{-1/4} \quad (2 \times 10^3 < \text{Re}_D < 10^5) \tag{1.28}$$

$$f \cong 0.046\text{Re}_D^{-1/5} \quad (2 \times 10^4 < \text{Re}_D < 10^6) \tag{1.29}$$

At sufficiently high Reynolds numbers, the friction factor depends only on the relative roughness, which means that for a given duct f is constant. This is known as the *fully rough* regime. This feature and Eq. (1.22) show that in fully rough turbulent flow ΔP is proportional to U^2, or to $\dot{m}^2$,

$$\frac{\Delta P}{\dot{m}^2} = \frac{2f}{\rho} \frac{L}{D_h A^2} \tag{1.30}$$

Compare this fully rough turbulent resistance with the fully developed laminar resistance (1.26), and you will see in another way the fact that the slope of the curves changes from left to right in Fig. 1.3.

1.2.2 Internal Flow: Local Losses

The relations between pressure drops and flow rates become more complicated when the ducts are too short to satisfy the negligible entrance assumption (1.15), and when they are connected with other ducts into larger flow networks (e.g., Fig. 1.4). The treatment of such cases is based on a formulation that begins with the generalized Bernoulli equation for irreversible flow through a duct as a control volume, from entrance (1) to exit (2), as shown in Eq. (1.5). The pressure loss ΔP is calculated by summing up all the flow imperfections,

$$\Delta P = \underbrace{\sum_d \left[f \frac{4L}{D_h} \frac{1}{2} \rho U^2 \right]_d}_{\text{distributed losses}} + \underbrace{\sum_l \left[K \frac{1}{2} \rho U^2 \right]_l}_{\text{local losses}} \tag{1.31}$$

The first summation refers to sections of long ducts, along which the pressure drops are due to fully developed flow, laminar or turbulent. Note the similarity between the expression written inside $[\;]_d$ and Eq. (1.22).

The second summation in Eq. (1.31) refers to *local losses* such as the pressure drops caused by junctions, fittings, valves, inlets, outlets, enlargements and contractions. Each local loss contributes to the total loss in proportion to $K\rho U^2/2$, where K is the respective local-loss coefficient. Experimental K data are available

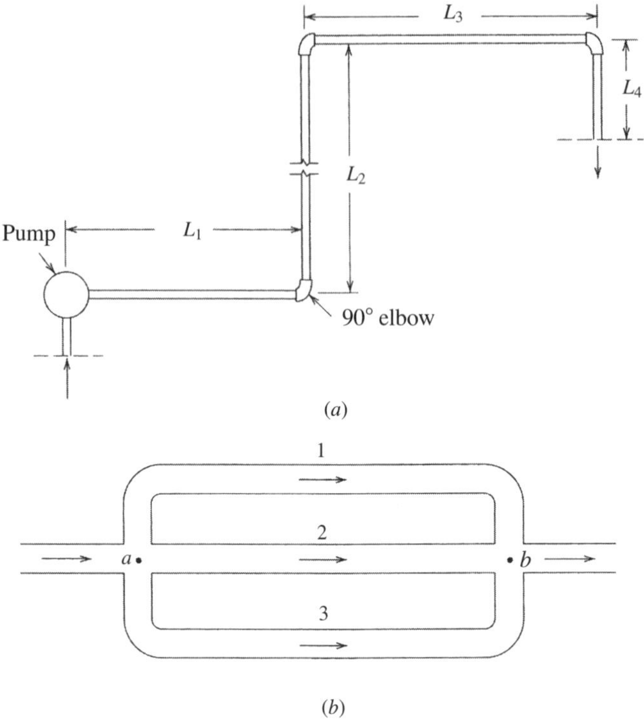

Figure 1.4 Examples of networks of pipes: (a) single flow path; (b) multiple paths [7].

in the fluids engineering literature: Table 1.1 shows a representative sample. Local losses can be non-negligible in the functioning of complex tree-shaped networks. How important, and when they can be neglected are questions tied closely to the concept of flow architecture svelteness, as we show in Example 1.1. It suffices to say that as the svelteness of the flow architecture increases from left to right in Fig. 1.1, all the lines of the drawing become thinner and, consequently, the distributed losses gain in importance relative to the local (junctions) losses.

Most of the fluid flow examples in this book have the objective to minimize in time the "energetic consumption" (exergy destruction) of the fluid network. We will now see that this objective can be attained by minimizing the overall pressure losses ΔP. We show this for an open system, and later for a closed system.

Assume that the open flow system is an urban hydraulic network. A certain volume of water V must be delivered per day to a city. This water is pumped from a river or a lake. Some water treatment may be needed to make the water drinkable for the users: this water treatment will occur at the level of the pump stations. The pumped water is not sent directly to the users in the city. The preferred solution consists of placing a water tank at a height between the pumping station and the city. In such a system, the tank has a double role. The first is to store water so that in case of pump malfunction the water supply (aqueduct) is decoupled from

Table 1.1 Local loss coefficients [7].

Resistance		K
	Changes in Cross-Sectional Area[a]	
Round pipe entrance 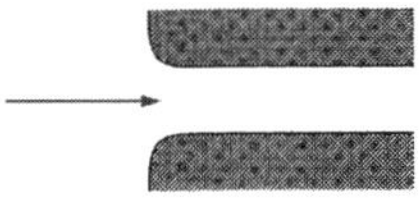		0.04–0.28
Contraction 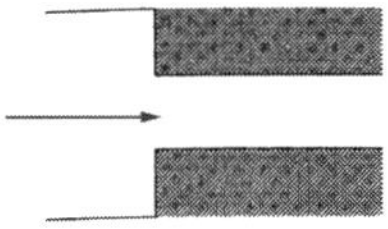	$\text{AR} = \dfrac{\text{smaller area}}{\text{larger area}}$	0.45(1−AR)
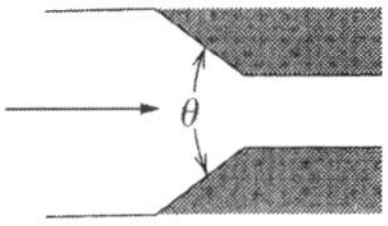	$\theta < 60°$ $0.1 \leq \text{AR} \leq 0.5$	0.04–0.08
Expansion 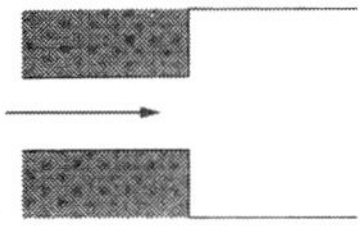		$\left(1 - \dfrac{1}{AR}\right)^2$
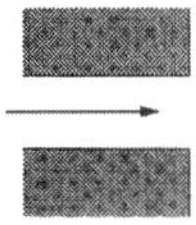		1.0
	Valves and Fittings	
Gate valve, open		0.2
Globe valve, open		6–10
Check valve (ball), open		70
45-degree elbow, standard		0.3–0.4
90-degree elbow, rounded		0.4–0.9

[a] The velocity used to evaluate the loss is the velocity upstream of the expansion. When AR = 0, $K = 1$.

the distribution network, and the city is not affected by accidents and continues to receive drinkable water. The second role of the tank is to provide a controlled water pressure to the city simply because of the altitude difference.

Figure 1.5 shows the main features of a water supply network. The total volume of water pumped per day and stored in the tank is fixed. The objective is to minimize the energetic consumption of the network C. The role of the pumping station (H_m) in this network is to compensate for the altitude difference between the tank and

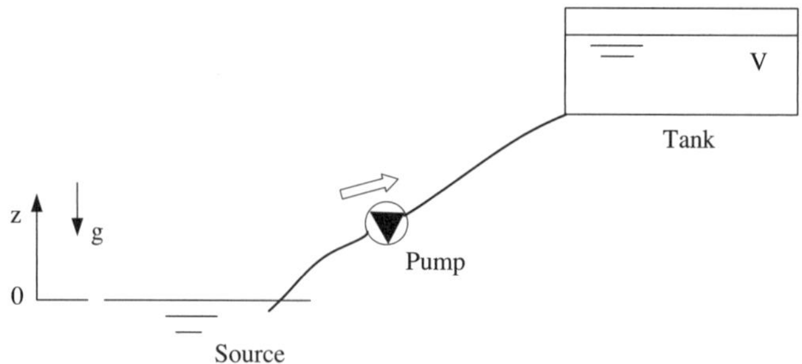

Figure 1.5 The route of city water supply from the source to the storage tank.

the source (river, lake) and the overall pressure loss,

$$H_m = z_{\text{tank}} - z_{\text{source}} + \xi \tag{1.32}$$

Here, ξ is the overall pressure loss expressed as head of water, that is, the height of a column of water, $\xi = \Delta P/\rho g$, where ΔP is the overall pressure loss. Typically, the pumps used in urban hydraulic applications are centrifugal pumps with blades curved backward. Their characteristic is given in Fig. 1.6 together with the network head curve [see the right-hand side of (1.32)]. The intersection of the two curves is called the *operating point*. The energetic consumption of the system is given by the product of the pumping power and the time of pumping,

$$C = \dot{W} t \tag{1.33}$$

where

$$\dot{W} = \dot{m} g H_m \tag{1.34}$$

$$C = \rho g H_m V \tag{1.35}$$

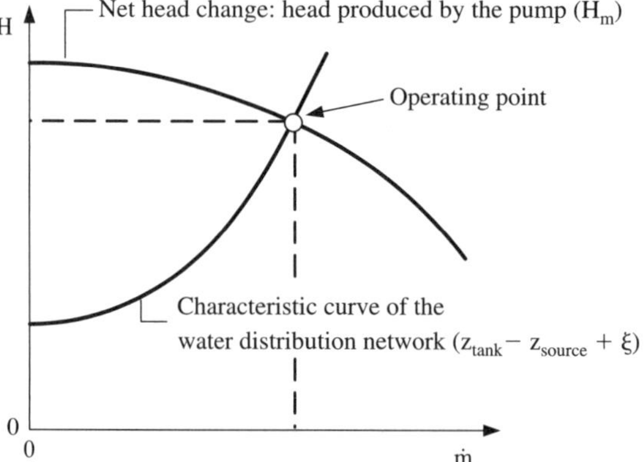

Figure 1.6 The operating point of the water distribution network, as the intersection between the characteristic curve of the pump and the characteristic curve of the water distribution network.

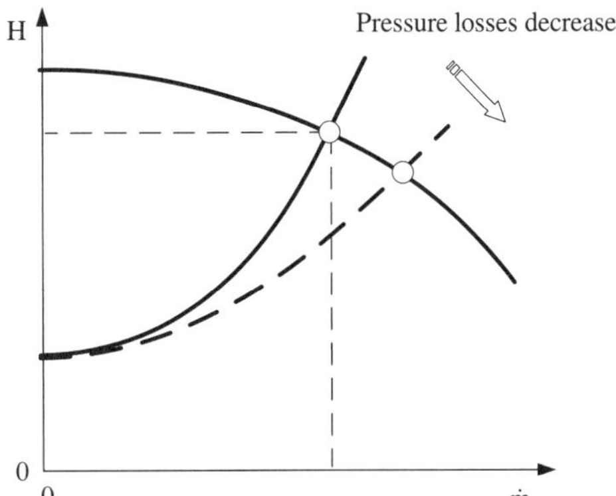

Figure 1.7 The change in the operating point as the pressure losses of the network decrease.

Because $H_m \sim \Delta P$ and the water volume V is constrained, we see that $C \sim \Delta P$. Therefore, to minimize the energetic consumption means to minimize the pressure losses. Figure 1.7 shows how the operating point shifts when the pressure losses are minimized. A higher flow rate is the result.

Next, we connect the tank to the users (Fig. 1.8). No pump is needed because the pressure required at the city level is maintained by the altitude difference between the tank and the city,

$$\rho g z_{\text{tank}} = P_{\text{required}} + \rho g z_{\text{city}} + \Delta P \tag{1.36}$$

Note that the kinetic energy term $\frac{1}{2}\rho U^2$ does not appear here. The reason is that in urban hydraulic applications the fluid velocity is in the range of 1 m/s, making the changes in kinetic energy negligible. Again, we see that the optimization of this flow system (a higher P_{required}) means that the pressure losses ΔP must be minimized.

An example of a closed system is presented in Fig. 1.9, which shows the principle of steady-flow operation of a heating network in a building. The network is made of basic components such as a water heater, a radiator, and a pump. The heater uses combustion in order to heat the water stream. The radiator releases heat into the space of the building. Because the system is closed, the gravity effect vanishes (the water that flows up must eventually flow down). Thus, we arrive at $H_m \sim \Delta P$.

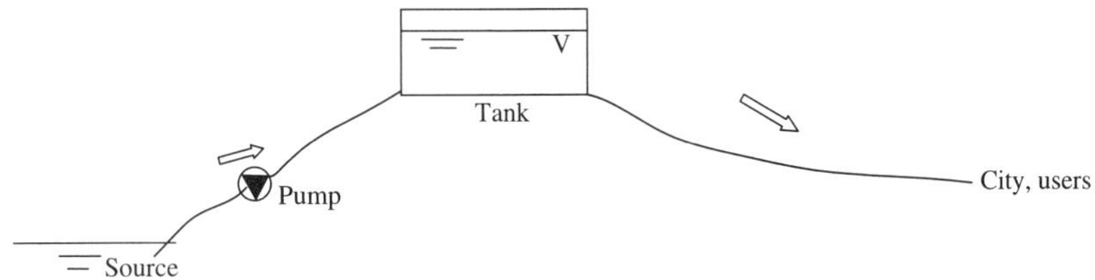

Figure 1.8 Complete network for water supply, storage, and distribution.

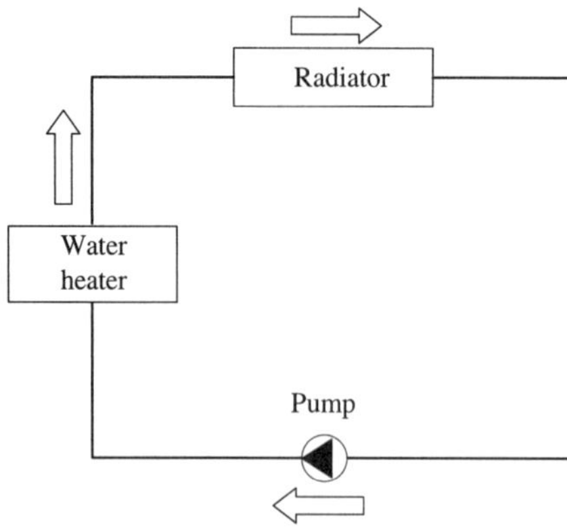

Figure 1.9 Closed loop circulation of water for heating a building.

Once again, the pumping power will be minimized when the pressure losses are minimized.

Example 1.1 Here, we show how to determine the local pressure loss associated with the sudden expansion of a duct (Fig. 1.2). Incompressible fluid flows through a pipe of diameter D_1 and length L_1, and continues flowing through a pipe of diameter D_2 and length L_2. Consider the cylindrical control volume indicated with dashed line in Fig. 1.10. The control-surface convention that we use is that there is no space between the dashed line and the wall (the solid line). The dashed line represents the *internal* surface of the wall. The control volume contains the enlargement of the stream, including the recirculation occurring after the backward-facing steps. We apply the momentum theorem in the longitudinal direction,

$$\dot{m}(V_1 - V_2) = A_2(P_2 - P_1) \tag{a}$$

where $\dot{m}$ is the mass flow rate

$$\dot{m} = \rho A_1 V_1 = \rho A_2 V_2 \tag{b}$$

and $A_{1,2} = (\pi/4)D_{1,2}^2$. See the lower part of Fig. 1.10. From Eq. (a) we learn that the pressure at the outlet of the control volume is

$$P_2 = P_1 + \rho V_2(V_1 - V_2) \tag{c}$$

This pressure can be compared with the value in the ideal (frictionless, reversible) limit where the stream expands smoothly from A_1 to A_2. According to the Bernoulli equation,

$$P_1 + \rho g z + \frac{1}{2}\rho V_1^2 = P_2 + \rho g z + \frac{1}{2}\rho V_2^2 + \Delta P_{\text{local}} \tag{d}$$

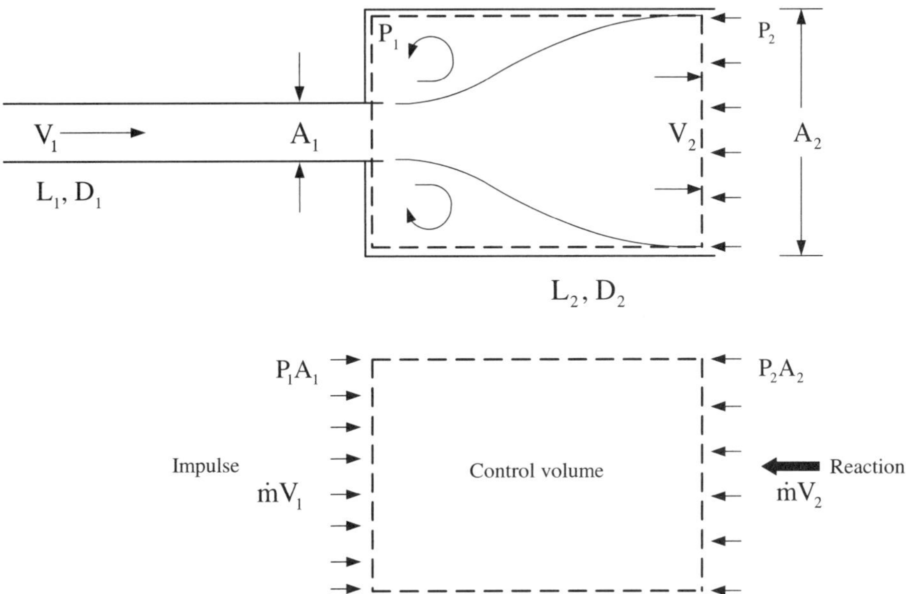

Figure 1.10 Control volume formulation for the duct with sudden expansion and local pressure loss, which is analyzed in Example 1.1.

the local pressure loss is

$$\Delta P_{\text{local}} = P_1 - P_2 + \frac{1}{2}\rho(V_1^2 - V_2^2) \tag{e}$$

which, after using Eq. (c) yields

$$\Delta P_{\text{local}} = \frac{1}{2}\rho(V_1 - V_2)^2 \tag{f}$$

Next, we eliminate V_2 using Eq. (b), and arrive at the local loss coefficient reported in Table 1.1 [see also Eq. (1.2)]:

$$K = \frac{\Delta P_{\text{local}}}{\frac{1}{2}\rho V_1^2} = \left(1 - \frac{A_1}{A_2}\right)^2 = \left[1 - \left(\frac{D_1}{D_2}\right)^2\right]^2 \tag{g}$$

To show analytically how the local loss becomes negligible as the svelteness of the flow system increases, assume that the small pipe is much longer than the wider pipe. The Sv definition (1.1) becomes

$$\text{Sv} \cong \frac{L_1}{\left(\frac{\pi}{4}D_1^2 L_1\right)^{1/3}} = \left(\frac{4}{\pi}\right)^{1/3}\left(\frac{L_1}{D_1}\right)^{2/3} \tag{h}$$

The distributed losses are due to fully developed flow in the L_1 pipe. According to Eq. (1.22), the pressure drop along L_1 is

$$\Delta P_1 = f_1 \frac{4L_1}{D_1}\frac{1}{2}\rho V_1^2 = \Delta P_{\text{distributed}} \tag{i}$$

Dividing Eqs. (g) and (i) we obtain

$$\frac{\Delta P_{\text{local}}}{\Delta P_{\text{distributed}}} = \frac{1}{4f}\frac{\left[1-(D_1/D_2)^2\right]^2}{L_1/D_1} \tag{j}$$

and, after using Eq. (h),

$$\frac{\Delta P_{\text{local}}}{\Delta P_{\text{distributed}}} = \frac{1}{4f}\frac{\left[1-(D_1/D_2)^2\right]^2}{(\pi/4)^{1/2}\mathrm{Sv}^{3/2}} \tag{k}$$

The ratio $\Delta P_{\text{local}}/\Delta P_{\text{distributed}}$ decreases as Sv increases. If the flow in the L_1 pipe is in the fully turbulent and fully rough regime, then f is a constant (independent of Re_1) with a value of order 0.01. For simplicity, we assume $f \cong 0.01$ and $(D_1/D_2)^2 \ll 1$, such that Eq. (k) yields the curve plotted for turbulent flow in Fig. 1.2:

$$\frac{\Delta P_{\text{local}}}{\Delta P_{\text{distributed}}} \cong \frac{25}{\mathrm{Sv}^{3/2}} \tag{l}$$

Noteworthy is the criterion for negligible local losses, $\Delta P_{\text{local}} \ll \Delta P_{\text{distributed}}$, which according to Eq. (l) means $\mathrm{Sv} \gg 8.5$.

If the flow in the L_1 pipe is laminar and fully developed, then in Eq. (l) we substitute $f = 16/\mathrm{Re}_1$. The Reynolds number ($\mathrm{Re}_1 = V_1 D_1/\nu$) is an additional parameter, which is known when $\dot{m}$ is specified. In place of Eq. (l) we obtain

$$\frac{\Delta P_{\text{local}}}{\Delta P_{\text{distributed}}} \frac{\mathrm{Re}_1/64}{\mathrm{Sv}^{3/2}} \tag{m}$$

The criterion $\Delta P_{\text{local}} \ll \Delta P_{\text{distributed}}$ yields in this case $\mathrm{Sv} \gg (\mathrm{Re}_1/64)^{2/3}$, which means $\mathrm{Sv} \gg 6.2$ when Re_1 is of order 10^3. The curves for $\mathrm{Re}_1 = 10^2$ and 10^3 are shown in Fig. 1.2.

Summing up, when the svelteness Sv exceeds 10 in an order of magnitude sense, the local losses are negligible regardless of flow regime.

1.2.3 External Flow

The fins of heat exchanger surfaces, the trunks of trees, and the bodies of birds are solid objects bathed all around by flows. Flow imperfection in such configurations is described in terms of the drag force (F_D) experienced by the body immersed in

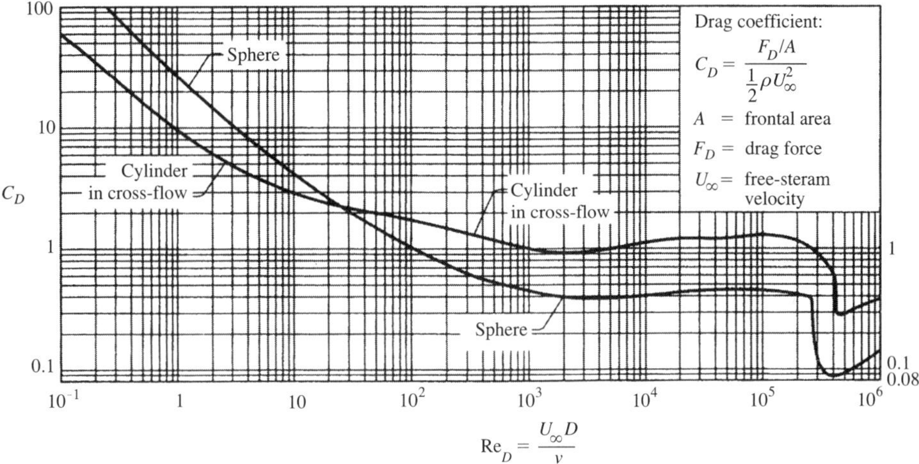

Figure 1.11 Drag coefficients for smooth sphere and smooth cylinder in cross-flow [6].

a flow with free-stream velocity U_∞,

$$F_D = C_D A_f \frac{1}{2} \rho U_\infty^2 \tag{1.37}$$

Here, A_f is the frontal area of the body, that is, the area of the projection of the body on a plane perpendicular to the free-stream velocity. The imperfection is also measured as the mechanical power used in order to drag the body through the fluid with the relative speed U_∞, namely $\dot{W} = F_D U_\infty$. We return to this thermodynamics aspect in Chapter 2.

The drag coefficient C_D is generally a function of the body shape and the Reynolds number. Figure 1.11 shows the two most common C_D examples, for the sphere and the cylinder in cross-flow. The Reynolds number plotted on the abscissa is based on the sphere (or cylinder) diameter D, namely $\mathrm{Re}_D = U_\infty D/\nu$. These two bodies are the most important because with just one length scale (the diameter D) they represent the two extremes of all possible body shapes, from the most round (the sphere) to the most slender (the long cylinder).

Note that in the Re_D range 10^2 – 10^5 the drag coefficient is a constant of order 1. In this range F_D is proportional to U_∞^2. In the opposite extreme, $\mathrm{Re}_D < 10$, the drag coefficient is such that the product $C_D \mathrm{Re}_D$ is a constant. This small-Re_D limit is known as Stokes flow: here, F_D is proportional to U_∞.

Compare side by side the flow resistance information for internal flow (Fig. 1.3) with the corresponding information for external flow (Fig. 1.11). All the curves on these log-log plots start descending (with slopes of -1), and at sufficiently high Reynolds numbers in fully rough turbulent flow they flatten out. Said another way,

f and C_D are equivalent nondimensional representations of flow resistance, f for internal flows, and C_D for external flows.

1.3 HEAT TRANSFER

Heat flows from high temperature to lower temperature in the same way that a fluid flows through a pipe from high pressure to lower pressure. This is the natural way. The direction of flow from high to low is the one-way direction of the second law (Chapter 2). For heat flow as a mode of energy interaction between neighboring entities, the temperature difference is the defining characteristic of the interaction: heating is the energy transfer driven by a temperature difference.

Heat flow phenomena are in general complicated, and the entire discipline of heat transfer is devoted to determining the q function that rules a physical configuration made of entities A and B [8]:

$$q = \text{function } (T_A, T_B, \text{time, thermophysical properties, geometry, fluid flow}) \tag{1.38}$$

In this section we review several key examples of the relationship between temperature difference $(T_A - T_B)$ and heat current (q), with particular emphasis on the ratio $(T_A - T_B)/q$, which is the thermal resistance. Examples are selected because they reappear in several applications and problems in this book. The parallels between this review and the treatment of fluid-flow resistance (section 1.2) are worth noting. One similarity is the focus on the simplest configurations, namely, steady flow, and materials with constant properties.

1.3.1 Conduction

Conduction, or thermal diffusion, occurs when the two bodies touch, and there is no bulk motion in either body. The simplest example is a two-dimensional slab of surface A and thickness L. One side of the slab is at temperature T_1 and the other at T_2. The slab is the space and material in which the two entities (T_1, T_2) make thermal contact. The thermal conductivity of the slab material is k. The total heat current across the slab, from T_1 to T_2, is described by the *Fourier law* of heat conduction,

$$q = k\frac{A}{L}(T_1 - T_2) \tag{1.39}$$

The group kA/L is the thermal conductance of the configuration (the slab). The inverse of this group is the thermal resistance of the slab,

$$R_t = \frac{T_1 - T_2}{q} = \frac{L}{kA} \tag{1.40}$$

There is a huge diversity of body-body thermal contact configurations, and thermal resistances are available in the literature (e.g. Ref. [8]). As in the discussion of the sphere and the cylinder of Fig. 1.11, we cut through the complications of diversity by focusing on the extreme shapes, the most round and the most slender. One such extreme is the *spherical shell* of inner radius r_i and outer radius r_o. We may view this shell as the wrapping of insulation of thickness $(r_o - r_i)$ on a spherical hot body of temperature T_i, such that the outer surface of the insulation is cooled by the ambient to the temperature T_o. The thermal resistance of the shell is

$$R_t = \frac{T_i - T_o}{q} = \frac{1}{4\pi k}\left(\frac{1}{r_i} - \frac{1}{r_o}\right) \tag{1.41}$$

where q is the total heat current from r_i to r_o, and k is the thermal conductivity of the shell material. Note that Eq. (1.41) reproduces Eq. (1.40) in the limit $r_o \to r_i$, where the shell is thin (i.e., like a plane slab). The *cylindrical shell* of radii r_i and r_o, length L and thermal conductivity k has the thermal resistance

$$R_t = \frac{\ln(r_o/r_i)}{2\pi k L} \tag{1.42}$$

Fins are extended surfaces that enhance the thermal contact between a base (T_b) and a fluid flow (T_∞) that bathes the wall. If the geometry of the fin is such that the cross-sectional area (A_c), the wetted perimeter of the cross-section (p), and the heat transfer coefficient [h, defined later in Eq. (1.56)] are constant, the heat transfer rate q_b through the base of the fin is approximated well by

$$q_b \cong (T_b - T_\infty)(kA_c hp)^{1/2} \tanh\left[\left(\frac{hp}{kA_c}\right)^{1/2}\left(L + \frac{A_c}{p}\right)\right] \tag{1.43}$$

In this expression L is the fin length measured from the tip of the fin to the base surface, and k is the thermal conductivity of the fin material. The heat transfer rate through a fin with variable cross-sectional area and perimeter can be calculated by writing

$$q_b = (T_b - T_\infty) h A_{exp} \eta \tag{1.44}$$

where A_{exp} is the total exposed (wetted) area of the fin and η is the fin efficiency, a dimensionless number between 0 and 1, which can be found in heat transfer books [8,9].

Equations (1.43) and (1.44) are based on the very important assumption that the conduction through the fin is essentially unidirectional and oriented along the fin. This assumption is valid when the *Biot number* Bi $= ht/k$ is small such that [8]

$$\left(\frac{ht}{k}\right)^{1/2} < 1 \tag{1.45}$$

where t is the thickness of the fin, that is, the fin dimension perpendicular to the conduction heat current q.

The Biot number definition should not be confused with the Nusselt number definition. The thermal conductivity k that appears in the Bi definition is the conductivity of the solid wall (e.g., fin) that is swept by the convective flow (h). In the Nu group defined later in Eq. (1.60), k is the thermal conductivity of the fluid in the convective flow.

Time-dependent conduction is also a phenomenon that we encounter and exploit for design in this book. For example, the temperature distribution in a semi-infinite solid, the surface temperature of which is raised instantly from T_i to T_∞, is

$$\frac{T(x,t) - T_\infty}{T_i - T_\infty} = \operatorname{erf}\left[\frac{x}{2(\alpha t)^{1/2}}\right] \tag{1.46}$$

The counterpart of this heating configuration is the surface with imposed heat flux. The temperature field under the surface of a semi-infinite solid that, starting with the time $t = 0$, is exposed to a constant heat flux q'' (heat transfer rate per unit area) is

$$T(x,t) - T_i = 2\frac{q''}{k}\left(\frac{\alpha t}{\pi}\right)^{1/2} \exp\left(-\frac{x^2}{4\alpha t}\right) - \frac{q''}{k} x \operatorname{erfc}\left[\frac{x}{2(\alpha t)^{1/2}}\right] \tag{1.47}$$

Note the arguments of erf, exp, and erfc: they all contain the group $x/(\alpha t)^{1/2}$, where x is the distance (under the surface) to which the thermal wave penetrates during the time t, and α is the thermal diffusivity of the material ($\alpha = k/\rho c$). In Eq. (1.46), for example, thermal penetration means that x is such that $T(x,t) - T_\infty \sim T_i - T_\infty$, which means that

$$\frac{x}{2(\alpha t)^{1/2}} \sim 1 \tag{1.48}$$

Therefore, a characteristic of all thermal diffusion processes is that the thickness of thermal penetration under the exposed surface grows as $(\alpha t)^{1/2}$, that is infinitely fast at $t = 0^+$, and more slowly as t increases. We return to this characteristic in the time-optimization of electrokinetic decontamination in Chapter 9.

Additional examples of time-dependent diffusion are the temperature fields that develop around concentrated heat sources and sinks. Common phenomena that can be described in terms of concentrated heat sources are underground fissures filled with geothermal steam, underground explosions, canisters of nuclear and chemical waste, and buried electrical cables. It is important to distinguish between instantaneous heat sources and continuous heat sources.

Consider first instantaneous heat sources released at $t = 0$ in a conducting medium with constant properties (ρ, c, k, α) and uniform initial temperature T_i. The temperature distribution in the vicinity of the source depends on the shape of the source:

$$T(x,t) - T_i = \frac{Q''}{2\rho c(\pi \alpha t)^{1/2}} \exp\left(-\frac{x^2}{4\alpha t}\right)$$

[instantaneous plane source, strength $Q''(\text{J/m}^2)$ at $x = 0$] (1.49)

$$T(r,t) - T_i = \frac{Q'}{4\rho c\pi \alpha t} \exp\left(-\frac{r^2}{4\alpha t}\right)$$

[instantaneous line source, strength $Q'(\text{J/m})$ at $r = 0$] (1.50)

$$T(r,t) - T_i = \frac{Q}{8\rho c(\pi \alpha t)^{3/2}} \exp\left(-\frac{r^2}{4\alpha t}\right)$$

[instantaneous point source, strength $Q(\text{J})$ at $r = 0$] (1.51)

The temperature distributions near continuous heat sources, which release heat at constant rate when $t > 0$, are

$$T(x,t) - T_i = \frac{q''}{\rho c}\left(\frac{t}{\pi \alpha}\right)^{1/2} \exp\left(-\frac{x^2}{4\alpha t}\right) - \frac{q''\,|x|}{2k}\operatorname{erfc}\left[\frac{|x|}{2\,(\alpha t)^{1/2}}\right]$$

[continuous plane source, strength q'' (W/m^2) at $x = 0$] (1.52)

$$T(r,t) - T_i = \frac{q'}{4\pi k}\int_{r^2/4\alpha t}^{\infty} \frac{e^{-u}}{u}\,\text{du} \cong \frac{q'}{4\pi k}\left[\ln\left(\frac{4\alpha t}{r^2}\right) - 0.5772\right], \text{ if} \frac{r^2}{4\alpha t} < 1$$

[continuous line source, strength $q'(\text{W/m})$ at $r = 0$] (1.53)

$$T(r,t) - T_i = \frac{q}{4\pi k r}\operatorname{erfc}\left[\frac{r}{2(\alpha t)^{1/2}}\right] \cong \frac{q}{4\pi k r} \quad \text{if} \frac{r^2}{2\,(\alpha t)^{1/2}} < 1$$

[continuous point source, strength $q(W)$ at $r = 0$] (1.54)

Equations (1.49) through (1.54) also describe the temperature fields near concentrated heat sinks. In such cases the numerical values of the source strengths (Q'', Q', Q, q'', q', q) are negative.

A related time-dependent configuration is plane melting and solidification. A semi-infinite solid that is isothermal and at the melting point (T_m) melts if its surface is raised to a higher temperature (T_0). In the absence of the effect of convection, the liquid layer thickness δ increases in time according to

$$\delta(t) \cong \left[2\frac{kt}{\rho h_{sf}}(T_0 - T_m)\right]^{1/2} \tag{1.55}$$

where ρ and k are the density and thermal conductivity of the liquid, and h_{sf} is the latent heat of melting. Equation (1.55) is valid provided that $c(T_0 - T_m)/h_{\text{sf}} < 1$, where c is the specific heat of the liquid.

The solidification of a motionless pool of liquid is described by Eq. (1.55), in which $T_0 - T_m$ is replaced by $T_m - T_0$ because the liquid is saturated at T_m, and the surface temperature is lowered to T_0. In the resulting expression δ is the thickness of the solid layer, and k, c and α are properties of the solid layer.

1.3.2 Convection

Convection is the heat transfer mechanism in which a flowing material (gas, fluid, solid) acts as a conveyor for the energy that it draws from (or delivers to) a solid wall. As a consequence, the heat transfer rate is affected greatly by the characteristics of the flow (e.g., velocity distribution, turbulence). To know the flow *configuration* and the regime (laminar vs. turbulent) is an important prerequisite for calculating convection heat transfer rates. Furthermore, the nature of the boundary layers (hydrodynamic and thermal) plays an important role in evaluating convection.

Convection is said to be external when a much larger space filled with flowing fluid (the free stream) exchanges heat with a body immersed in the fluid. According to Eq. (1.38), the objective is to determine the relation between the heat transfer rate (or the heat flux through a spot on the wall, q''), and the wall-fluid temperature difference ($T_w - T_\infty$). The alternative is to determine the convective heat transfer coefficient h, which for external flow is defined by

$$h = \frac{q''}{T_w - T_\infty} \tag{1.56}$$

where q'' is the heat flux, $q'' = q/A$, where A is the area swept by the flowing fluid. This means that Eq. (1.56) can also be written as

$$q = hA\,(T_w - T_\infty) \tag{1.57}$$

or that the convective thermal resistance is

$$R_t = \frac{T_w - T_\infty}{q} = \frac{1}{hA} \tag{1.58}$$

Figure 1.12 shows the order of magnitude of h for various classes of convective heat transfer configurations. Techniques for estimating h accurately are available (e.g., Ref. [6]). The most basic example is the boundary layer on a plane wall. When the fluid velocity U_∞ is uniform and parallel to a wall of length L, the hydrodynamic boundary layer along the wall is laminar over L if $\text{Re}_L \leq 5 \times 10^5$, where the Reynolds number is defined by $\text{Re}_L = U_\infty L/\nu$ and ν is the kinematic viscosity. The leading edge of the wall is perpendicular to the direction of the free stream (U_∞). The wall shear stress in laminar flow averaged over the length L is

$$\bar{\tau} = 0.664\rho U_\infty^2 \text{Re}_L^{-1/2} \quad (\text{Re}_L \leq 5 \times 10^5) \tag{1.59}$$

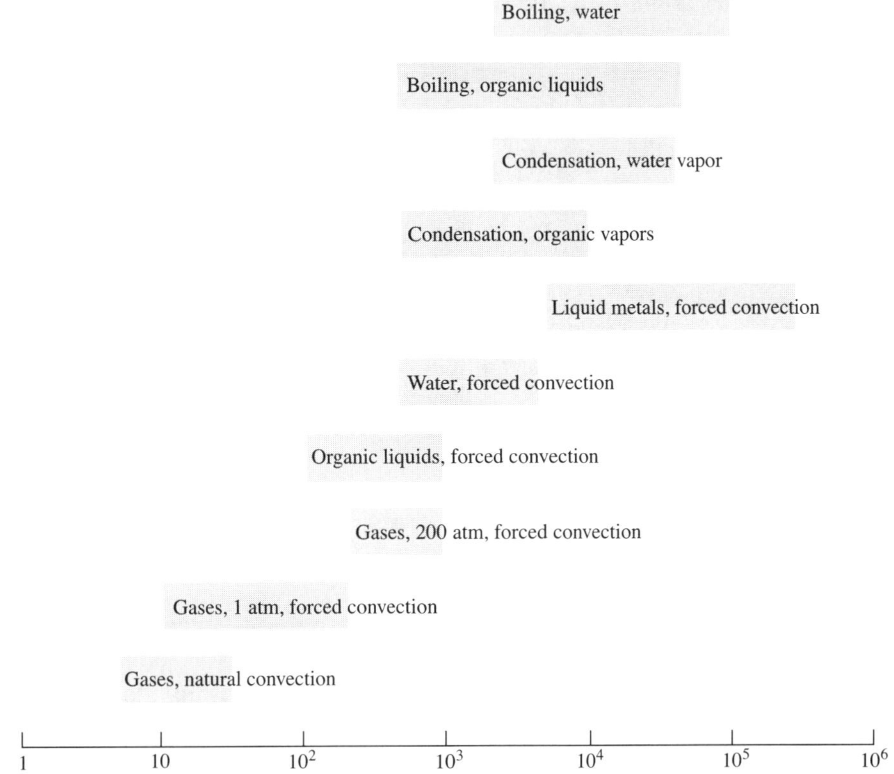

Figure 1.12 Convective heat transfer coefficients, showing the effect of flow configuration [8].

so that the total tangential force experienced by a plate of width W and length L is $F = \bar{\tau} LW$. The length L is measured in the flow direction. The thickness of the hydrodynamic boundary layer at the trailing edge of the plate is of order $L\mathrm{Re}_L^{-1/2}$.

If the wall is isothermal at T_w, the heat transfer coefficient $\bar{h}$ averaged over the flow length L is ($\mathrm{Re}_L \leq 5 \times 10^5$):

$$\mathrm{Nu} = \frac{\bar{h}L}{k} = \begin{cases} 0.664\,\mathrm{Pr}^{1/3}\,\mathrm{Re}_L^{1/2} & (\mathrm{Pr} \geq 0.5) \\ 1.128\,\mathrm{Pr}^{1/2}\,\mathrm{Re}_L^{1/2} & (\mathrm{Pr} \leq 0.5) \end{cases} \tag{1.60}$$

In these expressions k and Pr are the fluid thermal conductivity and the Prandtl number $\mathrm{Pr} = \nu/\alpha$. The free stream is isothermal at T_∞. The total heat transfer rate through the wall of area LW is $q = \bar{h}LW(T_w - T_\infty)$. The group $\mathrm{Nu} = \bar{h}L/k$ is the overall Nusselt number, where k is the thermal conductivity of the fluid.

When the wall heat flux q'' is uniform, the wall temperature T_w increases away from the leading edge ($x = 0$) as $x^{1/2}$:

$$T_w(x) - T_\infty = \frac{2.21 q'' x}{k\,\text{Pr}^{1/3}\,\text{Re}_x^{1/2}} \qquad (\text{Pr} \geq 0.5,\ \text{Re}_x \leq 5 \times 10^5) \tag{1.61}$$

Here, the Reynolds number is based on the distance from the leading edge, $\text{Re}_x = U_\infty x/\nu$. The wall temperature averaged over the flow length L, $\bar{T}_w$, is obtained by substituting, respectively, $\bar{T}_w$, Re_L, and 1.47 in place of $T_w(x)$, Re_x, and 2.21 in Eq. (1.61).

At Reynolds numbers Re_L greater than approximately 5×10^5, the boundary layer begins with a laminar section that is followed by a turbulent section. For $5 \times 10^5 < \text{Re}_L < 10^8$ and $0.6 < \text{Pr} < 60$, the average heat transfer coefficient $\bar{h}$ and wall shear stress $\bar{\tau}$ are

$$\text{Nu} = \frac{\bar{h}L}{k} = 0.037\,\text{Pr}^{1/3}(\text{Re}_L^{4/5} - 23{,}550) \tag{1.62}$$

$$\frac{\bar{\tau}}{\rho U_\infty^2} = \frac{0.037}{\text{Re}_L^{1/5}} - \frac{871}{\text{Re}_L} \tag{1.63}$$

Equation (1.62) is sufficiently accurate for isothermal walls (T_w) as well as for uniform-heat flux walls. The total heat transfer rate from an isothermal wall is $q = \bar{h}LW(T_w - T_\infty)$, and the average temperature of the uniform-flux wall is $\bar{T}_w = T_\infty + q''/\bar{h}$. The total tangential force experienced by the wall is $F = \bar{\tau}LW$, where W is the wall width. The Nu results for many other external flow configurations (sphere, cylinder in cross-flow, etc.) are provided in Refs. [6, 9].

In flows through ducts, the heat transfer surface surrounds and guides the stream, and the convection process is said to be internal. For internal flows the heat transfer coefficient is defined as

$$h = \frac{q''}{T_w - T_m} \tag{1.64}$$

Here, q'' is the heat flux (heat transfer rate per unit area) through the wall where the temperature is T_w, and T_m is the mean (bulk) temperature of the stream,

$$T_m = \frac{1}{UA}\iint_A uT\,dA \tag{1.65}$$

The mean temperature is a weighted average of the local fluid temperature T over the duct cross-section A. The role of weighting factor is played by the longitudinal fluid velocity u, which is zero at the wall and large in the center of the duct cross section [e.g., Eq. (1.6)]. The mean velocity U is defined by

$$U = \frac{1}{A}\iint_A u\,dA \tag{1.66}$$

The volumetric flow rate through the A cross-section is

$$Q = UA = \iint_A u \, dA \tag{1.66$'$}$$

In laminar flow through a duct, the velocity distribution has two distinct regions: the entrance region, where the walls are lined by growing boundary layers, and farther downstream, the fully developed region, where the longitudinal velocity is independent of the position along the duct. It is assumed that the duct geometry (cross-section A, internal wetted perimeter p) does not change with the longitudinal position. A measure of the length scale of the duct cross section is the hydraulic diameter D_h defined in Eq. (1.21). The hydrodynamic entrance length X for laminar flow can be calculated with the formula (1.15), or, more exactly,

$$\frac{X}{D_h} \sim 0.05 \mathrm{Re}_{D_h} \tag{1.67}$$

where the Reynolds number is based on mean velocity and hydraulic diameter, $\mathrm{Re}_{D_h} = UD_h/\nu$. As shown in section 1.2.1, when the duct is much longer than its entrance length, $L \gg X$, the laminar flow is fully developed along most of the length L, and the friction factor is independent of L [cf. Eq. (1.23) and Table 1.2].

The heat transfer coefficient in fully developed flow is constant, that is, independent of longitudinal position. Table 1.2 lists the h values for fully developed laminar flow for two heating models: duct with uniform heat flux (q'') and duct with isothermal wall (T_w). Using these h values is appropriate when the duct length L is considerably greater than the thermal entrance X_T over which the temperature distribution is developing (i.e., changing) from one longitudinal position to the next. The thermal entrance length for the entire range of Prandtl numbers is

$$\frac{X_T}{D_h} \sim 0.05 \mathrm{Re}_{D_h} \tag{1.68}$$

Note that in Table 1.2 the Nusselt number ($\mathrm{Nu} = hD_h/k$) is based on D_h, which is unlike in Eq. (1.60). Means for calculating the heat transfer coefficient for laminar duct flows in which X_T is not much smaller than L can be found in Ref. [6].

Turbulent flow becomes fully developed hydrodynamically and thermally after a relatively short entrance distance:

$$X \sim X_T \sim 10 D_h \tag{1.69}$$

In fully developed turbulent flow ($L \gg X$) the friction factor f is independent of L, as shown by the family of curves drawn for turbulent flow on the Moody chart (Fig. 1.3). The turbulent flow curves can be used for ducts with other cross-sectional shapes, provided that D is replaced by the appropriate hydraulic diameter of the duct, D_h. The heat transfer coefficient h is constant in fully developed turbulent flow and can be estimated based on the Colburn analogy between heat transfer and

Table 1.2 Friction factors (f) and heat transfer coefficients (h) for hydrodynamically fully developed laminar flows through ducts [6].

Cross-section shape	$\text{Po} = f\text{Re}_{D_h}$	hD_h/k	
		Uniform q''	Uniform T_w
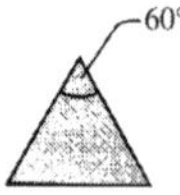 Equilateral triangle (60°)	13.3	3	2.35
Square	14.2	3.63	2.89
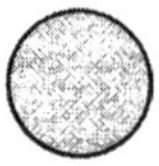 Circle	16	4.364	3.66
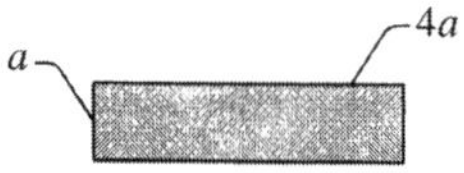 Rectangle (a × $4a$)	18.3	5.35	4.65
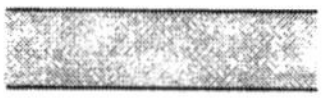 Parallel plates	24	8.235	7.54
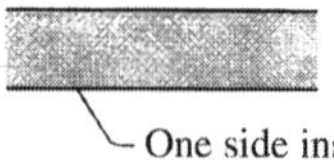 Parallel plates, one side insulated	24	5.385	4.86

momentum transfer:

$$\frac{h}{\rho c_P U} = \frac{f/2}{\text{Pr}^{2/3}} \tag{1.70}$$

This formula holds for $\text{Pr} \geq 0.5$ and is to be used in conjunction with Fig. 1.3, which supplies the f value. Equation (1.70) applies to ducts of various cross-sectional shapes, with wall surfaces having uniform temperature or uniform heat flux and various degrees of roughness. The dimensionless group $\text{St} = h/(\rho c_P U)$ is known as the Stanton number.

How does the temperature vary along a duct with heat transfer? Referring to Fig. 1.13, the first law can be applied to an elemental duct length dx to obtain

$$\frac{dT_m}{dx} = \frac{p}{A}\frac{q''}{\rho c_P U} \tag{1.71}$$

where p is the wetted internal perimeter of the duct and A is the cross-sectional area. Equation (1.71) holds for both laminar and turbulent flow. It can be combined with Eq. (1.64) and the h value furnished by Table 1.2 to determine the longitudinal variation of the mean temperature of the stream, $T_m(x)$. When the duct wall is

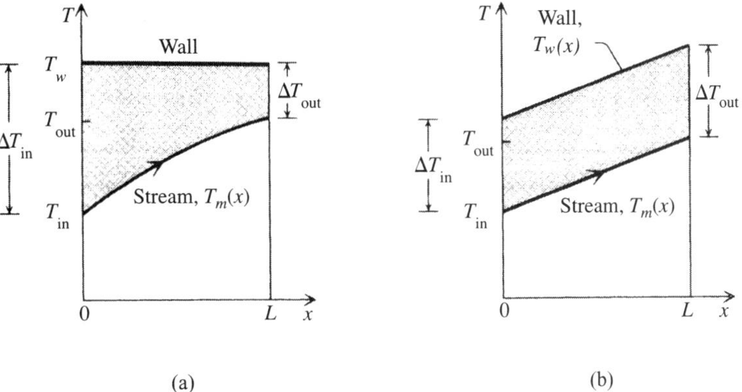

Figure 1.13 Distribution of temperature along a duct with heat transfer: (a) isothermal wall; (b) wall with uniform heat flux [8].

isothermal at T_w (Fig. 1.13a), the total rate of heat transfer between the wall and a stream with the mass flow rate $\dot{m}$ is

$$q = \dot{m} c_P \Delta T_{\text{in}} \left[1 - \exp \left(-\frac{h A_w}{\dot{m} c_P} \right) \right] \tag{1.72}$$

where A_w is the wall surface, $A_w = pL$. A more general alternative to Eq. (1.72) is

$$q = h A_w \Delta T_{\text{lm}} \tag{1.73}$$

where ΔT_{lm} is the log-mean temperature difference

$$\Delta T_{\text{lm}} = \frac{\Delta T_{\text{in}} - \Delta T_{\text{out}}}{\ln (\Delta T_{\text{in}} / \Delta T_{\text{out}})} \tag{1.74}$$

Equation (1.73) is more general than Eq. (1.72) because it applies when T_w is not constant, for example, when $T_w(x)$ is the temperature of a second stream in counterflow with the stream whose mass flow rate is $\dot{m}$. Equation (1.72) can be deduced from Eqs. (1.73) and (1.74) by writing $T_w =$ constant, and $\Delta T_{\text{in}} = T_w - T_{\text{in}}$ and $\Delta T_{\text{out}} = T_w - T_{\text{out}}$. When the wall heat flux is uniform (Fig. 1.13b), the local temperature difference between the wall and the stream does not vary with the longitudinal position x: $T_w(x) - T_m(x) = \Delta T$ (constant). In particular, $\Delta T_{\text{in}} = \Delta T_{\text{out}} = \Delta T$, and Eq. (1.74) yields $\Delta T_{\text{lm}} = \Delta T$. Equation (1.73) reduces in this case to $q = h A_w \Delta T$.

In natural convection, or free convection, the fluid flow is driven by the effect of buoyancy. This effect is distributed throughout the fluid and is associated with the tendency of most fluids to expand when heated. The heated fluid becomes less dense and flows upward, while packets of cooled fluid become more dense and sink.

One example is the natural convection boundary layer formed along an isothermal vertical wall of temperature T_w, height H, and width W, which is in contact with a fluid with far-field temperature T_∞. Experiments show that the transition from the laminar section to the turbulent section of the boundary layer occurs at the altitude y (between the leading edge, $y = 0$, and the trailing edge, $y = H$), where [8]

$$\mathrm{Ra}_y \sim 10^9\,\mathrm{Pr} \quad (10^{-3} < \mathrm{Pr} < 10^3) \tag{1.75}$$

In this expression Ra_y is the Rayleigh number based on temperature difference,

$$\mathrm{Ra}_y = \frac{g\beta(T_w - T_\infty)y^3}{\alpha\nu} \tag{1.76}$$

where β is the coefficient of volumetric thermal expansion, $\beta = (-1/\rho)\,(\partial\rho/\partial T)_P$. If the fluid behaves as an ideal gas, β equals $1/T$, where T is expressed in K. The heat transfer results shown next are valid when $|\beta(T_w - T_\infty)| \ll 1$. The boundary layer remains laminar over its entire height H when $\mathrm{Ra}_y < 10^9\,\mathrm{Pr}$. The average heat transfer coefficient (h) for a wall of height H with laminar boundary layer flow is [6]:

$$\frac{\bar{h}H}{k} = \begin{cases} 0.671\mathrm{Ra}_H^{1/4}, & \mathrm{Pr} \gg 1 \\ 0.8(\mathrm{Ra}_H\,\mathrm{Pr})^{1/4}, & \mathrm{Pr} \ll 1 \end{cases} \tag{1.77}$$

where $\mathrm{Ra}_H = g\beta\,(T_w - T_\infty)H^3/(\alpha\nu)$. The heat transfer rate from the wall is $q = \bar{h}HW(T_w - T_\infty)$, where W is the width (horizontal dimension) of the vertical wall.

A vertical wall that releases the uniform heat flux q'' into the fluid has a temperature that increases with altitude, $T_w(y)$. The relation between q'' and the wall-fluid temperature difference $(T_w - T_\infty)$ is represented adequately by Eq. (1.77), provided h is replaced by $q''/(\bar{T}_w - T_\infty)$. Note that the temperature difference $(\bar{T}_w - T_\infty)$ is averaged over the wall height H.

We end with examples of chimney flow: fully developed laminar flow driven by buoyancy through a vertical duct of hydraulic diameter D_h, height H, and inner surface temperature T_w. The top and bottom ends of the duct are open to a fluid at a temperature T_∞. It is assumed that the duct is sufficiently slender so that $H/D_h > \mathrm{Ra}_{D_h}$, where $\mathrm{Ra}_{D_h} = g\beta(T_w - T_\infty)D_h^3/\alpha\nu$. The average heat transfer coefficient h depends on the shape of the duct cross section:

$$\frac{hH/k}{\mathrm{Ra}_{D_h}} = \begin{cases} \dfrac{1}{192} & \text{(parallel plates)} \\ \dfrac{1}{128} & \text{(round)} \\ \dfrac{1}{113.6} & \text{(square)} \\ \dfrac{1}{106.4} & \text{(equilaterial triangle)} \end{cases} \tag{1.78}$$

The total heat transfer rate between the duct and the stream is approximately $q \cong \dot{m}c_P(T_w - T_\infty)$ when the group $\bar{h}A_w/\dot{m}c_p$ is greater than 1, where the mass flow rate is $\dot{m} = \rho AU$, and the duct cross-sectional area is A. The mean velocity U can be estimated using Eq. (1.22), in which $\Delta P/L$ is now replaced by $\rho g\beta(T_w - T_\infty)$. In other words, the chimney flow is Poiseuille flow driven upward by the effective vertical pressure gradient $\rho g\beta(T_w - T_\infty)$.

More examples of thermal resistance dominate the field of *radiation* heat transfer. For this body of thermal sciences we refer the reader to more complete treatments (e.g. Refs. [8, 9]).

Mass transfer is the transport of chemical species in the direction from high species concentrations to lower concentrations. Mass diffusion is analogous to thermal diffusion (section 1.3.1); therefore, it is not expanded in this section. In the simplest treatment, instead of the Fourier law (1.39), mass diffusion is based on the Fick law (Chapter 9), which proclaims a proportionality between species mass flux and concentration gradient. This topic is treated in detail in Chapter 9.

REFERENCES

1. A. Bejan, *Advanced Engineering Thermodynamics*, 2nd ed. (Ch. 13). New York: Wiley, 1997.
2. A. Bejan, *Shape and Structure, from Engineering to Nature*. Cambridge, UK: Cambridge University Press, 2000.
3. A. Bejan and S. Lorente, The constructal laws and the thermodynamics of flow systems with configuration. *Int J Heat Mass Transfer*, Vol. 47, 2004, pp. 3073–3083.
4. A. Bejan and S. Lorente, *La Loi Constructale*. Paris: L'Harmattan, 2005.
5. S. Lorente and A. Bejan, Svelteness, freedom to morph, and constructal multi-scale flow structures. *Int J Thermal Sciences*, Vol. 44, 2005, pp. 1123–1130.
6. A. Bejan, *Convection Heat Transfer*, 3rd ed. Hoboken, NJ: Wiley, 2004.
7. A. Bejan, G. Tsatsaronis, and M. Moran, *Thermal Design and Optimization*. New York: Wiley, 1996.
8. A. Bejan, *Heat Transfer*. New York: Wiley, 1993.
9. A. Bejan and A. D. Kraus, eds., *Heat Transfer Handbook*. Hoboken, NJ: Wiley, 2003.
10. A. Bejan, *Advanced Engineering Thermodynamics*, 3rd ed. Hoboken: Wiley, 2006.
11. A. Bejan and M. Almogbel, Constructal T-shaped fins. *Int J Heat Mass Transfer*, Vol. 43, 2000, pp. 141–164.
12. G. Lorenzini and S. Moretti, Numerical analysis of heat removal enhancement with extended surfaces. *Int J Heat Mass Transfer*, Vol. 50, 2007, pp. 746–755.
13. A. Bejan, How to distribute a finite amount of insulation on a wall with nonuniform temperature. *Int J Heat Mass Transfer*, Vol. 36, 1993, pp. 49–56.

PROBLEMS

1.1. *Flow strangulation* is not good for performance. Uniform distribution of strangulation is. Consider a long duct with Poiseuille flow (Fig. P1.1). The duct has two sections, a narrow one of length L_1 and cross-sectional area A_1, followed by a wider one of length L_2 and cross-sectional area A_2. The total length L, the mass flow rate, and the total duct volume are fixed. Show that the duct with minimal global flow resistance is the one with uniform cross-section ($A_1 = A_2$). To demonstrate this analytically, write that the pressure drop along each duct section is proportional to the flow length and inversely proportional to the cross-sectional area squared.

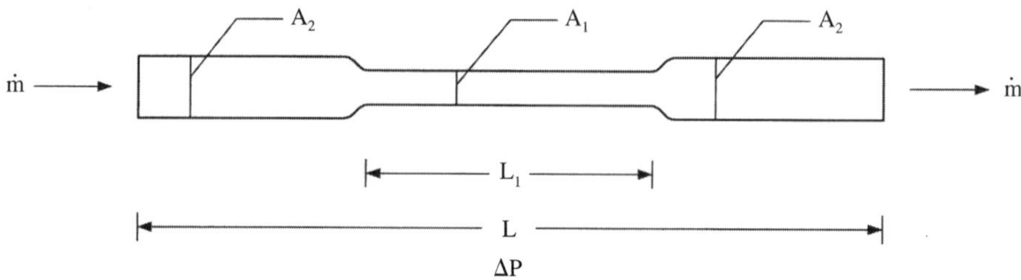

Figure P1.1

1.2. Consider the following heat conduction illustration (Fig. P1.2). A long bar of heat-conducting material (thermal conductivity k, length L) has two sections: one thin (length L_1, cross-section A_1) and the other thicker (length L_2, cross-section A_2). The total volume of conducting material (V) is fixed. Heat is conducted in one direction (along the bar) in accordance with the Fourier law. The global thermal resistance of the bar is $\Delta T/q$, where ΔT is the temperature drop along the distance L, and q is the end-to-end heat current. Show analytically that $\Delta T/q$ is minimum when $A_1 = A_2$, that is, *optimal distribution of imperfection* means *uniform distribution of flow strangulation*.

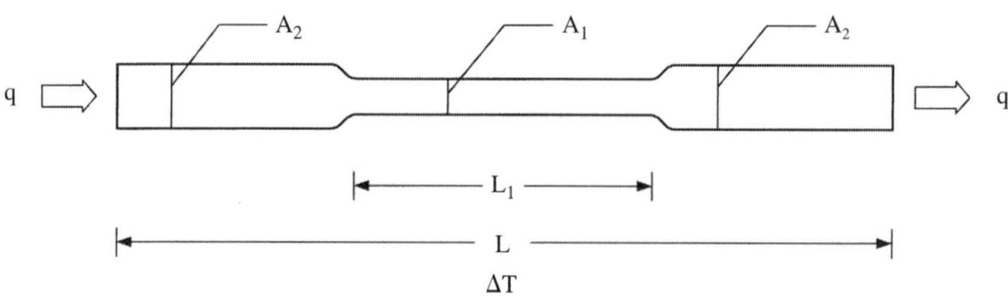

Figure P1.2

1.3. The phenomenon of *self-lubrication* is an illustration of the natural tendency of flow systems to generate configurations that provide maximum access for the things that flow. Consider the two-dimensional configuration shown in Fig. P1.3. The sliding solid body has an unspecified thickness D. The slot through which the body slides has the spacing $D + 2\delta$. The relative-motion gaps have the spacings $\delta + \varepsilon$ and $\delta - \varepsilon$, where ε represents the eccentricity of the body-housing alignment. The relative-motion gaps are filled with a

Newtonian liquid of viscosity μ. The body slides parallel to itself with the velocity V. Show that the body encounters minimum resistance when it centers itself in the slot ($\varepsilon = 0$).

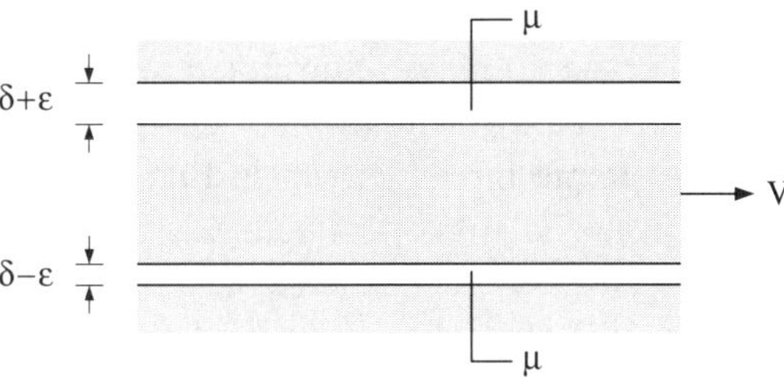

Figure P1.3

1.4. A round rod slides through a round housing as shown in Fig. P1.4. The motion of the rod is in the direction perpendicular to the figure. In general, the rod and housing cross-sections are not concentric. The eccentricity ε and the gap thickness $\delta(\theta)$ are much smaller than the rod radius R. The gap is filled with a lubricating liquid of viscosity μ. The rod slides with the velocity V. Show that the configuration that obstructs the movement of the rod the least is the concentric configuration ($\varepsilon = 0$). When cylindrical bodies slide with lubrication and coat themselves with a liquid film of constant thickness, they exhibit the natural phenomenon of *self-lubrication*. The concentric configuration, which is predictable from the maximization of flow access, is a constructal configuration.

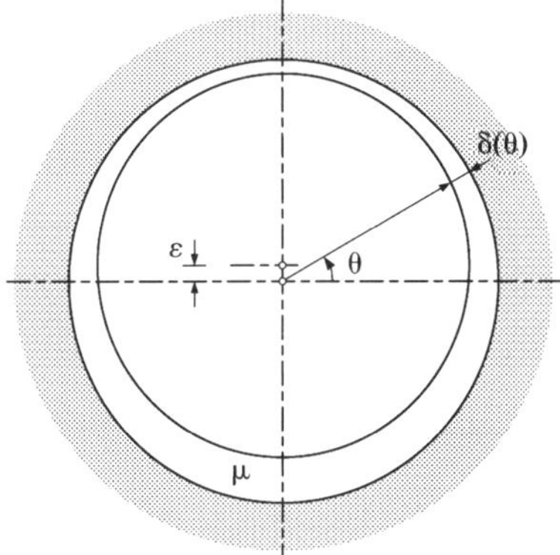

Figure P1.4

1.5. Two Newtonian liquids flow coaxially in the Poiseuille regime through a tube of radius R. The liquids have the same density ρ. The longitudinal pressure gradient $(-dP/dx)$ is specified. The liquids have different viscosities, $\mu_1 > \mu_2$. One liquid flows through a central core of radius R_i, and the other flows through an annular cross-section extending from $r = R_i$ to the wall ($r = R$). Show that the total flow rate is greater when the μ_2 liquid coats the wall. This phenomenon of self-lubrication is observed in nature and is demanded by the constructal law. Self-lubrication is achieved through the generation of flow configuration in accordance with the constructal law.

1.6. The modern impulse turbine developed based on the Pelton wheel invention has buckets in which water jets are forced to make 180° turns. According to the simple model shown in Fig. P1.6, nozzles bring to the wheel a stream of water of mass flow rate $\dot{m}$, room temperature, room pressure, and inlet velocity V_{in} (fixed). The bucket peripheral speed is V. The relative water velocity parallel to V, in the frame attached to the bucket, is V_r. The outlet stream velocity in the stationary frame is V_{out}. Determine the optimal wheel speed V such that the shaft power generated by the wheel is maximum.

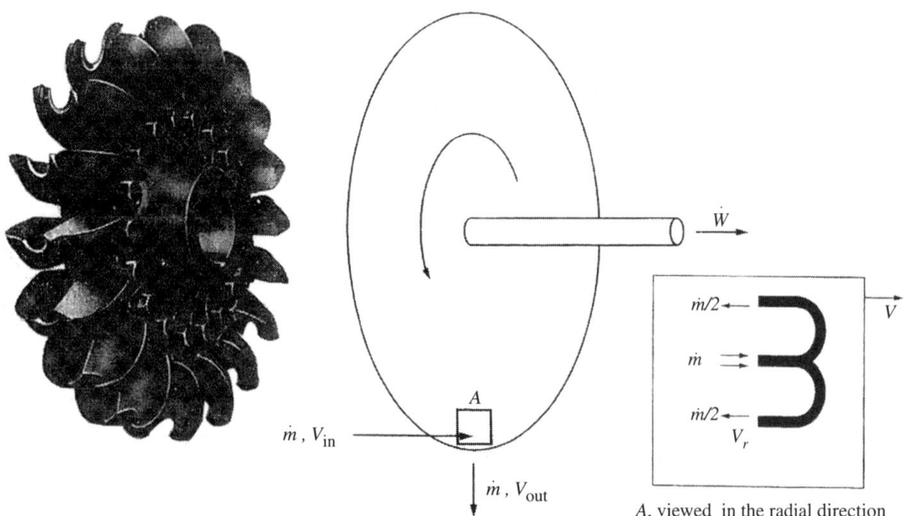

Figure P1.6

1.7. In this problem we explore the idea to replace the two-dimensional design of an airplane wing or wind mill blade with a three-dimensional design that has bulbous features on the front of the wing (Fig. P1.7) [10]. If the frontal portion of the wing is approximated by a long cylinder of diameter D_c, the proposal is qualitatively the same as replacing this material with several equidistant spheres of diameter D_s and spacing S. The approach air velocity is V. In the Reynolds number range $10^3 < VD_{c,s}/\nu < 10^5$, the sphere and cylinder drag coefficients are $C_s \cong 0.4$ and $C_c \cong 1$, (cf. Fig. 1.11). Evaluate the merit of this proposal by comparing the drag forces experienced by the spheres and cylinder segment, alone (i.e., by neglecting the rest of the wing).

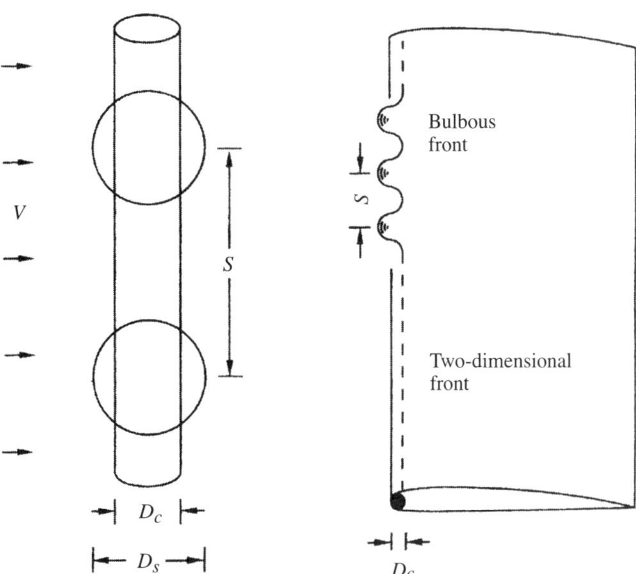

Figure P1.7

1.8. Show that the scale of the residence time of water in a river basin is proportional to the scale of the flow path length divided by the gradient (slope) of the flow path [10]. A simple and effective way to demonstrate this is to rely on constructal theory, according to which "optimal distribution of imperfection" in river basins means a balance between the high-resistance seepage along the hill slope and the low-resistance channel flow. Consequently, the global scales of the river basin are comparable with the scales of Darcy-flow seepage along the hill slope of length L, slope z/L, and water travel time t. Show that the proportionality between t and $L/(z/L)$ is $t \sim (\nu /gK)L/(z/L)$, where ν is the water kinematic viscosity and g is the gravitational acceleration. The permeability of soil K has values in the range 2.9×10^{-9} to 1.4×10^{-7} cm^2.

1.9. The time required to inflate a tire or a basketball is shorter than the time needed by the compressed air to leak out if the valve is left open. Model the inhaling-exhaling cycle executed by the lungs, and explain why the exhaling time is longer than the inhaling time, even though the two times are on the same order of magnitude. A simple one-dimensional model is shown in Fig. P1.9. The lung volume expands and contracts as its contact surface with the thorax travels the distance L. The thorax muscles pull this surface with the force F. The pressure inside the lung is P, and outside is P_{atm}. The flow of air into and out of the lung is impeded by the flow resistance r, such that $P_{atm} - P = r\dot{m}_{in}^n$ during inhaling and $P - P_{atm} = r\dot{m}_{out}^n$ during exhaling. The exponent n is a number between $n \cong 2$ (turbulent flow) and

$n = 1$ (laminar flow). For simplicity, assume that the density of air is constant and that the volume expansion rate during inhaling (dx/dt) is constant. The lung tissue is elastic and can be modeled as a linear spring that places the restraining force kx on surface A. Determine analytically the inhaling time (t_1) and the exhaling time (t_2), and show that $t_2/t_1 \geq 2$.

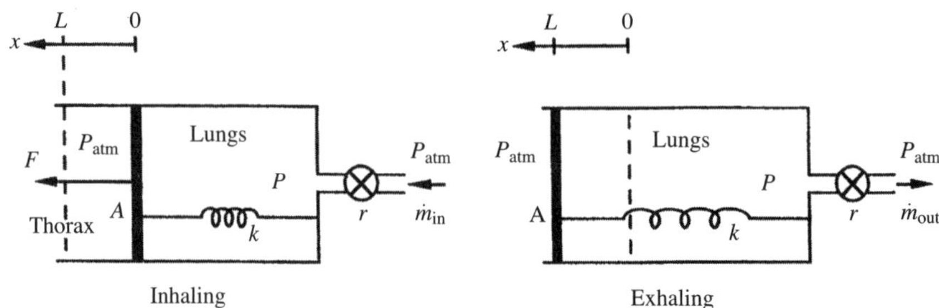

Figure P1.9

1.10. A domestic refrigerator and freezer have two inner compartments, the refrigerator or cooler of temperature T_r and the freezer of temperature T_f. The ambient temperature is T_H. This assembly is insulated with a fixed amount of insulation ($A_f L_f + A_r L_r =$ constant), where A_f and A_r are the surfaces of the freezer and the refrigerator enclosures. The thermal conductivity of the insulating material is k. Figure P1.10 shows this assembly schematically, in unidirectional heat flow fashion, with cold on the left, and warm on the right.

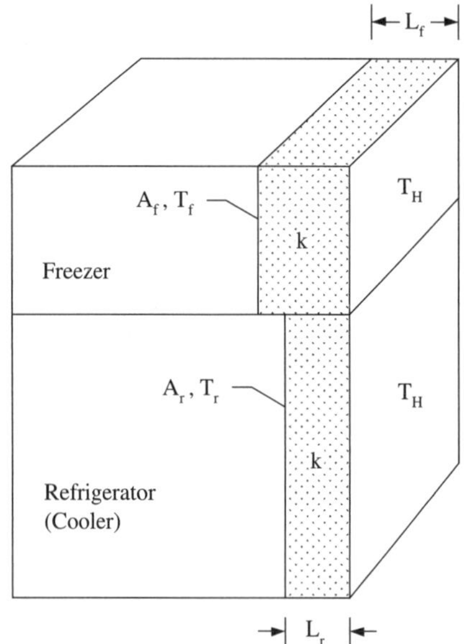

Figure P1.10

The two heat currents that penetrate the insulation and arrive at T_f and T_r are removed by a reversible refrigerating machine that consumes the power W. Determine the optimal insulation architecture, for example, the ratio L_f/L_r, such that the total power requirement W is minimum.

1.11. As shown in Fig. P1.11, a hot stream originally at the temperature T_h flows through an insulated pipe suspended in a space at temperature T_0. The stream temperature $T(x)$ decreases in the longitudinal direction because of the heat transfer that takes place from $T(x)$ to T_0 everywhere along the pipe. The function of the pipe is to deliver the stream at a temperature (T_{out}) as close as possible to the original temperature T_h. The amount of insulation is fixed. Determine the best way of distributing the insulation along the pipe.

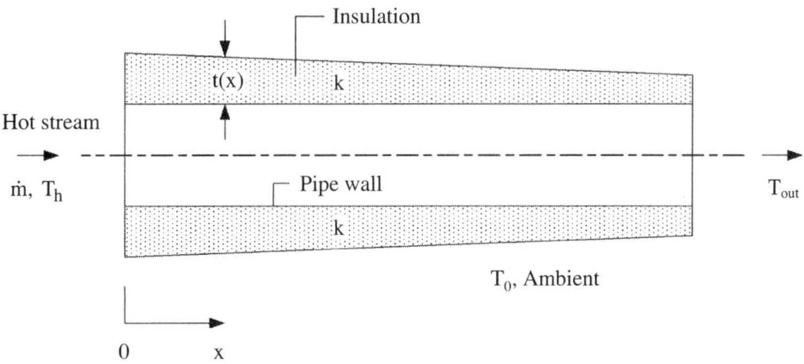

Figure P1.11

1.12. The temperature distribution along the two-dimensional fin with sharp tip shown in Fig. P1.12 is linear, $\theta(x) = (x/L)\theta_b$, where the local temperature difference is $\theta(x) = T(x) - T_\infty$ and $\theta_b = T_b - T_\infty$. The fin temperature is $T(x)$, base temperature is T_b, and the fluid temperature is T_∞. The fin width is W. The tip is at the temperature of the surrounding fluid: $\theta(0) = 0$. Show that the thickness δ of this fin must be $\delta = (h/k)x^2$. Derive an expression for the total heat transfer rate through the base of this fin, $\dot{Q}_b$.

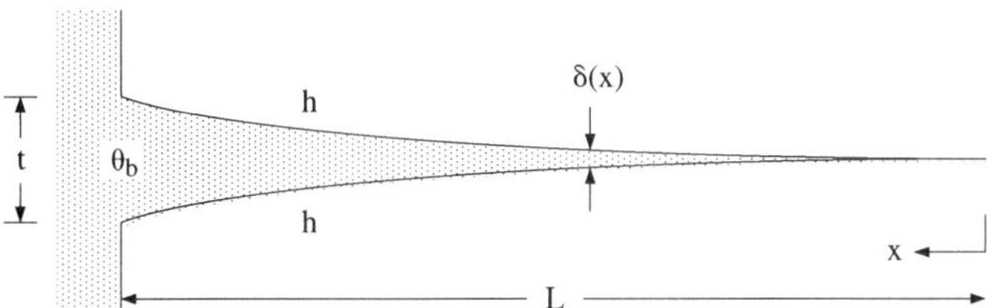

Figure P1.12

1.13. The sizing procedure for a fin generally requires the selection of more than one physical dimension. In a plate fin, Fig. P1.13, there are two such dimension, the length of the fin, L, and the thickness, t. One of these dimensions or the shape of the fin profile t/L can be selected optimally when the total volume of fin material is fixed. Such a constraint is often justified by the high cost of the high-thermal-conductivity metals that are employed in the manufacture of finned surfaces (e.g., copper, aluminum) and by the cost associated with the weight of the fin.

Consider a plate fin of constant thickness t, conductivity k, and length L extending from a wall of temperature T_b into a fluid flow of temperature T_∞ [8]. The plate-fin dimension perpendicular to $t \times L$ is W. The heat transfer coefficient h is uniform on all the exposed surfaces. The profile is slender ($t \ll L$), the Biot number is small ($ht/k \ll 1$), and the unidirectional fin conduction model applies. Assume that the heat transfer through the fin tip is negligible so that the total heat transfer rate through the fin base is $q_b = \theta_b (kA_c hp)^{1/2} \tanh(mL)$, where the notation is $\theta_b = (T_b - T_\infty)$, $m = (hp/kA_c)^{1/2}$, $A_c = tW$, and $p = 2W$. The global conductance of the fin is q_b/θ_b. Maximize this function subject to the constraint that the fin volume is fixed, $V = tLW$. Report analytically the optimal shape of the fin profile and the maximized global conductance.

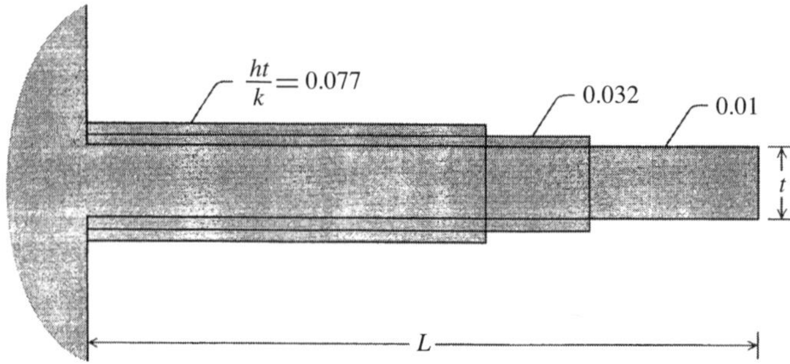

Figure P1.13

1.14. Consider the T-shaped assembly of fins [11, 12] sketched in Fig. P1.14. Two elemental fins of thickness t_0 and length L_0 serve as tributaries to a stem of thickness t_1 and length L_1. The configuration is two-dimensional, with the third dimension (W) sufficiently long in comparison with L_0 and L_1. The heat transfer coefficient h is uniform over all the exposed surfaces. The temperatures of the root (T_1) and the fluid (T_∞) are specified. The temperature at the junction (T_0) is one of the unknowns and varies with the geometry of the assembly.

The objective is to determine the optimal geometry (L_1/L_0, t_1/t_0) that is characterized by the maximum global thermal conductance $q_1/(T_1 - T_\infty)$, where q_1 is the heat current through the root section. The optimization is subjected to two constraints, the total-volume (i.e., front area) constraint,

$A = 2L_0L_1$, and the fin-material volume constraint, $A_f = 2L_0t_0 + t_1L_1$. The latter can be expressed in dimensionless form as the solid volume fraction $\phi_1 = A_f/A$, which is a specified constant considerably smaller than 1. The analysis that delivers the global conductance as a function of the assembly geometry consists of accounting for conduction along the L_0 ad L_1 fins and invoking the continuity of temperature and heat current at the T junction. The unidirectional conduction model is recommended for the analysis of each fin. Develop the global conductance as the dimensionless function $\tilde{q}_1(\tilde{L}_0, \tilde{t}_0, \tilde{L}_1, \tilde{t}_1, a)$, where

$$\tilde{q}_1 = \frac{q_1}{k_p W (T_1 - T_\infty)} \qquad a = \left(\frac{h A^{1/2}}{k_p}\right)^{1/2}$$

k_p is the fin material thermal conductivity and $(\tilde{L}_0, \tilde{t}_0, \tilde{L}_1, \tilde{t}_1) = (L_0, t_0, L_1, t_1)/A_1^{1/2}$. Set a = 1 and $\phi_1 = 0.1$. Use the dimensionless versions of the A and A_f constraints and recognize that the geometry of the assembly has only two degrees of freedom. Choose L_1/L_0 and t_1/t_0 as degrees of freedom. First, maximize numerically $\tilde{q}_1$ with respect to L_1/L_0 while holding t_1/t_0 constant, and your result will be the curve $\tilde{q}_{1,m}(t_1/t_0)$. Finally, maximize $\tilde{q}_{1,m}$ with respect to t_1/t_0 and report the twice-maximized conductance $\tilde{q}_{1,mm}$ and the corresponding architecture $(L_1/L_0)_{\text{opt}}$ and $(t_1/t_0)_{\text{opt}}$. Make a scale drawing of the T-shaped geometry that represents this design.

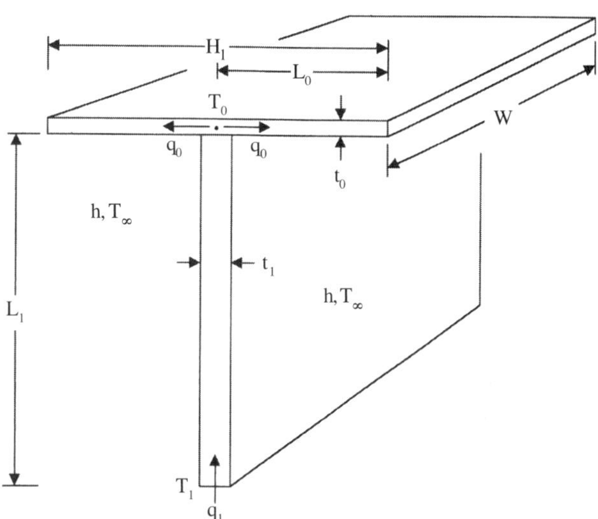

Figure P1.14

1.15. In this problem we consider the fundamental question of spreading optimally a finite amount of insulation on a nonisothermal surface. The simplest configuration in which this question can be examined is shown in Fig. P1.15. A plane wall of length L and width W (perpendicular to the figure) has the known temperature distribution $T(x)$. A layer of insulating material of low thermal conductivity k and unknown thickness $t(x)$ covers the wall. The temperature of the ambient is T_0. For simplicity, we assume that the thermal resistance (e.g., convection) between the outer surface of the insulation and the ambient is negligible relative to the resistance of the insulating layer. This is equivalent to assuming that the temperature of the outer surface of the insulating layer is essentially uniform and equal to T_0. When the layer is relatively thin ($t \ll L$), the flow of heat through the insulating material is unidirectional, perpendicular to the wall. Develop an integral expression for the total heat leak through the arbitrary-thickness design $t(x)$, and minimized it subject to the constraint that the total volume of insulation is fixed. Rely on variational calculus (see Appendix C) to show that the optimal insulation thickness function is

$$t_{\text{opt}}(x) = \frac{t_{\text{avg}} L [T(x) - T_0]^{1/2}}{\int_0^L [T(x) - T_0]^{1/2} dx}$$

where t_{avg} is the L-averaged thickness of the insulation. Determine the corresponding (minimal) heat leak through the insulation, q_{min}. The same $t_{\text{opt}}(x)$ distribution is obtained when the amount of insulation material is minimized subject to a fixed rate of heat loss to the ambient. Show that when $T(x)$ increases linearly from T_0 at $x = 0$ to $T_0 + \Delta T$ at $x = L$, the optimal thickness increases as $x^{1/2}$ along the wall. This problem was proposed and treated in detail in Ref. [13].

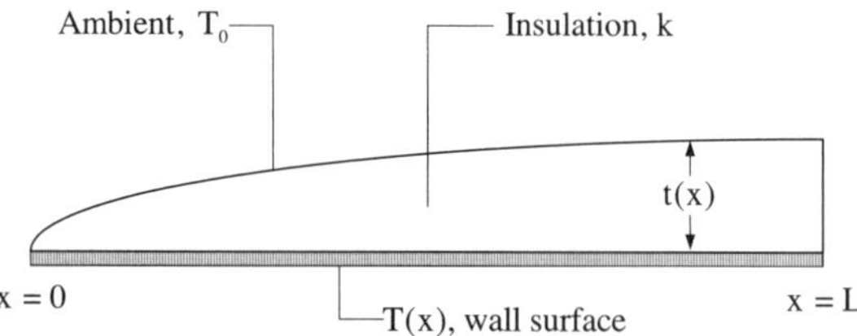

Figure P1.15

1.16. Consider again the problem of distributing a finite amount of insulation on a nonisothermal wall. In the preceding problem, we derived the general solution for the optimal distribution of insulation thickness and the corresponding total heat leak through the insulation, q_{min}. The example was for a wall with linear temperature distribution. In this problem, assume that the wall temperature $T(x)$ has the general form

$$T(x) = T_0 + \Delta T \left(\frac{x}{L}\right)^n$$

where the end excess temperature ΔT is fixed, and n is a positive constant. Determine $q_{\min}$ as a function of n. In addition, consider the case where insulation thickness is uniform ($t = t_{avg}$), and determine the corresponding heat leak (q), as a function of n. Show that the ratio $q_{\min}(n)/q(n)$ is equal to $(n+1)/(\frac{n}{2}+1)^2$, and that it decreases as n increases. Comment on the type of wall temperature distributions for which the use of the optimal thickness design is attractive.

2

IMPERFECTION

2.1 EVOLUTION TOWARD THE LEAST IMPERFECT POSSIBLE

Common sense and language are permanent reminders of the belief that obstacles (the effort to overcome resistances) are to be relegated to the past. They are sources of loss, imperfection, and "dissipation"—the irretrievable spreading of something into nothing. These concepts from the past (from our education before engineering) are clear and useful. They point to the future. It is good to avoid bottlenecks, from highway traffic to electrical wires that are too narrow for their currents.

It is the direction of the future to construct paths and structures, for example, the shortest path, the least expensive route, the lightest and strongest structural member, and so on. It is also *natural* for configurations to evolve in this way, and to succeed each other in time so that the greater system with the configuration "survives," that is, it persists in time. It is natural to morph *to go with the flow*.

In constructal theory, this natural behavior is seen as a phenomenon of generation of configuration in time. The configuration is the drawing of how and where the *channels* are distributed on the available area or volume. The territory is the *background*, the white on which the channels are the black. A flow system (e.g., river basin) covers the entire space, the white (seepage down hill slopes) and the black (rivulets and rivers). It takes at least two different ways of flowing, for example, diffusion (the white) and channeling (the black), in order to have configuration. In time, the imperfection is reduced by distributing channels to area elements so that the whole system flows more easily. To distribute black lines on white paper is to make the drawing. This natural tendency of flowing with better and better configuration is *optimal distribution of imperfection*, which is the essence of the constructal law (section 1.1).

Imperfection and perfection (ideality) are core concepts of thermodynamics. Sadi Carnot, the young engineer who kick-started thermodynamics, saw the direction of improvement for his configurations, the heat engines. According to him, future designs had to avoid fluid flow obstacles (friction, shocks) and heat flow jumps across finite temperature differences.

The future does not belong to the perfect, or the ideal. It belongs to the imperfect design that is *the least imperfect possible*. This has been the time arrow of design evolution in technology, biology, geomorphology, and social organization [1–4].

In this chapter we review the laws of thermodynamics in order to teach two ideas that are essential to design with constructal theory. One is the scientific meaning of imperfection: what it is, how to measure it, how to compare it with other imperfections, how costly it is, and how to reduce it. The other idea is the conceptual territory covered by classical thermodynamics, and how greatly the constructal law enlarges this territory. The conclusion will be that, with the constructal law, thermodynamics becomes a science of all nonequilibrium (flow) systems *with configuration* [5]. In this new domain, the constructal law is the self-standing principle that accounts for the time arrow of the phenomenon of generation of configuration everywhere.

2.2 THERMODYNAMICS

Thermodynamics describes the relationship between mechanical work, heating and other forms of energy transfer, or energy interactions. There are two facets of classical thermodynamics that must be stressed in an introductory chapter such as this. The first is the equivalence of *work* and *heat* as two possible forms of energy exchange. This facet is expressed by the first law of thermodynamics. The second aspect is the one-way character, or irreversibility of all flows that occur in nature. As expressed by the second law of thermodynamics, irreversibility or entropy generation is what prevents us from extracting the most possible work from various sources. It is also what prevents us from doing the most with the work that is already at our disposal.

In the first part of this chapter we review the first and second laws of thermodynamics and their implications in engineering, particularly with respect to imperfection in energy conversion and conservation. The analytical aspects (the formulas) of engineering thermodynamics are reviewed primarily in terms of the behavior of a pure substance, as would be the case of the working fluid in a heat engine or in a refrigeration machine. In the second part we introduce basic concepts of design—from objective and constraints to flow configuration—by using thermodynamics fundamentals and the common-sense objective of improving performance in processes and machines. This second part is an opening to the entire course, which teaches how to generate the configuration and how to achieve it strategically, with minimum cost.

Thermodynamic analysis requires the careful use of concepts and words that hold special, unambiguous meaning in the discipline. To talk about words may sound trivial, but without agreement on the meaning of words, any thermodynamic analysis is nonsense. Here are the most important entries in this glossary:

- *System*: The region or the collection of matter in space selected for analysis.
- *Boundary*: The real or imaginary surface delineating the thermodynamic system. The boundary separates the system from its environment. The boundary is an unambiguously defined surface. The boundary has zero thickness and zero volume.
- *Environment*: The thermodynamic system that is external to the thermodynamic system.
- *Closed system*: The thermodynamic system whose boundary is not crossed by mass flow.
- *Open system*: A thermodynamic system whose boundary is permeable to mass flow. Open systems have their own nomenclature: the system is usually referred to as the control volume, the boundary of the open system is the control surface, and the special regions of the boundary that are crossed by mass flows are the inlet and outlet ports.
- *State*: The condition (the being) of a thermodynamic system at a particular point in time, as described by an ensemble of quantities called thermodynamic properties (e.g., pressure, volume, temperature, energy, enthalpy, entropy). Properties are only those quantities that do not depend on the history (path) of the system between two different states. Quantities that depend on the system evolution (path) between states are not thermodynamic properties. Examples of nonproperties are the work, heat, and mass transfer; the entropy transfer; the entropy generation; and the destroyed exergy.
- *Extensive properties*: Properties whose values depend on the size of the system (e.g., mass, volume, energy, enthalpy, entropy).
- *Intensive properties*: Properties whose values do not depend on the size of the system (e.g., pressure, temperature).
- *Phase*: The collection of all system elements that have the same intensive state (e.g., the liquid droplets dispersed in a liquid-vapor mixture have the same intensive state, that is, the same pressure, temperature, specific volume, specific entropy, etc.).
- *Process*: The change of state from one initial state to a final state. In addition to the end states, knowledge of the process implies knowledge of the interactions experienced by the system while in communication with its environment (e.g., work transfer, heat transfer, mass transfer, and entropy transfer). To know the process also means to know the path (the history, or the succession of states) followed by the system from the initial to the final state.
- *Cycle*: The special process in which the final state coincides with the initial state.

2.3 CLOSED SYSTEMS

The first law of thermodynamics is a statement that brings together three concepts in thermodynamics: work transfer, heat transfer, and energy change. Of these concepts, only energy change or, simply, energy, is a thermodynamic property.

Consider the force F_x experienced by a system at a point on its boundary. The infinitesimal *work transfer* between system and environment is

$$\delta W = -F_x \, dx \tag{2.1}$$

where the boundary displacement dx is defined as positive in the direction of the force F_x. When the force $\boldsymbol{F}$ and the displacement of its point of application $d\boldsymbol{r}$ are not collinear, the general definition of infinitesimal work transfer is

$$\delta W = -\boldsymbol{F} \cdot d\boldsymbol{r} \tag{2.2}$$

According to the heat-engine sign convention, the work transfer is considered positive when the system does work on its environment—in other words, when $\boldsymbol{F}$ and $d\boldsymbol{r}$ are oriented in opposite directions. For example, if the closed system expands or contracts quasi-statically (i.e., slowly enough, in mechanical equilibrium internally and with the environment) so that at every point in time the pressure P is uniform throughout the system, then the work transfer term can be calculated as being equal to the work done by all the boundary pressure forces as they move with their respective points of application,

$$\int_1^2 \delta W = \int_1^2 P \, dV \tag{2.3}$$

The work transfer integral can be evaluated if the path of the quasi-static process, $P(V)$, is known. This is another reminder that the work transfer is path dependent (i.e., not a thermodynamic property).

In order for a system to experience work transfer, two things must occur: (1) a force must be present on the boundary, and (2) the point of application of this force (hence, the boundary) must move. The mere presence of forces on the boundary, without the displacement or the deformation of the boundary, does not mean work transfer. Likewise, the mere presence of boundary displacement without a force opposing or driving this motion does not mean work transfer. For example, in the free expansion of a gas into an evacuated space, the gas system does not experience work transfer because throughout the expansion the pressure at the imaginary system-environment interface is zero.

The physical evidence for a universal principle of energy conservation is as follows. If a closed system can interact with its environment only via work transfer (i.e., in the absence of heat transfer δQ discussed later), then measurements show that the work transfer during a change of state from state 1 to state 2 is the same for all processes linking states 1 and 2,

$$-\left(\int_1^2 \delta W\right)_{\delta Q=0} = E_2 - E_1 \tag{2.4}$$

In this special case the work transfer $(W_{1-2})_{\delta Q=0}$ is a property of the system because its value depends solely on the end states. This thermodynamic property is the *energy change* of the system, $E_2 - E_1$. The statement that preceded Eq. (2.4) is the first law of thermodynamics for closed systems that do not experience heat transfer.

Heat transfer is an energy interaction that can take place between a system and its environment. The transfer of heat is driven by the *temperature difference* established between the system and its environment. The distinction between δQ and δW is made by the second law of thermodynamics, discussed in the next section: heat transfer is the energy interaction accompanied by entropy transfer, whereas work transfer is the energy interaction taking place in the absence of entropy transfer.

Measurements also show that a closed system undergoing a change of state $1 \rightarrow 2$ in the absence of work transfer experiences a heat interaction whose magnitude depends solely on the end states:

$$\left(\int_1^2 \delta Q\right)_{\delta W=0} = E_2 - E_1 \tag{2.5}$$

In the special case of zero work transfer, the heat transfer is a thermodynamic property of the system because it is equal to the energy change experienced by the system in going from state 1 to state 2. Equation (2.5) is the first law of thermodynamics for closed systems incapable of experiencing work transfer. Heat transfer is considered positive when it increases the energy of the system.

Most thermodynamic systems do not manifest purely mechanical ($\delta Q = 0$) or purely thermal ($\delta W = 0$) behavior. Most systems manifest *coupled* mechanical and thermal behavior. The preceding first-law statements can be used to show that the first law of thermodynamics for a process executed by a closed system experiencing both work transfer and heat transfer is

$$\underbrace{\underbrace{\int_1^2 \delta Q}_{\substack{\text{heat}\\\text{transfer}}} - \underbrace{\int_1^2 \delta W}_{\substack{\text{work}\\\text{transfer}}}}_{\substack{\text{energy interactions}\\\text{(nonproperties)}}} = \underbrace{\underbrace{E_2 - E_1}_{\substack{\text{energy}\\\text{change}}}}_{\text{(property)}} \tag{2.6}$$

The first law means that the net heat transfer into the system equals the work done by the system on the environment, plus the increase in the energy of the system.

The first law of thermodynamics for a cycle or for an integral number of cycles executed by a closed system is

$$\oint \delta Q = \oint \delta W = 0 \tag{2.7}$$

The net change in the thermodynamic property energy is zero during a cycle or an integral number of cycles.

The energy change term $E_2 - E_1$ appearing on the right-hand side of the first law can be replaced by a more general notation that distinguishes between macroscopically identifiable forms of energy storage (kinetic $mV^2/2$, gravitational mgz) and energy stored internally (U),

$$\underbrace{E_2 - E_1}_{\text{energy change}} = \underbrace{U_2 - U_1}_{\text{internal energy change}} + \underbrace{\frac{mV_2^2}{2} - \frac{mV_1^2}{2}}_{\text{kinetic energy change}} + \underbrace{mgz_2 - mgz_1}_{\text{gravitational energy change}} \tag{2.8}$$

Here, m, V, and z are the mass, speed, and altitude of the center of mass of the system.

To review the second law as a summary of physical observations, consider a closed system executing a cycle or an integral number of cycles while in thermal communication with no more than one temperature reservoir. A temperature reservoir is a thermodynamic system that experiences only heat transfer and whose temperature remains constant during such interactions. Observe that the net work transfer during each cycle cannot be positive,

$$\oint \delta W \leq 0 \tag{2.9}$$

In other words, a closed system cannot deliver work during one cycle, while in communication with one temperature reservoir or with no temperature reservoir at all. An example of such cyclic operation is the vibration of a mass hanging from a spring or a ball bouncing on the pavement. In order for such systems to return to their respective initial heights, that is, in order for them to execute cycles, the environment (e.g., humans) must perform work on them.

The limit of frictionless cyclic operation is termed *reversible* because in this limit the system returns to its initial state without intervention (work transfer) from the environment. Therefore, the distinction between reversible and irreversible cycles executed by closed systems in communication with no more than one temperature reservoir is

$$\oint \delta W = 0 \quad \text{(reversible)} \tag{2.10}$$

$$\oint \delta W < 0 \quad \text{(irreversible)} \tag{2.11}$$

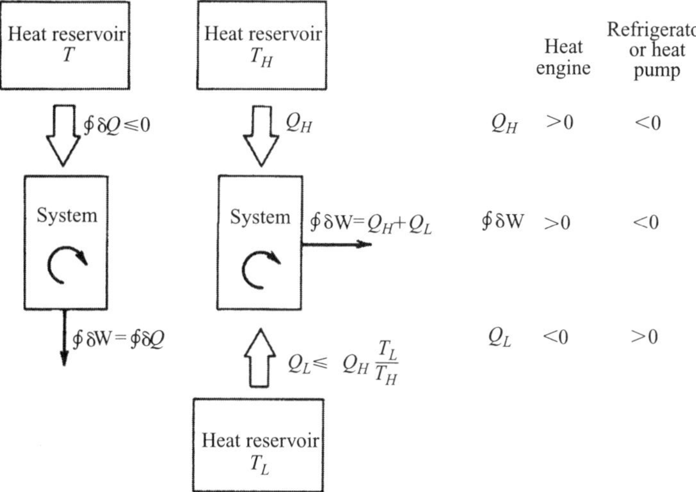

Figure 2.1 The first and second laws of thermodynamics for a closed system operating cyclically while in communication with one or two temperature reservoirs.

Put together, the first and second laws for closed systems operating cyclically in contact with no more than one temperature reservoir are (Fig. 2.1)

$$\oint \delta W = \oint \delta Q \leq 0 \tag{2.12}$$

This statement of the second law can be used to show [1] that in the case of a closed system executing one or an integral number of cycles while in communication with two temperature reservoirs, the following inequality holds (Fig. 2.1)

$$\frac{Q_H}{T_H} + \frac{Q_L}{T_L} \leq 0 \tag{2.13}$$

where H and L denote the high-temperature and the low-temperature reservoirs, respectively. Symbols Q_H and Q_L stand for the value of the cyclic integral $\oint \delta Q$, where δQ is in one case exchanged only with the T_H reservoir, and in the other with the T_L reservoir. In the reversible limit, the second law reduces to $T_H/T_L = -Q_H/Q_L$, which serves as definition for the thermodynamic temperature scale denoted by symbol T.

A heat engine is a special case of a closed system operating cyclically while in thermal communication with two temperature reservoirs. It is a system that during each cycle receives heat ($Q_H > 0$), rejects heat ($Q_L < 0$), and delivers work:

$$\oint \delta W = \oint \delta Q = Q_H + Q_L > 0 \tag{2.14}$$

The goodness of the heat engine can be described in terms of the first-law efficiency

$$\eta_I = \frac{\oint \delta W}{Q_H} \leq 1 - \frac{T_L}{T_H} \tag{2.15}$$

Alternatively, the second-law efficiency of the heat engine is defined as [6–8]

$$\eta_{II} = \frac{\oint \delta W}{\left(\oint \delta W\right)_{\substack{\text{maximum} \\ \text{(reversible case)}}}} = \frac{\eta_I}{1 - T_L/T_H} \tag{2.16}$$

A refrigerating machine or a heat pump operates cyclically between two temperature reservoirs ($Q_H < 0$), ($Q_L > 0$) such that during each cycle it receives work and delivers net heat to the environment,

$$\oint \delta W = \oint \delta Q = Q_H + Q_L < 0 \tag{2.17}$$

The goodness of such machines can be expressed in terms of a coefficient of performance (COP)

$$\text{COP}_{\text{refrigerator}} = \frac{Q_L}{-\oint \delta W} \leq \frac{1}{T_H/T_L - 1} \tag{2.18}$$

$$\text{COP}_{\text{heat pump}} = \frac{-Q_H}{-\oint \delta W} \leq \frac{1}{1 - T_L/T_H} \tag{2.19}$$

Generalizing the second law for closed systems operating cyclically, one can show that if during each cycle the system experiences any number of heat interactions Q_i with any number of temperature reservoirs whose respective absolute temperatures are T_i, then $\sum_i Q_i/T_i \leq 0$. Note that T_i is the absolute temperature of the boundary spot crossed by Q_i. Another way to write the second law in this case is

$$\oint \frac{\delta Q}{T} \leq 0 \tag{2.20}$$

where, again, T is the temperature of the boundary pierced by δQ.

Of special interest is the reversible cycle limit, in which the second law states that $(\oint \delta Q/T)_{\text{rev}} = 0$. According to the definition of thermodynamic property, the quantity $(\delta Q/T)_{\text{rev}}$ is the infinitesimal change in a property of the system. By definition, then, that property is the *entropy change*:

$$dS = \left(\frac{\delta Q}{T}\right)_{\text{rev}} \quad \text{or} \quad S_2 - S_1 = \left(\int_1^2 \frac{\delta Q}{T}\right)_{\text{rev}} \tag{2.21}$$

Combining this definition with the second law for a cycle, $\oint \delta Q/T \leq 0$, yields the second law of thermodynamics for any process executed by a closed system:

$$\underbrace{S_2 - S_1}_{\substack{\text{entropy}\\ \text{change}\\ \text{(property)}}} - \underbrace{\int_1^2 \frac{\delta Q}{T}}_{\substack{\text{entropy}\\ \text{transfer}\\ \text{(nonproperty)}}} \geq 0 \tag{2.22}$$

Note that any heat interaction (δQ) is accompanied by entropy transfer ($\delta Q/T$), whereas the work transfer δW is not. The entire left-hand side in the inequality (2.22) is by definition the *entropy generated* by the process:

$$S_{gen} = S_2 - S_1 - \int_1^2 \frac{\delta Q}{T} \tag{2.23}$$

Entropy generation is a measure of the inequality sign in the second law and hence a measure of the irreversibility of the process. As shown in Sections 2.6 and 2.7, entropy generation is also a measure of the imperfection, because it is proportional to the available work destroyed during the process.

2.4 OPEN SYSTEMS

If $\dot{m}$ represents the mass flow rate through a port in the control surface, the principle of mass conservation in the control volume requires that

$$\underbrace{\sum_{\text{in}} \dot{m} - \sum_{\text{out}} \dot{m}}_{\text{mass transfer}} = \underbrace{\frac{\partial M}{\partial t}}_{\text{mass change}} \tag{2.24}$$

Subscripts in and out refer to summation over all the inlet and outlet ports, respectively, while M stands for the instantaneous mass inventory of the control volume.

The first law of thermodynamics must account for the flow of energy associated with the $\dot{m}$ streams:

$$\underbrace{\sum_{\text{in}} \dot{m}\left(h + \frac{V^2}{2} + gz\right) - \sum_{\text{out}} \dot{m}\left(h + \frac{V^2}{2} + gz\right) + \sum_i \dot{Q}_i - \dot{W}}_{\text{energy transfer}} = \underbrace{\frac{\partial E}{\partial t}}_{\text{energy change}} \tag{2.25}$$

On the left side we have the energy interactions: heat, work, and the energy transfer associated with mass flow across the control surface. The specific enthalpy h, fluid velocity V, and height z are evaluated right at the boundary. On the right side, E is the instantaneous system energy integrated over the control volume.

The second law of thermodynamics for an open system has the form

$$\underbrace{\sum_{\text{in}} \dot{m}s - \sum_{\text{out}} \dot{m}s + \sum_{i} \frac{\dot{Q}_i}{T_i}}_{\text{entropy transfer}} \leq \underbrace{\frac{\partial S}{\partial t}}_{\text{entropy change}} \tag{2.26}$$

The specific entropy s is representative of the thermodynamic state of each stream right at the system boundary. The entropy generation rate is defined by

$$\dot{S}_{\text{gen}} = \frac{\partial S}{\partial t} + \sum_{\text{out}} \dot{m}s - \sum_{\text{in}} \dot{m}s - \sum_{i} \frac{\dot{Q}_i}{T_i} \tag{2.27}$$

The magnitude of the term $\dot{S}_{\text{gen}}$ is a measure of the irreversibility of open system operation.

The design importance of $\dot{S}_{\text{gen}}$ stems from the proportionality between it and the rate of destruction of available work. If the following parameters are fixed—all the mass flows ($\dot{m}$), the peripheral conditions (h, s, V, z), and the heat interactions (Q_i, T_i) except (Q_0, T_0)—then one can use the first law and the second law to show that the work transfer rate cannot exceed a theoretical maximum [1, 8]:

$$\begin{aligned} \dot{W} \leq & \sum_{\text{in}} \dot{m} \left(h + \frac{V^2}{2} + gz - T_0 s \right) \\ & - \sum_{\text{out}} \dot{m} \left(h + \frac{V^2}{2} + gz - T_0 s \right) - \frac{d}{dt}(E - T_0 s) \end{aligned} \tag{2.28}$$

The right side in this inequality is the maximum work transfer rate $\dot{W}_{\text{max}}$, which would exist only in the ideal limit of reversible operation. The rate of lost work, or the rate of exergy (availability) destruction, is defined as

$$\dot{W}_{\text{lost}} = \dot{W}_{\text{max}} - \dot{W} \tag{2.29}$$

Again, using both laws, one can show that lost work is directly proportional to entropy generation:

$$\dot{W}_{\text{lost}} = T_0 \dot{S}_{\text{gen}} \tag{2.30}$$

This result is known as the Gouy-Stodola theorem [6, 8]. Conservation of available work (exergy, useful energy) in thermodynamic systems can be achieved based only on the systematic *balance and distribution* of entropy generation in all the components of the system, that is, throughout the system.

2.5 ANALYSIS OF ENGINEERING COMPONENTS

This section contains a summary of the equations obtained by applying the first and second laws of thermodynamics to flow-system components encountered in most engineering systems, such as power cycles and refrigeration cycles. It is assumed that each component operates in steady state.

- *Valve (throttle)* or adiabatic duct with friction (Fig. 2.2a): <TB:vspace space="3pt"/>

$$\text{First law} \quad h_1 = h_2 \tag{2.31}$$

$$\text{Second law} \quad \dot{S}_{\text{gen}} = \dot{m}(s_2 - s_1) > 0 \tag{2.32}$$

- *Expander* or *turbine* with negligible heat transfer to the ambient (Fig. 2.2b):

$$\text{First law} \quad \dot{W}_T = \dot{m}(h_1 - h_2) \tag{2.33}$$

$$\text{Second law} \quad \dot{S}_{\text{gen}} = \dot{m}(s_2 - s_1) \geq 0 \tag{2.34}$$

$$\text{Efficiency} \quad \eta_T = \frac{h_1 - h_2}{h_1 - h_{2,\text{rev}}} \leq 1 \tag{2.35}$$

- *Compressor* or *pump* with negligible heat transfer to the ambient (Fig. 2.2c):

$$\text{First law} \quad \dot{W}_C = \dot{m}(h_2 - h_1) \tag{2.36}$$

$$\text{Second law} \quad \dot{S}_{\text{gen}} = \dot{m}(s_2 - s_1) \geq 0 \tag{2.37}$$

$$\text{Efficiency} \quad \eta_C = \frac{h_{2,\text{rev}} - h_1}{h_2 - h_1} \leq 1 \tag{2.38}$$

- *Nozzle* with negligible heat transfer to the ambient (Fig. 2.2d):

$$\text{First law} \quad \frac{1}{2}(V_2^2 - V_1^2) = h_1 - h_2 \tag{2.39}$$

$$\text{Second law} \quad \dot{S}_{\text{gen}} = \dot{m}(s_2 - s_1) \geq 0 \tag{2.40}$$

$$\text{Efficiency} \quad \eta_N = \frac{V_2^2 - V_1^2}{V_{2,\text{rev}}^2 - V_1^2} \leq 1 \tag{2.41}$$

- *Diffuser* with negligible heat transfer to the ambient (Fig. 2.2e):

$$\text{First law} \quad h_2 - h_1 = \frac{1}{2}(V_1^2 - V_2^2) \tag{2.42}$$

$$\text{Second law} \quad \dot{S}_{\text{gen}} = \dot{m}(s_2 - s_1) \geq 0 \tag{2.43}$$

$$\text{Efficiency} \quad \eta_D = \frac{h_{2,\text{rev}} - h_1}{h_2 - h_1} \leq 1 \tag{2.44}$$

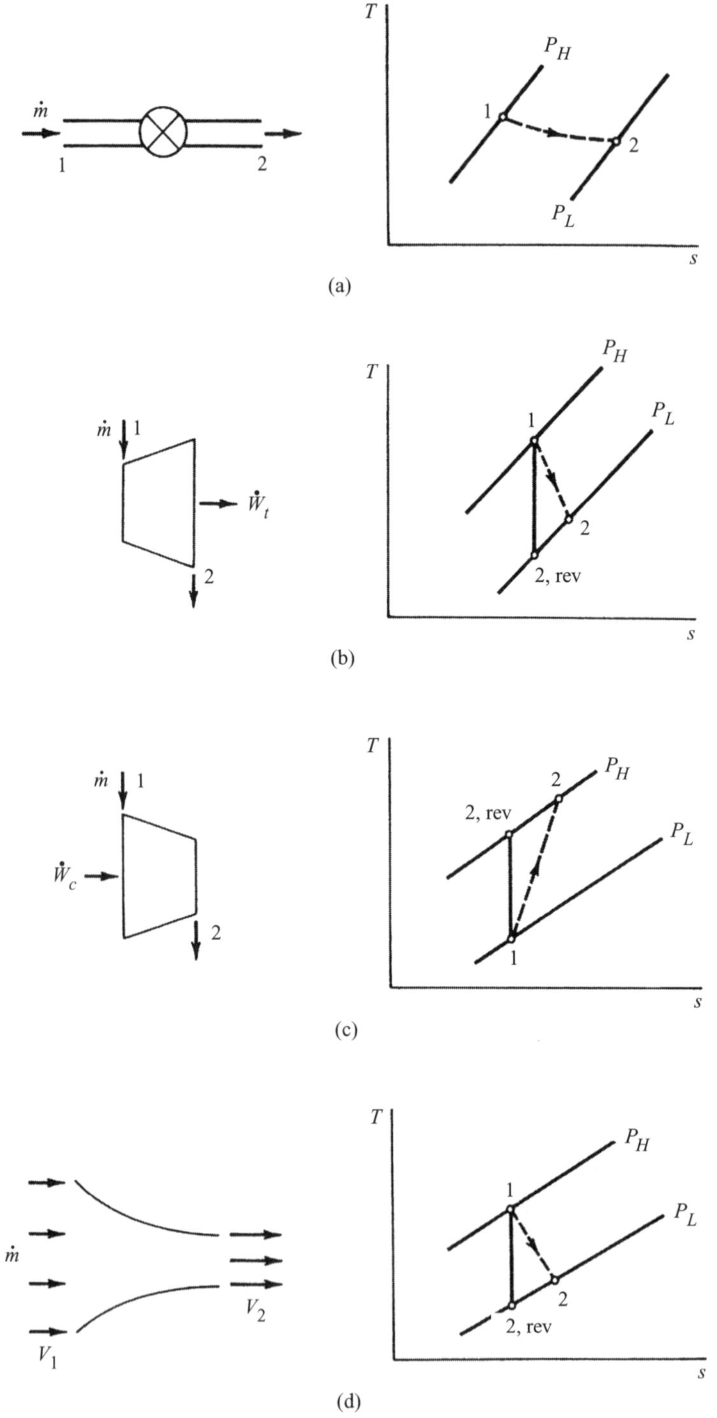

Figure 2.2 Engineering system components and their inlet and outlet states on the T-s plane, (P_H = high pressure; P_L = low pressure).

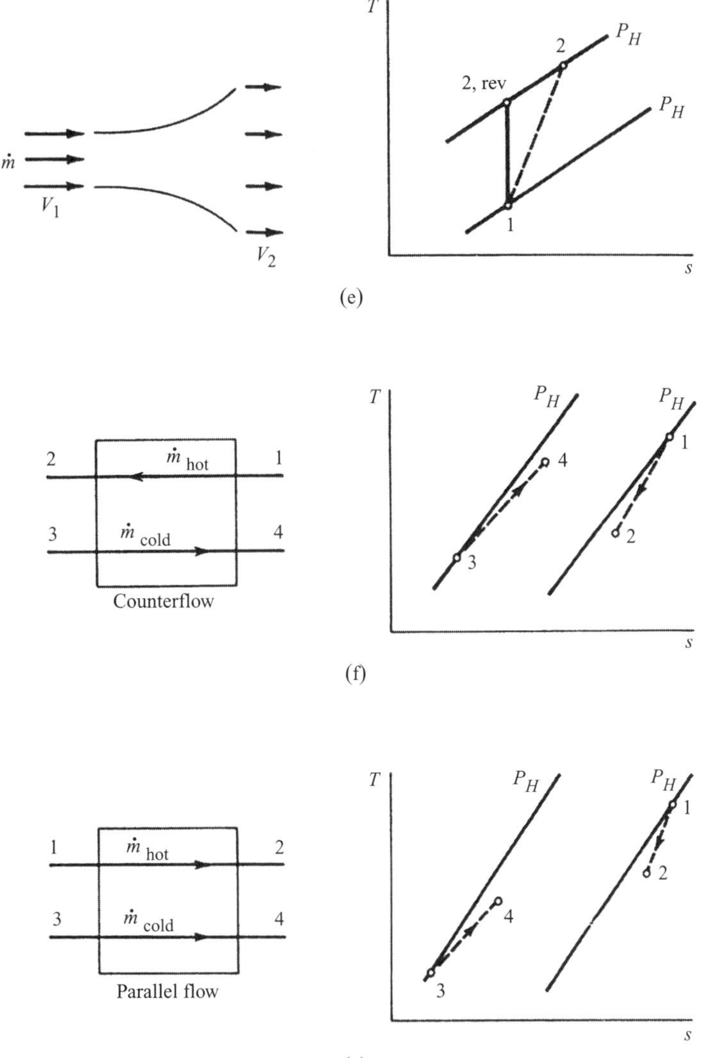

Figure 2.2 (*Continued*)

- *Heat exchangers* with negligible heat transfer to the ambient (Figs. 2.2f and 2.2g):

$$\text{First law} \qquad \dot{m}_{hot}(h_1 - h_2) = \dot{m}_{cold}(h_4 - h_3) \tag{2.45}$$

$$\text{Second law} \qquad \dot{S}_{gen} = \dot{m}_{hot}(s_2 - s_1) + \dot{m}_{cold}(s_4 - s_3) \geq 0 \tag{2.46}$$

Figures 2.2f and 2.2g, show that a pressure drop always occurs in the direction of flow, in any heat exchanger flow passage. Nature is such that, by themselves (i.e., if not forced), *all streams flow from high to low*.

2.6 HEAT TRANSFER IMPERFECTION

We are now in a position to make the connection between thermodynamic imperfection and the natural tendency of streams to keep on flowing by overcoming resistances (e.g., fluid flow and heat transfer; Chapter 1).

Consider the first drawing in Fig. 2.3, in which two systems of different temperatures (T_H, T_L) experience the heat transfer rate $\dot{Q}$. Note the two common notations for heat transfer rate: $\dot{Q}$, which is used in thermodynamics and this chapter, and q, which is used for heat current in the field of heat transfer and Chapter 1. Note that the two systems do not communicate directly, because if they did, they would have a common boundary (hence $T_H \equiv T_L$, which would contradict Fig. 2.3). Sandwiched between system T_H and system T_L is a third system: the temperature gap system. The heat current $\dot{Q}$ enters and leaves the temperature gap undiminished; that is, $\dot{Q}$ is conserved as it crosses the gap.

The second law for the temperature gap as a closed system in the steady state [cf. Eq. (2.27)] requires that

$$\dot{S}_{\text{gen}} = \frac{\dot{Q}}{T_L} - \frac{\dot{Q}}{T_H} = \frac{\dot{Q}}{T_L T_H}(T_H - T_L) \geq 0 \tag{2.47}$$

The imperfection exists (it is finite) when the product $\dot{Q}\,(T_H - T_L)$ is finite. Two features must be present at the same time for a temperature gap to be a source of imperfection—there must be a heat current, and the current must cross the abyss represented by the temperature difference. The stream ($\dot{Q}$) must fall from high (T_H) to low (T_L).

If T_L happens to be greater than T_H, the physical direction of $\dot{Q}$ must be the opposite of the direction shown in Fig. 2.3. The second law ($\dot{S}_{\text{gen}} \geq 0$) is always true.

A more palpable interpretation of this discussion is provided by the second drawing in Fig. 2.3, which is completely equivalent to the first drawing [6, 8]. This

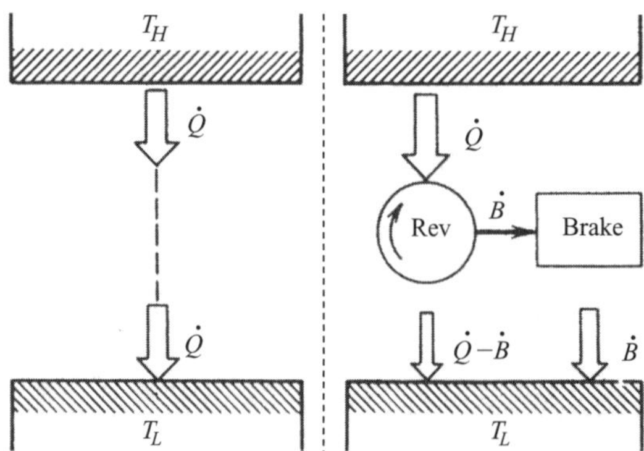

Figure 2.3 Imperfection of heat transfer: destruction of mechanical power in a temperature gap traversed by a heat current [6, 8].

time the temperature gap is occupied by an ideal heat engine and a brake. The heat engine is a closed system that operates steadily. For it, the first law and the second law require

$$\dot{Q}_H - \dot{Q}_L = \dot{W}_{\text{rev}} \tag{2.48}$$

$$\frac{\dot{Q}_H}{T_H} = \frac{\dot{Q}_L}{T_L} \tag{2.49}$$

In Fig. 2.3, the label for $\dot{Q}_H$ is $\dot{Q}$, the label for $\dot{Q}_L$ is $\dot{Q} - \dot{B}$, and instead of $\dot{W}_{\text{rev}}$ we wrote the mechanical power $\dot{B}$, which is transferred by the shaft of the engine to the brake. The brake is a closed system operating steadily and converting the mechanical power input $\dot{B}$ into an equal stream of heat ($\dot{B}$) dumped into the T_L reservoir. From the outside of the temperature-gap system, the observer sees only that $\dot{Q}$ enters and $\dot{Q}$ exits.

By eliminating $\dot{Q}_L$ between Eqs. (2.48) and (2.49), we find $\dot{W}_{\text{rev}}$, or, in the notation of Fig. 2.3 (right),

$$\dot{B} = \frac{T_H - T_L}{T_H}\dot{Q} \tag{2.50}$$

This brake power is lost. It is dissipated into heat, and dumped irretrievably into the ambient (T_L). Taken together, Eqs. (2.47) and (2.50) state that

$$\dot{B} = T_L \dot{S}_{\text{gen}} \geq 0 \tag{2.51}$$

which is the same as Eq. (2.30). In conclusion, the generated entropy, or the strength of the inequality sign in the second law is a measure of the imperfection.

2.7 FLUID FLOW IMPERFECTION

The story of fluid-flow imperfection is analogous to what we just presented for heat transfer. Consider a perfectly insulated duct through which the flow is steady. The control volume enclosed by the dashed boundary is an open adiabatic system with steady flow. It is the same sort of system as the control volume enclosing a valve (Fig. 2.2a), and this means that we can use Eqs. (2.31) and (2.32):

$$h_{\text{in}} = h_{\text{out}} \tag{2.52}$$

$$\dot{S}_{\text{gen}} = \dot{m}(s_{\text{out}} - s_{\text{in}}) \tag{2.53}$$

where $\dot{m} = \dot{m}_{\text{in}} = \dot{m}_{\text{out}}$. Next, we relate $\dot{S}_{\text{gen}}$ to the pressure drop $P_{\text{in}} - P_{\text{out}}$ by noting the differential relation between the changes in h, s, and P [1]:

$$dh = T\,ds + v\,dP \tag{2.54}$$

where the specific volume is $v = 1/\rho$. By integrating Eq. (2.54) from (out) to (in) along the duct, and noting that $dh = 0$ [cf. Eq. (2.52)], we obtain

$$\dot{S}_{\text{gen}} = \dot{m} \int_{\text{out}}^{\text{in}} \frac{dP}{(\rho T)}_{h=\text{constant}} \tag{2.55}$$

If the fluid is an ideal gas, then its two equations of state are

$$P = \rho R T \tag{2.56}$$

$$h = h(T) \tag{2.57}$$

The second equation of state shows that because $dh = 0$ along the stream, the temperature does not change from the inlet to the outlet. Replacing ρT with P/R in Eq. (2.55), we arrive at

$$\dot{S}_{\text{gen}} = \dot{m} R \ln \frac{P_{\text{in}}}{P_{\text{out}}} \tag{2.58}$$

which can be linearized in the limit $(P_{\text{in}} - P_{\text{out}}) \leq P_{\text{in}}$,

$$\dot{S}_{\text{gen}} \cong \dot{m} R \frac{P_{\text{in}} - P_{\text{out}}}{P_{\text{in}}} \tag{2.59}$$

If the fluid is incompressible, the corresponding result is [6, 8]

$$\dot{S}_{\text{gen}} = \frac{\dot{m}}{\rho T_{\text{in}}} (P_{\text{in}} - P_{\text{out}}) \tag{2.60}$$

In conclusion, Eqs. (2.59) and (2.60) show that the imperfection ($\dot{S}_{\text{gen}}$) is dictated by the product $\dot{m}(P_{\text{in}} - P_{\text{out}})$. It takes two features, flow ($\dot{m}$) and pressure drop, to have imperfection in fluid flow. This is the same as the conclusion drawn for heat transfer, where $\dot{S}_{\text{gen}}$ was dictated by the product $\dot{Q}\,(T_H - T_L)$. It is also the same as the conclusion for external fluid flow, for which the imperfection is proportional to the product $F_D U_\infty$ [6, 8]:

$$\dot{S}_{\text{gen}} = \frac{F_D U_\infty}{T_\infty} \tag{2.61}$$

As shown in section 1.2.3, the external flow configuration is characterized by the drag force F_D, the relative velocity U_∞, and the far-field fluid temperature T_∞. The mechanical power spent in order to drag the immersed body through the fluid is $\dot{W} = F_D U_\infty$, or after using Eq. (2.61),

$$\dot{W} = T_\infty \dot{S}_{\text{gen}} \tag{2.62}$$

This expression is the same as Eq. (2.51), and shows one more time that thermodynamic imperfection ($\dot{S}_{\text{gen}}$) means loss–the dissipation of useful work or mechanical power.

2.8 OTHER IMPERFECTIONS

Fluid flow and heat transfer are not the only sources of imperfection in nature and our designs. In this book we will deal with several other cases, which continue the analogy that springs out of sections 2.6 and 2.7: imperfection occurs whenever a current flows by overcoming a resistance. Mass diffusion down a concentration gradient, or across a finite concentration difference (ΔC), is another feature of imperfection (e.g., Chapter 9). The flow of an electric current through an electrical resistance is another example.

In results such as Eqs. (2.47), (2.60) and (2.61), we stressed the fact that the thermodynamic imperfection is ruled by the *product* of two characteristics of the current that flows. The two are the "across" quantity, which drives the flow, and the "through" quantity, which is the magnitude of the stream. Examples of across and through quantities mentioned in this chapter are put together in one drawing in Fig. 2.4. The existence of one (e.g., $\dot{m}$) without the other (e.g., ΔP), does not represent imperfection. For example, a heat current that flows across a vanishingly small temperature difference is accompanied by vanishingly small imperfection. This is the limit of *reversible* heat flow. Likewise, a finite temperature difference with zero heat leak across it does not represent imperfection. A perfectly insulated wall is a configuration of this kind.

In most flow configurations, the imperfection is due to more than one mechanism. Mass diffusion and electrical conduction occur together in electrodiffusion (e.g., section 9.1). The turbulent mixing of two dissimilar fluids is a can of worms of imperfection—currents (eddies) of momentum, heat, and species, each driven by local differences of pressure, temperature, and concentration. Heat exchangers

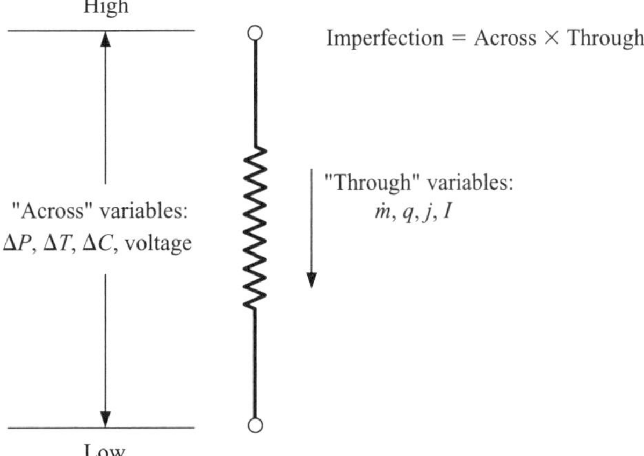

Figure 2.4 Thermodynamic imperfection can be thought of as the collaboration (the product) of *two* features of the flow configuration, the current and the forces that drive it (the "through" and the "across"). The two, together, generate imperfection.

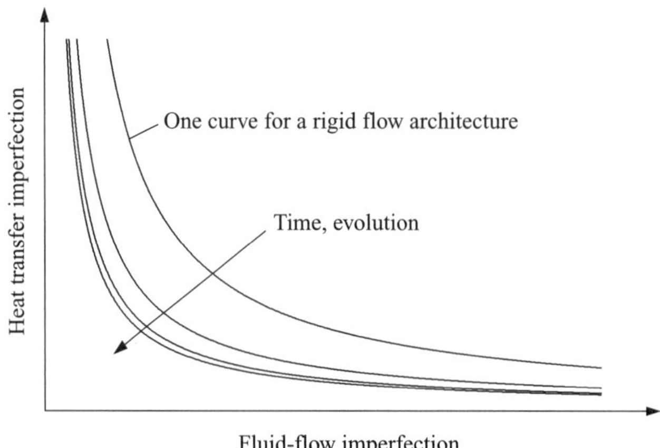

Figure 2.5 The time evolution of a flow configuration with imperfection of two kinds, fluid flow and heat transfer.

are flow configurations with fluid flow and heat transfer imperfections. Mass exchangers have two imperfections of their own, fluid flow and mass transfer. Such flow structures exist because they evolve in time by becoming less imperfect in all the features that plague them. This evolution is illustrated in Fig. 2.5 for a flow system with fluid-flow (x) and heat transfer (y) imperfection. A given (rigid) flow configuration is represented by one x-y curve. The future points toward the origin, in the direction of small x and small y. Survival is possible only by evolution, that is, *by changing the drawing* in ways that make the x-y curve migrate toward small x and small y. See again the constructal law (p. 2).

Even in simple flow systems with only one or two types of imperfection, there is an additional imperfection that is associated with the weight of the flow system. In the flowing nature, bodies must be carried. Organs inside animals, and flow components on vehicles, require the expenditure of useful energy (exergy, fuel, food) in order to be transported. From this important observation comes the discovery that a flow system must evolve not only toward a better configuration but also toward a better size [9]. This discovery is pursued in sections 2.9 and 4.2.

The thermodynamic meaning of flow imperfection, and the new place that the constructal law occupies in thermodynamics, form the subject of several segments of this course. Along the way, it is important to keep in mind that we consider all the imperfections important, and that we consider them *together*. Imperfections are not reduced to zero. They are not eliminated, alone or together. They are arranged, rearranged, and arranged again so that their combined contribution is the least imperfect whole. Out of this evolutionary process emerges the drawing, and the thermodynamics of systems with structure and configuration [5, 10].

2.9 OPTIMAL SIZE OF HEAT TRANSFER SURFACE

We end this chapter with an illustration of the importance and use of the concept of heat transfer imperfection (section 2.6). The additional lesson taught by this example is the demonstration of the existence of an "optimal organ size" [9] for use in a larger and more complex installation. This lesson is pursued further in section 4.2, because its implications stretch across the board, from engineering installations of all scales to animal design of all scales. The simplest way to demonstrate the existence of an optimal size for a component in a larger installation is to consider the total thermodynamic imperfection associated with the heat transfer configuration shown in Fig. 2.6. The heat transfer area of size A and temperature T_A has the function of transferring the heat current q to a fluid flow of temperature T_f. The heat transfer coefficient based on A is h; therefore,

$$q = hA(T_A - T_f) \tag{2.63}$$

There are at least two mechanisms of imperfection (destruction of useful energy) in this configuration. One is the flow of heat from T_A to T_f, which according to Eq. (2.47) is accompanied by the generation of entropy at the rate

$$\dot{S}_{\text{gen}} = \frac{q}{T_f} - \frac{q}{T_A} \tag{2.64}$$

Combining Eqs. (2.63) and (2.64), and assuming that $(T_A - T_f) \ll (T_A, T_f)$, we obtain

$$\dot{S}_{\text{gen}} = \frac{q^2}{T^2 hA} \tag{2.65}$$

where T stands for the thermodynamic temperature level represented by T_A and T_f. We may convert Eq. (2.65) into a rate of destruction of useful energy $\dot{W}_1$ by invoking the Gouy-Stodola theorem (2.30), where it is assumed that the temperature level T also represents the environment,

$$\dot{W}_1 = \frac{q^2}{ThA} \tag{2.66}$$

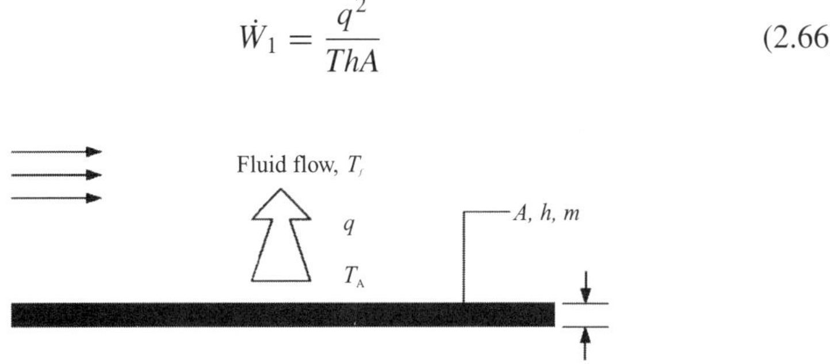

Figure 2.6 Finite-size surface with convective heat transfer to an external flow [9].

The second destruction mechanism is the mechanical power destroyed in order to carry the heat transfer surface A on the installation (e.g. vehicle or animal). According to the constructal theory of animal and engineered locomotion [8, 11], the useful energy needed in order to fly an incremental mass (m) is proportional to that mass:

$$\dot{W}_2 = 2\,mgV \tag{2.67}$$

Here, V is the thermodynamically optimized speed of the larger installation (vehicle). We treat V as a known parameter, which is dictated by the evolution of constructal locomotion in a global sense, by seeing the total mass of the vehicle as being much larger and fixed. In other words, the mass of the heat transfer device A is negligible relative to the mass of the vehicle. In the simplest model, the mass m is equal to $\rho_w\, tA$, where ρ_w and t are the density and thickness of the wall of heat transfer area A.

Adding $\dot{W}_1$ and $\dot{W}_2$, we obtain the total rate of useful-energy destruction associated with A,

$$\dot{W}_1 + \dot{W}_2 \cong \frac{q^2}{ThA} + 2gV\rho_w tA \tag{2.68}$$

The area size for which the total rate of useful-energy destruction is minimum is

$$A_{\text{opt}} = \frac{q}{(2gV\rho_w t\,Th)^{\,1/2}} \tag{2.69}$$

The optimal area size is proportional to the heat-transfer duty q, and inversely proportional to other parameters such as $V^{1/2}$ and $h^{1/2}$. In other words, there is an optimal heat flux, q/A_{opt}, which increases in proportion with $(V\rho_w th)^{1/2}$.

This demonstration of the optimal surface size for convection was based on the simplest model. More complex heat-transfer surfaces, such as the surfaces employed in heat exchangers, can be sized optimally based on the same argument—the trade-off between reduced heat-transfer imperfection and reduced mass. In the case of a heat exchanger, the imperfection is due to three mechanisms, heat transfer, fluid friction, and the transportation of the heat exchanger. In an animal organ such as the human lung, the imperfection is due to a more complicated combination of mechanisms: mass transfer, fluid friction, heat transfer, and the transportation of the organ. It is worth contemplating the question of optimal organ sizes in animals from the point of view of this section and Ref. [11]. It is also useful to consider it in the context of multiscale and multiobjective flow structures presented in Chapters 6 and 7.

REFERENCES

1. A. Bejan, *Advanced Engineering Thermodynamics*, 2nd ed. New York: Wiley, 1997.
2. A. Bejan and S. Lorente, Constructal theory of configuration generation in nature and engineering. *J Appl Phys*, Vol. 100, 2006, 041301.
3. A. Bejan, S. Lorente, A. F. Miguel, and A. H. Reis, *Along with Constructal Theory*. Lausanne, Switzerland: University of Lausanne, Faculty of Geosciences and the Environment, 2006.

4. A. Bejan and G. W. Merkx, eds., *Constructal Theory of Social Dynamics*. New York: Springer, 2007.
5. A. Bejan and S. Lorente, The constructal law and the thermodynamics of flow systems with configuration. *Int J Heat Mass Transfer*, Vol. 47, 2004, pp. 3203–3214.
6. A. Bejan, *Entropy Generation through Heat and Fluid Flow*. New York: Wiley, 1982.
7. A. Bejan, *Entropy Generation Minimization*. Boca Raton, FL: CRC Press, 1996.
8. A. Bejan, *Advanced Engineering Thermodynamics*, 3rd ed. (Ch. 13). Hoboken, NJ: Wiley, 2006.
9. A. Bejan and S. Lorente, Thermodynamic optimization of flow architecture: Dendritic structures and optimal sizes of components, Paper IMECE2002-33158, ASME International Mechanical Engineering Congress & Exposition, November 17–22, 2002, New Orleans, LA.
10. L. Wang, An approach to thermodynamic reasoning. *Int J Modern Phys B*, Vol. 10, No. 20, 1996, pp. 2531–2551.
11. A. Bejan and J. H. Marden, Constructing animal locomotion from new thermodynamics theory. *American Scientist*, Vol. 94, July–August 2006, pp. 342–349.

PROBLEMS

2.1. A cylinder contains air at a constant pressure P maintained by a frictionless piston, which is pushed from the outside with a constant force. The cylinder pressure P is greater than the atmospheric pressure P_0. There is a crack in the cylinder wall, and air leaks slowly through it. The piston travels slowly and without friction from right to left until all the compressed air has left the cylinder ($V_2 = 0$). During this motion, the air that is inside the cylinder is in thermal equilibrium with the atmosphere of temperature T_0. The system is defined by the dashed boundary shown in Fig. P2.1. The process starts from state 1 (volume V_1) and ends at state 2 (volume V_2).

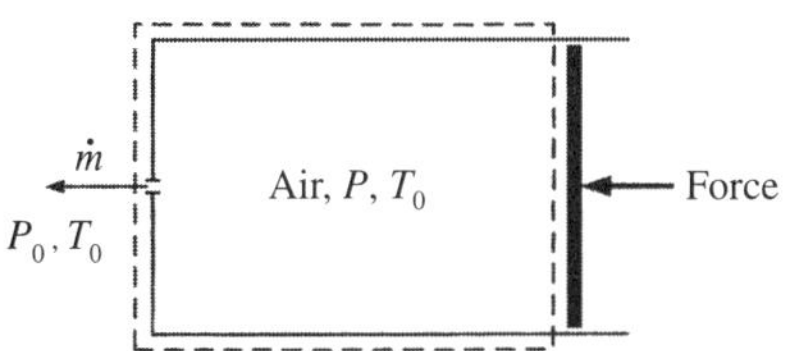

Figure P2.1

(a) Is the system open or closed?
(b) Is the process steady or unsteady?
(c) Write the per-unit-time statements of mass conservation and the first and second laws.
(d) Determine the size and direction of the total heat and work interactions experienced by the system, Q_{1-2} and W_{1-2}.
(e) Determine the total entropy generated by the system during the process, $S_{gen,1\text{-}2}$. Is the process $1 \rightarrow 2$ reversible or irreversible?

2.2. A rigid and perfectly insulated vessel contains air at atmospheric conditions (T_1, P_1). The volume of the vessel is V. At the time $t = t_1$, a stream of air $(\dot{m})$ at atmospheric temperature (T_1) is forced to flow into the vessel until $t = t_2$, when the pressure inside the container reaches $P_2 = 2P_1$. The objective is to determine the final temperature and mass of the air inside the vessel, relative to their respective initial values, namely T_2/T_1 and m_2/m_1. Solve the problem in this order:

(a) Select as "thermodynamic system" the air that resides instantaneously inside the vessel V. Is the system closed or open?

(b) Is the process from t_1 to t_2 steady or unsteady?

(c) Write the per-unit-time statement for mass conservation in the system.

(d) Write the per-unit-time statement for the first law for the system.

(e) Integrate (c) and (d) from t_1 to t_2, and combine these equations so that you obtain a single relation between the two unknowns, T_2 and m_2. [Hints: $h_1 = h(T_1)$ for ideal gas, $du = c_v dT$ for ideal gas, and $h_1 = u_1 + P_1 v_1$].

(f) Write the P-V-T equation of state for the air inside the system, at state 1 and at state 2. Combine the two statements, and obtain a second relation between the unknowns T_2 and m_2.

(g) Solve the system (e) + (f), and report analytically T_2/T_1 and m_2/m_1. Your expressions should show that $T_2 > T_1$ and $1 < m_2/m_1 < 2$.

2.3. Two masses (m_1, m_2) travel in the same direction at different velocities (V_1, V_2). They happen to touch and rub against each other and after a sufficiently long time they acquire the same velocity (V_∞). Consider the system composed of m_1 and m_2, and also consider the process (a) – (b) illustrated in Fig. P2.3. There are no forces between the system and its environment. Determine the ratio $y = KE_b/KE_a$ as a function of m_2/m_1 and V_2/V_1, where KE is the kinetic energy inventory of the system. Show that $\eta < 1$ when $V_2 \neq V_1$ and that η is of order 1 when m_2/m_1 is of order 1. Next, assume that the masses m_1 and m_2 are incompressible substances and that (a) and (b) are states of thermal equilibrium with the ambient of temperature T_0. Determine an expression for the heat transfer between the system and the environment.

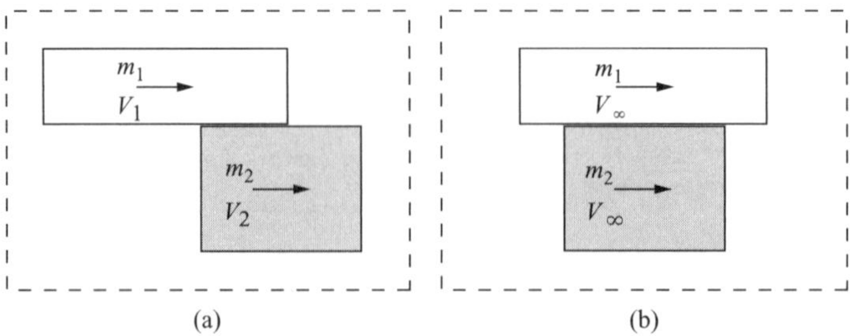

Figure P2.3

2.4. The brake of a vehicle can be modeled as a solid mass m, specific heat c, and spatially uniform temperature T, which is initially equal to the ambient temperature T_0. During a braking process, the brake receives the work input W, its temperature rises to T, and it experiences heat transfer (cooling) in contact with the ambient. Consider the system composed of the brake material and the adjacent air that is heated by the brake. In other words, the system boundary is a surface of temperature T_0.

(a) Determine the entropy generated by the system during the braking process and show how S_{gen} varies with W, mc, T, and T_0.

(b) Show that S_{gen} is generally less than W/T_0. How do you explain this inequality?

(c) Following the braking process, the brake material is cooled further, and its temperature returns to T_0. Determine the total entropy generated from the start of the braking process until the brake cools down to T_0. How do you explain the fact that the total entropy generation is greater than the value calculated in part (b)?

2.5. In the Middle Ages, hydraulic power was produced near waterfalls and fast (inclined) rivers by using waterwheels. A waterwheel has any buckets on its rim. On the heavy side of the wheel, the buckets face upward and are filled with water from the natural stream. On the opposite side of the wheel, the buckets are upside down (empty), the half-rim is light, and the wheel turns with finite torque and delivers power to its user, who is connected to the shaft.

An even simpler version of this invention is the "air wheel" shown in Fig. P2.5. The wheel is immersed in a stationary body of water. On its rim it has a number of equidistant cylinders filled with air and with pistons, which are frictionless and leak proof. The figure shows only a pair of diametrically opposed air cylinders. On the left side, the cylinder faces upward, the air charge is compressed by the weight of the piston, and consequently, the air displaces less of the surrounding water. On the right side, the cylinder faces downward, the air charge expands, and the volume of displaced water is greater. The shaft of the air wheel is connected to a flour mill, which needs mechanical power to turn. How will the air wheel turn?

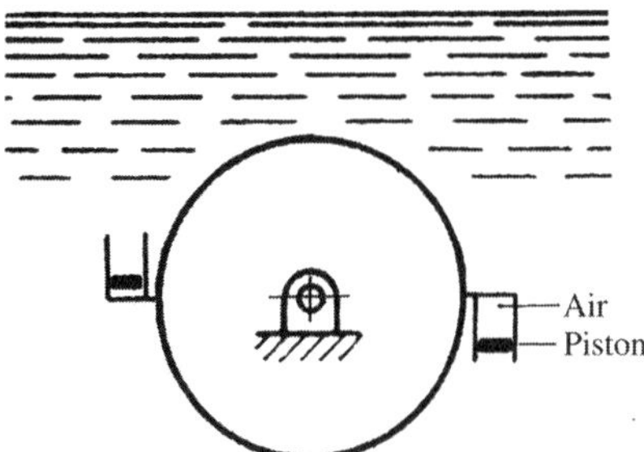

Figure P2.5

2.6. The work produced by the irreversible power plant (P) shown in Fig. P2.6 drives the irreversible refrigeration plant (R). The second-law efficiencies of the power and refrigeration plant are $\eta_{II}^{(P)}$ and $\eta_{II}^{(R)}$. Now consider the aggregate system ($P + R$), which operates between T_H and T_L. Show that the second-law efficiency of the aggregate system is

$$\eta_{II}^{(P+R)} = \eta_{II}^{(P)}\eta_{II}^{(R)}$$

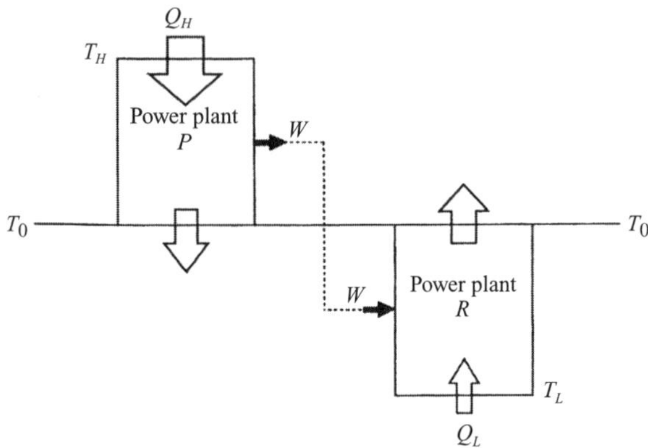

Figure P2.6

2.7. It is contemplated to solve simultaneously the heating requirements of an office building and the refrigeration requirements of an ice-storage facility by installing the refrigerator/heat pump system shown in Fig. P2.7. The temperature of the ice-storage facility (T_1) and the temperature of the office building interior (T_2) are fixed by design. The steady-state heat leak from the ambient to the ice-storage facility (Q_1) and the heat leak from the building interior to the ambient (Q_2) are both proportional to the respective temperature differences that drive the leak, $\dot{Q}_1 = C_1(T_0 - T_1)$ and $\dot{Q}_2 = C_2(T_2 - T_0)$, where T_0 is the ambient temperature, and heat conductances C_1 and C_2 are known constants. The refrigerator/heat pump system is in thermal contact with the ambient and experiences the heat transfer input $\dot{Q}_0$ (note the positive direction of the energy interaction arrows and the vertical alignment of temperatures on the figure). Finally, it is assumed that the refrigerator/heat pump system operates cyclically and reversibly.

(a) Determine the mechanical power requirement $\dot{W}$ as a function of T_0, T_1, T_2, C_1, and C_2.

(b) The ambient temperature T_0 can vary daily or seasonally while T_1 and T_2 remain fixed. Determine the ambient temperature $T_{0,\text{opt}}$ that would minimize the power requirement $\dot{W}$.

(c) Show that the optimum condition described in part (b) corresponds to the special case where the refrigerator/heat pump system does not exchange heat with the ambient ($\dot{Q}_1 = 0$).

(d) What is the entropy-generation rate $\dot{S}_{gen}$ of the aggregate system that includes the refrigerator/heat pump apparatus, the ice storage facility, and the office building?

(e) Show that regardless of weather conditions (T_0), the power requirement $\dot{W}$ is directly proportional to $\dot{S}_{gen}$.

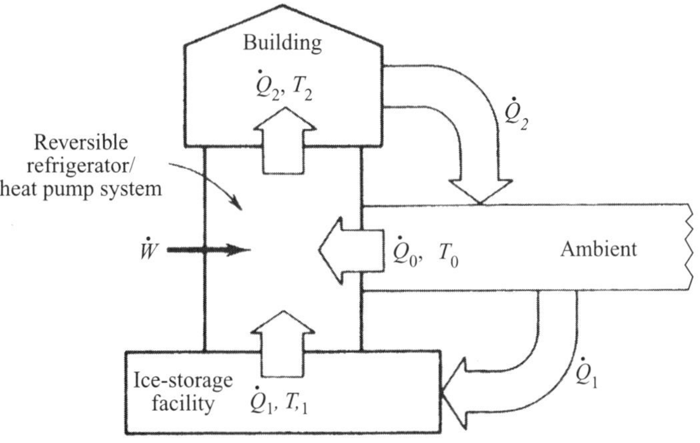

Figure P2.7

2.8. This problem is about the fundamentals of global circulation and natural convection in general. We explore the fundamentals in two parts, in accordance with Fig. P2.8.

(a) Consider a stream of ideal gas with the flow rate $\dot{m}$, which flows isothermally and reversibly through the system shown in Fig. P2.8a. The temperature T is constant throughout the system. The inlet and outlet pressures are P_{in} and P_{out}. Invoke the first and second laws, the ideal gas model, and the isothermal and reversible model, and show that the heat input rate $\dot{Q}$ and work output rate $\dot{W}$ are equal and given by

$$\dot{Q} = \dot{W} = \dot{m}RT \ln\frac{P_{in}}{P_{out}}$$

(b) Next, the circulation of the atmosphere can be modeled as a heat engine that functions in a cycle of our processes (Fig. P2.8b): 1–2, isothermal heating and expansion at T_H, 2–3, isobaric cooling at P_L, 3–4,

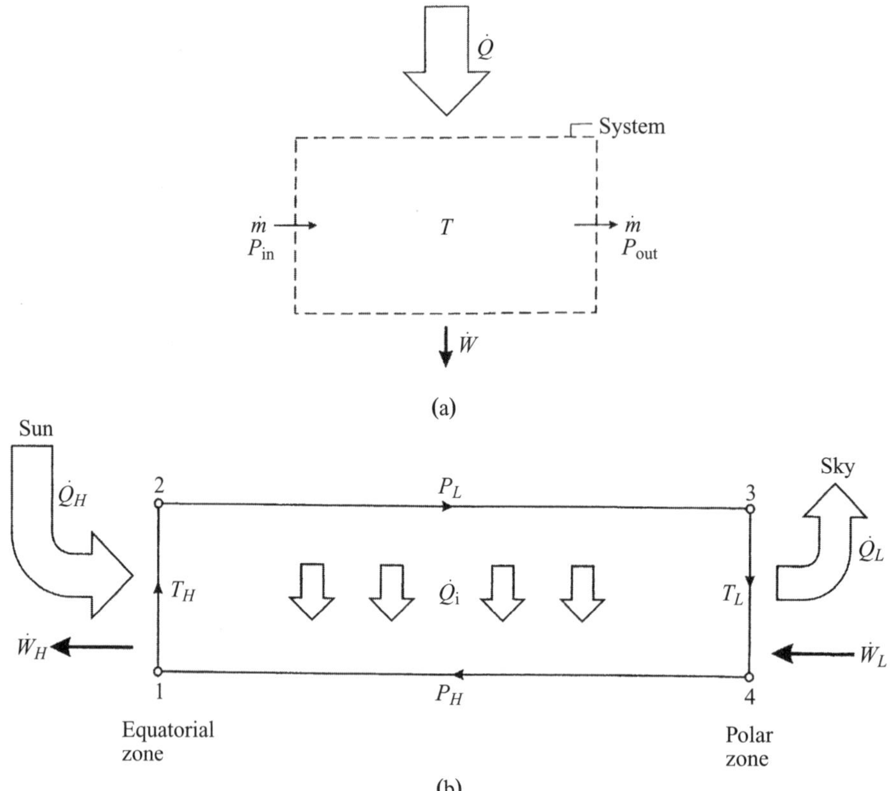

Figure P2.8

isothermal cooling and compression at T_L, and 4–1, isobaric heating at P_H. The cycle is executed reversibly: There are no pressure drops from 2 to 3 and from 4 to 1, and locally, there is no temperature difference between the 2–3 and 4–1 streams. The internal (regenerative) heat transfer $\dot{Q}_i$ occurs across a zero temperature difference. The heating and expansion process is a model for how the air warms up and rises to higher altitudes (lower pressures) over the equatorial zone (T_H). The cooling and compression is a model for the sinking of the same airstream over the polar zones (T_L). The counterflow formed by the 4–1 and 2–3 streams is a model for the circulation of the atmosphere in the meridional direction.

Use the results of part (a) to calculate the net power output of the atmospheric heat engine $\left(\dot{W}_{\text{net}} = \dot{W}_H - \dot{W}_L\right)$ and the energy conversion efficiency $\eta = \dot{W}_{\text{net}}/\dot{Q}_H$. Does your resulting expression for η look familiar? Why?

2.9. For the design of a new power plant, an engineer proposes to change the configuration in such a way that the power plant achieves two objectives independently, maximum power output $\dot{W}$, and minimum entropy generation rate $\dot{S}_{gen}$. The engineer realizes that a larger $\dot{W}$ and a smaller $\dot{S}_{gen}$ mean a larger ratio $\dot{W}/\dot{S}_{gen}$, and on this basis he decides to pursue a design that maximizes the ratio $\dot{W}/\dot{S}_{gen}$. Is this decision correct? If not, then for what special class of power plants does the maximization of $(\dot{W}/\dot{S}_{gen})$ mean that the $\dot{W}$ maximum and $\dot{S}_{gen}$ are achieved at the same time? Can the thermodynamic performance of this special class of power plants be improved?

2.10. A heat engine communicating with two temperature reservoirs (T_H, T_L) can be modeled as a Carnot engine connected in parallel with a conducting strut. In this model the conducting strut accounts for the leakage of heat from T_H to T_L around the power-producing components of the engine. The bypass heat leak Q_C is proportional to the temperature difference across it, $Q_C = C(T_H - T_L)$, where C is a measured constant.

(a) Is the heat engine operating reversibly? Explain.

(b) If T_H is a design variable, determine the optimal temperature $T_{H,\text{opt}}$ such that the work output (or heat engine efficiency W/Q_H) is maximum.

2.11. A combined-cycle power plant is configured as a gas-turbine power plant at high temperatures and a steam-turbine power plant at lower temperatures. Model the two power plants as closed systems that execute cycles. The gas-turbine cycle receives the heat input Q_1, produces the work W_1, and rejects the heat transfer Q_2. The steam-turbine cycle is driven by Q_2 and produces the work W_2. The two cycles are not reversible. Show that the first-law efficiency of the combined-cycle power plant $[\eta = (W_1 + W_2)/Q_1]$ is given by

$$\eta = \eta_1 + \eta_2(1 - \eta_1)$$

where η_1 and η_2 are the first-law efficiencies of the gas-turbine cycle and steam-turbine cycle, respectively.

2.12. In the simple model shown in Fig. P2.12, the irreversibilities of a refrigeration plant are assumed to be located in the two heat exchangers, $(UA)_H$ above room temperature (T_H, fixed) and $(UA)_L$ below the refrigeration load temperature (T_L, fixed). A reversible compartment between T_{HC} and T_{LC} completes the model. The heat transfer rates $\dot{Q}_H$ and $\dot{Q}_L$ are given by $\dot{Q}_H = (UA)_H\,(T_{HC} - T_H)$ and $\dot{Q}_L = (UA)_L\,(T_L - T_{LC})$. With reference to the right-hand drawing of Fig. P9.1, minimize $UA = (UA)_H + (UA)_L$ while regarding the refrigeration load $\dot{Q}_L$ as specified. Show that in this case $(UA)_{\min}$ is divided equally between the two heat exchangers. Derive an expression for $\dot{W}/\dot{Q}_L$ and comment on the effect of $(UA)_{\min}$.

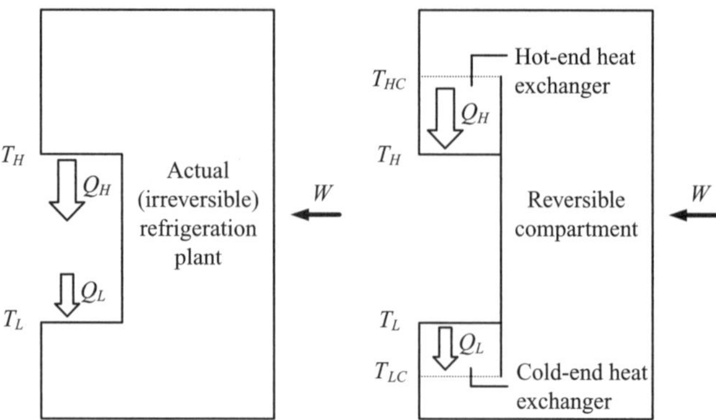

Figure P2.12

2.13. It is widely recognized that fluid flow friction is one of the sources of "imperfection," which must be distributed optimally during the configuring of the flow architecture. In this problem we illustrate the thermodynamic meaning of such imperfection. On the left side of Fig. P2.13, we see an adiabatic turbine that expands a stream of ideal gas from the high pressure P_1 to the low pressure P_2. The expansion is not ideal (not isentropic): this imperfection is accounted for by the turbine isentropic efficiency η_t, which is less than 1. On the right side of the figure, we see an equivalent drawing of the imperfect turbine. It is an ideal (isentropic) expansion from P_1 to the intermediate pressure P_i, followed by the flow with friction through a flow resistance, from P_i to P_2. Show analytically that "imperfection" means that the traditional inequality $\eta_t < 1$ is equivalent to $P_i > P_2$ across the flow resistance.

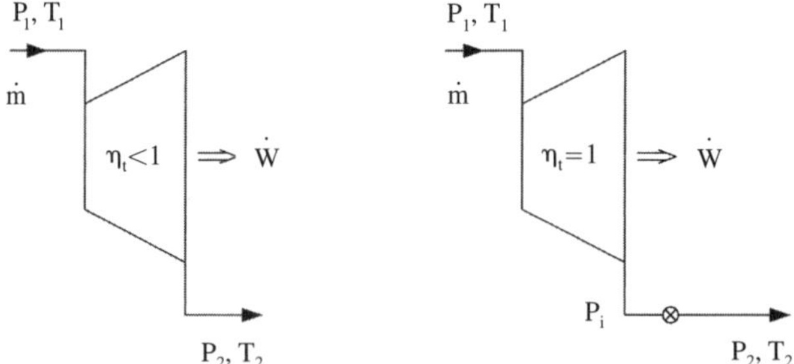

Figure P2.13

2.14. Another imperfection in thermofluid design is the flow of heat across a finite temperature difference, that is, against a thermal resistance. In this problem we show the thermodynamic meaning of heat-transfer imperfection by analyzing the work of a heat engine (closed system) that executes an irreversible cycle. On the left side of Fig. P2.14, the work output W can be expressed relative to the work output of a reversible cycle W_{rev} subjected to the same conditions (Q_H, T_H, T_L), namely $\eta = W/W_{rev} < 1$. On the right side of the figure, the same work output is produced by a reversible cycle between T_H and the intermediate temperature T_i, sandwiched with a temperature gap (thermal resistance) traveled by Q_L, from T_i to T_L. Show that the traditional way of accounting for imperfection ($\eta < 1$) is equivalent to the inequality $T_i > T_L$, in other words, the heat transfer across the finite temperature difference is responsible for the imperfection.

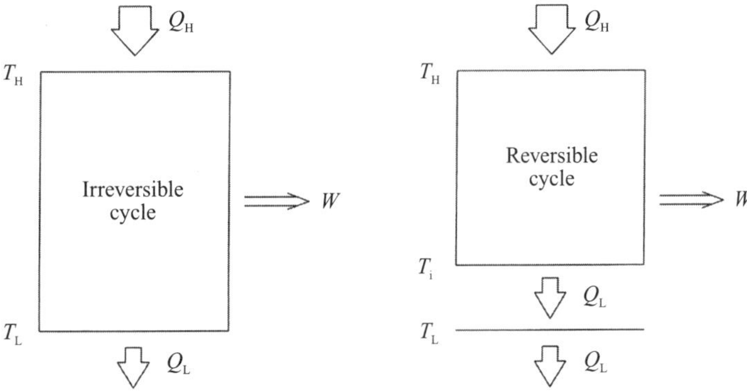

Figure P2.14

2.15. The power plant of a vehicle cruising at the constructal speed V is represented by the simple model with two compartments shown in Fig. P2.15. The power plant operates between temperatures T_H and T_L, and receives the specified heating rate $\dot{Q}_H$. The power plant is a closed system that operates in steady state. The compartment delineated between T_H and the internal temperature T_i operates reversibly, therefore, $\dot{W} = \dot{Q}_H (1 - T_i/T_H)$. The compartment sketched between T_i and T_L is the temperature gap necessary for driving the rejected heat $\dot{Q}_L$ out of the system, $\dot{Q}_L = hA(T_i - T_L)$, where h is the heat-transfer coefficient between T_i and T_L. The wall of temperature T_i has the variable surface area A, fixed thickness t, and density ρ.

Determine the optimal size A such that the total power destruction (i.e., fuel use) associated with A is minimum. This problem was proposed in Ref. [9]. There are two losses due to the finiteness of A:

(i) The loss due to the irreversibility of the power plant. Write the expression for the power plant output $\dot{W}$ and the power output in the reversible limit, $\dot{W}_{rev}$. Calculate the difference (the lost power), $\dot{W}_1 = \dot{W}_{rev} - \dot{W}$, and show that $\dot{W}_1$ is proportional to $1/A$.

(ii) The power consumed by a vehicle (or animal) is proportional to the total mass of the vehicle [9, 10]. Deduce from this the relation between A and the power ($\dot{W}_2$) used in order to transport A. Minimize $\dot{W}_1 + \dot{W}_2$ with respect to A, and find the optimal size A.

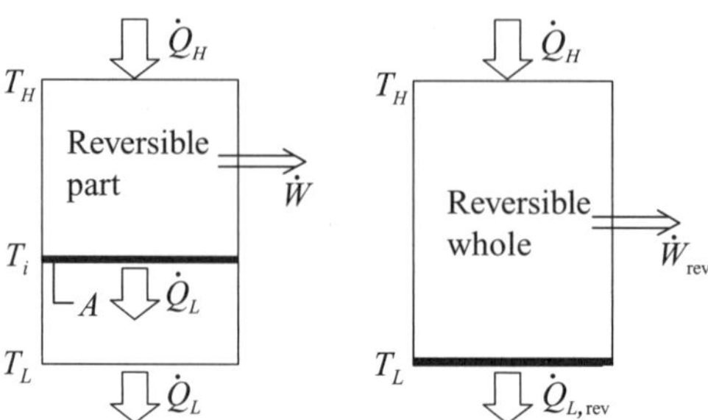

Figure P2.15

2.16. An ice storage facility is maintained at the low temperature T_L (below the ambient T_0) by a refrigerator modeled as a closed system operating in cycles. The coefficient of performance (COP) of the refrigerator is specified. The ice storage facility is an enclosure with the surface area A_L, which absorbs a heat leak from the ambient, at the rate $Q_L = A_L\, U_L\, (T_0 - T_L)$. The proportionality factor U_L is called overall heat-transfer coefficient, and it is specified.

A former student of this course proposes to kill two birds with the same stone, and use the heat rejected by the refrigerator in order to heat a building. The temperature inside the building is T_H, the surface of the building enclosure is A_H, and the rate at which heat leaks from the building to the ambient is $Q_H = A_H\, U_H\, (T_H - T_0)$, where U_H is another specified heat-transfer coefficient.

It turns out that, as a good approximation, $U_H = U_L$, and that T_0 is half way between T_H and T_L. Show that the warm building (A_H) must be larger than the cold building (A_L).

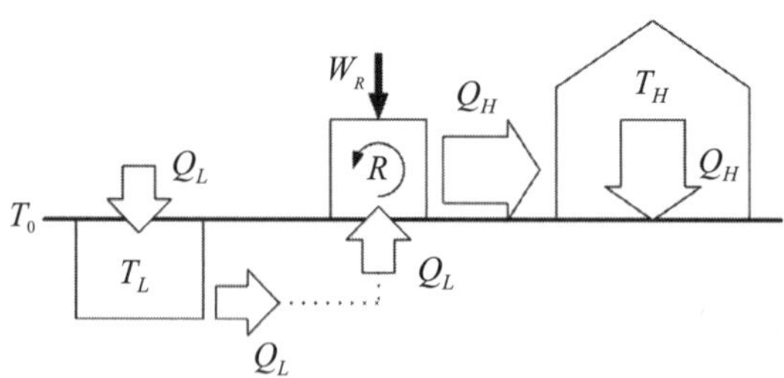

Figure P2.16

3

SIMPLE FLOW CONFIGURATIONS

In this chapter we offer an introduction to how to generate flow configuration from principle. We teach the simplest first: duct cross-sections, river cross-sections, and spacings between plates and other components in packages bathed by forced convection and natural convection. These lessons are simple (single scale) and can be taught analytically with simple methods such as scale analysis (Appendix A) and the intersection of asymptotes (section 3.5).

3.1 FLOW BETWEEN TWO POINTS

3.1.1 Optimal Distribution of Imperfection

We start with one of the simplest examples of how the collision between global objective and global constraints generates the complete architecture of the flow system. Consider the flow between two points (Fig. 3.1), where we will see that simplicity characterizes only the optimal and near-optimal architectures [1, 2]. This makes the example easy to present graphically. The rest of the world of possible point-point designs is conceptually as vast and complicated as in any other example. When the flow architecture is free to morph, the design space is infinite. There is an infinity of flow architectures that can be chosen to guide a fluid stream ($\dot{m}$) from one point to another point.

Constructal theory begins with (1) the tendency (the phenomenon) to flow with configuration, and (2) the finiteness of the flow system. These concepts are represented by the global objectives and the global constraints of the flow system, and the fact that in the beginning the geometry of the flow is missing. *Geometry is the big unknown.*

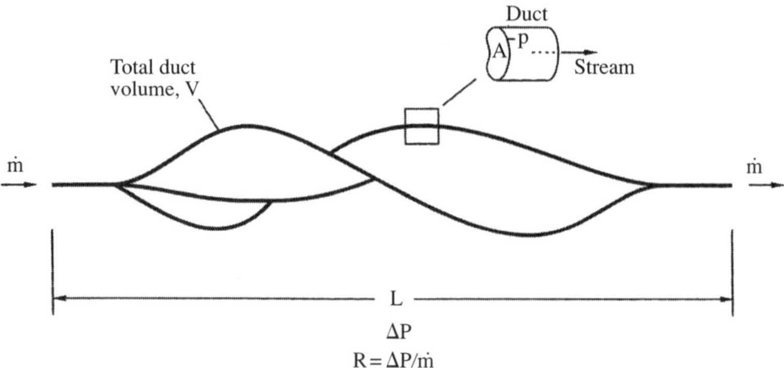

Figure 3.1 General flow architecture for guiding a stream from one point to another point [1].

In Fig. 3.1 the objective is to guide the single-phase fluid stream $\dot{m}$ to flow from one point to another point with maximum access, in this case, by using minimal pumping power. When $\dot{m}$ is fixed, this objective is the same as seeking flow architectures with minimal pressure overall difference (ΔP, cf. section 1.2.2), minimal overall flow resistance (R), or minimal rate of entropy generation by fluid friction (cf. section 2.7).

There are two *global* constraints, one external and the other internal. The external constraint is the "system size," which is represented by the distance between the two points, L. The internal global constraint is the space invested in making the flow architecture. In Fig. 3.1 that amount is the total volume (V) of all the ducts of the flow structure, or the specified (finite) svelteness $\mathrm{Sv} = L/V^{1/3}$. Without such an investment, there is no flow—not even a drawing that would show the flow. A flow must be guided. Flow means direction, geometry, and architecture (finite Sv), in addition to flow rate.

Why is there an infinity of eligible flow architectures that meet the global objective and global constraints? There are many reasons, that is, many directions in which the number of possible architectures increases:

(i) The flow pattern may be two-dimensional (in the plane of Fig. 3.1), or three-dimensional.
(ii) There may be any number of ducts connected in parallel between the two points.
(iii) Any duct may have any number of branches or tributaries at any location between the two points.
(iv) A single duct may have any length.
(v) The cross-sectional shape may vary along the duct.
(vi) A single duct may have any cross-sectional shape.

We will see that the best configuration (the straight round tube) is far from being alone on the podium of maximal performance. This podium and the design world under it are the physics domain covered by the constructal law.

How do we identify the geometric features that bring a flow architecture to the highest level of global performance? There are many lessons of this type throughout engineering, and, if remembered, they constitute *strategy*. They shorten dramatically the search for the geometry in which all the features are "useful" in serving the global objective. Constructal design is about strategy, the strategy learned from seeing and applying the constructal law in basic flow configurations. It is about the compact lessons of optimal shape and structure, which are fundamental and universally applicable.

Here are some of the older lessons that accelerate the search through the broad categories listed as (i) through (vi). Again, for simplicity assume that ducts are slender, and the flows are slow so that in each cross-section the regime is laminar and fully developed. This means that the svelteness Sv of all the possible flow architectures is greater than 10 (cf. section 1.1). Each lesson is identified by the symbol, (i) through (v) of the geometric feature listed above. The lessons are:

(i)–(iii) One single duct with large cross-section offers a smaller flow resistance than two ducts with smaller cross-sections connected in parallel.

(iv) The lowest resistance belongs to the shortest duct, that is, the straight duct between the two points.

(v) The duct with cross-sectional geometry that does not vary longitudinally has a lower resistance than the duct with variable cross-section (cf. Problem 1.1). If the cross-section is the throttle that represents the imperfection (resistance) of the flow path, then, in accord with constructal theory, the duct with constant cross-section is the duct with *optimal distribution of imperfection* [3].

3.1.2 Duct Cross Sections

Summing up, out of the infinity of designs represented by (i) through (v), we have selected a single straight duct with a cross-sectional shape that does not vary from one end of the duct to the other. According to (vi), however, there is still an infinite number of possible cross-section shapes: symmetric vs. asymmetric, smooth vs. polygonal, and so on. Which impedes the flow the least?

The answer becomes visible if we assume cross-sections with polygonal shapes. Start with an arbitrary cross-section shaped as a triangle. The area of the cross-section A is fixed because the total duct volume V and the duct length L are fixed, namely $A = V/L$. Triangular cross-sections constrict the flow when one of the angles is much smaller than the other two. The least resistance is offered by the most open triangular cross-section. The most open is the equilateral triangle. Once again, if one very small angle and two larger ones represent a nonuniform distribution of geometric features of imperfection (i.e., features that impede the flow), then the equilateral triangle represents the architecture with optimal distribution of imperfection.

The same holds for any other polygonal shape. The least resistance is offered by a cross-section shaped as a *regular* polygon. In conclusion, out of the infinity of flow architectures recognized in class (vi) we have selected an infinite number of candidates. They are ordered according to the number of sides (n) of the regular polygon, from the equilateral triangle ($n = 3$) to the circle ($n = \infty$). The flow resistance in Poiseuille flow through a straight duct with polygonal cross-section can be written as

$$\frac{\Delta P}{\dot{m}} = \frac{\nu L}{8V^2} \frac{\text{Po}\, p^2}{A} \tag{3.1}$$

where p is the perimeter of the cross-section. Equation (3.1) comes from the general pressure drop formula, Eq. (1.17) (see also Ref. 4, p. 105),

$$\Delta P = f \frac{pL}{A} \frac{1}{2} \rho U^2 \tag{3.2}$$

where U is the mean fluid velocity, $U = \dot{m}/(\rho A)$. As shown in Table 3.1, the dimensionless perimeter $p/A^{1/2}$ depends only on n. The same is true about the factor Po (the Poiseuille constant), which appears in the solution for friction factor in Poiseuille flow,

$$f = \frac{\text{Po}}{\text{Re}} \tag{3.3}$$

where $\text{Re} = UD_h/\nu$ and $D_h = 4A/p$.

In conclusion, the group Po p^2/A depends only on n and accounts for how this last geometric degree of freedom influences global performance. The group Po p^2/A is the dimensionless global flow resistance of the flow system. The smallest Po p^2/A value is the best, and, as expected, the best is the round cross-section.

Figure 3.2 is a plot of the flow resistance data of Table 3.1. The flow structure with minimal global resistance is approached gradually (with diminishing decrements)

Table 3.1 The laminar flow resistances of straight ducts with regular polygonal cross-sections with n sides [5].

n	Po	$p/A^{1/2}$	Po p^2/A
3	40/3	4.559	277.1
4	14.23	4	227.6
5	14.74	3.812	214.1
6	15.054	3.722	208.6
8	15.412	3.641	204.3
10	15.60	3.605	202.7
∞	16	$2\pi^{1/2}$	201.1

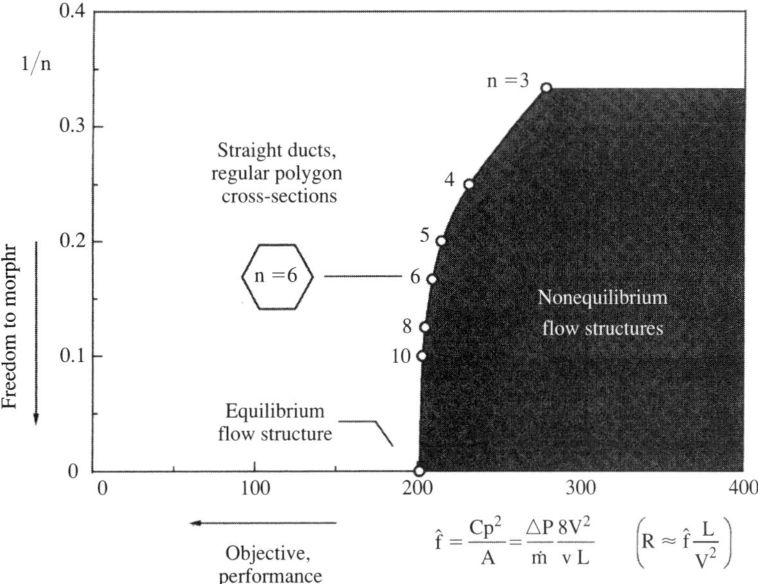

Figure 3.2 The performance-freedom world of designs: the approach to the minimal global flow resistance between two points when the number of sides of the regular-polygon cross-section (n) increases (data from Table 3.1) [1].

as n increases. The polygonal cross-section with $n = 10$ performs nearly as well as the round cross-section ($n = \infty$). The "evolution" of the cross-sectional shape stops when the number of features of (n) has become infinite, that is, when the structure has become the most free. This design domain where changes in global performance have stopped is the *equilibrium flow architecture* [1, 2]. Here, the performance does not change, even though the configuration is the most free to change.

The curve plotted in Fig. 3.2 was generated by calculations for regular-polygon cross-sections. This curve is actually a sequence of discrete points, one point for each n value. We drew a continuous line through these points in order to teach an additional idea. For any n, the regular polygon and straight duct with constant cross-section is already the "winner" from an infinitely large group of competing architectures. Qualitatively, this means that the global flow resistances of all the designs that are not covered by Table 3.1 fall to the right of the curve sketched in Fig. 3.2.

In sum, the immense world of possible designs occupies only a portion of the two-dimensional domain illustrated in Fig. 3.2. This domain can be described qualitatively as a *performance versus freedom* space, when global constraints such as L and V are specified. The boundary of the domain is formed by a collection of the better flow structures. The best is achieved by putting more freedom in the geometry of the flow structure (e.g., a larger n). The best performance belongs to the structure that was most free to morph—the equilibrium configuration. In its immediate vicinity, however, we find many configurations that are different (they have finite n values), but have practically the same global performance level. These are the *near-equilibrium flow structures*.

3.2 RIVER CHANNEL CROSS-SECTIONS

Freedom is good for design. Performance vs. freedom design worlds are inhabited by all the flow systems with configuration and freedom to morph. In this section we consider a class of flow systems as simple as that of Fig. 3.1, and as ubiquitous. A river channel guides a point-to-point flow on the landscape. It has the width W and the depth d. In spite of the many complications (e.g., geological properties) that account for the diversity of river channels, one simple geometric feature stands out: large (wide) rivers are deeper than tiny rivers. This begs for a theoretical explanation that should be as simple as the scaling law itself:

$$W \sim d \tag{3.4}$$

From lessons (i) through (v) that ended section 3.1.1, we know that the best point-to-point connection is a straight channel with constant cross-section. Which is the best cross-section? The cross-section of an open channel is characterized by an upper straight segment (the free surface), which is shear free, and the rest of the perimeter (the bottom), which is labeled p and is characterized by shear (Table 3.2). Assume that the channel is straight. The cross-sectional area A and the mass flow rate $\dot{m}$ are known constants. The area A is the result of the slow development of the river bed, and, in this sense, it describes the age of the river bed. By fixing the age, we focus our inquiry on the present, or, in view of the slow development, on the channel as a system in quasi-steady state (with fixed A).

The mass flow rate is fixed by the global constraint represented by the rainfall on the drainage surface situated upstream of A. Several important conclusions follow from the A and $\dot{m}$ constraints. First, the mean flow velocity is fixed, $V = \dot{m}/\rho A$, where ρ is the water density. Next the longitudinal shear stress along the bottom is also fixed, $\tau = \frac{1}{2} C_f \rho V^2$, because in turbulent flow over a very rough surface the average skin friction coefficient C_f is practically constant (by analogy with f for fully rough turbulent flow, Fig. 1.3).

The total force per unit of channel length in the flow direction ($p\tau$) decreases in proportion with the bottom perimeter p. This conclusion can also be expressed in thermodynamic terms. The rate of work destruction $p\tau V$ at the longitudinal station where A is located is also proportional to p. The same holds for the rate of entropy generation in A, namely $p\tau V/T_0$, where T_0 is the ambient temperature.

Table 3.2 Constructal cross-sectional shapes of open channels [5].

	Constructal Shape	$(W/d)_{opt}$	$p_{min}/A^{1/2}$
Rectangle	A	2	2.828
Triangle	A	2	2.828
Parabola	A	2.056	2.561
Circle	W, A, d, p	2	2.507

We arrive at a very simple geometric problem that accounts for the constructal design of the river flow through the cross-section A:

> Find the cross-sectional shape that has the minimum ground perimeter p, which, along with a straight (free-surface) segment, encloses the fixed area A.

Table 3.2 shows four steps in the search for the cross-sectional shape. In each step, the shape of the bottom had to be assumed. In the second example, the bottom shape is rectangular, and its perimeter ($p = W + 2d$) can be minimized subject to the area constraint ($A = Wd$). The result is the proportionality:

$$W = 2d \tag{3.5}$$

In conclusion, the existence of a natural scaling law of type (3.4) is deducible from the constructal law. The other shapes (triangular, parabolic, circular) reveal the same scaling law. The minimized bottom perimeter decreases slightly from one example to the next.

Table 3.2 also shows the route to the mathematically optimal channel cross-section. The optimal drawing has two parts, or 2 degrees of freedom: (1) the shape, or bottom profile, which had to be assumed in each of the examples of Table 3.2, and (2) the slenderness ratio W/d, which was optimized analytically in each example.

(1) The optimal shape of the bottom can be determined based on variational calculus (see Appendix C). We assume an arbitrary function $y(x)$ for the bottom profile, where y is the depth corresponding to the coordinate x measured along the free surface (in the plane of A: see the bottom sketch in Table 3.2). The objective

is to minimize the total length of the bottom perimeter:

$$p = \int_{-W/2}^{W/2} \left[1 + \left(\frac{dy}{dx} \right)^2 \right]^{1/2} dx \tag{3.6}$$

where, for the time being, the width W is considered fixed. The function $y(x)$ for which the integral of Eq. (3.6) is minimized subject to the cross-sectional area constraint,

$$A = \int_{-W/2}^{W/2} y dx \tag{3.7}$$

is readily obtained when the associated Euler equation is solved (Appendix C). The answer is that the optimal $y(x)$ is the arc of a circle, or that A is the segment of a circle such that W is the chord. There is an infinite number of such solutions, from the very thin segment of the circle to the full circle, depending on how W compares with the length scale $A^{1/2}$. Three examples are shown in the upper part of Fig. 3.3.

(2) The slenderness ratio W/d of the optimal shape is determined next by minimization of the ground contact perimeter subject to fixed A. The result of this operation is the half-circle [$W/d = 2$, $p_{\min} = (2\pi A)^{1/2}$], which is shown as the fourth case in Table 3.2.

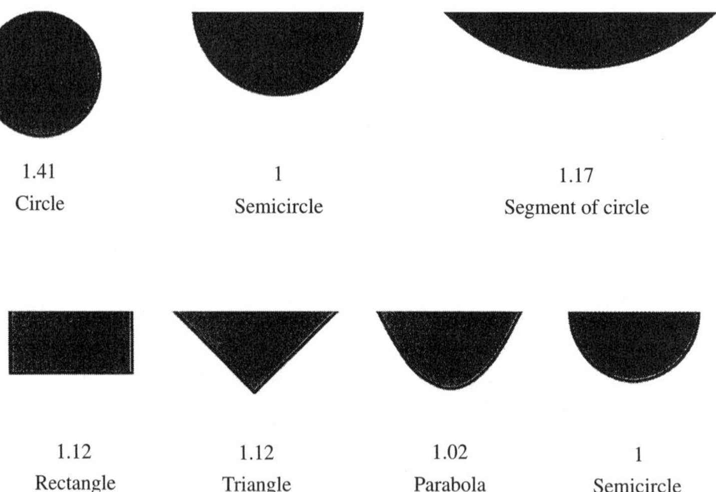

Figure 3.3 Channels with the same cross-sectional area, different shapes, and different wetted perimeters. The top row shows three examples with bottoms shaped as arcs of circles. The number under each drawing represents the flow resistance divided by the flow resistance of the channel with semicircular cross-section. The four cross sections shown on the bottom row have had their depth/width ratios optimized for minimum resistance. Note that the resistance does not vary appreciably from one shape to the next. Channel cross-sectional designs are robust [4].

In summary, the scaling law between river width and maximum depth can be anticipated in nakedly simple terms by maximizing the flow access, or minimizing the dissipation (the mechanical power destroyed) in the entire cross-sectional area A. In this way, the optimal geometry of the channel is associated with a *global* optimization subject to imposed, global constraints (A, $\dot{m}$).

The march toward the equilibrium flow structure (the semicircle) in Table 3.2 can be traced on a performance vs. freedom plane just like Fig. 3.2. The performance is indicated by the value of $p_{min}/A^{1/2}$, which would be plotted on the abscissa. One could argue that the freedom to morph increases downward through the table, because from the triangle to the semicircle the number of bendable "corners" of the rim increases. The four points plotted based on Table 3.2 would define a ragged line just as the points of Table 3.1 traced in Fig. 3.2. To the right of this line would lie all the open channels with prescribed shape (triangle, rectangle, etc.) that are not optimal.

The examples aligned in Table 3.2 reveal two additional features of the optimal channel geometry. First, in the optimal shape (the half-circle) the river banks extend vertically downward into the water and are likely to crumble under the influence of erosion (drag on particles) and gravity. This will decrease the slopes of the river bed near the free surface and, depending on the bed material, will increase the slenderness ratio W/d.

The second feature revealed by Table 3.2 and the lower part of Fig. 3.3 is that, in practical engineering terms, the four designs have nearly the same performance (the same $p_{min}/A^{1/2}$ value). The rectangular and triangular channels perform identically, whereas the parabolic and semicircular channels agree to within 2.2%. The two most extreme examples given in Table 3.2 are separated by only 12%, even though the rectangular shape is strikingly "rough" in comparison with the parabolic and the round shapes. This high level of agreement with regard to performance represents design *robustness*. It accounts for the significant scatter in the data on river bottom profiles. Global flow performance is what matters, not local shape.

This situation is completely analogous to the robustness exhibited by other highly complex systems in the field of engineering design. Quite often, in the end it turns out that the chosen configuration of the optimized system performs *nearly* as well as the truly optimal configuration, even though, visually, the nearly optimal and the optimal configurations may be different.

3.3 INTERNAL SPACINGS FOR NATURAL CONVECTION

3.3.1 Learn by Imagining the Competing Extremes

In this section and the next, we pursue two teaching objectives at the same time. First, we show how to determine (with pencil and paper) the optimal spacings for virtually all the possible channel configurations with natural and forced convection. Second, we teach a very powerful method for solving such problems: *the intersec-*

tion of asymptotes method. The basic idea is one of common sense: if the solution to the problem at hand is not intuitively obvious (i.e., if the problem is "opaque"), then, perhaps, the solution to the same problem will be visible if we imagine the problem in one extreme. Next, perhaps the solution is also easy to see in the opposite extreme. With the two extremes visible in our minds, we begin to see more clearly the opaque: the important solution, which resides between the two extremes. Such solutions can be located by literally intersecting the extremes—the asymptotes.

This kind of thinking pays dividends in many domains of science. The wise learn to imagine two different scenarios, in order to develop a feel for the realistic scenario. Closer to our engineering (e.g., Chapter 2), a feel for the complicated properties of a real substance can be acquired by learning the properties of the same substance in its two limits of behavior, the most compressible (the ideal gas model) and the least compressible (the incompressible substance model). We return to the essential steps and value of the intersection of asymptotes method in section 3.5.

Consider designing a package of electronic components, which must fit inside a given volume—the *global* constraint. The *objective* of the design is to install as much circuitry as possible in this volume. Because electrical components generate heat, this objective is equivalent to installing as much heat generation rate (q, or $\dot{Q}$) as possible. The highest temperatures in the package (the hot spots) must not exceed a specified value, T_{max}: this is the *local* constraint, or the allowable peak of imperfection. If the local temperature (T) rises above this allowable ceiling, the electronic functioning of the local component is threatened.

The reader should note that the packaging of electronics into a fixed volume is the "icon" for the general design problem of high-density *stacking*. We find it not only in packages of electronics, but also in packages of fuel cells, mass exchangers, and so on, all the way to gills of fish and swarms of honey bees [3].

In sum, the designed configuration is better (more compact) when q is larger. In other words, the design is better when the global thermal conductance ratio $q/(T_{max} - T_0)$ is larger, where T_0 is the initial (reference, sink) temperature of the coolant that absorbs q. Alternatively, the problem of discovering the configuration can be formulated by fixing the amount of electronics (q) and minimizing the global volume. Another route is to fix q and the global volume and to minimize the hot-spot temperature T_{max}. All these objectives can be pursued based on the method presented next, with the same results for the recommended flow geometry.

Internal structure (graininess, internal spacings) results from the pursuit of a greater global conductance while the flow structure is free to morph. To illustrate, we assume that the heat generation rate q is spread almost uniformly over the given volume, and that the flow "happens"—the flow is driven by natural convection. The engine that drives all "natural" flows is another example constructal design in nature (the sun-earth-universe engine), which is described in Refs. [6] and [7].

The electronic circuitry is mounted on equidistant vertical boards of height H, filling a space of height H, and horizontal dimensions L and W. For simplicity,

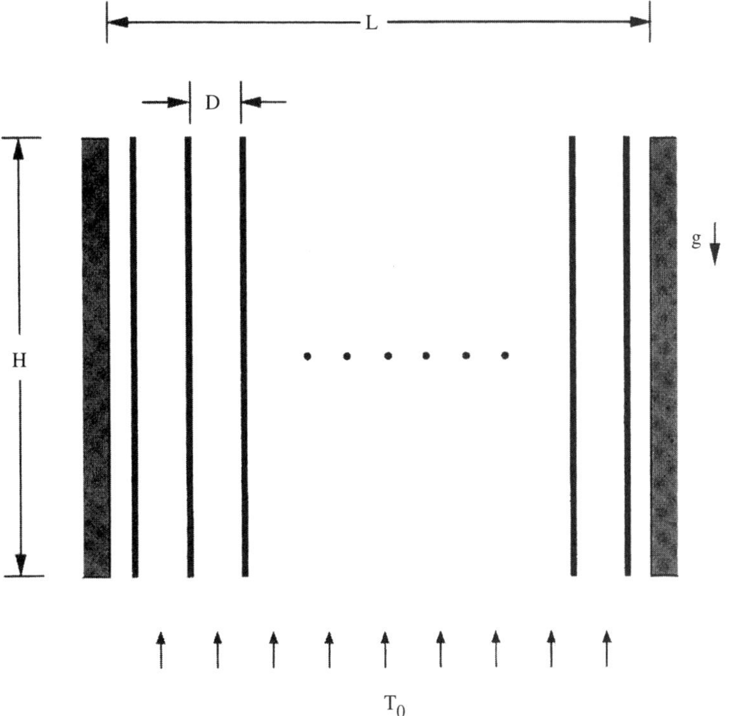

Figure 3.4 Stack of parallel plates that generates heat and is cooled by natural convection.

assume that the configuration is two-dimensional with respect to the ensuing flow, as shown in Fig. 3.4. Not specified is the board-to-board spacing D, or the number of vertical parallel-plate channels,

$$n = \frac{L}{D} \tag{3.8}$$

Determining the spacing D that maximizes q can be a laborious task. We will show that with a little ingenuity the problem can be solved on the back of an envelope. The method is the intersection of asymptotes described in section 3.5, which was proposed in 1984 [8]. The method was used extensively in Refs. [3] and [4]. In this chapter, it is based on three assumptions:

1. The flow is laminar.
2. The board surfaces are sufficiently smooth to justify the use of heat transfer results for natural convection over vertical smooth walls.
3. The maximum temperature $T_{\max}$ is closely representative of the temperature at every point on the board surface.

The method has two steps. In the first, we identify the two possible extremes: the small-D limit and the large-D limit. In the second step, the two extremes are intersected to locate the D value for which q is maximum.

3.3.2 Small Spacings

When D becomes sufficiently small, the channel formed between two boards becomes narrow enough for the flow and heat transfer to be fully developed. According to the first law of thermodynamics we have

$$q_1 = \dot{m}_1 c_P (T_{max} - T_0) \tag{3.9}$$

for the heat-transfer rate extracted by the coolant from one of the channels of spacing D, where T_0 is the inlet temperature and T_{max} is the outlet temperature. The mass flow rate is

$$\dot{m}_1 = \rho D W U \tag{3.10}$$

where the mean velocity U can be estimated by replacing $\Delta P/L$ with $\rho g \beta (T_{max} - T_\infty)$ in the pressure drop formula, Eq. (1.22):

$$\rho g \beta (T_{max} - T_0) = f \frac{4}{D_h} \frac{1}{2} \rho U^2 \tag{3.11}$$

This substitution comes from the following argument [4]. In the package of Fig. 3.4, there are two vertical fluid columns of height H: the one on the outside, which is cold (T_0), and the internal column the temperature of which is T_{max} when D is small enough. The bottom-top hydrostatic pressure difference is $\rho(T_0)gH$ along the cold fluid column, and $\rho(T_{max})gH$ along the warm columns. This means that the pressure difference measured horizontally between the bottom ends of the cold and warm columns is $\Delta P \sim [\rho(T_0) - \rho(T_{max})]\, gH$. The same ΔP scale, but with the opposite sign, is estimated between the two top ends of the cold and warm columns.

The conclusion is that the effective pressure difference that pushes the bottom cold fluid into the bottom of the warm columns is ΔP. Similarly, the pressure deficit that sucks the fluid out of the top of the warm columns is ΔP. It means that the pressure difference scale ΔP plays the same role along the H-tall channel as the ΔP along the L-long pipe in Eq. (1.22). This is why the effective longitudinal pressure gradient in the vertical channels of Fig. 3.4 is $\Delta P/H \sim [\rho(T_0) - \rho(T_{max})]\, g \sim \rho g \beta (T_{max} - T_0)$, where ρ is the scale of both $\rho(T_0)$ and $\rho(T_{max})$, and β is the coefficient of thermal expansion (e.g., $\beta = 1/T$ for an ideal gas).

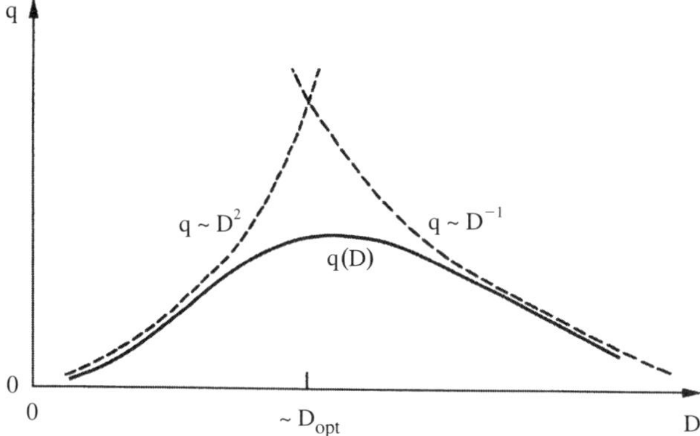

Figure 3.5 The intersection of asymptotes method: the maximization of the global thermal conductance of a stack of parallel plates cooled by natural convection.

For laminar flow between parallel plates we insert $f = 24/(UD_h/\nu)$ and $D_h = 2D$ in Eq. (3.11), and we obtain

$$U = g\beta\,(T_{\max} - T_0)\, D^2/(12\nu) \tag{3.12}$$

$$\dot{m}_1 = \rho D W U = \rho W g\beta\,(T_{\max} - T_0)\, D^3/(12\nu) \tag{3.13}$$

The rate at which heat is removed from the entire package is $q = nq_1$, or

$$q = \rho c_P W L g\beta\,(T_{\max} - T_0)^2\, D^2/(12\nu) \tag{3.14}$$

In conclusion, in the $D \to 0$ limit the total heat transfer rate varies as D^2. This trend is indicated by the small-D asymptote plotted in Fig. 3.5.

3.3.3 Large Spacings

The other limit is where D is large enough that it exceeds the thickness of the thermal boundary layer that forms on each vertical surface: $\delta_T \sim H\,\mathrm{Ra}_H^{-1/4}$, where $\mathrm{Ra}_H = g\beta H^3(T_{\max} - T_0)/(\alpha\nu)$ and $\mathrm{Pr} \geq 0.5$. In other words, the spacing is considered large when [4]

$$D > H\left[\frac{g\beta H^3\,(T_{\max} - T_0)}{\alpha\nu}\right]^{-1/4} \tag{3.15}$$

In this limit the boundary layers are distinct. They are thin compared with D, and the central plane of the board-to-board spacing is occupied by fluid of temperature T_0. The number of distinct boundary layers is $2n = 2L/D$ because there are two boundary layers for each D spacing. The heat-transfer rate through one boundary layer is $\bar{h}HW(T_{\max} - T_0)$, for which $\bar{h}$ is furnished by the solution for pure boundary layer natural convection [4], cf. Eq. (1.77) for Pr = 0.7 (air),

$$\frac{\bar{h}H}{k} = 0.517\,\mathrm{Ra}_H^{1/4} \tag{3.16}$$

The rate of heat transfer extracted from the entire package is $2n$ times larger than $\bar{h}HW(T_{\max} - T_0)$:

$$q = 2\frac{L}{D}HW\,(T_{\max} - T_0)\,\frac{k}{H}0.517\,\mathrm{Ra}_H^{1/4} \tag{3.17}$$

Equation (3.17) shows that in the large-D limit the total heat transfer rate decreases as D^{-1} when the board-to-board spacing changes. This second asymptote has been added to Fig. 3.5.

3.3.4 Optimal Spacings

So far we have determined the two asymptotes of the actual (unknown) curve of q versus D. Figure 3.5 shows that the asymptotes intersect above what would be the peak of the actual $q(D)$ curve. The trade-off between the two extremes is the balance between two really different imperfections: the fluid flow resistance and the heat-flow resistance. One may call these "apples and oranges," but they do compete, and they decide the configuration that is best.

All designs are combinations of some apples and some oranges, *in the right amounts*. When D is smaller than D_{opt} the coolant cannot flow easily through the parallel-plates channels; that is, the fluid-flow resistance is too large. When D is larger than D_{opt} the solid surfaces are too few, and the generated heat current has a difficult time flowing from the plates to the coolant. In this second limit the heat-flow resistance is too large.

It is not necessary to determine the exact $q(D)$ relation. The optimal spacing D_{opt} for maximum q can be estimated as the D value where Eqs. (3.14) and (3.17) intersect [8]

$$\frac{D_{opt}}{H} \cong 2.3\left[\frac{g\beta\,(T_{\max} - T_0)\,H^3}{\alpha\nu}\right]^{-1/4} \tag{3.18}$$

This D_{opt} estimate is within 20% of the optimal spacing deduced based on much lengthier methods, such as the maximization of the $q(D)$ relation [9] and the finite-difference simulations of the complete flow and temperature fields in the package.

An estimate of the maximum heat-transfer rate can be obtained by substituting D_{opt} into Eq. (3.17) [or Eq. (3.14)],

$$q_{\max} \approx 0.45k\,(T_{\max} - T_0)\,\frac{LW}{H}\mathrm{Ra}_H^{1/2} \tag{3.19}$$

The approximation sign is a reminder that the peak of the actual $q(D)$ curve falls under the intersection of the two asymptotes (Fig. 3.5). This result also can be expressed as the maximum volumetric rate of heat generation in the $H \times L \times W$ volume:

$$\frac{q_{max}}{HLW} \approx 0.45 \frac{k}{H^2} (T_{max} - T_0) \, \mathrm{Ra}_H^{1/2} \tag{3.20}$$

In conclusion, if the heat-transfer mechanism is laminar natural convection, the maximum volumetric rate of heat generation is proportional to $(T_{max} - T_0)^{3/2}$ $H^{-1/2}$, and the group of properties $k(g\beta/\alpha\nu)^{1/2}$.

How does the lesson of Fig. 3.5 fit on a performance-freedom plane such as Fig. 3.2? The performance measure $(T_{max} - T_0)/q$ would be plotted on the abscissa of Fig. 3.2, in which migration to the left means "better." The freedom to morph the design would be plotted on the ordinate. In the present example, there was not much freedom: just one degree of freedom (D). The optimal design (D_{opt}, q_{max}) is one point in the performance-freedom plane. This point would occupy a position like that of the point $n = 3$ in Fig. 3.2. To the right of this point would lie all the nonoptimal designs ($D \neq D_{opt}$) aligned on the abscissa of Fig. 3.5. The march toward greater performance, culminating with the discovery of the equilibrium flow structure consists of repeating the morphing done in Fig. 3.5 for stacks of vertical plates with increasing numbers of degrees of freedom. This evolution toward multiscale flow structures forms the subject of Chapter 6.

3.3.5 Staggered Plates and Cylinders

Structures with a single degree of freedom have attracted considerable attention. For example, when the fixed volume is filled with many equidistant parallel staggered plates, which are considerably shorter than the volume height ($b << H$ in Fig. 3.6a), there exists an optimal horizontal spacing (D) between neighboring plates [10]. In laminar natural convection, the optimal spacing scales with the height of the fixed volume, not with the height of the individual plate:

$$\frac{D_{opt}}{H} \cong 0.63 \left(\frac{Nb}{H}\right)^{1.48} \mathrm{Ra}_H^{-0.19} \tag{3.21}$$

where N is the number of plate surfaces that face one elemental channel (e.g., $N = 4$ in Fig. 3.6a). The dimensionless group (Nb/H) is of the order of 1 and represents the relative contact area present along the boundaries of the elemental channel.

The similarities between Eq. (3.21) and (3.18) are important, because they point to the *robustness* of the design principle that generates internal spacings. The corresponding correlation for the maximum thermal conductance is [10]

$$\frac{q_{max}}{k\,(T_{max} - T_0)\,LW/H} \cong 1.92 \exp\left(-0.7\frac{Nb}{H}\right) \mathrm{Ra}_H^{0.43} \tag{3.22}$$

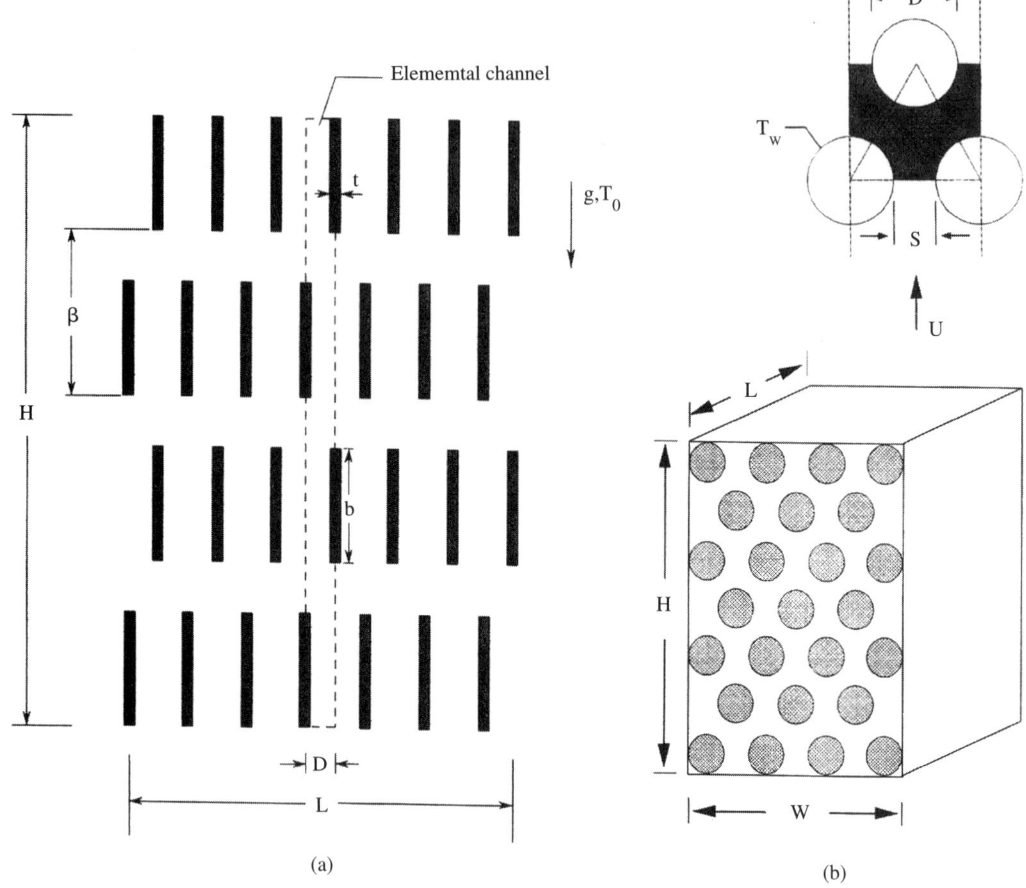

Figure 3.6 Volumes heated uniformly and cooled by natural convection: (a) array of staggered plates [10], and (b) array of horizontal cylinders [11].

where W is the dimension perpendicular to the plane of Fig. 3.6a. The form and numerical estimates based on Eq. (3.22) are similar to those of Eq. (3.20). Experiments and numerical simulations show that the correlations (3.21) and (3.22) for vertical staggered plates in natural convection, are accurate within 6% in the range $10^3 \leq \text{Ra}_H \leq 5 \times 10^5$ and $0.4 \leq (Nb/H) \leq 1.2$.

When the volume is heated by an array of horizontal cylinders of diameter D [Fig. 3.6(b)], the optimal cylinder-to-cylinder spacing S scales with the overall height of the volume [11]. This is the same as in the other optimal spacings discussed so far. In laminar natural convection, the optimal spacing is correlated closely by a formula that resembles Eq. (3.18):

$$\frac{S_{\text{opt}}}{H} \cong 2.72 \left(\frac{H}{D}\right)^{1/12} \text{Ra}_H^{-1/4} + 0.263 \frac{D}{H} \tag{3.23}$$

where the second term on the right side is a small correction factor. The correlation for the corresponding maximum thermal conductance is

$$\frac{q_{max} D^2}{k\,(T_{max} - T_0)\, HLW} \cong 0.448 \left[\left(\frac{H}{D} \right)^{1/3} \mathrm{Ra}_D^{-1/4} \right]^{-1.6} \tag{3.24}$$

where $\mathrm{Ra}_D = g\beta\,(T_{max} - T_0)D^3/(\alpha\nu)$. Equations (3.23) and (3.24) agree within 1.7% with numerical simulations and experiments performed in the range $\mathrm{Pr} = 0.72$, $350 \le \mathrm{Ra}_D \le 10^4$ and $6 \le H/D \le 20$.

In conclusion, the maximization of the overall thermal conductance of a flow system that morphs freely subject to the global and local constraints generates a characteristic internal structure—an optimal relative positioning of the solid components that fill the given volume. This is the principle that generates internal geometric form.

3.4 INTERNAL SPACINGS FOR FORCED CONVECTION

Consider next the analogous problem of installing the optimal number of heat-generating plates in a space cooled by forced convection [12]. As shown in Fig. 3.7, the swept length of each board is L, and the transverse dimension of the entire

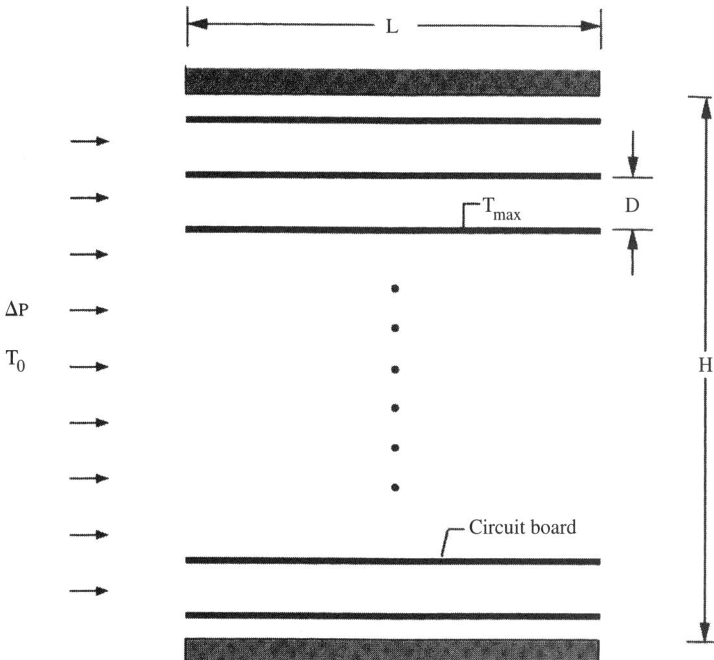

Figure 3.7 Forced convection cooling of a stack of parallel plates that generate heat.

package is H. The width of the stack W is perpendicular to the plane of the figure. We retain the assumptions 1 through 3 listed under Eq. (3.8). The thickness of the individual board is again negligible relative to the board-to-board spacing D, so that the number of channels is

$$n = \frac{H}{D} \tag{3.25}$$

The pressure difference across the package, ΔP, is assumed constant and known. This is a good model for electronic systems in which several packages and other features (e.g., channels) receive their coolant *in parallel*, from the same plenum. The plenum pressure is maintained by a fan, which may be located upstream or downstream of the package. This way of bathing a volume is different than the bathing provided by a single stream that is distributed as a tree throughout the volume (see Chapter 4).

3.4.1 Small Spacings

When D becomes sufficiently small, the channel formed between two boards becomes narrow enough for the flow and heat transfer to be fully developed. The mean outlet temperature of the fluid approaches the board temperature T_{max}. The total rate of heat transfer from the $H \times L \times W$ volume is

$$q = \dot{m} c_P (T_{max} - T_0), \tag{3.26}$$

where $\dot{m} = \rho HWU$. The mean velocity through the channel, U, is known from the Poiseuille flow solution [4]:

$$U = \frac{D^2}{12\mu} \frac{\Delta P}{L} \tag{3.27}$$

The corresponding expression for the total heat transfer rate is

$$q = \rho HW \frac{D^2}{12\mu} \frac{\Delta P}{L} c_P (T_{max} - T_0) \tag{3.28}$$

In this way we conclude that the total heat-transfer rate decreases as D^2 when D decreases. This trend is illustrated by the small-D asymptote in Fig. 3.8.

3.4.2 Large Spacings

Consider next the limit in which D is large enough that it exceeds the thickness of the thermal boundary layer that forms on each horizontal surface. In this case it is necessary to determine the free-stream velocity U_0 that sweeps these boundary layers. Because the pressure drop ΔP is fixed, a force balance on the $H \times L \times W$ control volume reads

$$\Delta PHW = 2nLW\bar{\tau} \tag{3.29}$$

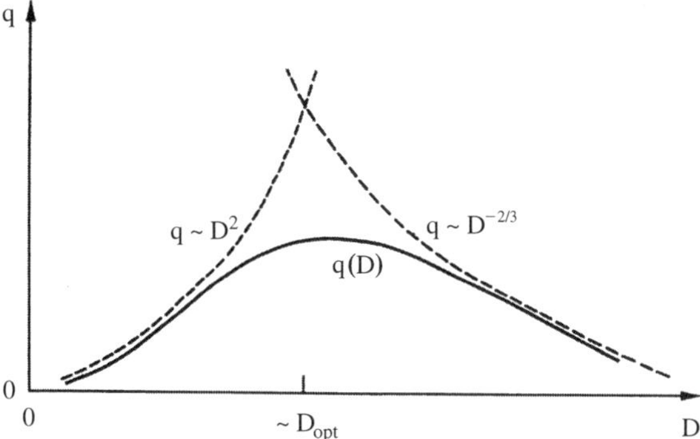

Figure 3.8 The intersection of asymptotes method: the maximization of the global thermal conductance of a stack of parallel plates cooled by forced convection.

where $\bar{\tau}$ is the wall shear stress averaged over L, namely $\bar{\tau} = 0.664\ \rho U_0^2 \mathrm{Re}_L^{-1/2}$ for $\mathrm{Re}_L \leq 5 \times 10^5$ [4]. Equation (3.29) yields

$$U_0 = \left(\frac{\Delta P H}{1.328 n L^{1/2} \rho \nu^{1/2}} \right)^{2/3} \tag{3.30}$$

The heat-transfer rate through a single board surface is

$$q_1 = \bar{h} L W (T_{\max} - T_0), \tag{3.31}$$

where the heat-transfer coefficient averaged over L is

$$\bar{h} = 0.664 \frac{k}{L} \mathrm{Pr}^{1/3} \mathrm{Re}_\mathrm{L}^{1/2} \tag{3.32}$$

when $\mathrm{Pr} \geq 0.5$ [4]. The total heat-transfer rate from the entire package is $q = 2nq_1$ or, after the $\bar{h}$ and U_0 expressions given above are used:

$$q = 1.21 kHW (T_{\max} - T_0) \left(\frac{\mathrm{Pr}\, L \Delta P}{\rho \nu^2 D^2} \right)^{1/3} \tag{3.33}$$

In conclusion, in the large-D limit the total heat-transfer rate decreases as $D^{-2/3}$ as the board-to-board spacing increases. This trend is shown in Fig. 3.8.

3.4.3 Optimal Spacings

The intersection of the two $q(D)$ asymptotes, Eqs. (3.28) and (3.33), yields an estimate for the board-to-board spacing for maximum global thermal conductance:

$$\frac{D_{\mathrm{opt}}}{L} \cong 2.7 \left(\frac{\Delta P L^2}{\mu \alpha} \right)^{-1/4} \tag{3.34}$$

This spacing increases as $L^{1/2}$ and decreases as $\Delta P^{-1/4}$ with increasing L and ΔP, respectively. Equation (3.34) underestimates by 12% the more exact value obtained by locating the maximum of the actual $q(D)$ curve [12] and is adequate when the board surface is modeled either as uniform flux or isothermal. It was shown later in Ref. [13] that Eq. (3.34) holds even when the board thickness is not negligible relative to the board-to-board spacing. At this stage we introduce the dimensionless pressure difference number defined in Refs. [14] and [15]:

$$\mathrm{Be} = \frac{\Delta P L^2}{\mu \alpha} \tag{3.35}$$

so that Eq. (3.34) becomes

$$\frac{D_{\mathrm{opt}}}{L} \cong 2.7 \mathrm{Be}^{-1/4} \tag{3.36}$$

The manner in which the design parameters influence the maximum volumetric rate of heat removal from the package can be expressed as

$$\frac{q_{\max}}{HLW} \approx 0.6 \frac{k}{L^2} (T_{\max} - T_0)\, \mathrm{Be}^{1/2} \tag{3.37}$$

which is obtained by setting $D = D_{\mathrm{opt}}$ in Eq. (3.28) or Eq. (3.33). Once again, the approximation sign is a reminder that the actual $q_{\max}$ is as much as 20% smaller because the peak of the $q(D)$ curve is situated under the point where the two asymptotes cross in Fig. 3.8.

The similarity between the forced convection results [Eqs. (3.34) and (3.37)] and the corresponding results for natural convection cooling [Eqs. (3.18) and (3.20)] is worth noting. The role played by the Rayleigh number Ra_H in the free convection case is played in forced convection by the pressure drop group Be [15].

3.4.4 Staggered Plates, Cylinders, and Pin Fins

The optimal internal spacings belong to the volume as a whole, with its global purpose and constraints, not to the individual element (L-long plate in Fig. 3.7) on which heat is being generated. The generality of this conclusion becomes clear when we look at other elemental shapes for which optimal spacings have been determined. A volume heated by an array of staggered plates in forced convection (Fig. 3.9a) is characterized by an internal spacing D that scales with the swept length of the volume, L [16]:

$$\frac{D_{\mathrm{opt}}}{L} \cong 5.4\, \mathrm{Pr}^{-1/4} \left(\mathrm{Re}_L \frac{L}{b} \right)^{-1/2} \tag{3.38}$$

In this relation the Reynolds number is $\mathrm{Re}_L = U_\infty L/\nu$. The range in which this correlation was developed based on numerical simulations and laboratory experiments is $\mathrm{Pr} = 0.72$, $10^2 \le \mathrm{Re}_L \le 10^4$ and $0.5 \le (Nb/L) \le 1.3$.

When the elements are cylinders in cross-flow (Fig. 3.9b), the optimal spacing S is influenced the most by the longitudinal dimension of the volume. The

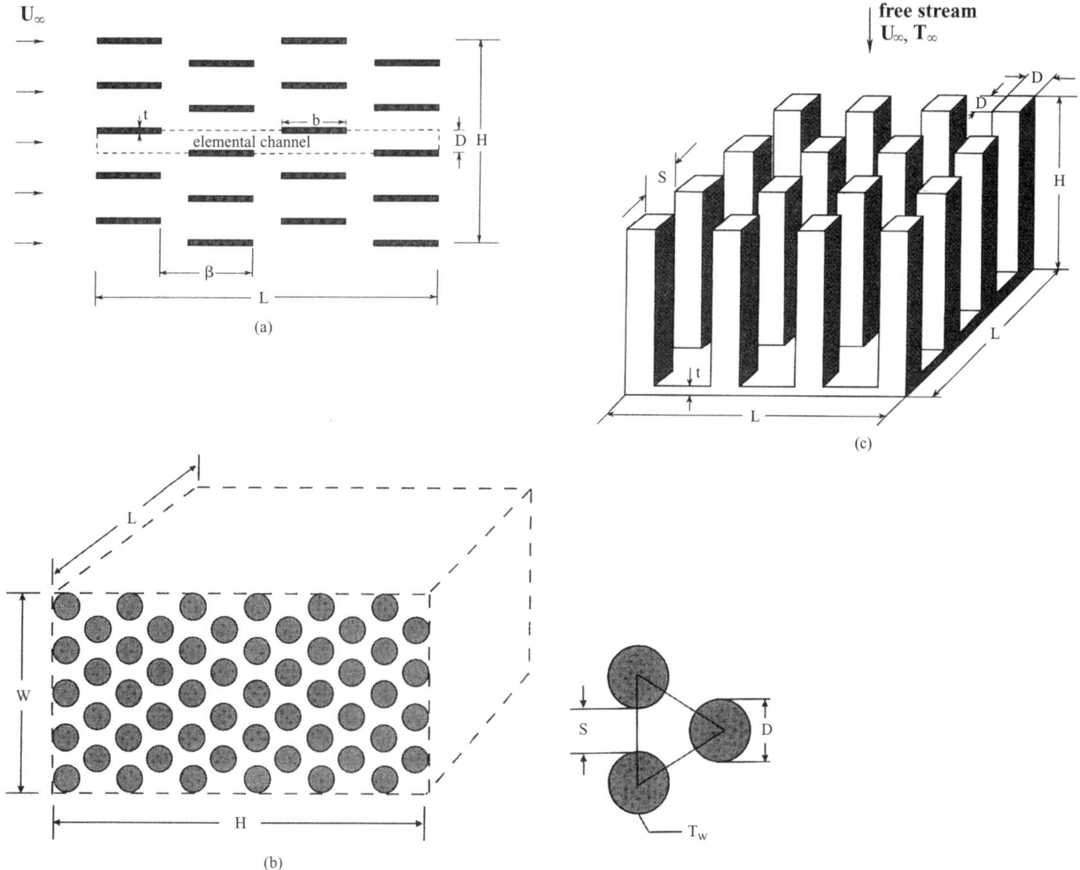

Figure 3.9 Volumes heated uniformly and cooled by forced convection: (a) array of staggered plates [16], (b) array of horizontal cylinders [17, 18], and (c) square pins with impinging flow [19].

optimal spacing was determined based on the method of intersecting the asymptotes [17, 18]. The asymptotes were derived from the large volume of empirical data accumulated in the literature for single cylinders in cross-flow (the large-S limit) and for arrays with many rows of cylinders (the small-S limit). In the range $10^4 \leq \tilde{P} \leq 10^8$, $25 \leq H/D \leq 200$ and $0.72 \leq \mathrm{Pr} \leq 50$, the optimal spacing is correlated within 5.6% by

$$\frac{S_{\mathrm{opt}}}{D} \cong 1.59 \frac{(H/D)^{0.52}}{\tilde{P}^{0.13}\,\mathrm{Pr}^{0.24}} \tag{3.39}$$

where $\tilde{P}$ is a dimensionless pressure drop based on D [proportional to Be, see Eq. (3.35)]:

$$\tilde{P} = \frac{\Delta P D^2}{\mu \nu} \tag{3.40}$$

When the free stream velocity U_∞ is specified (instead of ΔP), Eq. (3.39) may be transformed by noting that, approximately, $\Delta P \sim (1/2)\rho U_\infty^2$:

$$\frac{S_{\text{opt}}}{D} \cong 1.7 \frac{(H/D)^{0.52}}{\text{Re}_D^{0.26}\,\text{Pr}^{0.24}} \tag{3.41}$$

This correlation is valid in the range $140 < \text{Re}_D < 14{,}000$, where $\text{Re}_D = \text{U}_\infty D/\nu$. The minimized global thermal resistance that corresponds to this optimal spacing is

$$\frac{T_D - T_\infty}{qD/(kLW)} \cong \frac{4.5}{\text{Re}_D^{0.9}\,\text{Pr}^{0.64}} \tag{3.42}$$

where T_{D} is the cylinder temperature, and q is the total rate of heat transfer from the *HLW* volume to the coolant (T_∞). If the cylinders are arranged such that their centers form equilateral triangles (see Fig. 3.9b), the total number of cylinders present in the bundle is $HW/[(S + D)^2 \cos 30°]$. This number and the contact area based on it may be used in order to deduce from Eq. (3.42) the volume-averaged heat-transfer coefficient between the array and the stream.

Optimal spacings also emerge when the flow is three-dimensional, as in an array of pin fins with impinging flow [Fig. 3.9(c)]. The flow is initially aligned with the fins, and later makes a 90-degree turn to sweep along the base plate and across the fins. The optimal spacings are correlated within 16% by [19]

$$\frac{S_{\text{opt}}}{L} \cong 0.81\text{Pr}^{-0.25}\text{Re}_L^{-0.32} \tag{3.43}$$

which is valid in the range $0.06 < D/L < 0.14$, $0.28 < H/L < 0.56$, $0.72 < \text{Pr} < 7$, $10 < \text{Re}_D < 700$ and $90 < \text{Re}_L < 6{,}000$. Note that the spacing S_{opt} is controlled by the linear dimension of the volume, L. The corresponding minimum global thermal resistance between the array and the coolant is given within 20% by

$$\frac{T_D - T_\infty}{q/(kH)} \cong \frac{(D/L)^{0.31}}{1.57\text{Re}_L^{0.69}\,\text{Pr}^{0.45}} \tag{3.44}$$

The global resistance refers to the entire volume occupied by the array (HL^2), and is the ratio between the fin-coolant temperature difference ($T_D - T_\infty$) and the total heat current q generated by the HL^2 volume. In the square arrangement of Fig. 3.9(c) the total number of fins is $L^2/(S + D)^2$.

3.5 METHOD OF INTERSECTING THE ASYMPTOTES

The intersection of asymptotes method illustrated in sections 3.3 and 3.4 should not be confused with the method of *matched* asymptotic expansions. Matched asymptotes is an old technique of applied mathematics, in which the asymptotes are indeed fused (aligned, spliced) together through the appropriate choice of free parameters. Classical examples from fluid mechanics and heat transfer are the boundary layer solutions of Blasius [20] and Pohlhausen [21].

The intersection of asymptotes is quite different. It is the opposite. It is the competition, clash, or collision of the asymptotes, not the match. It is the sharp intersection of two vastly dissimilar trends (e.g., Figs. 3.5 and 3.8) that determines the spacings for optimal internal structure subject to global constraints. This method provides a much more direct route to structure optimization than the questionable (model-dependent) development of the complete behavior of the system versus its varying internal structure.

Not every possible design is important enough to be simulated and compared with other designs. Not every point of the rounded solid-line curves sketched in Figs. 3.5 and 3.8 deserves to be determined. The important designs are in the close vicinity of the intersection of the two competing trends. Optimal architecture is the visible (physical) statement made by the *balance*, equipartition, or equilibrium between competing geometries. This balance is the *optimal distribution of imperfection*. We will see this many times in constructal theory and design, for example, in the next section.

Additional work on the discovery of optimal spacings for natural and forced convection is available in the current literature. The optimal spacings for stacks of parallel plates with *mixed convection* (forced and natural) are reported and correlated in Ref. [22]. An analogous set of results exists for packages of bodies (plates, spheres) where the interstices are occupied by fluid-saturated porous media subjected to natural or forced convection [4, 23]. Such configurations are equivalent to Figs. 3.4, 3.6, 3.7, and 3.9, in which the fluid space is replaced by a fluid-saturated porous medium.

To optimize the spacings in such systems means to discover the structure of the skeleton of the multiscale solid that participates in the fluid-solid composite. The constructal composite uncovered by this procedure is a *designed porous medium* [23, 24], and it offers maximum flow access and volumetric density of heat transfer or mass transfer.

Natural porous flow structures exhibit multiple pore scales (diameters, lengths) and nonuniform distribution of scales through the available space. In Ref. [25] we showed that such heterogeneous flow structures can be derived from the constructal law of generation of flow configuration for greater flow access. The predicted porous medium has tree-shaped labyrinths with multiple scales that are distributed nonuniformly. We return to this important advance in Chapter 8.

The heterogeneous porous medium idea has the engineering-design applications sketched in Figs. P3.6 and P3.7. A solid wall is bombarded by intense heat flux from the side, and cooled by fluid flowing through internal channels parallel to the heated surface. The total volume occupied by fluid is fixed, because it must be small in order to preserve the mechanical strength of the wall. It was proposed in Ref. [7] (pp. 832–834) to distribute the channel volume in such a way that the hot-spot temperatures created on the heating surfaces are minimum. This led to the discovery that the best flow configuration has multiple channel scales that are positioned optimally in the body, hence nonuniformly. This design concept was validated numerically by Robbe et al. [26].

3.6 FITTING THE SOLID TO THE "BODY" OF THE FLOW

We conclude with a retrospective on the road from traditional design to the constructal method used in this chapter. The traditional approach to the design of convective flow structures such as heat exchangers and electronics cooling starts with the channels and the ducts. Solid features (walls, fins, etc.) are first assumed. Later, they are connected and assembled into larger constructs that fill the space allocated to the device. The flows, which are many and diverse, are forced (stuffed) into regular and rigid spaces.

The traditional approach is so common that we do not even think of questioning it. The very teaching of thermal engineering consists of "results" for "typical" configurations that confine convective flows: flat plates, tubes parallel plates, and the like. The typical configurations are assumed.

To force the flow to perform inside a standard, prescribed geometry is like forcing your foot into a standard (one-size) shoe [27]. It is not a good idea. The foot has its own "body"—its own size and shape [28]. The foot and the shoe go together.

A convective flow is the same. For example, the "body" of a thermally convective flow near a flat plate parallel to a stream is the thermal boundary layer region. This body has a natural shape, with a blunt nose and a characteristic thickness that increases downstream.

All the constructal spacings presented in this chapter illustrate a proposal to move away from the traditional approach. Start with the body of the flow, and build the confining walls so that they mate with the body as smoothly as possible. In this new philosophy, *the foot comes before the shoe*.

The idea is not limited to convective flows. In fluid flow, momentum is transferred between walls and the body of the flow (e.g., the boundary layer regions), and the flow of momentum is helped if the walls are fitted to the body of the flow. Such design work is common, but it is not recognized as such. All drag-reduction developments, morphing, and adaptable structures are in line with this philosophy. We demonstrate this by means of a simple morphing example in Problem 3.10.

To summarize this proposal, consider the problem of cooling a heat-generating line (or narrow strip) with a stream that flows parallel to the strip (Fig. 3.10, top). The global objective is to fit this convective flow into the smallest volume, or to pack the largest heat-transfer rate in a volume of fixed size and variable shape. The natural body of the convective flow can be anticipated based on boundary layer theory. As shown in Fig. 3.10, we expect a convective region shaped as a body of revolution, which becomes thicker in the downstream direction.

Next comes the task of shaping the volume so that it mimics the geometry of the convective body. The ideal way to proceed is to give the fixed volume V a

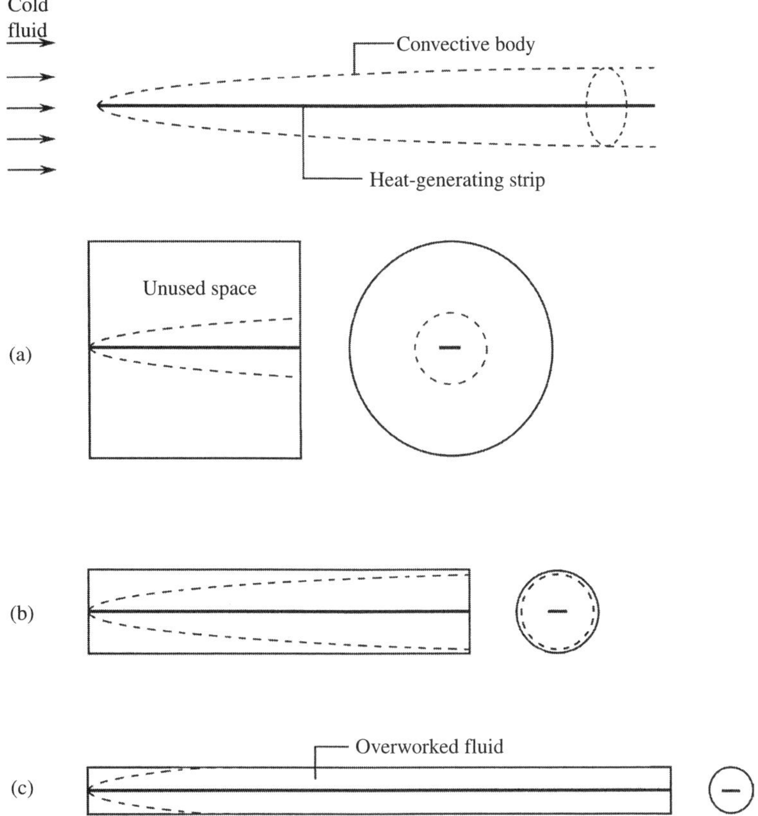

Figure 3.10 The body of a convective flow, and how to fit a duct around it [27].

large number of degrees of freedom, and to optimize the heat-transfer rate density with respect to each degree. The simplest version of this approach is to endow the volume V with a single degree of freedom, so that we still have access to an infinity of V shapes that could be fitted around the convective body. This was the approach taken in sections 3.3 and 3.4.

In Fig. 3.10, it was assumed that V is a cylinder positioned coaxially with the heat strip, and that the slenderness of the cylinder is free to vary. Three competing designs are shown. When the shape of V is robust, Fig. 3.10a, a large portion of V is wasted because it is bathed by fluid that does not "work." The isothermal fluid that is outside the convective body is useless.

In the opposite extreme, Fig. 3.10c, the V shape is considerably more slender than the convective body, and the heat-transfer fluid is "overworked." Most of the downstream position of the heat-generating strip is cooled by fluid that has been used already. The stream warms up as it flows downstream and becomes a poorer cooling agent.

The best V shape is in-between (Fig. 3.10b). An enclosure that hugs the contour of the convective body promises to be filled to the maximum with cold fluid that interacts thermally with the heat source. We exploit this body-fitting principle further in Chapter 6, where we endow the solid structure with more degrees of freedom.

The early work on constructal-theory spacings for forced and natural convection was reviewed in Refs. [3, 4]. New work is being contributed in this area at a significant pace (e.g. Refs. [29] through [36]). Noteworthy is Ref. [32], in which the optimal spacings for forced convection (section 3.4), are validated experimentally at microscales.

3.7 EVOLUTION OF TECHNOLOGY: FROM NATURAL TO FORCED CONVECTION

By comparing the optimized structures for natural convection (section 3.3) and forced convection (section 3.4), we have an opportunity to see the evolution of technology, as flow structures that morph in the same time direction as all other animate and inanimate flow architectures. This universal migration in the design space, from nonequilibrium flow structures, toward equilibrium structures [1] forms the subject of section 11.7.

Here we compare the volumetric cooling densities for natural convection [$q'''_{NC} = q_{max}/HLW$, Eq. (3.20)], with the corresponding result for forced convection [$q'''_{FC} = q_{max}/HLW$, Eq. (3.37)]. Their ratio is

$$\frac{q'''_{FC}}{q'''_{NC}} \sim \frac{\text{Be}^{1/2}/L^2}{\text{Ra}_H^{1/2}/H^2} \sim \frac{(\Delta P/\mu\alpha)_{FC}^{1/2}/L}{(g\beta\Delta T/\nu\alpha)_{NC}^{1/2}/H^2} \tag{3.45}$$

According to constructal theory, the future calls for greater and greater access to the flow of q out of the space HLW. This means greater and greater densities, q'''_{FC} and q'''_{NC}. Both quantities increase as their respective flow lengths (L, H) decrease. This means that the future calls for *miniaturization*. This is an important lesson from this design course: a technological trend (miniaturization) is *predicted*, and the prediction matches observations across the board and across all scales.

Another predicted trend is the replacement of natural convection with forced convection in cooling technologies. The reason is that q'''_{FC} increases faster than q'''_{NC} as the scale of the swept length (L, or H) decreases. Note that q'''_{FC} increases as L^{-1}, and q'''_{NC} as $H^{-1/2}$. The transition is from forced to natural convection (and not from natural to forced), and it occurs when $q'''_{FC} > q'''_{NC}$, or, if we write $L = H$ for the swept length in both designs, when

$$L\frac{(\Delta P/\mu\alpha)_{FC}}{(g\beta\Delta T/\nu\alpha)_{NC}} \tag{3.46}$$

where $\Delta T = T_{max} - T_0$. Below this length scale, forced convection cooling accommodates greater heat transfer densities than natural convection cooling.

It is possible to use the same method as in this section to locate the even smaller length scales below which forced convection cooling will be replaced by even more effective mechanisms. This philosophy is applied to much smaller scales in solid conduction (section 5.4) and in the realm of *designed porous media* in Ref. [23].

REFERENCES

1. A. Bejan and S. Lorente, The constructal law and the thermodynamics of flow systems with configuration. *Int J Heat Mass Transfer*, Vol. 47, 2004, pp. 3203–3214.
2. A. Bejan and S. Lorente, *La Loi Constructale*. Paris; L'Harmattan, 2005.
3. A. Bejan, *Shape and Structure, from Engineering to Nature*. Cambridge, UK: Cambridge University Press, 2000.
4. A. Bejan, *Convection Heat Transfer*, 3rd ed. Hoboken, NJ: Wiley, 2004.
5. A. Bejan, *Advanced Engineering Thermodynamics*, 2nd ed., (ch. 13). New York: Wiley, 1997.
6. A. H. Reis and A. Bejan, Constructal theory of global circulation and climate. *Int J Heat Mass Transfer*, Vol. 49, 2006, pp. 1857–1875.
7. A. Bejan, *Advanced Engineering Thermodynamics*, 3rd ed. (pp. xxi, 126, 472–484). Hoboken, NJ: Wiley, 2006.
8. A. Bejan, *Convection Heat Transfer* (p. 157, Problem 11). New York: Wiley, 1984.
9. A. Bar Cohen and W. M. Rohsenow, Thermally optimum spacing of vertical, natural convection cooled, parallel plates. *J Heat Transfer*, Vol. 106, 1984, pp. 116–123.
10. G. A. Ledezma and A. Bejan, Optimal geometric arrangement of staggered vertical plates in natural convection. *J Heat Transfer*, Vol. 119, 1997, pp. 700–708.
11. A. Bejan, A. J. Fowler, and G. Stanescu, The optimal spacing between horizontal cylinders in a fixed volume cooled by natural convection. *Int J Heat Mass Transfer*, Vol. 38, 1995, pp. 2047–2055.
12. A. Bejan and E. Sciubba, The optimal spacing of parallel plates cooled by forced convection. *Int J Heat Mass Transfer*, Vol. 35, 1992, pp. 3259–3264.
13. S. Mereu, E. Sciubba, and A. Bejan, The optimal cooling of a stack of heat generating boards with fixed pressure drop, flow rate or pumping power. *Int J Heat Mass Transfer*, Vol. 36, 1993, pp. 3677–3686.
14. S. Bhattacharjee and W. L. Grosshandler, The formation of a wall jet near a high temperature wall under microgravity environment. *ASME HTD*, Vol. 96, 1988, pp. 711–716.
15. S. Petrescu, Comments on the optimal spacing of parallel plates cooled by forced convection. *Int J Heat Mass Transfer*, Vol. 37, 1994, p. 1283.
16. A. J. Fowler, G. A. Ledezma, and A. Bejan, Optimal geometric arrangement of staggered plates in forced convection. *Int J Heat Mass Transfer*, Vol. 40, 1997, pp. 1795–1805.

17. A. Bejan, The optimal spacings for cylinders in crossflow forced convection. *J Heat Transfer*, Vol. 117, 1995, pp. 767–770.
18. G. Stanescu, A. J. Fowler, and A. Bejan, The optimal spacing of cylinders in free-stream cross-flow forced convection. *Int J Heat Mass Transfer*, Vol. 39, 1996, pp. 311–317.
19. G. Ledezma, A. M. Morega, and A. Bejan, Optimal spacing between pin fins with impinging flow. *J Heat Transfer*, Vol. 118, 1996, pp. 570–577.
20. H. Blasius, Grenzschichten in Flüssigkeiten mit kleiner Reibung. Z *Math Phys*, Vol. 56, 1908, p. 1; also NACA TM 1256.
21. E. Pohlhausen, Der Wärmeaustausch zwischen festen Körpern und Flüssigkeiten mit kleiner Reibung und kleiner Wärmeleitung. *Z Angew Math Mech*, Vol. 1, 1921, pp. 115–121.
22. T. Bello-Ochende and A. Bejan, Optimal spacings for mixed convection. *J Heat Transfer*, Vol. 126, 2004, pp. 956–962.
23. A. Bejan, Designed porous media: maximal heat transfer density at decreasing length scales. *Int J Heat Mass Transfer*, Vol. 47, 2004, pp. 3073–3083.
24. A. Bejan, I. Dincer, S. Lorente, A. F. Miguel, and A. H. Reis, *Porous and Complex Flow Structures in Modern Technologies*. New York: Springer, 2004:
25. S. Lorente and A. Bejan, Heterogeneous porous media as multiscale structures for maximum flow access. *J Appl Phys,* Vol. 100, 2006, 114909.
26. M. Robbe, E. Sciubba, A. Bejan, and S. Lorente, Numerical analysis of a tree-shaped cooling structure for a 2-D slab: A validation of a "constructally optimal" configuration, ESDA 2006, 8th Biennial ASME Conference on Engineering Systems Design and Analysis, Turin, July 4–7, 2006.
27. T. Bello-Ochende and A. Bejan, Fitting the duct to the "body" of the convective flow. *Int J Heat Mass Transfer*, Vol. 46, 2003, pp. 1693–1701.
28. D. A. Fried, *Shoes That Don't Hurt*, Victoria, BC, Canada: Trafford Publishing, 2006.
29. S. K. W. Tou, C. P. Tso, and X. Zhang. 3-D numerical analysis of natural convective cooling of a 3×3 heater array in rectangular enclosures. *Int J Heat Mass Transfer*, Vol. 44, 1999, pp. 3231–3244.
30. A. Yilmaz, O. Buyukalaca, and T. Yilmaz, Optimum shape and dimensions of channel for convective heat transfer in laminar flow at constant wall temperature. *Int J Heat Mass Transfer*, Vol. 43, 2000, pp. 767–775.
31. K. C. Toh, X. Y. Chen, and J. C. Chai, Numerical computation of fluid flow and heat transfer in microchannels. *Int J Heat Mass Transfer*, Vol. 45, 2002, pp. 5133–5141.
32. M. Favre-Marinet, S. Le Person, and A. Bejan, Maximum heat transfer rate density in two dimensional minichannels and microchannels. *Microscale Thermophysical Engineering*, Vol. 8, 2004, pp. 225–237.
33. Y. S. Muzychka, Constructal design of forced convection cooled micro-channel heat sinks and exchangers. *Int J Heat Mass Transfer*, Vol. 48, 2005, pp. 3119–3124.
34. D.-K. Kim and S. J. Kim, Closed-form correlations for thermal optimization of microchannels. *Int J Heat Mass Transfer*, Vol. 50, 2007, pp. 5318–5322.
35. L. Gosselin, Fitting the flow regime in the internal structure of heat transfer systems. *Int Comm Heat Mass Transfer*, Vol. 33, 2006, pp. 30–38.
36. Y. S. Muzychka, Constructal multi-scale design of compact micro-tube heat sinks and heat exchangers. *Int J Thermal Sciences*, Vol. 46, 2007, pp. 245–252.
37. S. Lorente, W. Wechsatol, and A. Bejan, Tree-shaped flow structures designed by minimizing path lengths. *Int J Heat Mass Transfer*, Vol. 45, 2002, pp. 3299–3312.

PROBLEMS

3.1. Points A and B are connected by the broken line ARB, such that point R is free to move along the base of the drawing, Fig. P3.1. In the principle of Heron of Alexandria, AR is the incident ray of light, RB is the reflected ray, and the base is the mirror. Show that the path ARB is the shortest when $x = 0$, that is, when the angle of incidence (α_i) is equal to the angle of reflection (α_r).

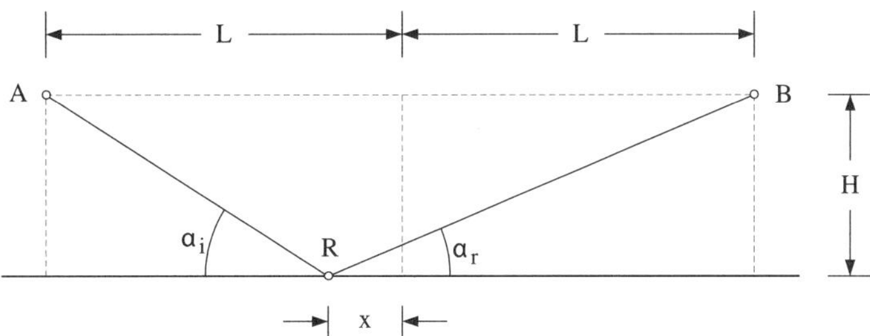

Figure P3.1

3.2. Consider the travel along the route ARB shown in Fig. P3.2. The speeds are V_0 along the AR segment and V_1 along the RB segment. Break point R is free to slide along the line that separates the V_0 and the V_1 domains. In the law of Fermat, the two domains are two different media through which a ray of light connects A to B; the point of refraction (R) is situated on the interface that separates the two media. Show that the time of travel between A and B is minimum when the break in the path is such that

$$\frac{\sin \beta_{\mathrm{i}}}{\sin \beta_{\mathrm{r}}} = \frac{V_0}{V_1}$$

The angles β_i and β_r are measured relative to the normal to the interface. In optics, the above relation is known as Snell's law. See also Problem 3.15.

3.3. Water is being distributed to a large number of users by using a round duct of length L and inner diameters $D(x)$, where x is the longitudinal coordinate. Each user consumes the same fraction of the initial flow rate of the water stream. The users are distributed equidistantly along the duct. The initial flow rate is $\dot{m}$ and enters the pipe through the $x = L$ end. The last user served by the stream is located near the $x = 0$ end. Assume that the users are sufficiently numerous, and their individual demands sufficiently small, so that the flow rate varies continuously along the duct, $\dot{m}(x)$. The water flow is in the fully turbulent regime, with a friction factor that is independent of flow rate (the fully rough limit). Find the optimal duct shape $D(x)$ that minimizes the total power required to pump the stream through the entire system. Assume that the resistance to water flow is due mainly to the flow through

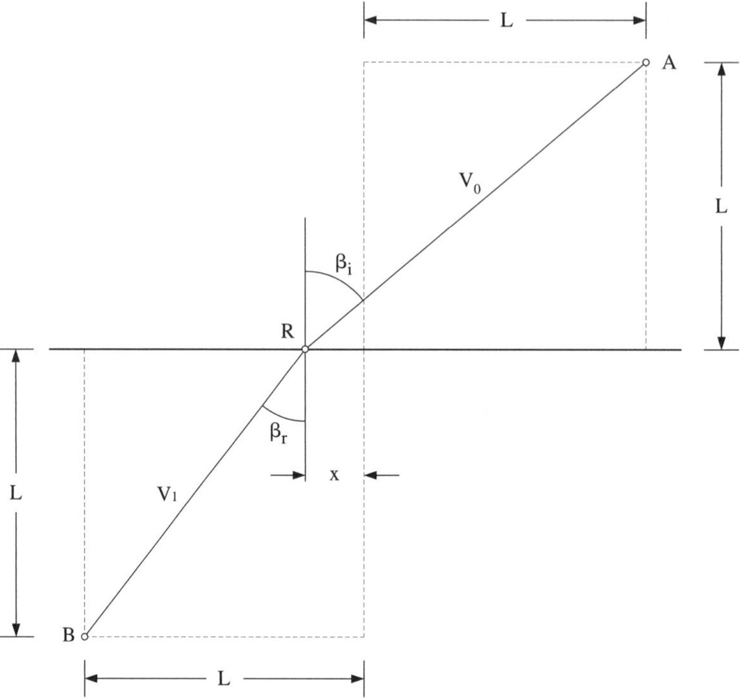

Figure P3.2

the duct. Assume also that the duct wall thickness is small in comparison with D, and that the total amount of duct wall material is fixed.

3.4. Determine the optimal shape of a duct that must carry a certain stream ($\dot{m}$) between two pressure levels (P_0 and $P_L = P_0 - \Delta P$) separated by a fixed distance (L). The objective is to minimize the overall flow resistance $\Delta P/\dot{m}$. The duct is straight with the flow pointing in the x direction; however, the cross-sectional area $A(x)$ and wetted perimeter $p(x)$ may vary with the longitudinal position. Two constraints must be taken into account. One is the total duct-volume constraint:

$$\int_0^L A(x)dx = V, \quad \text{constant}$$

The volume constraint is important in the design of compact heat exchangers. Another possible constraint refers to the total amount of duct wall material used:

$$\int_0^L p(x)dx = M, \quad \text{constant}$$

The material constraint is important in designs in which the unit cost of the material is high or in which the weight of the overall heat exchanger is constrained, as in aerospace applications. We can calculate the overall flow resistance by noting that regardless of whether the flow is laminar, turbulent, or fully developed, the pressure gradient is given by

$$\frac{dP}{dx} = -\tau_w(x)\frac{p(x)}{A(x)}$$

where τ_w is the wall shear stress. To calculate the global resistance to flow, we substitute the friction factor $f = \tau_{\mathrm{w}}/\left(\frac{1}{2}\rho U^2\right)$, the Reynolds number $\mathrm{Re}_{D_h} = D_h U/\nu$, and the mass flow rate $\dot{m} = \rho UA$, and integrate from the inlet to the outlet:

$$\frac{\Delta P}{\dot{m}} = \int_0^L f\mathrm{Re}_{D_h}\frac{p\nu}{2A^2 D_h}dx$$

Consider as an example the hydraulic entrance region to a duct of round cross-section [diameter $D(x)$] through which the flow is sufficiently isothermal such that ν may be regarded as constant. Show that when the overall resistance is minimized subject to the volume constraint, the optimal diameter $D(x)$ varies as $x^{-1/12}$, which means that the duct is shaped like a trumpet. Show also that when the optimization is subjected to the wall material constraint, the optimal duct shape varies as $D \sim x^{-1/10}$. Derive the corresponding shapes for a duct with the cross-section shaped like a very flat rectangle of spacing $D(x)$ and width W, such as $D(x) \ll W$.

3.5. The temperature of gas-turbine blades is controlled by circulating cool air through channels made longitudinally in the body of the blade. Figure P3.5 shows a two-dimensional model of the technique (Ref. [7], pp. 832–833). The heat input received from the hot gases of combustion can be modeled as a curtain of uniform heat flux q''. The channels are round with diameter D and center-to-center spacing S. The blade is a two-dimensional conducting slab of thickness H (fixed) and thermal conductivity k. The hole centers are on the axis of symmetry of the slab cross-section. The hole surface is isothermal at temperature T_{min}. The hot spots (T_{max}) occur on the external surface at the points situated the farthest from the holes. The objective is to minimize T_{max} by selecting the best flow configuration (S, D) subject to the constraint that the volume fraction (ϕ) occupied by all the channels is fixed. Assume that to minimize $T_{\mathrm{max}} - T_{\mathrm{min}}$ is approximately the same as minimizing the length L of the straight path from T_{max} to T_{min}. This approach is known as the minimal length method [37] (see also Problem 4.1) and represents the near-optimal allocation of one flow path length (L) to elemental area $H \times S/2$. Determine analytically the optimal channel diameter and spacing in the range $\phi \ll 1$.

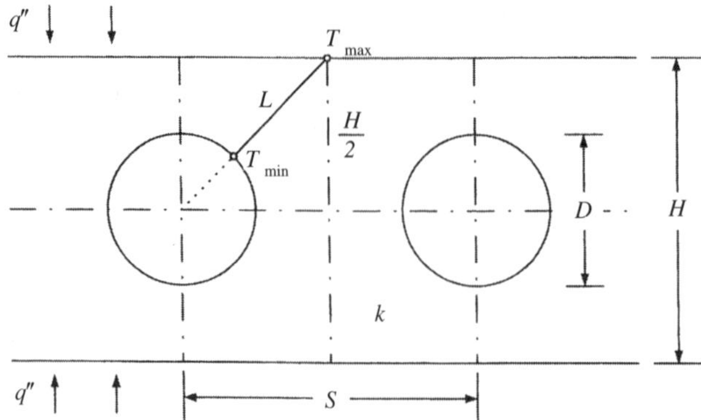

Figure P3.5

3.6. The hot-spot temperature of the configuration shown in Fig. P3.5 can be reduced further by installing two smaller D_2 channels halfway between two of the original D_1 channels. The new configuration is shown in Fig. P3.6. In every elemental volume of cross-section $H \times S$ there are two D_2 channels and one D_1 channel. The volume fraction (ϕ) occupied by all the channels is fixed. The new hot spots occur at a distance x from the transversal plane in which the D_2 centers are located. The heat flux that enters the hot spot splits into two equal currents as it flows toward the two heat sinks, D_1 and D_2. Approximate the respective flow paths as the straight segments L_1 and L_2: consequently, the equipartition of the heat flux means that $L_1 = L_2$ $(= L)$. The minimization of $(T_{\text{max}} - T_{\text{min}})$ is the same as minimizing L. Nondimensionalize all the dimensions with respect to H (fixed), and determine numerically the optimal configuration in dimensionless terms $(D_1/H, D_2/H, S/H, x/H, y/H)$.

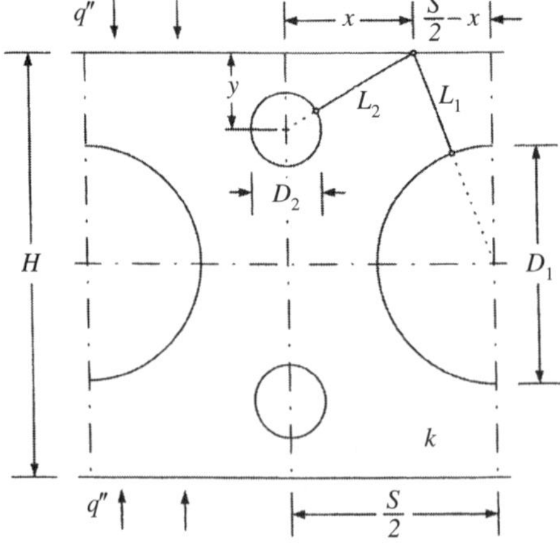

Figure P3.6

3.7. In this problem we ask whether ducts with square cross-sections (s) are better than parallel-plate channels (p) for achieving greater heat-transfer densities in volumes packed with such channels. The comparison is shown in Fig. P3.7. The flow is from left to right, and the pressure difference is ΔP in both designs. Each channel is the best that it can be: its boundary layers merge just as they exit the channel.

According to section 3.4, the optimal spacing between parallel plates with fixed length (L_p) and pressure drop ΔP is $D_p/L_p = a_p \mathrm{Be}_p^{-1/4}$. The corresponding maximum heat-transfer rate (q_p) packed in a stack of thickness H_p and width H_p is $q_p L_p/(H_p^2 k \Delta T) = b_p\ \mathrm{Be}_p^{1/2}$, where $\mathrm{Be}_p = \Delta P L_p^2/(\mu\alpha)$. There are $n_p = H_p/D_p$ plates in the stack. Each plate has the area $H_p L_p$ and negligible thickness.

The numerical factors a_p and b_p are dimensionless and of order 1. We assume that these factors do not change much when the shape of the channel cross section changes. Thus, for a stack of channels with square cross-sections (square side D_s, length L_s), we write similarly $D_s/L_s = a_s\ \mathrm{Be}_s^{-1/4}$ and $q_s L_s/(H_s^2 k \Delta T) = b_s\ \mathrm{Be}_s^{1/2}$, where $\mathrm{Be}_s = \Delta P L_s^2/(\mu\alpha)$, $a_s \simeq a_p$ and $b_s \simeq b_p$. For simplicity, assume that $a_s = a_p$ and $b_s = b_p$.

Two global constraints that apply to both configurations, parallel plates and squares. The total volume of the stack is the same, and so is the total amount of channel wall material. Show that the (p) and (s) configurations have the relative dimensions drawn in Fig. P3.7:

$$\frac{D_p}{D_s} = \frac{1}{2}\frac{L_p}{L_s} = \frac{1}{4}\frac{H_p}{H_s} = 2$$

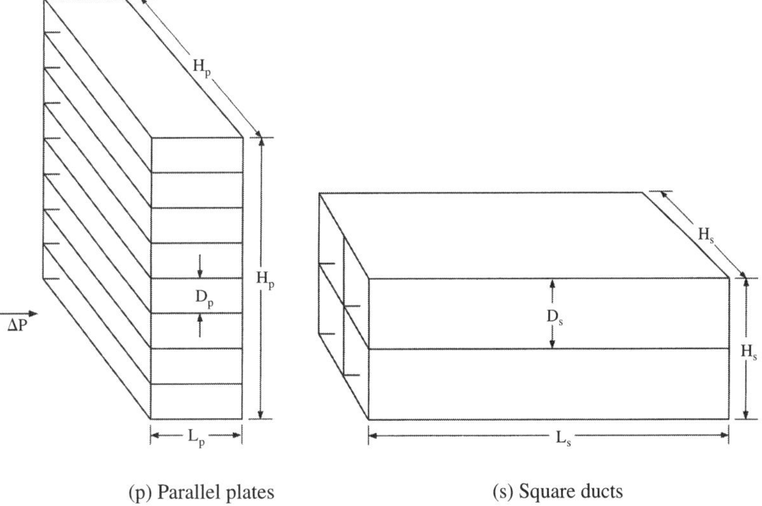

Figure P3.7

and that the stack of parallel plates packs four times more heat transfer than the stack of square channels, $q_p/q_s = 4$.

This conclusion will be approximate, and its accuracy can be assessed by repeating the analysis using exact values of $(a, b)_{p,s}$. Furthermore, stacks with channels that fill the H^2L volume with wall material that is comparable to that of square channels (e.g., triangular, round and hexagonal cross-sections), will have a total heat-transfer rate similar to q_s, not q_p. The general conclusion then is that stacks with parallel plates are four times "more compact" than other stacks.

This brings up the future research question of how to stiffen each plate in a stack of parallel plates. The final stack configuration will have parallel plates with plate-to-plate stiffeners: the cross-sections of the resulting flow channels will have rectangular cross-sections with aspect ratios to be determined from the competition between maximum heat transfer and maximum stiffness (Chapter 10).

3.8. In this problem we question the treatment of the large-D analysis of Eqs. (3.29) through (3.33). Imagine D spacings much larger than the wall boundary layer thicknesses. Most of the fluid driven by the imposed ΔP flows without friction through the channels. Its velocity (U) is dictated by the Bernoulli equation applied on the center line of the vein that flows through the D-wide channel, $U \sim (\Delta P/\rho)^{1/2}$. This scale replaces Eq. (3.30). Repeat the analysis of section 3.4, and show that the intersection of asymptotes leads to the transition spacing shown in Eq. (3.36). This means that when D increases above the scale $(\Delta P/\rho)^{1/2}$ the flow rate *stagnates* at this level. Larger increases in D are useless. This makes the transition D scale (3.36) much more important, because it marks the D scale above which increases in D do not yield increases in U.

3.9. Nature shows that moving bodies that are faster are also more slender. This is true of fishes, birds, and falling raindrops. Demonstrate analytically that faster bodies must be slender, in accordance with the constructal law of greater flow access through the evolution of configuration. As shown in Fig. P3.9, assume that the body is two-dimensional (a slab) with the thickness H and swept length L. The fixed body size is indicated by the fixed area of the profile, $A = HL$, or the fixed body length scale $A^{1/2}$. The relative speed between body and medium (V) is fixed.

The total drag force aligned with V is F' [N/m] $= F'_L + F'_H$, where F'_L is the friction force along the two L faces, $F'_L = \bar{\tau}2L$. The F'_H contribution is due to the blunt shape of the body, and can be approximated as $F'_H \sim \frac{1}{2}\rho V^2 H$, where $\frac{1}{2}\rho V^2$ is the scale of the stagnation pressure felt by the leading surface. Estimate $\bar{\tau}$ in two ways:

(a) Turbulent flow: $\bar{\tau} = 0.037\rho V^2 \mathrm{Re}_L^{-1/5}$, valid for $\mathrm{Re}_L = 10^5 - 10^7$, where $\mathrm{Re}_L = VL/\nu$, cf. Ref. [4], p. 353.

(b) Laminar flow: $\bar{\tau} = 0.064\rho V^2 \mathrm{Re}_L^{-1/2}$, valid for $\mathrm{Re}_L = 10 - 10^5$, cf. Ref. [4], p. 54.

Minimize the total drag force F' by varying the aspect ratio H/L subject to fixed $A^{1/2}$ and V, and show that the optimal aspect ratio must decrease as V increases:

(a) $\left(\frac{H}{L}\right)_{\mathrm{opt}} = 0.093\mathrm{Re}_{A^{1/2}}^{-2/9} = 0.144\mathrm{Re}_L^{-2/11}$

(b) $\left(\frac{H}{L}\right)_{\mathrm{opt}} = 1.46\mathrm{Re}_{A^{1/2}}^{-2/3} = 1.76\mathrm{Re}_L^{-1}$

Here, $\mathrm{Re}_{A^{1/2}}$ is the fixed Reynolds number based on body length scale, $\mathrm{Re}_{A^{1/2}} = VA^{1/2}/\nu = (H/L)^{1/2}\mathrm{Re}_L$.

A similar analysis can be performed for a body of revolution of length L and thickness H. Consider a water droplet dripping from a faucet ($y = 0$). Assume laminar flow (b), and that the effect of drag is still negligible so that the droplet can be modeled as falling freely. The falling distance y is measured downward from the faucet ($y = 0$). Consider two distances, $y_2 > y_1$ such that $y_2 = 2y_1$. In the absence of an analysis for a body of revolution, assume that $(H/L)_{\mathrm{opt}}$ formula (b) is approximately valid. Show that (H/L) at level y_2 must be smaller than the slenderness at altitude y_1.

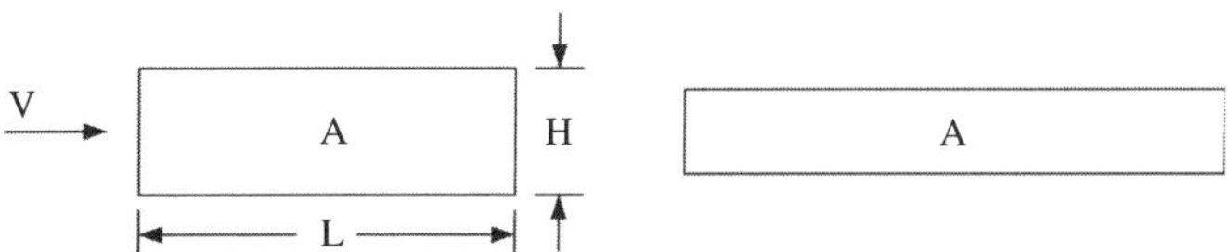

Figure P3.9

In the study of fish swimming, it is observed that the transition from swimming by side flippers to swimming by undulating the body and tail occurs when $\mathrm{Re}_L \sim 10$. Use the solution for part (b) and show that the expected aspect ratio of the smallest undulating fish is $(H/L)_{\mathrm{opt}} \sim 1/6$. Larger fish swim faster and they must be more slender.

3.10. Consider again the problem of Fig. P3.9, and recognize W as the third dimension of the slab: W is perpendicular to the plane of the figure. Assume that $W >> H$. The boundary layers are turbulent, but instead of model (a) of Problem 3.9, assume the model where the average wall shear stress is simply $\bar{\tau} = c\rho V^2$, in which c is nearly constant, with values between 10^{-3} and 10^{-2}. Minimize the drag force subject to fixed profile area ($A = HL$), and show that the optimal configuration is represented by $H/L = 4c$, $L = [A/(4c)]^{1/2}$ and $H = (4cA)^{1/2}$, for which the minimized drag force is $F = \rho V^2(4cA)^{1/2}W$.

3.11. If the width W of the plate shown in Fig. P3.10 is not much larger than the thickness H, then the friction force along the longitudinal surfaces is $\bar{\tau} = 2(W + H)$. Minimize this force subject to the assumption that the

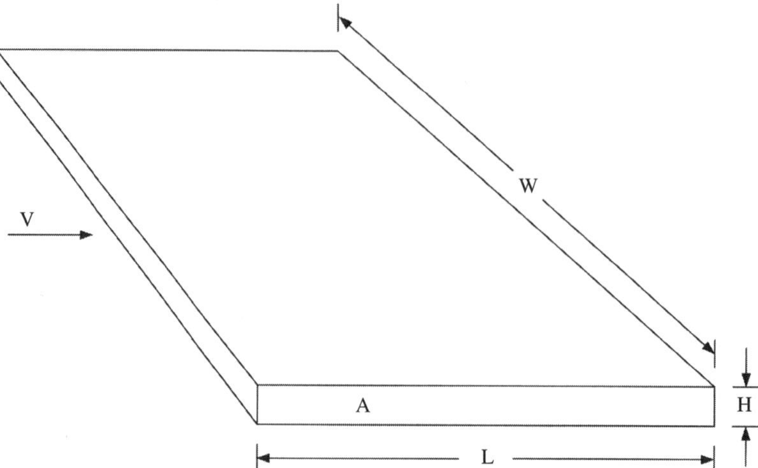

Figure P3.10

frontal area is fixed (WH = constant), and conclude that the constructal configuration must have a square cross-section, $H = W$.

3.12. The "smooth" version of the design uncovered in the preceding problem is a cylinder of length L and diameter D. The fluid free-stream velocity is perpendicular to the frontal area, which is a disc with diameter D. The volume of the cylinder ($B = D^2L$) is fixed. For simplicity assume (as in Problem 3.11) that the wall averaged shear stress is $\bar{\tau} = c\rho V^2$, where c is nearly constant with values in the range 0.001 – 0.01. Vary the aspect ratio D/L, and show that when $D/L = 4c$, $D = (4cB)^{1/3}$ and $L = [B/(4c)^2]^{1/3}$ the total drag force is minimal, $F = (3\pi/8)\rho V^2(4cB)^{2/3}$.

3.13. In problems 3.10 and 3.12 we discovered the shapes that endow a slab and a cylinder with minimal drag. The shapes and minimal drag forces are listed at the end of the problem statements. Assume that these two objects are assembled as in Fig. P3.13. The wing span W is a known parameter. The total volume of the assembly is fixed, $V_T = WA + B'$, where WA is the volume of the wing and B' is the volume of the fuselage. Use the results of Problems 3.10 and 3.12, in which c is a dimensionless factor of order $10^{-3} - 10^{-2}$. Minimize the total drag force on the assembly by varying the way in which V_T is divided between WA and B'. Express your results graphically as $\tilde{A}(\tilde{W}, c)$, $\tilde{B}'(\tilde{W}, c)$ and $\tilde{F}_T(\tilde{W}, c)$, where $\tilde{B}' = B'/V_T$, $\tilde{A} = A/V_T^{2/3}$, $\tilde{W} = W/V_T^{1/3}$ and $\tilde{F}_T = F_T/(\rho V^2 V_T^{2/3})$.

3.14. A dog and its owner stand on the beach (at water's edge) at point A. The owner throws a stick that falls in the water at point B. There are no currents in the water. The distance from B to point P on shore is H. The distance from A to P is L. See Fig. P3.14.

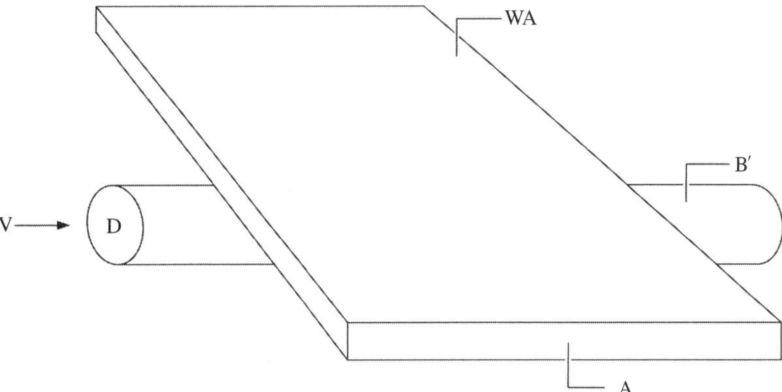

Figure P3.13

The dog must travel from A to B by combining running with swimming. He must decide at what location (J) to jump in the water. The dog knows constructal theory, and on that basis he selects point J. He looks at the problem in two ways:

(a) According to the constructal theory of animal locomotion, the minimum work spent per distance traveled is proportional to the body weight (Mg) times a constant of order 1, which depends on the surrounding medium, namely c_r for running, and c_s for swimming, where $c_s > c_r$. Derive an expression for the total work spent on the route AJB, minimize this expression with respect to the distance a, and report the optimal location of point J as $a_{opt}/H = \text{function}\ (c_s/c_r)$.

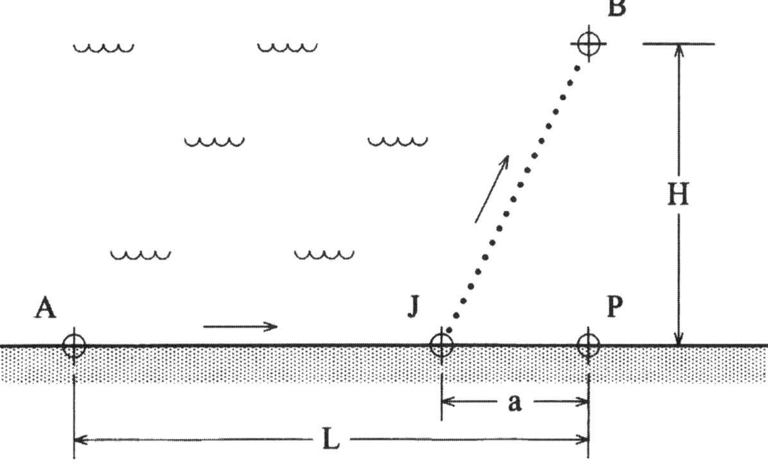

Figure P3.14

(b) According to measurements, the dog running speed V_r is roughly ten times greater than the swimming speed V_s. Derive an expression for the total time required for travel on the route AJB, minimize this time with respect to the length a, and report the location of J as $a_{opt}/H =$ function (V_r/V_s).

(c) The value calculated for a_{opt}/H in part (b) turns out to be the same as the value of a_{opt}/H determined in part (a). What is the value of the ratio c_s/c_r?

4

TREE NETWORKS FOR FLUID FLOW

This chapter is an introduction to the generation of tree-shaped layouts for fluid flow, so that the flow structures use the available space for maximum global benefit. The focus is on generating the *drawing*, that is, the configuration, geometry, and geography of the flow system. We start with the simplest class of tree-shaped flows: T- and Y-shaped constructs. We devote more space to trees that connect a circle with its center, because these quasi-radial structures will be used in applications for heating and cooling (Chapters 5, 7, and 8). We also teach methods for decreasing the cost and time needed to generate tree architectures, for example, by minimizing path lengths everywhere, and optimizing the angles of confluence of two branches.

The purpose of a tree network is to make a flow connection between one point (source, sink) and a *continuum*—an *infinity* of points (line, area, volume). In civil engineering applications such as the distribution of water and electricity, the area is approximated by a large number of evenly or unevenly distributed points (the consumers [1, 2]). Tree-shaped flows over continua are the rule in nature, not the exception. They define the design of animate and inanimate flow systems: lungs, vascularized tissues, nervous systems, river basins and deltas, lightning, snowflakes, vegetation, and so on.

Dendritic flows have been studied extensively in physiology and geophysics. The empirical view inherited from the natural sciences is that tree-shaped flows are examples of spontaneous self-organization and self-optimization. In contrast with this, the constructal view is theory: flow architectures such as the tree are the results of a process of evolution toward greater global flow access. Tree-shaped flows are deducible from the constructal law.

Trees emerge in the same way as the flow systems with optimized internal structure (spacings) presented in Chapter 3. Tree-shaped flows persist in time—in nature, as well as in engineering—because they are efficient and use the available space to the maximum. They are important fundamentally in engineering, because the relation between efficiency and compactness is key to the progress toward design and integration of increasing numbers of smaller and smaller functioning components. Tree-shaped architectures are the new weapon in the miniaturization and vascularization revolution.

One common feature of tree-shaped flows is that they consist of two dissimilar flow mechanisms, or flow paths: one with low resistivity (channels, streams), and the other with high resistivity (diffusion across interstices between channels). Maximum flow access is achieved when the many streams of the tree are organized such that the flows with high resistivity inhabit the smallest scales of the flow structure, whereas the flows with the lower resistivity are assigned to the larger scales. The occurrence of two flow mechanisms to maximize access is common in nature (cf. section 11.2).

In the lung, for example, the high-resistivity flow is the diffusion of O_2 and CO_2 through the tissue that separates the smallest flow spaces of the structure (the alveoli). An entire sequence of air-flow scales is distributed in a very special way over the larger scales: tubes become wider, and longitudinal length scales increase. In the conduction cooling trees of Chapter 5, the high resistivity belongs to the interstices configured between low-resistivity inserts.

Flow resistances cannot be eliminated. They can be arranged or assembled—forced to coexist with each other in a finite volume—so that their global impact on performance is minimal. Flow diversity conspires with optimized organization, and the result is construction, or drawing.

4.1 OPTIMAL PROPORTIONS: *T*- AND *Y*-SHAPED CONSTRUCTS

The generation of geometry is the key to achieving maximum global performance under global constraints, and to predicting along the way the emergence of flow configuration in natural systems. This becomes clearer if we consider a simple case of tree flow, namely Fig. 4.1. To start with, the maximum-access geometry for fluid flow between two points is the straight duct with round cross-section (cf. section 3.1). The round shape holds for both laminar and turbulent flow. It is a very robust design feature: nearly round shapes are almost as effective as the perfectly round shape.

The simplest tree is the T-shaped construct of round tubes shown in Fig. 4.1. The flow connects one point (source or sink) with two points. There are two global constraints: (1) the total duct volume,

$$V = \frac{\pi}{4}\left(D_1^2 L_1 + 2D_2^2 L_2\right), \text{ constant} \tag{4.1}$$

and (2) the total space (area) occupied by the construct,

$$A = 2L_2 L_1, \text{ constant} \tag{4.2}$$

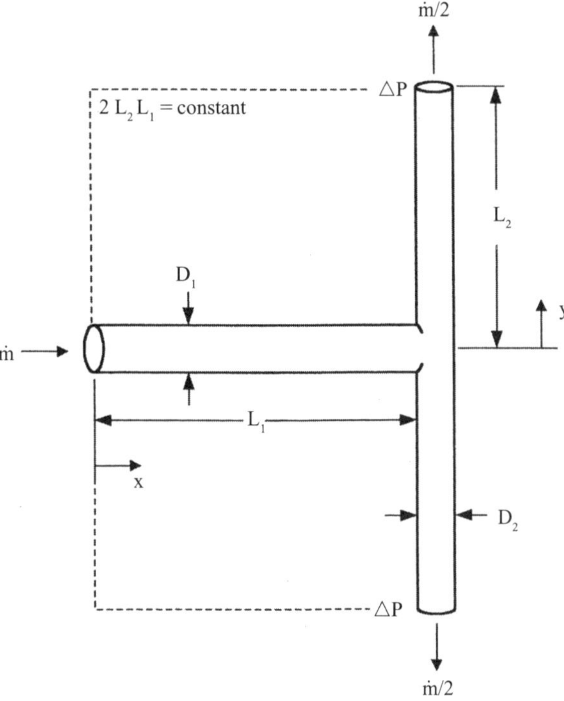

Figure 4.1 T-shaped construct of round tubes [3].

Large or small, the image of Fig. 4.1 is decided by selecting the values of two dimensionless ratios, D_1/D_2 and L_1/L_2. The latter represents the shape of the area A. Both are free to vary; therefore, they represent the 2 *degrees of freedom* of the drawing (flow architecture). In the following analysis, we determine the two aspect ratios by minimizing the total flow resistance posed by the T-shaped construct on area A.

The optimal ratio D_1/D_2 turns out to be independent of the shape of A, and is determined as follows. Assume that the flow through every tube is in the Poiseuille regime, that is, that local pressure losses (entrance, junction) are negligible in comparison with the frictional pressure loss that is distributed uniformly along the length of the tube. This means that the svelteness of the construct (Sv) is large, namely, $\mathrm{Sv} = A^{1/2}/V^{1/3}$ (cf. section 1.1). The longitudinal pressure drops for the two tube sizes shown in Fig. 4.1 are

$$\Delta P_1 = C\dot{m}_1 \frac{L_1}{D_1^4} \tag{4.3}$$

$$\Delta P_2 = C\dot{m}_2 \frac{L_2}{D_2^4} \tag{4.4}$$

where $C = 128\nu/\pi$, and ν is the kinematic viscosity of the fluid. The two mass flow rates are related by

$$\frac{\dot{m}_1}{\dot{m}_2} = 2 \tag{4.5}$$

because the assumed T configuration is symmetric: symmetry rules out a flow imbalance between the streams that flow through the two L_2 tubes. Without symmetry, there is no equipartitioning of the $\dot{m}_1$ streams into two $\dot{m}_2$ streams.

The overall pressure drop is $\Delta P = \Delta P_1 + \Delta P_2$. After using Eqs. (4.3) through (4.5), we find the global flow resistance:

$$\Delta P = C \dot{m}_1 R_{\text{lam}} \tag{4.6}$$

where R_{lam} is a factor that depends solely on the geometry of the T:

$$R_{\text{lam}} = \frac{L_1}{D_1^4} + \frac{L_2}{2D_2^4} \tag{4.7}$$

To minimize R_{lam}, our thoughts guide us to increasing both D_1 and D_2. This idea is not workable because the diameters D_1 and D_2 are related through the volume constraint (4.1): D_1 and D_2 cannot be increased indefinitely and independently.

The correct path to the optimal D_1 and D_2 values is to minimize R_{lam} subject to keeping V constant. This can be done analytically in several ways; however, we use this as an opportunity to illustrate the use of the method of Lagrange multipliers (see Appendix B). To minimize the expression shown in Eq. (4.7) subject to the constant expressed in Eq. (4.1) is equivalent to seeking the extremum of the aggregate function:

$$\Phi = \frac{L_1}{D_1^4} + \frac{L_2}{2D_2^4} + \lambda \left(D_1^2 L_1 + 2 D_2^2 L_2 \right) \tag{4.8}$$

which was obtained by combining linearly the two expressions. The factor λ is a constant (the Lagrange multiplier), which can be determined from the constraint (4.1) later, after optimization. The extremum of Φ is found by solving the system of equations:

$$\frac{\partial \Phi}{\partial D_1} = \frac{-4L_1}{D_1^5} + \lambda 2 D_1 L_1 = 0 \tag{4.9}$$

$$\frac{\partial \Phi}{\partial D_2} = \frac{-4L_2}{2D_2^5} + \lambda 4 D_2 L_2 = 0 \tag{4.10}$$

from which L_1 and L_2 will drop out. This is how we discover that the layout (L_1, L_2) does not influence the diameter ratio D_1/D_2 (this is true only if the T or Y junction is symmetric: see Problem 4.4). Solving Eqs. (4.9) and (4.10) for $D_1(\lambda)$ and $D_2(\lambda)$, and eliminating λ we find

$$\frac{D_1}{D_2} = 2^{1/3} \tag{4.11}$$

This ratio was first reported in physiology by Hess [4, 5] and 12 years later by Murray [6]; therefore, we shall refer to it as the Hess-Murray rule. This result is remarkable for its robustness: the optimal D_1/D_2 ratio is independent of the assumed tube lengths. It is also independent of the relative positions of the three tubes; hence, it is independent of geometry.

Another remarkable consequence of Eq. (4.11) is that the wall shear stress along the D_1 tube is the same as along the D_2 tube (Problem 4.11). The design distributes stresses uniformly through the solid that houses the bifurcating flow structure. The uniform distributing of stresses is a constructal design aspect that is treated in detail in Chapter 10.

The Hess-Murray rule holds for perfectly symmetric bifurcations. When the mother channel (D_1, L_1) splits into two dissimilar daughter channels, (D_2, L_2) and (D_3, L_3), the optimal size ratios D_1/D_2 and D_2/D_3 depend on the ratio L_2/L_3, which accounts for the lack of symmetry. This new development is the subject of Problem 4.4.

It is important to capitalize on an optimization result right after it is obtained. According to Ref. [3], if we substitute Eq. (4.11) into Eqs. (4.1) and (4.7), we obtain

$$\frac{2V}{\pi D_2^2} = 2^{-1/3} L_1 + L_2 \tag{4.12}$$

$$2 D_2^4 R_{\text{lam}} = 2^{-1/3} L_1 + L_2 \tag{4.13}$$

Eliminating D_2 between these two equations we find

$$\frac{8}{\pi^2} V^2 R_{\text{lam}} = \left(2^{-1/3} L_1 + L_2\right)^3 \tag{4.14}$$

in which V is constant. The global resistance R_{lam} decreases when both L_1 and L_2 decrease, but the two tube lengths cannot be varied independently because of the area constraint (4.2). The best that we can do is to minimize the expression $(2^{-1/3} L_1 + L_2)$ subject to constraint (4.2), which is the same as finding the extremum of another aggregate function:

$$\Psi = 2^{-1/3} L_1 + L_2 + \mu L_1 L_2 \tag{4.15}$$

where μ is a new Lagrange multiplier. The Ψ extremum is located where its two first derivatives are zero:

$$\frac{\partial \Psi}{\partial L_1} = 2^{-1/3} + \mu L_2 = 0 \tag{4.16}$$

$$\frac{\partial \Psi}{\partial L_2} = 1 + \mu L_1 = 0 \tag{4.17}$$

Eliminating μ between Eqs. (4.16) and (4.17), we discover that the optimal shape of the A rectangle that houses the T construct is

$$\frac{L_1}{L_2} = 2^{1/3} \tag{4.18}$$

Combining this ratio with the A construct (4.2), we find the optimal lengths $L_1 = 2^{-1/3}A^{1/2}$ and $L_2 = 2^{-2/3}A^{1/2}$, in which $A^{1/2}$ plays the role of length scale.

Combining Eqs. (4.11) and (4.18), we find that the pressure drop along the D_1 tube is the same as along the D_2 tube (Problem 4.11). The uniform distributing of pressure drop from tube to branches is the merit of coupling the Hess-Murray diameter ratio with the lengths ratio determined from the search for the best layout of tubes [3] in a constrained space.

We capitalize on this latest result [Eq. (4.18)], and by substituting the optimal L_1 and L_2 expressions into Eq. (4.14), we arrive at the smallest of all possible resistances, which is

$$R_{\text{lam}} = \frac{\pi^2 A^{3/2}}{4V^2} \tag{4.19}$$

This corresponds to the best T-shaped architecture, which is represented by Eqs. (4.11) and (4.18). The R_{lam} expression (4.19) makes sense: this is the best (the smallest) that can be achieved by morphing the flow configuration subject to the V and A constraints. Further reductions in R_{lam} can be achieved only by changing the constraints, namely, by increasing V and/or decreasing A.

The integer 2 in the diameter ratio $2^{1/3}$ [Eq. (4.11)] comes from the assumption of dichotomy (pairing, bifurcation) in the T configuration of Fig. 4.1. If the L_1 tube splits into n identical tubes (L_2, D_2), then $2^{1/3}$ is replaced by $n^{1/3}$ (Problem 4.10). The global resistance of a junction with fixed total tube volume and one mother tube and n identical daughter tubes increases monotonically with n. This means that dichotomy ($n = 2$) is the best way to configure a junction with Poiseuille flow.

The exponent 1/3 in Eq. (4.11) is a reflection of the assumption of fully developed laminar flow. If in Fig. 4.1 the Poiseuille regime is replaced by fully developed turbulent flow in the fully rough regime, then the exponent 1/3 is replaced by 3/7 and Eq. (4.11) becomes [3]

$$\frac{D_1}{D_2} = 2^{3/7} \tag{4.20}$$

The corresponding forms of Eqs. (4.18) and (4.19) are

$$\frac{L_1}{L_2} = 2^{1/7} \tag{4.21}$$

$$R_{\text{turb}} = \frac{\pi^{5/2}}{4^{3/2}} \frac{A^{7/4}}{V^{5/2}} \tag{4.22}$$

As shown in Problem 4.8, these results are obtained by performing the analysis of Eqs. (4.1) through (4.19) after substituting in place of Eqs. (4.3) and (4.4) the corresponding relations for fully developed, fully rough duct flow, namely,

$$\Delta P_1 = C'\dot{m}_1^2 \frac{L_1}{D_1^5} \tag{4.23}$$

$$\Delta P_2 = C'\dot{m}_2^2 \frac{L_2}{D_2^5} \tag{4.24}$$

where C' is a constant factor. In place of Eq. (4.7), we find that the global flow resistance $\Delta P/\dot{m}_1^2$ is proportional to the geometric expression

$$R_{\text{turb}} = \frac{L_1}{D_1^5} + \frac{L_2}{4D_2^5} \tag{4.25}$$

The smallest of all the values of R_{turb} is given in Eq. (4.22), and it corresponds to the aspect ratios optimized in Eqs. (4.20) and (4.21).

The smallest R_{lam} and R_{turb} [Eqs. (4.19) and (4.22)] are surprisingly close to each other, even though their respective flow regimes are drastically different. The roles played by the global constraints (A, V) are clear. Flow resistances are smaller when the bathed territories are smaller and when the tube volumes are larger. Equations (4.19) and (4.22) can also be written as

$$R_{\text{lam}} = \frac{\pi^2}{4} \frac{\text{Sv}^3}{V} \tag{4.26}$$

$$R_{\text{turb}} = \frac{\pi^{5/2}}{4^{3/2}} \frac{\text{Sv}^{7/2}}{V^{4/3}} \tag{4.27}$$

where, again, the svelteness of the T construct is defined as $\text{Sv} = $ (external length scale)/(internal length scale) $= A^{1/2}/V^{1/3}$ [cf. Eq. (1.1)].

In accordance with the performance-freedom design space shown in Fig. 3.2, the construct of three tubes (Fig. 4.1) can be optimized further by giving the morphing geometry more degrees of freedom. One option is to allow the angle of confluence to vary. This alternative is outlined in Fig. 4.2, where the total space constraint is a disc-shaped area with specified radius (r). We found [3] that when the flow is fully developed and laminar, the optimized flow architecture is represented by $D_1/D_2 = 2^{1/3}$, $\alpha = 0.654$ rad, and $L_1 = L_2 = r$. In this configuration the tubes are connected in the center of the disc, and the angle between the two L_2 tubes is very close to 75 degrees.

More recent work has shown that the 75 degrees angle appears as an optimized feature in much more complex tree-shaped flow architectures [7], and that it is nearly insensitive to the change from symmetric Y to asymmetric Y [8, 9]. This angle, like the Hess-Murray rule (4.11), provides a useful shortcut in the development of effective strategies to design equilibrium or near-equilibrium tree flow architectures [10]. We will focus on these aspects of tree design in sections 4.5 and 4.6.

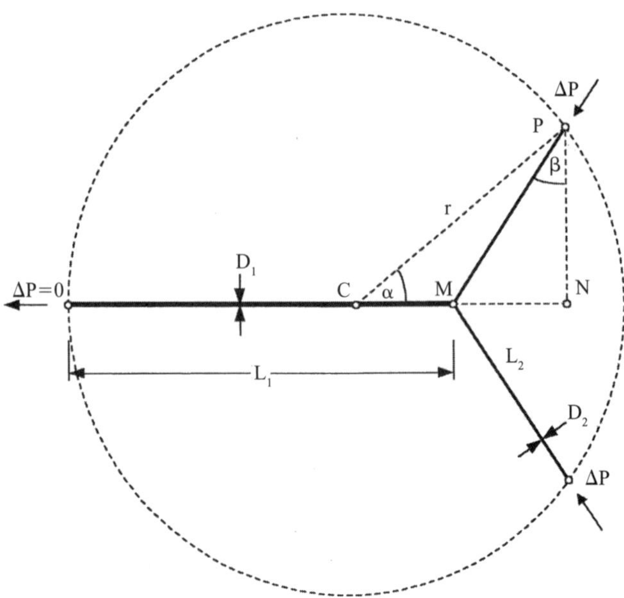

Figure 4.2 *Y*-shaped construct of round tubes [3].

Example 4.1 In Chapter 3 we learned that a channel packs maximum heat-transfer rate volume when the channel houses only the "entrance region" of the developing flow and temperature fields. In such a region the boundary layers grow and meet at the exit. For simplicity, we assume that the fluid has Prandtl number ($\mathrm{Pr} = \nu/\alpha$) or Schmidt number ($\mathrm{Sc} = \nu/D$) comparable with 1, so that the boundary layers sketched in Fig. 3.10 represent all the boundary layers that coat the walls (velocity, temperature, concentration).

In this example we teach how to use the above principle for the purpose of sizing the terminal (smallest scales) ramifications of a bifurcated assembly of three flow channels (e.g., Fig. 4.1) such that the flow volume occupied by the assembly packs maximum heat-transfer rate or mass-transfer rate. We are interested in the ratios D_1/D_2 and L_1/L_2. The first ratio follows from the minimization of pressure drop across the entire *Y*-shaped construct, subject to fixed total flow volume [the Hess-Murray rule, Eq. (4.11)]:

$$\frac{D_1}{D_2} = 2^{1/3} \tag{1}$$

The ratio L_1/L_2 is determined from the observation that in all the tubes the boundary layers meet at the exit,

$$\delta_1 \sim \frac{D_1}{2} \qquad \delta_2 \sim \frac{D_2}{2} \tag{2}$$

where δ_1 and δ_2 are the boundary layer thicknesses for laminar flow

$$\delta_1 = CL_1 \mathrm{Re}_1^{-1/2} \qquad \delta_2 = CL_2 \mathrm{Re}_2^{-1/2} \tag{3}$$

with

$$\mathrm{Re}_1 = \frac{U_1 L_1}{\nu} \qquad \mathrm{Re}_2 = \frac{U_2 L_2}{\nu} \tag{4}$$

The average velocities through the mother tube (U_1) and daughter tube (U_2) are defined as

$$U_1 = \frac{\dot{m}_1}{\rho \frac{\pi}{4} D_1^2} \qquad U_2 = \frac{\dot{m}_2}{\rho \frac{\pi}{4} D_2^2} \tag{5}$$

Dividing Eq. (3) and using Eq. (2), we obtain

$$\frac{D_1}{D_2} = \left(\frac{L_1}{L_2} \frac{U_2}{U_1} \right)^{1/2} \tag{6}$$

Next, we eliminate U_2/U_1 by using Eq. (5) and the assumption of *symmetric* bifurcations such that

$$\dot{m}_2 = \frac{1}{2} \dot{m}_1 \tag{7}$$

and Eq. (6) becomes independent of the ratio D_1/D_2:

$$\frac{L_1}{L_2} = 2 \tag{8}$$

In conclusion, for packing a maximum rate of transport (heat, mass) inside the flow volume, one must use rules (1) and (8), which, combined, read

$$\frac{L_1/D_1}{L_2/D_2} = 2^{2/3} \tag{9}$$

Another conclusion, then, is that the slenderness ratio L/D decreases in going toward smaller channels. These conclusions change somewhat if one accounts for the volume of the tissue that surrounds each tube and is penetrated by diffusion during the times L_1/U_1 and L_2/U_2, whether the Y configuration is asymmetric (Problem 4.4), and for whether each channel has a flat (parallel-plate) cross-section instead of the round cross-section assumed here.

4.2 OPTIMAL SIZES, NOT PROPORTIONS

The optimal aspect ratios (D_1/D_2, L_1/L_2) determined in the preceding section hold for T-shaped constructs of all sizes. These proportions represent the image (the drawing) of the best T configuration. They do not tell us the size of every tube in the T-shaped construct. Can we identify the most appropriate T size out of an infinity of T configurations that are geometrically similar?

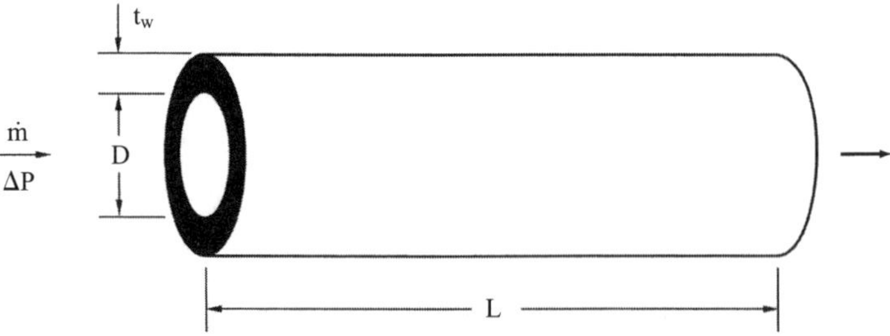

Figure 4.3 Round duct with specified flow rate [11].

We are in a position to answer this basic question on fundamental grounds [2, 11, 12]. Consider the simplest model of a flow passage, or a pore and solid combination: the straight tube of inner diameter D and prescribed mass flow rate $\dot{m}$, which is shown in Fig. 4.3. This could be one of the tubes shown in Fig. 4.1. We make the additional assumption that this simplest flow system is an element in a greater flow system, for example, an aircraft or an animal (flyer, runner, or swimmer). This global use assumption is absolutely essential. It means that the purpose of the elemental flow passage is to help the global flow system move on earth with maximum access or dissipation or minimal destruction of useful mechanical power (review sections 2.6 through 2.8).

There are two loss mechanisms in the configuration of Fig. 4.3: the flow with friction through the duct, and the power destroyed for the purpose of carrying the duct mass as part of a flow installation on a vehicle such as aircraft or animal. If the flow regime is laminar and fully developed, the pressure loss per unit of duct length is

$$\frac{\Delta P}{L} = \frac{32\mu}{D^2} U, \tag{4.28}$$

where U is the mean fluid velocity, $\dot{m}/(\rho\pi D^2/4)$. The first loss is the pumping power required to force $\dot{m}$ to flow through the tube:

$$\frac{\dot{W}_1}{L} = \frac{\dot{m}\Delta P}{\eta_p \rho L}, \tag{4.29}$$

where η_p is the pump isentropic efficiency, ρ is the density of the single-phase fluid (section 2.5), and it is assumed that the tube cross-section is constant (no acceleration/deceleration) and that there are no changes in altitude.

The second loss is the power spent in order to maintain the mass of the duct in locomotion (e.g., flight, Ref. [13], p. 239), which can be expressed per unit of duct length as

$$\frac{\dot{W}_2}{L} \cong 2\frac{m}{L} gV. \tag{4.30}$$

Here, m/L is the duct mass per unit length, and V is the cruising speed. Assume that the competing designs have duct cross-sections that are geometrically similar; that is, the thickness of the duct wall is a fraction of the duct diameter, $t_w = \varepsilon D$, where $\varepsilon \ll 1$, constant. The duct mass per unit length is $m/L = \rho_w \pi \varepsilon D^2$. If many tubes of this type fill a space, then the porosity of that medium is fixed.

The total loss is the sum of Eqs. (4.29) and (4.30):

$$\frac{\dot{W}_1 + \dot{W}_2}{L} \cong \frac{c_1}{D^4} + c_2 D^2 \tag{4.31}$$

where $c_1 = 128\mu\dot{m}^2/(\pi\eta_p\rho^2)$ and $c_2 = 2\pi\varepsilon\rho_w g V$. We see the competition between the two loss mechanisms: a large D is attractive from the point of view of avoiding large flow friction irreversibility, but it is detrimental because of the large duct mass. The optimal tube diameter is

$$D_{\text{opt}} = \left(2\frac{c_1}{c_2}\right)^{1/6} = \left(\frac{128\mu\dot{m}^2}{\pi^2\eta_p\rho^2\rho_w g V}\right)^{1/6} \tag{4.32}$$

The optimal diameter increases with the flow rate through the tube as $\dot{m}^{1/3}$ and decreases weakly as the flying speed increases. We will show that the competition between two losses similar to Eq. (4.31) dictates the optimal locomotion of every animal and vehicle [14–16].

As shown in the T-shaped example of section 4.1, the optimization of tube diameters in complex flow networks is based routinely on two statements:

I. The minimization of flow resistance.

II. The constrained (fixed) tube volume.

Although (I) can be rationalized as an invocation of the constructal law (p. 2), constraint (II) has been lacking a theoretical basis. That basis is made clear now by Eq. (4.31). The flow resistance (I) is represented by c_1/D^4, and the tube volume (II) by $c_2 D^2$. According to the method of Lagrange multipliers [e.g., Eq. (4.8)], to minimize (I) subject to fixed (II) is to minimize the sum formed on the right side of Eq. (4.31).

In conclusion, all the flow network optimization efforts seen in physiology and engineering can be justified theoretically on the basis of the constructal law: the global maximization of performance, which led to Eq. (4.31) and all the flow architectures that follow in this chapter.

This simple analysis is just the start of a series of key design problems: the size of every flow component that works inside a complex flow system. For example, the size of the round tube can be optimized for operation in the turbulent flow regime and for geometries where the tube wall thickness is not necessarily a fraction of the tube diameter. Fittings, junctions, bends, and all the other geometric features that impede flow and add mass to an animal or vehicle can have their sizes optimized in the same manner. The flow passages of heat exchangers are additional examples, although in their case the losses are due to three mechanisms: heat transfer, flow friction, and the air-lifting of the heat exchange mass.

In sum, the method illustrated in this section and in Refs. [2], [11], and [12] promises the identification of optimal shapes and sizes for flow "organs" optimized for the benefit of the greater system (vehicle, or animal). If, for example, the elemental tube in Fig. 4.3 represents a large blood vessel in a flying bird, then the tube mass per unit length is $m/L = \rho_B \pi D^2/4$, where ρ_B is the blood density. The optimal tube diameter continues to be given by Eq. (4.32), where $c_2 = (\pi/4)\, \rho_B g V$.

The main conclusion is that the sizes of flow components can be optimized, such that the aggregate flow system performs at the highest level possible. This conclusion may seem counterintuitive. For example, the total weight of the aircraft dictates its power requirement and limits its range. A small total weight is better. Smaller components in every subsystem of the aircraft may appear to be preferable. There is, however, a competing trend, which stalls the drive toward smaller sizes. Flow systems and their components function less efficiently when their sizes decrease. Their flow resistances increase when sizes decrease (Fig. 4.4). In a heat exchanger, for example, the heat-transfer area and the fluid-flow cross-sections decrease when the total mass and volume decrease. Large flow resistances lead to thermodynamic imperfection (exergy destruction) and, globally, to the requirement of installing more fuel on board. More fuel means more weight.

This conflict is summarized in general terms in Fig. 4.4. The total fuel (or food) required by a flow system is the sum of the fuel required to transport the system and the fuel that must be used in order to produce the exergy that is ultimately destroyed by the system. This trade-off is absolutely fundamental. We can expect it in every flow system—in every vehicle and living system, no matter how complex. Therefore, we must exploit it.

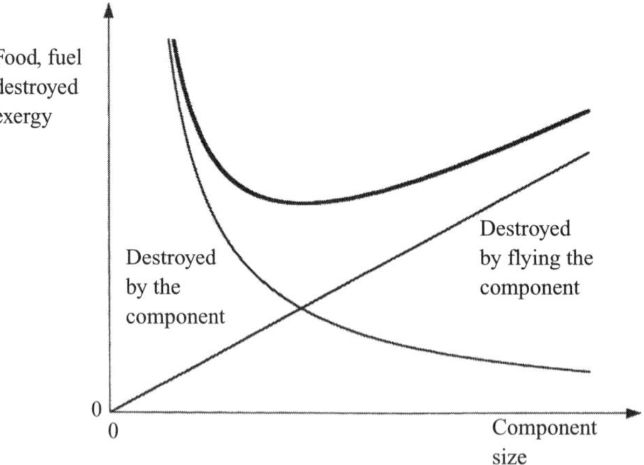

Figure 4.4 Minimization of the total loss associated with one flow component in a complex flow system [17].

4.3 TREES BETWEEN A POINT AND A CIRCLE

In this section we turn our attention to trees that are considerably more complicated than the *T*s and *Y*s of section 4.1. These new trees connect a circle with its center, and, as a basic configuration, serve as model for a large number of the flow distribution and collection problems that dominate natural and engineered flow systems.

Trees "happen" because they provide maximum access between one or a few special points (sources, sinks) and an infinity of points (curve, area, volume). In Fig. 4.5 the curve is the circle, and because the curve is simpler to describe than an area or volume, the point-circle trees are simpler than the point-area and point-volume trees that we will discuss later in this book. To make the point-circle tree problem even easier to attack, we approximate the circle as a string of N equidistant points (sinks, sources) separated by the small distance d, which is fixed.

In this section and the next two we use this simplified point-circle flow access problem in order to teach three lessons:

1. How trees emerge as architectures for maximum access.
2. How to discover such architectures faster and less expensively.
3. How the freedom to change (to morph) the drawing has positive effect on performance.

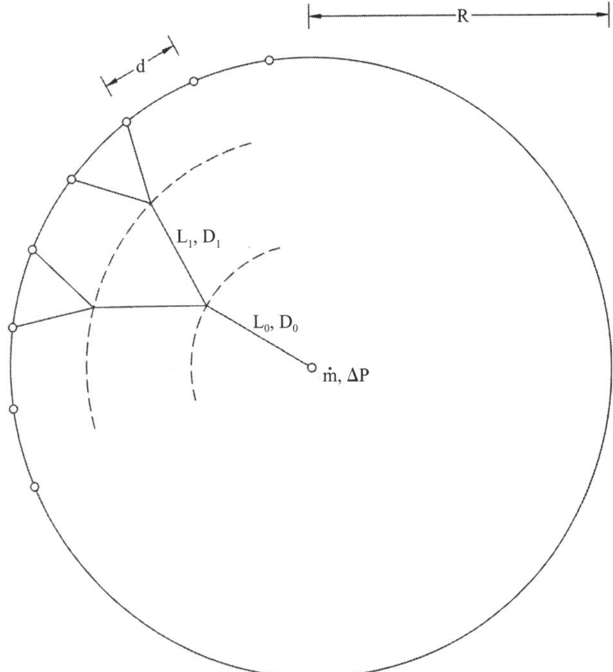

Figure 4.5 Dendritic layout of tubes connecting the center and the rim of a circular area [7].

We wish to discover the dendritic paths of least flow resistance (or minimum ΔP) between the center of a disc-shaped area of radius R and points on its perimeter. The flow is between a point and a circle, or vice versa. It is not between one point and another point (section 3.1). The ducts are round tubes of several diameters (D_i) and lengths (L_i; $i = 0, 1, \ldots$). The volume occupied by all the tubes is fixed. The flow regime in each tube is laminar and fully developed (Poiseuille). We seek to optimize all the geometric details of a structure shaped as a tree or, better, as a disc-shaped river basin or delta.

For more clarity, assume that the flow rate $\dot{m}$ enters the flow structure through the center of the disc, and flows almost radially through tubes that become more numerous toward the rim of the disc. Outlet ports are positioned equidistantly along the rim. The main features of the structure are illustrated in Fig. 4.5, where dichotomy (pairing, or bifurcation at each node in the network) was used because dichotomy is a constructal design feature for Poiseuille flow (cf. section 4.1).

4.3.1 One Pairing Level

The simplest setting for studying this problem is shown in Fig. 4.6. Several tubes (D_0, L_0) are positioned radially and equidistantly around the center port. The angle between two L_0 tubes is $\alpha = 2\pi/n_0$, where n_0 is the number of L_0 tubes (e.g., $n_0 = 4$ in Fig. 4.6). The flow rate through one L_0 tube is $\dot{m}_0 = \dot{m}/n_0$. Pairing means that there are $n_1 = 2n_0$ peripheral tubes of size (D_1, L_1). The flow rate through each peripheral tube is $\dot{m}_1 = \dot{m}/n_1$.

To select the geometry of the flow structure means to select the aspect ratios (L_1/L_0, D_1/D_0) and the tube numbers such that the global resistance $\Delta P/\dot{m}$ is

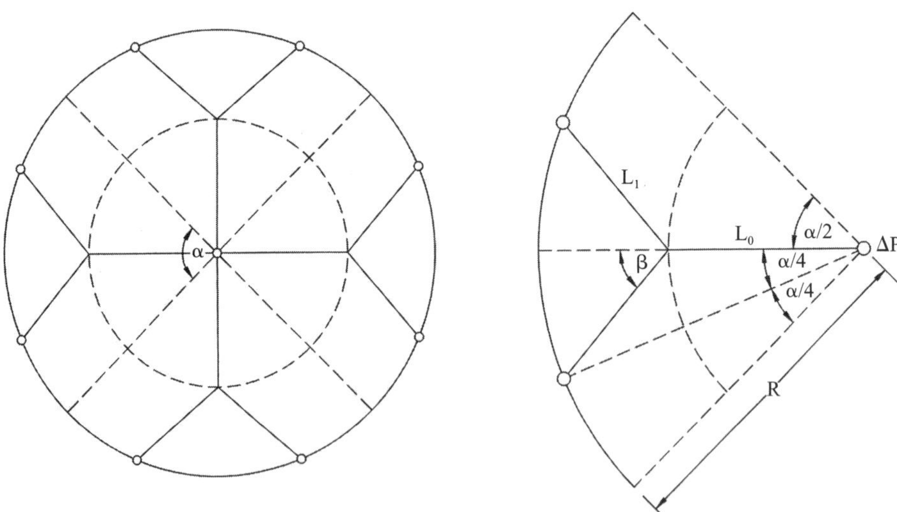

Figure 4.6 Point-circle trees with only one level of pairing [7].

minimum. In Poiseuille flow the resistance of tube (L_i, D_i) is

$$\frac{\Delta P_i}{\dot{m}_i} = \frac{128\,\nu}{\pi}\,\frac{L_i}{D_i^4} \tag{4.33}$$

We continue to aim for the best architecture, and assume the optimal step change in diameter at each pairing node such that the global flow resistance is minimized (the Hess-Murray rule):

$$\frac{D_{i+1}}{D_i} = 2^{-1/3} \tag{4.34}$$

In this notation, as in Figs. 4.5 and 4.6, the index i counts the tube sizes in the radial direction ($i = 0$ are the tubes that touch the center, and $i = p$ are the tubes that touch the perimeter). The optimal diameter ratio of Eq. (4.34) is very robust: it does not depend on the lengths and geometric layout of the respective tubes, provided that each bifurcation is symmetric (Problem 4.4).

The newer aspect of the present problem is that the area over which the tubes may be arranged is constrained. In the simpler case of T- and Y-shaped arrangements of tubes, it was shown that the ratio of successive tube lengths can also be optimized (section 4.1). In the present problem, the Y-shaped construct of two L_1 tubes and one L_0 tube occupies the fixed area of the circle sector of angle α (Fig. 4.6, right). The pressure drop between the center and the ports on the rim is

$$\Delta P = \Delta P_0 + \Delta P_1 = \dot{m}_0 \frac{128\,\nu}{\pi}\left(\frac{L_0}{D_0^4} + \frac{L_1}{2D_1^4}\right) \tag{4.35}$$

The volume occupied by the three tubes in the sector α is

$$V_\alpha = \frac{\pi}{4}\left(D_0^2 L_0 + 2\,D_1^2 L_1\right) \tag{4.36}$$

Using $D_1 = 2^{-1/3} D_0$, and eliminating D_0 between Eqs. (4.35) and (4.36), we arrive at

$$\Delta P = \dot{m}_0 \frac{8\pi\nu}{V_\alpha^2}\left(L_0 + 2^{1/3} L_1\right)^3 \tag{4.37}$$

So far, we have made no assumptions regarding the orientation of the L_1 ducts for the purpose of minimizing ΔP. The geometry of the Y-shaped construct depends on the radial position of the node (i.e., the length L_0), or the angle β. Both L_0 and L_1 vary with β when R is fixed:

$$L_0 = R\cos\left(\frac{\alpha}{4}\right) - R\,\frac{\sin(\alpha/4)}{\tan\beta} \tag{4.38}$$

$$L_1 = R\frac{\sin(\alpha/4)}{\sin\beta} \tag{4.39}$$

To minimize the flow resistance (4.37) means to minimize the expression ($L_0 + 2^{1/3}L_1$) by varying β in accordance with Eqs. (4.38) and (4.39). The minimum resistance occurs when

$$\beta = 0.654 \text{ rad}(37.47^\circ), \text{ or } 2\beta \cong 75^\circ \tag{4.40}$$

It is remarkable that this angle does not depend on the sector angle α. It does not depend on the aspect ratio of the area that houses the construct. This is in agreement with the observations made at the end of section 4.1. The minimized pressure drop that corresponds to this angle is

$$\Delta P = \dot{m}_0 \frac{8\pi\nu}{V_\alpha^2} R^3 \left[\cos\frac{\alpha}{4} + \sin\frac{\alpha}{4}\left(\frac{2^{1/3}}{\sin\beta} - \frac{1}{\tan\beta}\right)\right]^3 \tag{4.41}$$

The effect of the number of tubes becomes visible if we use $\alpha = 2\pi/n_0$, $\dot{m}_0 = \dot{m}/n_0$ and $V_\alpha = V/n_0$, where V is the total volume occupied by all the tubes. Equation (4.41) becomes

$$\Delta P = \dot{m}\frac{8\pi\nu}{V^2} R^3 f\,(n_0) \tag{4.42}$$

where

$$f\,(n_0) = n_0\left[\cos\frac{\pi}{2n_0} + \sin\frac{\pi}{2n_0}\left(\frac{2^{1/3}}{\sin\beta} - \frac{1}{\tan\beta}\right)\right]^3 \tag{4.43}$$

The effect of n_0 is such that the global resistance increases as n_0 increases: $f(2) = 3.897, f(3) = 5.849, f(4) = 7.213, f(5) = 8.381, f(6) = 9.471$. The smallest number of central tubes ($n_0 = 2$) is not realistic because its corresponding length (L_0) is negative. The smallest possible number is $n_0 = 3$, for which the corresponding lengths are $L_0 = 0.214R$ and $L_1 = 0.822R$. These aspect ratios and the optimal angle β are evident in the scale drawing shown in Fig. 4.7.

The smallest number of central tubes ($n_0 = 3$) is a result as fundamental as dichotomy for two-dimensional Y-shaped junctions and the Hess-Murray rule for successive diameter ratios (section 4.1). It means that the easiest way for a stream to emerge from one source and spread on a plane is as three equal streams. Similar rules can be discovered for junctions and source streams in three dimensions (Problem 4.9).

The imposed, global length scales of the flow pattern are not only the total extent (R, Fig. 4.5) but also the smallest distance between the points (ports) serviced by the flow structure (d, Fig. 4.5). According to constructal theory [13], the fixed smallest length scale is a characteristic of all dendritic flows, engineered or natural. Diffusion governs the flow around the densest points, that is, at length scales smaller than d. In this way, the flow of Fig. 4.5 connects the central point to the entire rim area of length $2\pi R$ and radial thickness d. In an engineered structure such as the supply of water or other goods from one central point [1], the rim represents the

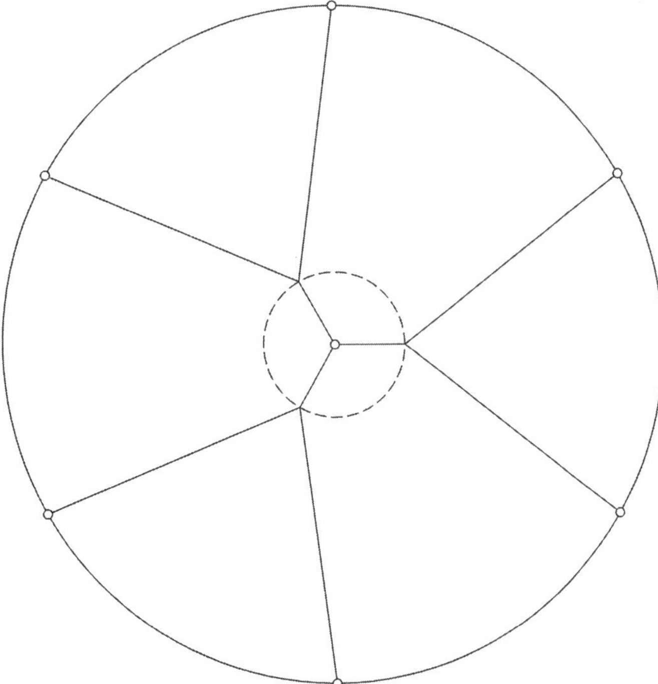

Figure 4.7 The minimum-resistance tree architecture when there is only one level of pairing ($p = 1$) and $n_0 = 3$ [7].

circle of consumers, where the elemental area represented by one consumer is of order Rd.

When d is smaller than but comparable with R, the best structure has only one level of pairing or branching (three inner nodes), as in Fig. 4.7. When d is larger, say between R and $2R$, nodes are not even necessary: the best way is to connect the central point to three equally spaced points around it. The angle between two adjacent ducts would be 120 degrees. We return to this observation in Fig. 4.10.

4.3.2 Free Number of Pairing Levels

Much more challenging is the case of a territory so large that the diffusion scale d is much smaller than R. This case is essential because all natural flows have the tendency to expand their flow territory, in the same way that they tend to provide greater access with more compactness, or svelteness [18, 19] (see also section 11.5). Consequently, most radial dendrites are characterized by $R \gg d$, where d is fixed, and R increases in time. In such cases the number of levels of pairing (p) is not fixed. This number is an additional degree of freedom that can be selected such that the overall flow resistance is minimum. We can expect the optimal number of pairings to increase as R increases.

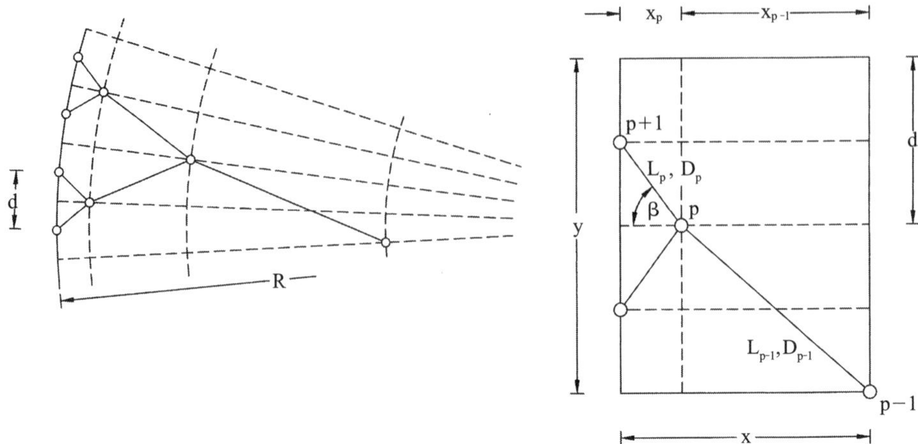

Figure 4.8 Rectangular element near the rim of the dendritic pattern when $R \gg d$ [7].

On the right side of Fig. 4.6 we saw that the area element that houses one Y-shaped construct (pairing, bifurcation) is a circular segment (i.e., a curvilinear triangle). When the number of pairing levels is greater than one, the Y-shaped constructs that do not touch the center (source, sink) are housed by curvilinear rectangles. When R is considerably larger than d, the curvilinear rectangles that are located closer to the rim look progressively more like true rectangles.

This trend is illustrated in Fig. 4.8. We focus on one area element near the periphery and assume that it is a rectangle. We seek the optimal architecture (the aspect ratios) such that the flow resistance of the element is minimum when the elemental area ($A = xy$) and tube volume are fixed. We do not know the best shape of A (i.e., the aspect ratio):

$$\xi = \frac{x}{y} \tag{4.44}$$

The second degree of freedom of the rectangular structure is the relative position of the internal node. This variable is represented by the angle β.

The third degree of freedom is the ratio of tube diameters D_p/D_{p-1}. According to Eq. (4.34), the optimal value is $2^{-1/3}$, regardless of the tube lengths (L_p, L_{p-1}) and their relative positions. Here we use $2^{-1/3}$ as an approximation, because in Fig. 4.8 the bifurcations (the Ys) are not symmetric (see again Problem 4.4). The minimized flow resistance of the rectangular construct is obtained through a change in notation in Eqs. (4.36) and (4.37),

$$\Delta P_p = \dot{m}_{p-1}\frac{8\pi\nu}{V_p^2}\left(L_{p-1} + 2^{1/3}L_p\right)^3 \tag{4.45}$$

$$V_p = \frac{\pi}{4}\left(D_{p-1}^2 L_{p-1} + 2D_p^2 L_p\right) \tag{4.46}$$

The pressure drop ΔP_p is measured between node $(p-1)$ and the periphery $(p+1)$, i.e., ΔP_p is centered on node p. In this formulation, $\dot{m}_{p-1}$ is the total flow rate through node p. The tube lengths can be expressed in terms of the degrees of freedom (ξ, β) and constraints (A, V_p):

$$L_p = \frac{y}{4 \sin \beta} \tag{4.47}$$

$$L_{p-1} = \left[\left(\frac{y}{2}\right)^2 + \left(x - \frac{y/4}{\tan \beta}\right)^2 \right]^{1/2} \tag{4.48}$$

where $x = (\xi A)^{1/2}$ and $y = (A/\xi)^{1/2}$. Equation (4.45) becomes

$$\Delta P_p = \dot{m}_p \frac{8 \pi \nu}{V_p^2} A^{3/2} g^3 \tag{4.49}$$

where

$$g(\xi, \beta) = \left[\frac{1}{4\xi} + \left(\xi^{1/2} - \frac{1}{4\xi^{1/2} \tan \beta} \right)^2 \right]^{1/2} + \frac{2^{1/3}}{4\xi^{1/2} \sin \beta} \tag{4.50}$$

The function g can be minimized with respect to both ξ and β:

$$g_{\min} = 1.324 \text{ at } \xi = 0.7656,\ \beta = 0.927 \text{ rad (53.1 degree)} \tag{4.51}$$

In summary, these values represent the optimized architecture of the peripheral area element highlighted in Fig. 4.8. This design is valid when the element is approximately rectangular. The ξ and β values can be used to calculate other geometric ratios of the element, for example, the distances between successive pairings and the successive tube lengths:

$$\frac{x_p}{x_{p-1}} = a = 0.325 \qquad \frac{L_p}{L_{p-1}} = b = 0.409 \tag{4.52}$$

All the elemental scales are proportional to the elemental spacing d; for example:

$$\frac{x_p}{d} = 0.375 \qquad \frac{L_p}{d} = 0.625 \tag{4.53}$$

As we proceed from the periphery toward the center, and if we approximate all the area constructs as rectangles, the distance from the periphery to the center is

$$R = \sum_{i=0}^{p} x_i = \frac{x_p(1 - a^{p+1})}{(1-a)\, a^p} \tag{4.54}$$

This equation dictates the approximate number of pairing levels (p) when the elemental scale (d, or x_p) and the global scale (R) of the structure are known. The overall flow resistance encountered by the total stream $\dot{m}_0$ between the periphery and the center of the disc is obtained by summing $(p + 1)$ pressure drops of the type shown in Eq. (4.33):

$$\Delta P = \sum_{i=0}^{p} \Delta P_i = \dot{m}_0 \frac{128\nu}{\pi} \frac{L_p}{D_p^4} (2^{-4/3} b)^p \frac{1 - (2^{1/3} b)^{p+1}}{1 - 2^{1/3} b} \tag{4.55}$$

The total volume of all the tubes in the structure is

$$V = \sum_{i=0}^{p} 2^i \frac{\pi}{4} D_i^2 L_i = \frac{\pi}{4} D_p^2 L_p \frac{1 - (2^{1/3} b)^{p+1}}{(2^{-2/3} b)^p (1 - 2^{1/3} b)} \tag{4.56}$$

The elemental tube diameter (D_p) is assumed fixed, along with the other dimensions of the peripheral element. In this case Eq. (4.56) establishes a one-to-one relationship between V and p.

The constant ratios x_p/x_{p-1} and L_p/L_{p-1} indicate that when there are many levels of branching or confluence, the optimized area elements become smaller in sizeable steps as we approach the periphery. This trend contradicts the features of Fig. 4.7, which shows that when there is only one branching level, the peripheral length scale (L_1) is greater than the central scale (L_0). To decide which trend is correct when the number of branching levels is moderately greater than 1, Wechsatol et al. [7] performed the analysis of Eqs. (4.33) through (4.43) for structures with two levels of pairing. Figure 4.9 shows the resulting optimized structure when the central region has only three ducts ($n_0 = 3$). This optimized geometry is represented by the lengths and angles reported in Refs. [7] and [10] for $p = 1, 2, \ldots, 7$. The overall flow resistance is

$$\Delta P = 8\pi \nu \dot{m} \frac{R^3}{V^2} f \tag{4.57}$$

where the f value is the minimum of the function

$$f = n_0 \left(\hat{L}_0 + 2^{1/3} \hat{L}_1 + 2^{2/3} \hat{L}_2 \right)^3 \tag{4.58}$$

and where $\hat{L}_i = L_i/R$. The flow rate $\dot{m}$ is the total flow rate, and $\dot{m}_0 = \dot{m}/n_0$ is the flow rate through a single L_0 tube.

The structure of Fig. 4.9 begins to bridge the gap between Fig. 4.7 and the construction rules written in Eqs. (4.51) through (4.53). Note that in Fig. 4.9 the central duct (L_0) is sensibly shorter than the next, post-bifurcation duct (L_1). This is in qualitative agreement with the features of Fig. 4.7. The next two ducts show that $L_1 > L_2$. This feature represents a shift from increasing lengths ($L_1/L_0 > 1$, as in Fig. 4.7) to decreasing lengths [$L_p/L_{p-1} < 1$, as in Eq. (4.52)].

The work detailed in this section was continued numerically to higher levels of complexity in Refs. [7] and [10]. The minimized dimensionless flow resistances are summarized as $f(N, p)$ in Fig. 4.10. Which dendritic pattern is better? The answer

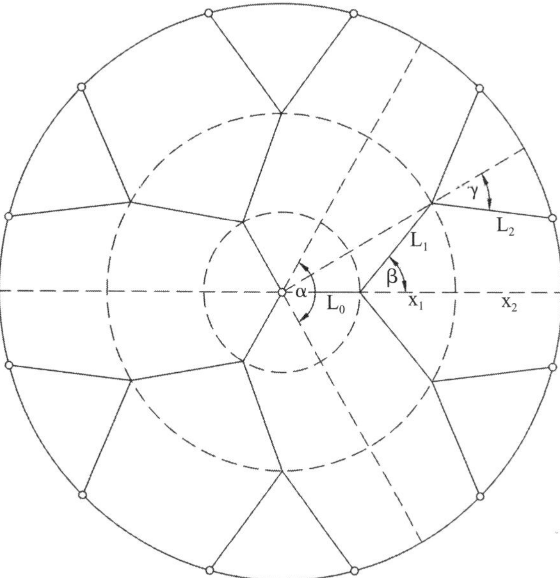

Figure 4.9 The minimum-resistance tree architecture with two levels of pairing and $n_0 = 3$ [7].

depends on what is held fixed. We continue to rely on the view that the smallest length scale of the flow pattern—the elemental scale—is known and fixed. This length scale is the distance (d) between two adjacent points on the circle. Let us also assume that the radius of the circle (R) is fixed. This means that the number of points on the circle (N) is fixed. We also assume that the total volume of the tubes installed on the R-fixed disc is also a constant.

Under these circumstances, formulas such as Eq. (4.57) show that the global flow resistance of the tree construct ($\Delta P/\dot{m}$) varies proportionally with f, while the other factors are constant. The flow pattern with less global resistance is the one with the lower f value. These values are plotted in Fig. 4.10, where for clarity we used continuous curves for each class (number of levels of pairing) even though N is an integer. The leftmost curve corresponds to purely radial flow (no pairings): one can show that this curve represents $f = N$. Each of the subsequent $f(N)$ curves is almost a straight line when plotted on a graph with linear scales in f and N.

The smallest f corresponds to $N = 1$, or a single radial tube between the center and a point on the circle. The next highest f belongs to $N = 2$. These banal cases are point-point flows and fall outside the class of flows that form the subject of this section (point-circle flows).

Figure 4.10 is instructive for additional reasons. Read this figure vertically, at $N =$ constant. One conclusion is that pairing is a useful feature only if N is

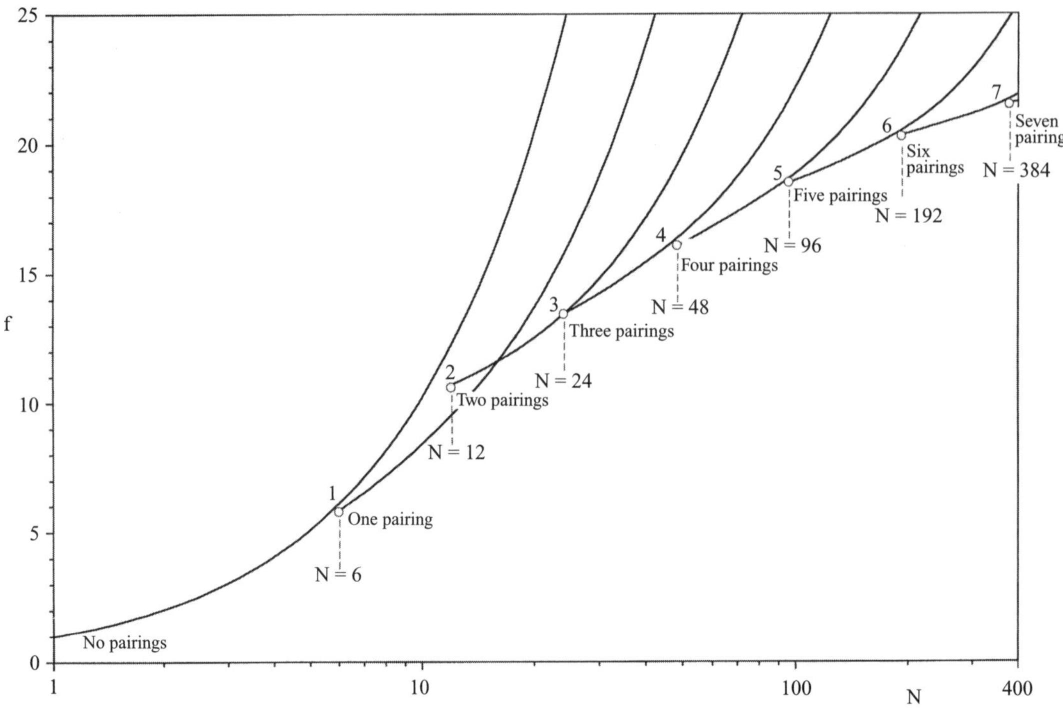

Figure 4.10 The effect of the number of pairing levels (p) on the minimized global flow resistance (f) when the number of ports on the circle (N) is specified, and the disc radius (R) and total tube volume (V) are fixed [7].

large enough (larger than 6, otherwise radial ducts are best). The larger the N value, the more likely the need to design more levels of pairings into the flow structure. If the number of points on the rim of the structure (N) increases, then the flow structure with minimal flow resistance becomes more complex. Complexity increases because N increases, and because the number of pairing levels increases. For example, if $N = 192$, the recommended tree has six levels of pairing, not an infinite number of pairings.

The discovery of complexity is the design principle. The resulting *finite* complexity must not be confused with "maximized complexity." Children can draw trees that are considerably more complex and fuzzy than the ones discovered in this section.

If we think of fixed rims (R) with more and more points (N), then the search for minimal flow resistance between the rim and the center requires discrete changes in the structure that covers the disc. To start with, N has to be large enough for an optimized structure with one or more pairings to exist. These starting N values (6, 12, 24, ...) are indicated with little circles in Fig. 4.10. As the structures become more complex, these circles suggest a "curve" in Fig. 4.10. When there are three or more levels of pairing, the circles indicate the *transition* from one type of structure to the next type with one more level of branching. This transition, or competition between different flow structures, is analogous to the transition and flow pattern selection in Bénard convection. In the vicinity of each circle in Fig. 4.10, the choice is between two structures, as both have nearly the same resistance. These choices are illustrated in Fig. 4.11, which shows the two dendritic structures that compete in the vicinity of points 3 and 4 on Fig. 4.10.

The abrupt change from one architecture ($p = 2$) to a totally different architecture ($p = 4$) in Fig. 4.11 illuminates the origin of abrupt changes (mutations) in configuration throughout nature, engineering, and social organization. Such changes are due to growth (internal growth or density increase) subject to fixed external constraints. For example, Fig. 4.11 can be understood as a prediction of how urban design will evolve. Assume that each "wheel" represents the optimized water distribution network from one source (the center) to N users on the disc perimeter. When $N = 24$ users, the best configuration is $p = 2$, and it will continue to be the best as N increases slowly above 24. But when the population doubles ($N = 48$), a totally different design ($p = 4$) emerges, and it erases the old design from the map. This change is "revolution." Urban design evolves slowly, but when the population doubles, the design must change abruptly and dramatically.

4.4 PERFORMANCE VERSUS FREEDOM TO MORPH

The apparent envelope of the curves in Fig. 4.10 divides the $f - N$ field into two distinct regions. Along the solid lines reside the optimized tree-shaped flow architectures (e.g., Fig. 4.9). Above each $p =$ constant curve (in the white space) reside the untold alternatives of inferior (larger f) tree configurations that have the same number of pairing levels (p) as the structures plotted on $p =$ constant curve.

In sum, above the envelope reside all the possible tree designs that have the same global constraints (R, V), hence the same svelteness ($\mathrm{Sv} = R/V^{1/3}$). The envelope is "apparent" because it is not a curve: it is a string of distinct points, which represent the frontier in the march toward the best performance possible. The white space situated below the envelope is inaccessible to flow drawings that have the specified R and V.

The world of possible (competing) designs is more evident if we limit the discussion to a fixed population of users on the perimeter of the circular area (e.g., $N = 192$). This means that we cut Fig. 4.10 with the vertical plane $N = 192$. In this plane we see the performance-freedom domain shown in Fig. 4.12. This mental viewing is the same as what we presented in Fig. 3.2 for straight ducts with various (competing) cross-sectional shapes. Every class of flow configurations lives on a map of performance versus freedom that is analogous to Figs. 3.2 and 4.12. Such maps are waiting to be drawn for other classes of flow architectures.

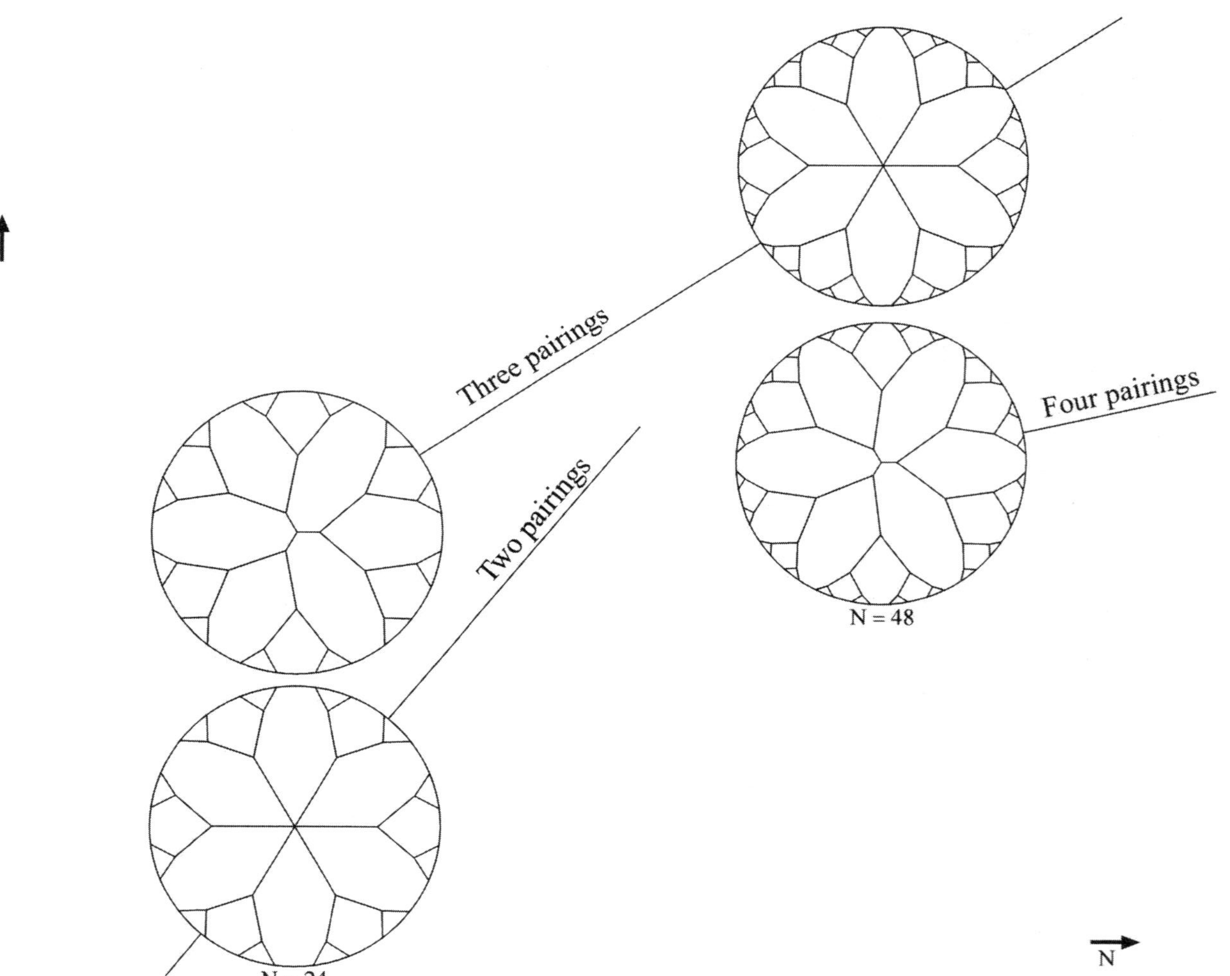

Figure 4.11 The abrupt transition from a tree with two pairing levels to a tree with four pairing levels, as the smallest global resistance (f) is sought while N increases [7]. See also Color Plate 3.

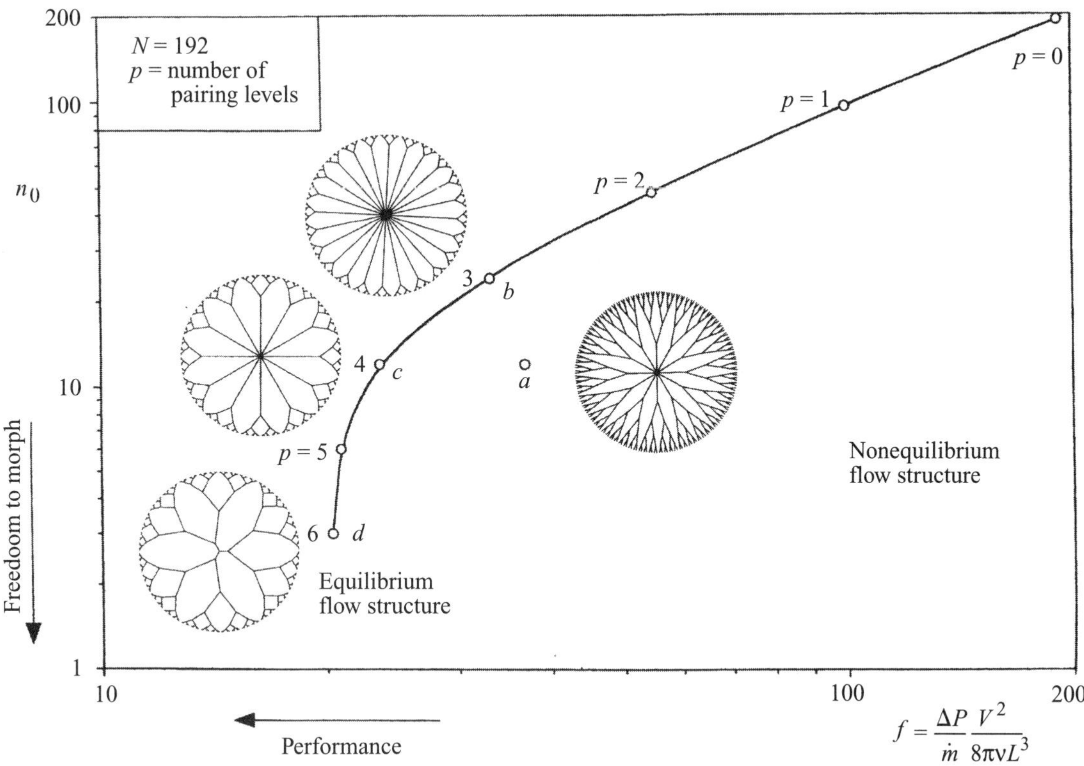

Figure 4.12 The performance-freedom domain of flows that connect the center with a number (N) of equidistant points on a circle [20].

Point-circle trees were first drawn by postulating a "fractal algorithm"—a fixed ratio between successive tube lengths [21]. This layout is shown as inset a in Fig. 4.12. If plotted on Fig. 4.10, the (f, N) point of such a design falls well above the envelope of the solid curves. The corresponding point a falls well inside the domain of nonequilibrium flow structures in Fig. 4.12, that is, flow architectures that have a way to go (to morph, to migrate in the performance-freedom plane) before they can be the least imperfect that they can be.

The abscissa of Fig. 4.12 indicates the dimensionless global resistance between the center and the rim of the disc:

$$f = \frac{\Delta P}{\dot{m}} \frac{V^2}{8\pi \nu L^3} \tag{4.59}$$

which is the same as Eq. (4.57) except that we used L for the external length scale of the architecture (the disc radius).

According to the constructal law, the evolution of the point-circle tree configurations in Fig. 4.12 must be proceeding toward the left. This migration is made possible by allowing the tube lengths to vary freely. The configurations shown as *b*, *c* and *d* were obtained by optimizing every tube length (see section 4.3 and Refs. [7] and [10]). The number of pairing or bifurcation levels is p, and the number of tubes that touch the center is n_0. Designs *a* and *c* have $p = 4$ as an additional constraint, and show how the freeing of the tube lengths leads to a significant improvement in performance (*a* smaller f), from *a* to *c*.

Additional improvements are possible if p is also allowed to vary. In this way, the optimal morphing of the tube layout brings the structure to configuration *d*, which has $p = 6$. We called this the *equilibrium flow structure* [18, 19], because in its vicinity of the design space the performance does not change even though it is here that the configuration is the most free to change.

Structures such as *b* through *d* outline a ragged boundary that divides the performance-freedom design space into two domains. To the right are the possible (suboptimal) designs, such as the assumed structure *a*. To the left of the *b*–*d* boundary, it is impossible to find flow configurations that have the same L, V, and Sv as configurations *a* through *d*.

Note once again that the best flow structure has finite complexity ($p = 6$), not maximum complexity. It would be easy to draw configurations much more complex than *d*, and their performance would be significantly inferior to that of *d*. Maximization of performance and morphing freedom in time (the constructal law) does not lead to maximum complexity.

The crowding of near-optimal designs near the equilibrium flow structure speaks of the *robustness* of tree-shaped architectures. Trees that do not look like the best trees perform practically as well as the best. This is an attractive feature that drives the generation of tree flows in nature and engineering. Robustness is increased further by installing loops [22] at the small and intermediate scales of the canopy, so that if one duct is damaged the fluid can flow the other way around the loop and preserve the global performance of the tree flow. We illustrate the merits of loops in section 5.2.

4.5 MINIMAL-LENGTH TREES

Powerful strategies are emerging for accelerating the search for optimal tree-shaped flow configurations. Examples are the Hess-Murray rule for successive duct diameters, dichotomy for duct with Poiseuille flow, and quadrupling for channels with fully developed turbulent flow (see Fig. 13.39 in Ref. [23]). Another shortcut is to recognize that in configuration *d* of Fig. 4.12 the optimized angles of confluence are approximately 75 degrees. This quasi-invariant result was obtained by optimizing Y-shaped constructs of tubes [3, 7, 8], and can be used as a rule for rapidly constructing tree structures that are situated very close to the *b*–*d* boundary in Fig. 4.12.

One highly effective shortcut is to select all the duct lengths by minimizing the length of each duct on the area element allocated to it [24]. From this follows the shape of every area, from the smallest to the largest, and the construction of the entire tree canopy in two or three dimensions.

4.5.1 Minimal Lengths in a Plane

Consider the problem of connecting all the points of a straight line (e.g., line P) with a single point (S) situated off the line, Fig. 4.13. The single point may be the source of a stream (e.g., water) and the line may represent a large number of users of the water stream. We approximate the line with a sequence of equidistant points. The distance between two consecutive points is d. The rectangular area around each point is finite and fixed, $A_0 = cd$, constant. This area represents the smallest element of the flow structure that will be designed. The link is the segment PQ, where Q is a corner on the side opposite the side for which P is the midpoint. An interesting property of the elemental system is that the length of the segment PQ,

$$L_{PQ} = [(d/2)^2 + c^2]^{1/2} \tag{4.60}$$

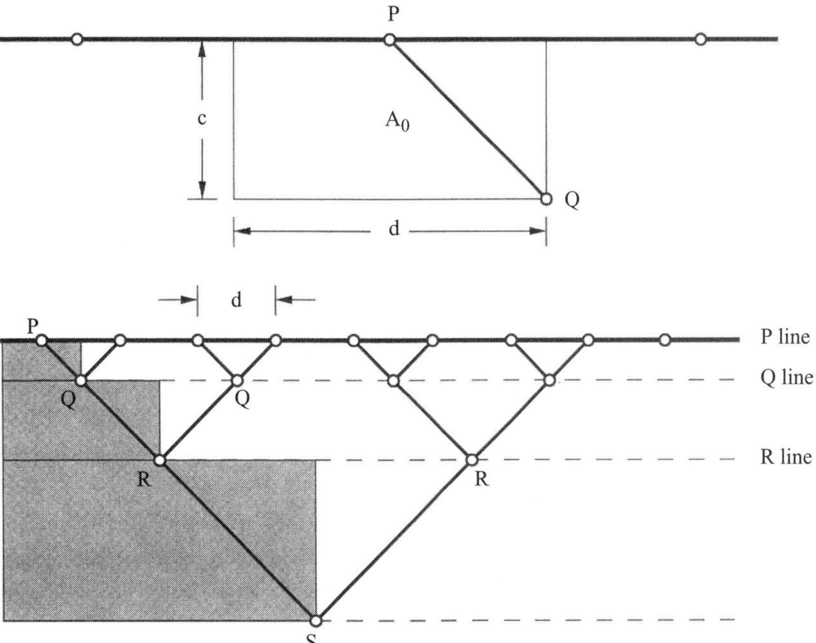

Figure 4.13 Elemental system with one flow segment (top), and the construction of the minimal-length tree between a line and a point (bottom) [24].

can be minimized by varying the element aspect ratio c/d subject to the A_0 constraint. The PQ segment is the shortest when the aspect ratio is $c/d = 1/2$. This special shape was drawn in the upper part of Fig. 4.13. The link PQ makes a 45-degree angle with the line of points of type P. This is already a sign that the tree structure that emerges is near optimal, not optimal, because we learned from Fig. 4.2 that the best angle of bifurcation is roughly 75 degrees, not 90 degrees.

Following this first optimization step, the links of type PQ occupy a strip of thickness $c = d/2$. Points of type Q occupy the lower boundary of this strip (the Q line), and the distance between them is $2d$. Each point Q is the junction of two PQ segments. Access from Q to other parts of the flow structure is effected along segments of type QR. The rectangle that houses the QR segment is completely analogous to the elemental rectangle that houses the PQ segment. These rectangles are stacked (shaded) in the lower part of Fig. 4.13. The length of the QR segment can be minimized by selecting the aspect ratio of the rectangle that houses it. The optimal aspect ratio is the same as at the elemental level, and the minimized QR segment makes a 45-degree angle with the Q line.

The rest of Fig. 4.13 outlines the subsequent steps of the construction. Each step is analogous to the preceding one, while the length scale of each optimized rectangle is the double of the preceding length scale. Two tubes are joined together at a 90-degree angle at each level of assembly. The total length from the source (point S) to each point P is the same. In other words, the tree structure generated by minimizing each segment is also the structure that maintains constant the distance between the point (S) and each of the points of line P. If we use the words *flow resistance* instead of *length*, then the structure designed in Fig. 4.13 distributes uniformly the flow resistance between one point and one line. Because the resistance to flow represents the thermodynamic imperfection of the flow system, the structure derived in Fig. 4.13 has the merit of distributing the imperfection uniformly throughout the flow system. *Optimal distribution of imperfection* is a recurring feature in constructal design and throughout this book.

The tree structure may be viewed as the result of fitting together rectangular flow elements, the shapes of which have been optimized. Although this construction sequence may resemble graphically the first method by which conduction trees were deduced [25], the present construction is different. First, all the rectangular building blocks shown in Fig. 4.13 are geometrically similar. Second, at the elemental level the flow path PQ is not aligned with one of the sides of the elemental rectangle. Third, the present construction is not concerned with the "diffusion" flow that would occupy the white regions of the A_0 area in order to connect Q with the infinity of points of the side on which P is situated. The entire tree architecture of Fig. 4.13 is a consequence of the geometric property that the length L_{PQ} can be minimized.

4.5.2 Minimal Lengths in Three Dimensions

The same geometric property can easily be put to use in three dimensions. The problem is to connect with minimum-length paths one point to the infinity of points of a plane. We approximate the latter with a patchwork of square area elements of side d, centered at one discrete point (P). In other words, the area $d \times d$ is represented by one point in the plane. The elemental volume is fixed, $V_0 = cd^2$, constant. The connection between P and the rest of the flow structure is made along the segment PQ, where Q is the middle of one of the sides of the square situated at the distance c relative to the plane of P. The length of the PQ segment is the same as in Eq. (4.60), for which d and c are defined in Fig. 4.14. This length is minimal when the aspect ratio of the parallelepiped $d \times d \times c$ is $c/d = 2^{-3/2}$. This aspect ratio was drawn to scale in the elemental volumes shown at the top of Fig. 4.14.

The next volume to be optimized is $2d \times d \times e$. The link QR is from the top rectangular face (plane Q) to the middle of one of the $2d$-long sides in plane R. The length QR is minimal when the volume aspect ratio is $e/d = 2^{-3/2}$. This aspect ratio is the same as at the smallest scale.

In summary, after two steps of volume shape optimization, we have covered a total volume of size $2d \times 2d \times (c + e)$. The shape of this volume is the same as that of the elemental volume, $(c + e)/2d = c/d = 2^{-3/2}$. Point R, which is the center of the bottom ($2d \times 2d$) square, is analogous to point P (the center of square $d \times d$), which started the construction. The construction continues toward planes situated successively under plane R, and the steps are the same two steps that we just outlined, from P to Q, and from Q to R. This sequence is indicated by the side view of the construct, which is shown in the lower part of Fig. 4.14. Again, the ragged structure that emerges is full of hints that a better tree can be obtained by taking the tree of Fig. 4.14 and "stretching" it so that each Y becomes plane (most likely with a 75 degree angle of bifurcation).

4.5.3 Minimal Lengths on a Disc

The same geometry-discovery opportunity can be exploited to derive minimal-length flow paths for considerably more complicated configurations. In Fig. 4.15 we propose to construct the connection between the points of a circle and the center of the circle. This configuration is similar to taking the point–line structure of Fig. 4.13, and curving the P line around point S, so that S becomes the center of a circle. The rectangular building blocks used in Fig. 4.13 become deformed (curvilinear) rectangles. The deformed rectangle is closer to a true rectangle if it is far from the center. The degree of closeness is associated with the distance (r) between the rectangular element and the center.

Consider the elemental curvilinear rectangle of radial distance r, angle θ, and radial thickness δ, which is shown in the upper part of Fig. 4.15. The area of this

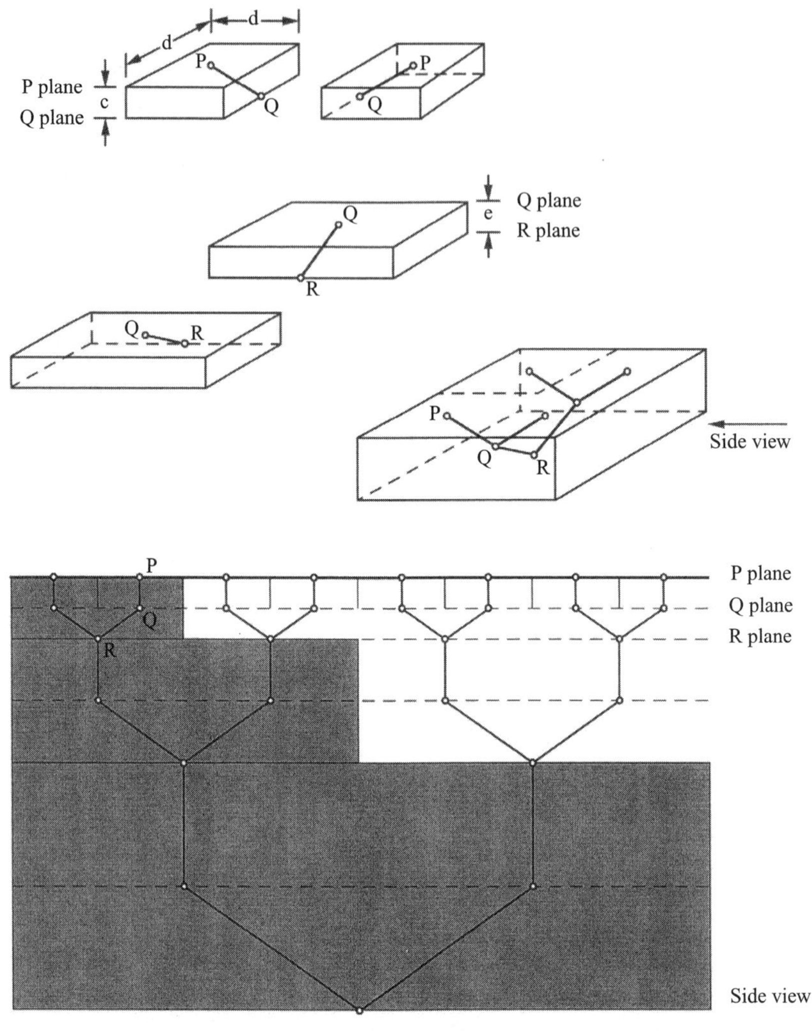

Figure 4.14 The construction of the minimal-length tree between one plane and one point (top), and side view of the emerging tree (bottom) [24].

element is fixed:

$$A = \frac{1}{2}\theta\left(2r\delta - \delta^2\right), \text{ constant} \tag{4.61}$$

The role of segment PQ of Fig. 4.13 is played by segment l. The objective is to minimize l subject to constraint (4.61), that is, to minimize

$$l^2 = a^2 + b^2 = \left[r\left(1 - \cos\frac{\theta}{2}\right) + \delta\cos\frac{\theta}{2}\right]^2 + (r - \delta)^2 \sin^2\frac{\theta}{2} \tag{4.62}$$

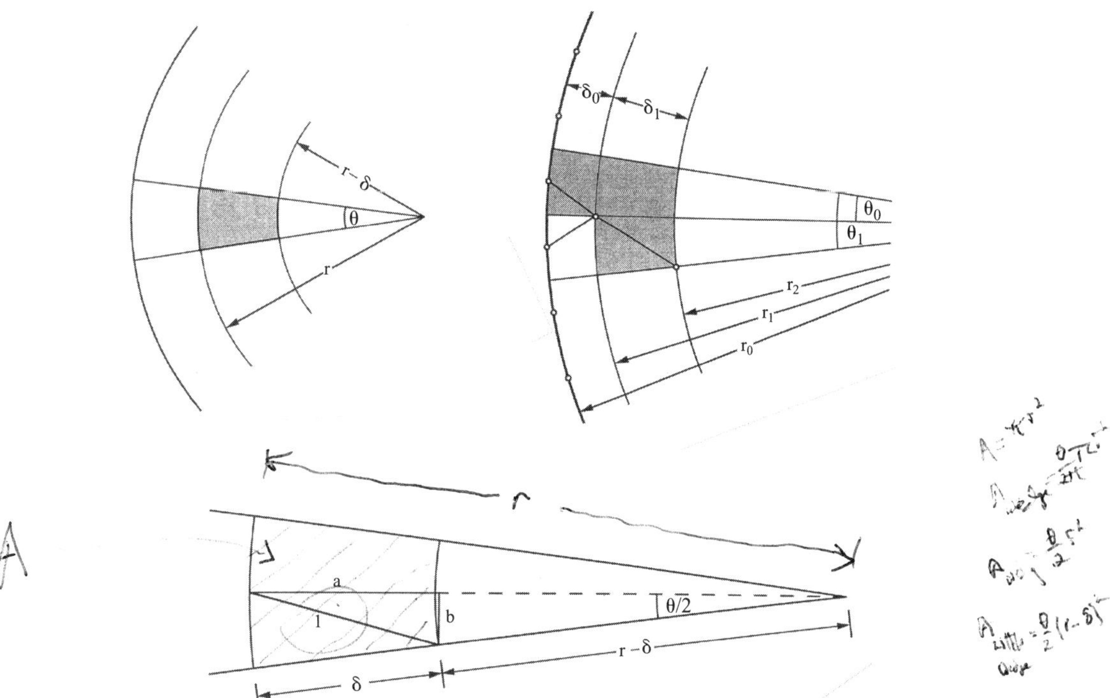

Figure 4.15 Curvilinear rectangle defined by the intersection of radial lines and concentric circles, and the construction of the minimal-length tree between a circle and its center [24].

Next we nondimensionalize (4.62) and (4.61) by using r as length scale

$$\left(\frac{l}{r}\right)^2 = (1-x)^2 - 2(1-x)\cos\frac{\theta}{2} + 1 \tag{4.63}$$

$$\frac{A}{r^2} = \frac{\theta}{2}\left(2x - x^2\right) \tag{4.64}$$

where $x = \delta/r$. According to the method of Lagrange multipliers, the problem of finding the extremum of function (4.63) subject to constraint (4.64) is equivalent to finding the extremum of the aggregate function (cf. Appendix B):

$$\Phi = (1-x)^2 - 2(1-x)\cos\frac{\theta}{2} + 1 + \lambda\theta\left(x - \frac{1}{2}x^2\right) \tag{4.65}$$

Solving the system $\partial\Phi/\partial x = 0$ and $\partial\Phi/\partial\theta = 0$, and eliminating the Lagrange multiplier λ, we obtain the relation that pinpoints the optimal aspect ratio of the curvilinear rectangle:

$$\frac{x}{\theta} = \frac{(1-x)^2 \sin\frac{\theta}{2}}{(2-x)\left(x - 1 + \cos\frac{\theta}{2}\right)} \tag{4.66}$$

To see how Eq. (4.66) leads to optimal shapes and later to point–circle trees, assume that $\theta \ll 1$, such that $\cos(\theta/2) \cong 1$ and $\sin(\theta/2) \cong \theta/2$. Then Eq. (4.66) becomes

$$\frac{x}{\theta} \cong \frac{1 - x}{2^{1/2}(2 - x)^{1/2}}. \tag{4.67}$$

where x/θ is the aspect ratio of the curvilinear rectangle, $x/\theta = \delta/(r\theta)$. When the radial dimension of the rectangle (δ) is small in comparison with its distance to the center (r) (i.e., when $x \ll 1$), the optimal aspect ratio is $\delta/(r\theta) \cong 1/2$. Closer to the center of the circle, where x is progressively larger, the optimal aspect ratio is progressively greater than 1/2. We can expect the same trends from the exact solution, Eq. (4.66).

To construct the minimal-length path between the outer circle (radius r_0) and the center (O), we start from the outer circle and approximate it as a string of equidistant points (P). See the lower part of Fig. 4.15. The distance between two consecutive points is d. The angle sustained by the outermost (elemental) rectangle of peripheral length d is $\theta_0 = d/r_0 \ll 1$. The value of θ_0 must be selected at the start of construction, e.g., $\theta_0 = 0.1$. Substituting θ_0 for θ in Eq. (4.66), we calculate x_0, or the aspect ratio of the elemental rectangle (x_0/δ_0), or the radial thickness of the element, $\delta_0 = r_0 x_0$. The radius of the inner circle that borders the elemental rectangle is $r_1 = r_0 - \delta_0$.

The next curvilinear rectangle subtends the angle $\theta_1 = 2\theta_0$, and has the radial position r_1. Eq. (4.66) delivers x_1, the aspect ratio x_1/δ_1, and the radial dimension $\delta_1 = r_1 x_1$. The radius of the next circle is $r_2 = r_1 - \delta_1$. This algorithm can be applied a sufficient number of times, marching toward the center of the circle, and drawing the resulting tree network. The construction must stop at a certain step (i) if

$$\theta_i > 2\pi \tag{4.68}$$

or

$$r_i < 0 \tag{4.69}$$

whichever occurs first. During the numerical implementation of the algorithm, we found that in this construction, thresholds (4.66) and (4.67) are reached simultaneously [24]. We expect the constructed tree to be imperfect approximate near the center, because the rectangle approximation becomes poorer in the limit $r \to 0$.

To evaluate the effectiveness of the minimal-length structures, we compared the point-circle construction with a tree-shaped network obtained by optimizing every geometric detail of the flow network. The optimization of every detail is reported in section 4.3 (see also Ref. [7]). The comparison is made on a common basis: the circle of fixed radius R, the use of tube pairing, and the fixed total volume of the tubes with Poiseuille flow, V. The step change in tube diameters is optimized in accordance with Eq. (4.34). The number of points arranged equidistantly on the circle is $N = 24$.

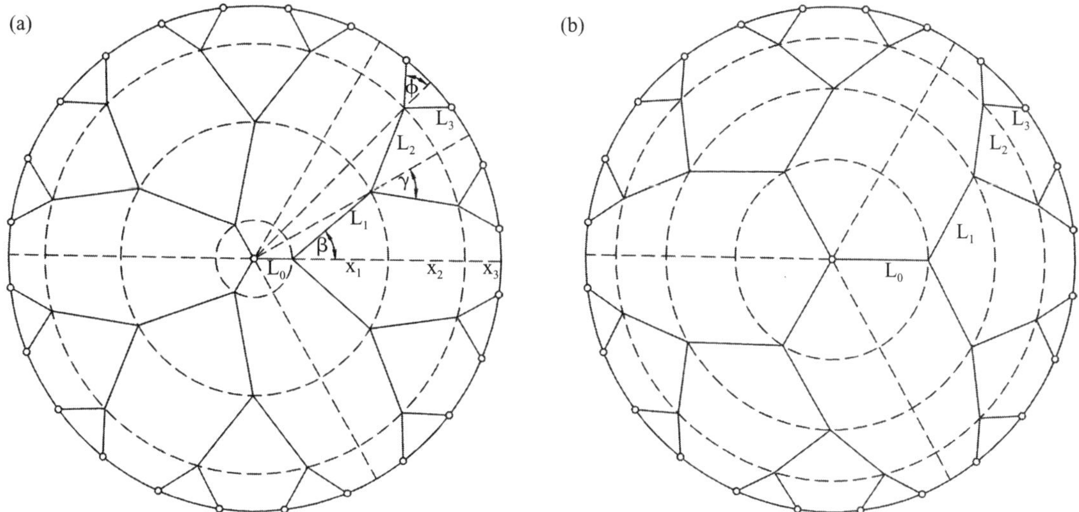

Figure 4.16 Tree flows between a circle and its center: (a) construct generated by optimizing every geometric detail; and (b) construct generated by minimizing the length of every duct on its allocated area [24].

Figure 4.16a shows the flow structure derived after optimizing features such as $L_0, L_1, L_2, L_3, \beta, \gamma, \phi, x_1, x_2$ and x_3, where $R = L_0 + x_1 + x_2 + x_3$. The overall flow resistance is given by Eq. (4.57), where ΔP is the pressure difference between the center and the circle, $\dot{m}$ is the total mass flow rate, and f is the dimensionless flow resistance

$$f = n_0 \left(\sum_{j=0}^{p} 2^{j/3} \frac{L_j}{R} \right)^3 \qquad (4.70)$$

where n_0 is the number of tubes that reach the center ($n_0 = 3$), and p is the number of levels of pairing ($N = 3 \times 2^p$). The structure of Fig. 4.16a is the result of minimizing f by exploiting all the degrees of freedom of the flow architecture. The minimized f value is plotted in Fig. 4.17 not only for $N = 24$ but also for other numbers of points on the circular perimeter (N).

Figure 4.16b shows the corresponding structure designed based on the length-minimization algorithm outlined in this section. Visually, there is relatively little difference between the two constructions, Figs. 4.16a and 4.16b. The same can be said about the f values of the competing designs. The squares plotted on Fig. 4.17 show that although consistently inferior, the performance of minimal-length structures resembles closely the performance of fully optimized structures. The closeness documented in Figs. 4.16 and 4.17 shows again that optimized tree flow structures are robust.

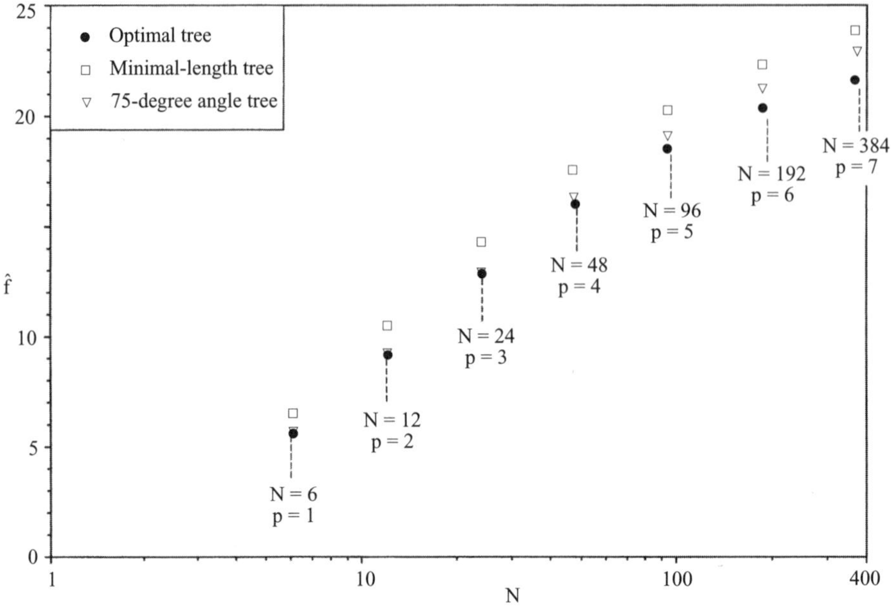

Figure 4.17 The lowest flow resistances of tree designs constructed by using three different methods.The global flow resistance for: (a) constructs generated by optimizing every geometric detail; (b) constructs generated by minimizing the length of every duct on its allocated area; and (c) constructs by setting the bifurcation angles equal to 75 degrees [10, 24]

4.6 STRATEGIES FOR FASTER DESIGN

4.6.1 Miniaturization Requires Construction

Engineering is racing toward smaller scales and more complex multiscale systems. When the design of a complex flow system starts, the flow architecture is not known and is malleable. Strategies are needed for the conceptual design of complex systems that employ progressively smaller scales. This step is particularly timely in view of the computational means that are available today. No matter how successful we are in discovering and understanding small-scale phenomena and processes, we are forced to face the challenge of assembling and connecting numerous components into palpable devices. The challenge is to *construct*, that is, to assemble and to optimize while assembling.

For this, we need principles of construction: assembly and optimization at every scale, in the pursuit of performance when space is at a premium. The macroscopic flow structure that emerges is a complex one, with many tree-shaped paths for the flowing currents, and with elemental systems filling the interstices—the spaces between the smallest branches of the flow trees. This structure is characterized by *multiple scales*—entire arpeggios of length, time, and force scales.

Traditionally, design begins with assuming one configuration, building a model, and optimizing its performance. If time and money permit, one or two competing configurations are modeled and optimized, so that in the end the designer may select the best configuration from the assumed few. As an alternative, the strategies illustrated in this chapter liberate the designer from the straitjacket of assuming a certain macroscopic structure. The physical configuration is the chief unknown in design, and the optimization of configuration is the route to maximal global system performance, under global constraints such as size, cost, and computational time.

The generation of geometry makes flow systems achieve their best. In the beginning there is no drawing, and the design domain is vast. The search for the architecture with minimal global resistance between one point and an area begins with recognizing the many ways in which the flow geometry can be varied. A growing body of work is showing that fluid flows that connect one point with large or infinite numbers of points, the best configuration is a tree.

The example shown in Fig. 4.18 and Table 4.1 suggests that even if we start the search by postulating a tree-shaped structure with straight round tubes and Poiseuille flow, there are many geometric features that must still be selected: the lengths (L_i) and diameters (D_i) of all the ducts, the number of central ducts (n_0), the distance between outlets (d), or the number of outlets on the rim ($N = 2\pi L/d$), and the number of pairing (or bifurcation) levels, which in Fig. 4.18 are indicated with dashed circles. Although the number of tributaries that form a larger stream is free to vary, the number two (dichotomy, pairing, bifurcation) is the best such number for Poiseuille flow.

4.6.2 Optimal Trees versus Minimal-Length Trees

Because the overall size (L) is fixed, the architecture of Fig. 4.18 has $2p + 1$ degrees of freedom, where p is the number of pairing levels. Degrees of freedom are the number of central tubes n_0, the dimensionless lengths $\hat{L}_i = L_i/L$, and the diameter ratios D_i/D_{i-1} ($i = 1, \ldots, p$). A classical result is the Hess-Murray rule, according to which the diameter ratios are all equal, $D_i/D_{i-1} = 2^{-1/3}$. This ratio is independent of the tube lengths and their layout. The global performance of the flow architecture—the global flow resistance—is expressible as in Eq. (4.59), where L and V are global constraints, and the dimensionless resistance factor f depends solely on the geometry of the flow structure.

As reference for the cost of generating optimal tree configurations, we use the computational cost required for generating the structure of the type illustrated in Fig. 4.18, which is the case c ($p = 5$) of Fig. 4.19. The structures with six and seven levels of pairing are detailed in Table 4.1.

The computations compared in Fig. 4.17 and Table 4.1 were performed on a PC with a Pentium 3 processor at 1.2 GHz, with 512 Mb of memory [10]. The black

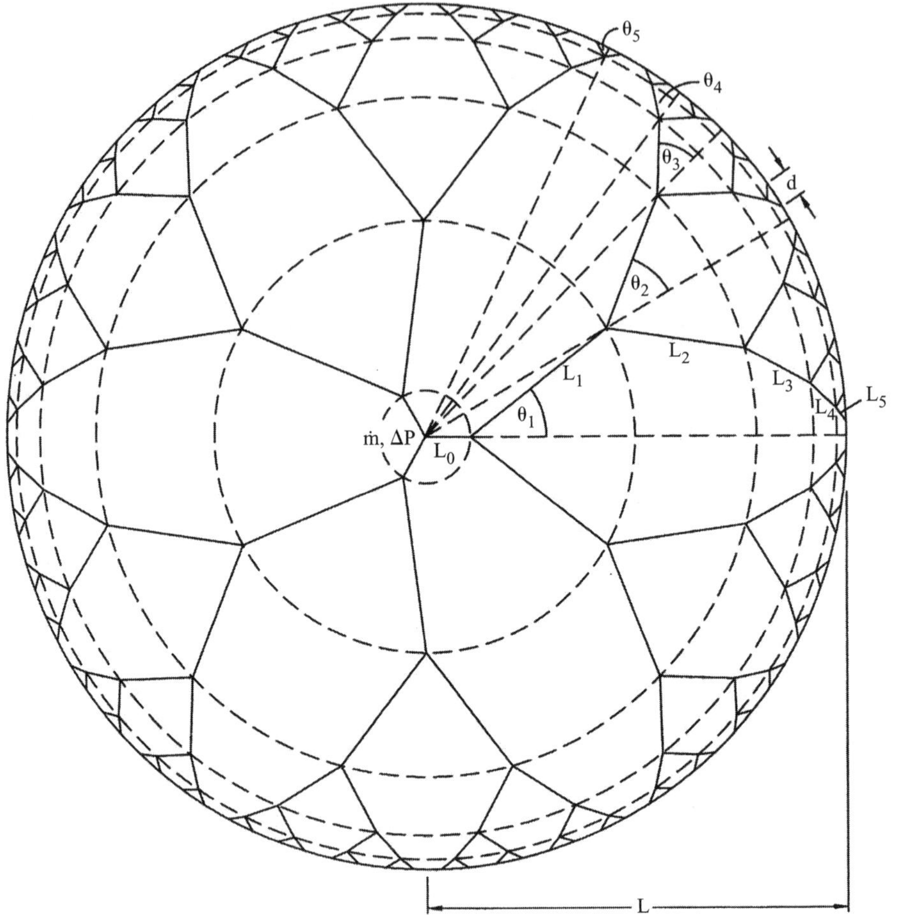

Figure 4.18 Geometric parameters of point-circle trees with five levels of pairing [10].

circles plotted in Fig. 4.19 indicate the computational times required for generating the $n_0 = 3$ optimal trees of Fig. 4.17. The inclined line connects the designs that were generated by using the same algorithm. This line shows that cost increases exponentially with complexity (p).

Table 4.2 and the vertical line (a–b–c) drawn at $p = 2$ in Fig. 4.19 describe the algorithm. As indicated in Fig. 4.18 and Table 4.1, the layout of the tree can be described and optimized in terms of tube lengths (L_i) or pairing angles ($\theta_1, \theta_2, \ldots$). In Ref. [10] the layout was described by using the angles. Each angle could be varied between 0 degrees and 180 degrees, but because of dichotomy, the search was reduced to the range 0 degrees to 90 degrees. Each half-angle was varied in steps of ε degrees during each of the loops in the ladder of embedded loops that covered all the possible tree configurations.

Table 4.1 The optimized geometric features of flow structures with six and seven levels of pairing [10]. The angles are expressed in degrees.

n_0	θ_1	θ_2	θ_3	θ_4	θ_5	θ_6	$\hat{L}_0$	$\hat{L}_1$	$\hat{L}_2$	$\hat{L}_3$	$\hat{L}_4$	$\hat{L}_5$	$\hat{L}_6$	f
3	37.5	38.1	43.1	49.8	56.6	62.7	0.106	0.407	0.327	0.175	0.083	0.039	0.018	20.3
4	37.5	39.9	45.9	52.7	59.3	65.0	0.269	0.398	0.258	0.129	0.060	0.028	0.014	20.9
5	37.5	41.6	48	54.8	61.2	66.5	0.394	0.365	0.209	0.101	0.047	0.022	0.011	21.3
6	37.5	42.8	49.6	56.5	62.6	67.9	0.487	0.329	0.175	0.083	0.039	0.018	0.009	21.8
7	37.5	43.8	50.9	57.7	63.7	68.9	0.557	0.297	0.150	0.070	0.033	0.016	0.008	22.3
8	37.5	44.6	51.9	58.7	64.7	69.6	0.610	0.269	0.131	0.061	0.029	0.014	0.007	22.9

n_0	θ_1	θ_2	θ_3	θ_4	θ_5	θ_6	θ_7	$\hat{L}_0$	$\hat{L}_1$	$\hat{L}_2$	$\hat{L}_3$	$\hat{L}_4$	$\hat{L}_5$	$\hat{L}_6$	$\hat{L}_7$	f
3	37.5	38.1	43.1	49.8	56.6	62.8	68	0.106	0.405	0.325	0.174	0.083	0.039	0.018	0.009	21.6
4	37.5	39.9	45.9	52.7	59.2	65.0	69.9	0.269	0.398	0.257	0.128	0.060	0.028	0.014	0.007	21.9
5	37.5	41.5	48	54.8	61.2	66.7	71.3	0.393	0.364	0.209	0.101	0.047	0.022	0.011	0.005	22.2
6	37.5	42.8	49.7	56.5	62.6	67.9	72.1	0.486	0.329	0.175	0.083	0.039	0.018	0.009	0.004	22.6
7	37.5	43.8	50.9	57.7	63.7	68.7	72.9	0.556	0.297	0.150	0.070	0.033	0.016	0.008	0.004	23.1

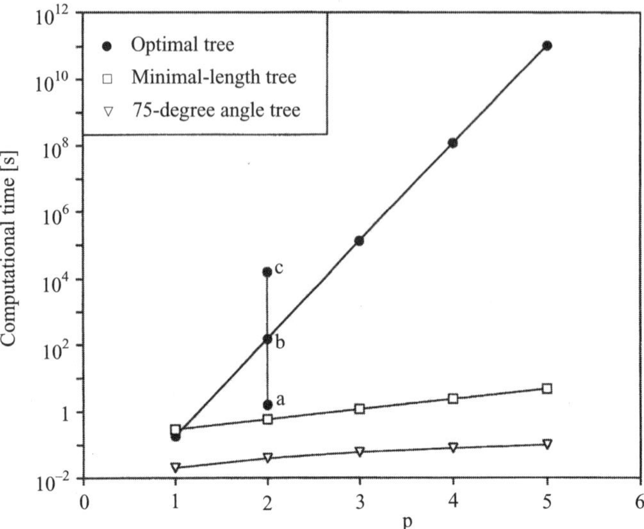

Figure 4.19 The lowest flow resistances of tree designs constructed by using three different methods. The computational times required for determining the designs of Fig. 4.17 when $n_0 =$ 3 [10].

Table 4.2 shows how the step-size in changing the angle influences the computational time and the accuracy of the final result (f). The computational time increases almost as ε^{-2} as ε decreases. The effect of ε on the optimized layout (f) is essentially zero if ε is smaller than 1°. The main conclusion is that the cost of generating complex flow structures increases dramatically with the complexity of the structure. Can this increase in computational time be arrested?

The squares plotted in Fig. 4.17 show that the global performance of the minimal-length trees is consistently inferior to that of the fully optimized structures (the black circles). The relative difference between their respective f values decreases to 10% as the complexity (N) increases.

The computational times required for constructing minimal-length trees are reported in Fig. 4.19. They increase with p, but are orders of magnitude smaller than the times required for generating fully optimized trees.

Table 4.2 The relation between the refinement of pairing angles (ε), accuracy (f), and computational time in the optimization of tree architectures with $n_0 = 3$ and $p = 2$ [10].

	ε	f	Time (s)
a	1°	9.3399	1.6
b	0.1°	9.3395	150
c	0.01°	9.3395	15,731

4.6.3 75 Degree Angles

Another shortcut in the construction of nearly optimal tree architectures is the observation that in fully optimized trees that are sufficiently complex (e.g., Fig. 4.18), the angles between two pairing tubes near the center tend to vary little from one architecture to the next. For example, Table 4.1 shows that the innermost angle is nearly the same ($2\theta_1 \cong 75$ degrees) for structures with six and seven levels of pairing and central tubes varying from 3 to 8. This observation—the constancy of the 75-degree angle—was made earlier in Eq. (4.40), and leads to the strategy to fix all the angles at 75 degrees, and to construct the tree network from the rim to the center. The distance between two adjacent ports on the rim (d) is fixed when L, p, and n_0 are fixed. The number of peripheral ports N depends on p and n_0. The triangles plotted in Fig. 4.17 show that the global resistance of the 75-degree-angle trees is comparable with that of the fully optimized structures. The performance of the 75-degree-angle trees deteriorates relative to that of the optimal trees as the complexity increases. The computational times required by the 75-degree-angle trees increase linearly with p, and are one order of magnitude smaller than the times required by the minimal-length trees.

We conclude that on both accounts, accuracy and cost, the 75-degree-angle method outperforms the minimal-length method. Both methods produce tree architectures that perform at levels comparable with the top performance. Both methods lead to dramatic reductions in computational cost, and drive home the message that principle-based strategy is highly recommended in the optimization of multiscale complex flow structures.

4.7 TREES BETWEEN ONE POINT AND AN AREA

In section 4.5 we constructed minimum-length paths for connecting one point source (or sink) to a large number of point sinks (or sources) situated on a line or on a plane. In the case of the plane, the point source S was situated at a distance away from the plane (Fig. 4.14). The space between the point source and the plane did not contain any sources or sinks—it contained the connecting network.

In this section we consider the more challenging problem of connecting the point source to a large number of point sinks that cover an area uniformly. The point source and the many sinks are in the same plane. We may think of this configuration as a river basin (the area) that collects rain water at a uniform rate per unit area, and which uses its links (rivulets and rivers) to channel the collected water out of the territory, as a single stream. We may also think of this problem as a modification of the line-point flow network of Fig. 4.13, where the added feature is the presence of uniformly distributed points (sources, or sinks) on the area between the P line and the common source or sink (S).

How the construction starts depends on how the points are distributed over the given area. The simplest construction is sketched at the top of Fig. 4.20, where

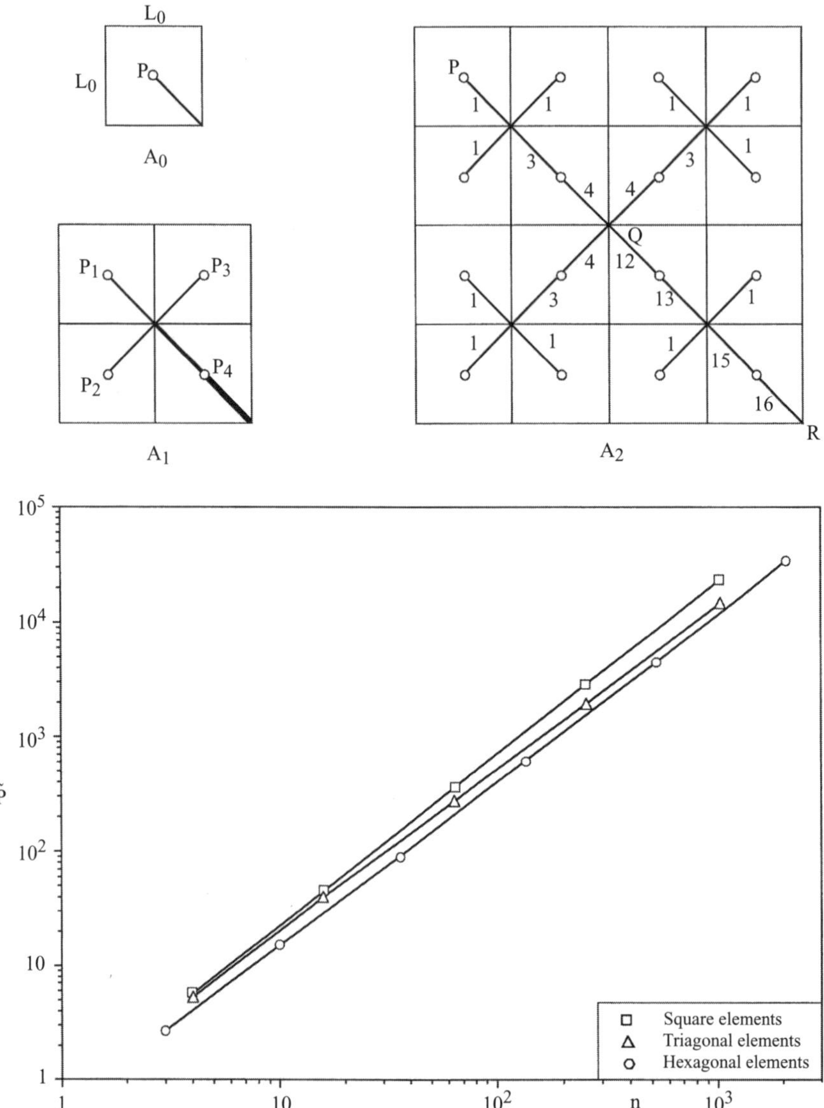

Figure 4.20 Top: The construction of the tree between one point (R) and an area covered uniformly by points (P) arranged in a square pattern. Bottom: The overall flow resistance between the point (source, sink) and the most distant point of the area that is covered uniformly with points [24].

the assumption is that the points are distributed in a square pattern. The start is the elemental area of size A_0, which is allocated to one of the many points P. By performing a length-minimization analysis that is analogous to the analysis of section 4.5, we find that the shape of A_0 must be square. In this case the distance from P to the corner of A_0 (the exit from the elemental area) is the shortest. The side of the elemental square is $L_0 = A_0{}^{1/2}$.

Square elements can be assembled tightly into square constructs of progressively larger sizes. Each new construct, A_i, consists of four constructs of the preceding size, $A_i = 4A_{i-1} = 4^i A_0$ $(i = 1,2,\ldots)$. The first construct (A_1, Fig. 4.20) has the purpose of connecting four points (P_1, P_2, P_3, P_4) to the single exit point Q. The way out of area A_1 is a cross in which three half-diagonals meet in the center of A_1, and the stem touches point P_4 before reaching Q. In the drawing of A_1 we have employed lines of increasing thicknesses, to suggest that elemental streams combine into larger streams en route to the exit (Q).

The second construct is itself a cross-shaped assembly of first constructs. Figure 4.20 shows that the exit points of three A_1-size constructs meet in the Q center of the second construct. The fourth A_1 construct is connected to the network, as the diagonal stem travels from Q to R. The flow rate increases in steps, from P to R. The numbers attached to each link indicate the relative magnitude of each local flow rate, where "1" stands for the flow rate that reaches each elemental point of type P.

As figure of merit for the resulting tree network we evaluate the pressure difference between the farthest point on the territory (e.g., point P on A_2) and the exit point (point R on A_2). For simplicity, we assume that all the pipes have the same diameter, and the flow is in the Poiseuille regime through every pipe. The pressure drop across one link (ΔP_k) is proportional to the mass flow rate through the link ($\dot{m}_0$) and the length of the link (L_k). The mass flow rate changes from one link to the next, where L_k is constant and equal to the half-diagonal of the elemental square, $L_k = 2^{-1/2} A_0{}^{1/2}$. If $\dot{m}_0$ is the mass flow rate through one elemental link (the link that touches point P), then the ratio $\dot{m}_k/\dot{m}_0$ is the relative flow rate indicated by the number attached to each link of the A_2 construct in Fig. 4.20.

The pressure drop across link k is

$$\Delta P_k = C \dot{m}_k L_k \tag{4.71}$$

where the C coefficient is a constant dictated by the kinematic viscosity of the fluid and the pipe diameter D, namely, $C = 128\nu/(\pi D^4)$. The pressure drop from the most distant point P (the upper-left corner of A_2) to the exit point (R) is obtained by summing along the descending diagonal the link pressure drop contributions ΔP_k:

$$\Delta P = \sum_{\text{from } P \text{ to } Q, R, \ldots} \Delta P_k. \tag{4.72}$$

This quantity can be nondimensionalized as

$$\tilde{P} = \frac{\Delta P}{C \dot{m}_0 A_0^{1/2}} = 2^{-1/2}(1 + 3 + 4 + 12 + \overset{Q}{13} + 15 + \overset{R}{16} + 48 + \cdots + \overset{S}{64} + \cdots). \tag{4.73}$$

where the numbers in the brackets represent the relative mass flow rates through the links of the descending diagonal. The pressure drop increases stepwise as the size of the construct increases. The latter is also indicated in Eq. (4.73) by the label of the exit corner of each construct (Q, R,. . .), which is placed above the relative flow rate of the stream that leaves through the lower-right corner of each construct (4, 16, 64,. . .). The way in which the overall pressure drop increases as the covered territory grows is indicated by square symbols in the lower part of Fig. 4.20, where n is the number of elements of size A_0 contained in the total area covered by the construct.

Another way to cover an area uniformly with points is to assign an approximately round area A_0 to each point, and to fit the A_0 elements so that they cover the given territory completely. A tight packing is obtained when each A_0 is squeezed into the shape of a regular hexagon with $L_0 = bA_0{}^{1/2}$ as the distance between the opposing sides of the hexagon, $b = (2/3)^{1/2}(\tan 30°)^{-1/2}$. The point P is the center of the hexagon, as shown in the first drawing of Fig. 4.21. The path out of the element

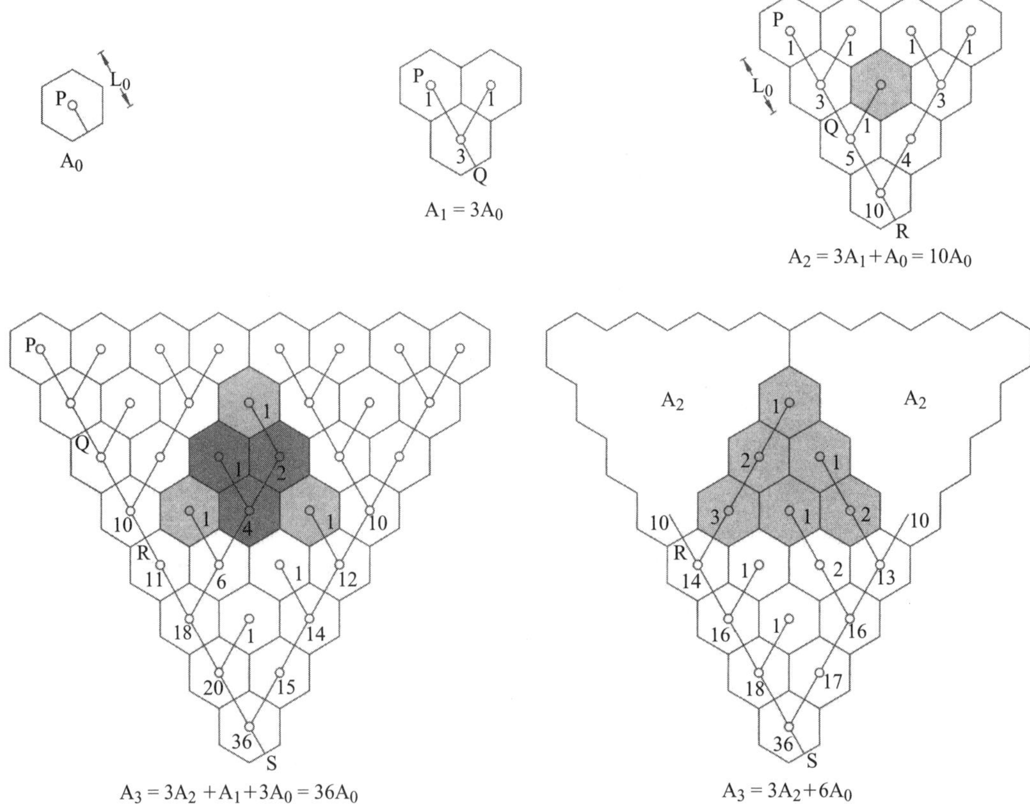

Figure 4.21 The construction of the tree between one point (Q, R,. . .) and an area covered uniformly by hexagonal area elements. The bottom row shows two designs for covering the central area of the third construct [24].

is the line of length $L_0/2$, from P to the midpoint of one side of the hexagon. This packing places the points of type P into a pattern of equilateral triangles.

The construction sequence is described in Fig. 4.21. The first construct consists of only three elements. The distance between adjacent points is L_0. The three points are connected by an asymmetric Y-shaped path with the root at the exit Q. We find that *asymmetry* is a necessary feature of assembly when the A_0 elements are the roundest (i.e., regular hexagons). We return to this observation in section 4.8. Asymmetry distinguishes the tree-shaped paths of Fig. 4.21 from the trees connecting square constructs (Fig. 4.20). Dichotomy, or pairing of two links to form a stem, is another feature that distinguishes Fig. 4.21 from Fig. 4.20.

The second construct (A_2) brings to light another feature that was not present in Fig. 4.20. A hole of size A_0 is left uncovered when first constructs are assembled into a second construct. In Fig. 4.21 the hole is in the center of A_2, and is covered by one A_0 element (shown darker). The drawing also shows the relative mass flow rates along each link. Asymmetry is accentuated at the second-construct level, because the central A_0 element must be connected to one of the two sides of the large V that contains the tree network. The flow rates collected by the two sides of the V are slightly unbalanced (5 vs. 4 mass flow rate units) as they reach the center of the bottom element. We will show that minimizing this imbalance is important from the point of view of generating the network, that is, for the purpose of distributing the imperfections of the network optimally.

These distinguishing features are even more striking in larger constructs. The first drawing in the bottom row of Fig. 4.21 shows how a third construct can be formed by assembling three second constructs. A hole of size $6A_0$ is left uncovered in the middle of the third construct. The central part of this hole was filled with one first construct, which is dark. The remaining corners of the hole (lighter shadow) are connected to the nearest points. The numbers indicate the relative mass flow rate through each link. Asymmetry and flow imbalance are evident. The two branches of the large V, which meet at the bottom of the third construct, show a flow rate imbalance of 20 units versus 15.

In addition to asymmetry and imbalance, in Fig. 4.21 we see the emergence of increasingly more freedom in how to connect the elements of the inner hole to the structure that delineates each new V-shaped construct. The rule of assembly applied from A_0 to A_1, from A_1 to A_2, from A_2 to A_3, and so on, concerns only the corner areas of each triangular construct. The new construct is made of three constructs of the preceding size, plus filler area in the middle. There is more than one way of connecting the central area to the peripheral building blocks of the structure. The lower-left drawing of Fig. 4.21 shows one way. Better ways can be found by analyzing the global performance of the tree-shaped flow network.

The calculation of the overall pressure drop between P and subsequent exit points ($Q, R, S, \ldots$) in Fig. 4.21 follows the steps described in conjunction with Eqs. (4.71) through (4.73). We employ the same notation, and assume that the elemental area A_0 of Fig. 4.21 is the same as the area A_0 of Fig. 4.20. This means that the length scale L_0 of Fig. 4.21 is not exactly the same as the L_0 scale of Fig. 4.20. With reference to the left side of the V-shaped A_3 construct shown in the lower-left drawing of Fig. 4.21, the overall pressure drop is

$$\tilde{P} = \frac{\Delta P}{C\dot{m}_0 A_0^{1/2}} = b\left(1 + 3 + 5 + \overset{Q}{10} + 11 + \overset{R}{18} + 20 + \frac{1}{2}\overset{S}{36}\ldots\right). \tag{4.74}$$

A smaller $\tilde{P}$ value is found if the calculation proceeds along the right side of the V:

$$\tilde{P} = b\left(1 + 3 + 5 + \overset{Q}{10} + 12 + \overset{R}{14} + 15 + \frac{1}{2}\overset{S}{36}\ldots\right). \tag{4.75}$$

The higher $\tilde{P}$ value prevails, and measures the overall imperfection (resistance) of the flow path. Clearly, a smaller flow rate at the root of the side along which $\tilde{P}$ is calculated is good for the purpose of decreasing $\tilde{P}$. A smaller flow rate is achieved by balancing as closely as possible the two sides of each V. The lower-right drawing of Fig. 4.21 shows a more balanced way of connecting the inner hole (dark) to the three A_2 constructs that make up A_3. The flow imbalance at the bottom of the V is 18 units versus 17. The overall pressure drop along the left side of the V is

$$\tilde{P} = b\left(1 + 3 + 5 + \overset{Q}{10} + 14 + \overset{R}{16} + 18 + \frac{1}{2}\overset{S}{36}\ldots\right). \tag{4.76}$$

which is less than in Eq. (4.74). Balancing resistances, or improving the distribution of imperfection is the route to better global performance.

The $\tilde{P}$ values calculated based on Fig. 4.21 (lower right) and Eq. (4.76) are plotted versus n in Fig. 4.20, where n is the number of A_0 elements in each construct. These points are indicated by circles, and fall below the apparent line of squares associated with the upper part of Fig. 4.20. The conclusion is that the construction of Fig. 4.21 is more effective for the purpose of connecting one point to the large number of points distributed uniformly over an area.

To see even more clearly why the assembly of regular hexagons is a more effective flow structure, consider the only option that is still available for covering an area with elements that are regular polygons. That option is to use elements shaped as equilateral triangles, Fig. 4.22. The elemental area A_0 is the same as in the case of square elements (Fig. 4.20) and hexagonal elements (Fig. 4.21). Figure 4.22 shows that when a large area is covered by equilateral triangles, the centers of mass of the triangles (P) arrange themselves in a pattern of regular hexagons. The side of the hexagon, or the shortest distance between two points of type P is $L_0 = b' {A_0}^{1/2}$, where $b' = (2/3)(\tan 30 \text{ degrees})^{-1/2}$. The length L_0 is not the same as the L_0 dimension used earlier.

Figure 4.22 shows the first steps in the construction based on equilateral triangles. The first construct (A_1) consists of four elements, and is itself an equilateral triangle.

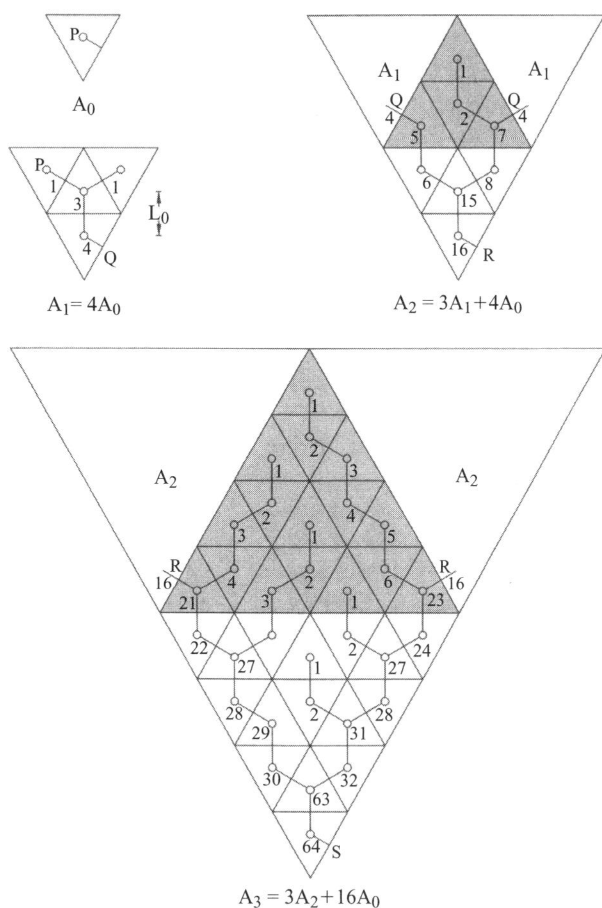

Figure 4.22 The construction of the tree between one point ($Q, R, \ldots$) and an area covered by area elements shaped as equilateral triangles [24].

The four center points (P) are linked into a symmetric Y-shaped flow path. The exit from the first construct (Q) is along the shortest path from the Y structure to the boundary of A_1. The number attached to each link indicates the relative flow rate, where 1 represents the elemental flow rate arriving at or departing from a center point P.

The second construct (A_2) is formed by three A_1 constructs and an inner area of size $4A_0$. The A_2 drawing shows the only way in which the centers of the four inner elements can be connected to the rest of the network, so that the resulting flow structure is without loops. The result is *asymmetry*. Note the asymmetric connections that cover the shaded area. Also note the flow rate imbalance (5 vs. 7) between the streams that enter the bottom A_1 construct. The construction of Fig. 4.22 can be extended to cover larger areas, as shown by the A_3 design.

The performance of the structure based on triangular elements can be evaluated by using the same pressure drop model and calculation procedure as in (4.73) and (4.75). For the structure of Fig. 4.22, we obtain

$$\tilde{P} = \frac{\Delta P}{C\dot{m}_0 A_0^{1/2}} = b'\left(1 + 3 + \overset{Q}{\frac{1}{2}4} + \frac{1}{2}4 + 7 + 8 + 15 + \overset{R}{\frac{1}{2}16} + \frac{1}{2}16 + \ldots\right). \tag{4.77}$$

Note the relative flow rate indicated by 7 in the fifth term inside the round brackets.

This means that at the A_2 level we chose to estimate the largest pressure drop, which occurs between the extremities of the upper-right A_1 area and the root R. Had we chosen the upper-left A_1 construct as the start of the ΔP calculation, we would have used 5 in place of 7, and we would have underestimated the actual overall pressure difference sustained by the second construct.

The $\tilde{P}$ values generated based on Fig. 4.22 and Eq. (4.77) are indicated by triangles in Fig. 4.20. These values fall between the preceding two sets (squares, circles), and reconfirm the conclusion that the most effective flow structure is the one based on hexagonal elements (the circles). Structures with hexagonal and triangular elements perform nearly the same as their complexity (n) increases. They perform better than structures with square elements, and this suggests that structures with pairings perform better than structures with three tributaries at each point of confluence (cf. section 4.1).

From triangular area elements to square and hexagonal elements, the angles decrease from 120 degrees to 90 degrees and 60 degrees. The smallest angle corresponds to the best structure, and it comes from "packing" needs—how to cover an area completely by using the shortest links between the centers of elements. The 60 degree angle of confluence or branching, which looks so much like the natural angles of the trees known from physiology and geophysics, is a consequence of packing, or the struggle to cover an area with identical elements.

4.8 ASYMMETRY

In sections 4.3 through 4.6 we pursued in great detail the morphing and generating of progressively better tree-shaped flow networks. We did this by focusing on trees that connect a circle with its center. The summary given in Fig. 4.12 shows that when there are 192 equidistant points on the circle, the equilibrium flow architecture has a certain (optimized, finite) degree of complexity, which is represented by six levels of bifurcation or pairing.

More recently, we discovered [9] that the circle-point tree architectures can be morphed and improved further, so that their points migrate even farther to the left on the performance versus freedom domain of Fig. 4.12. The true equilibrium flow architecture lies slightly to the left of point d indicated in Fig. 4.12. The path to

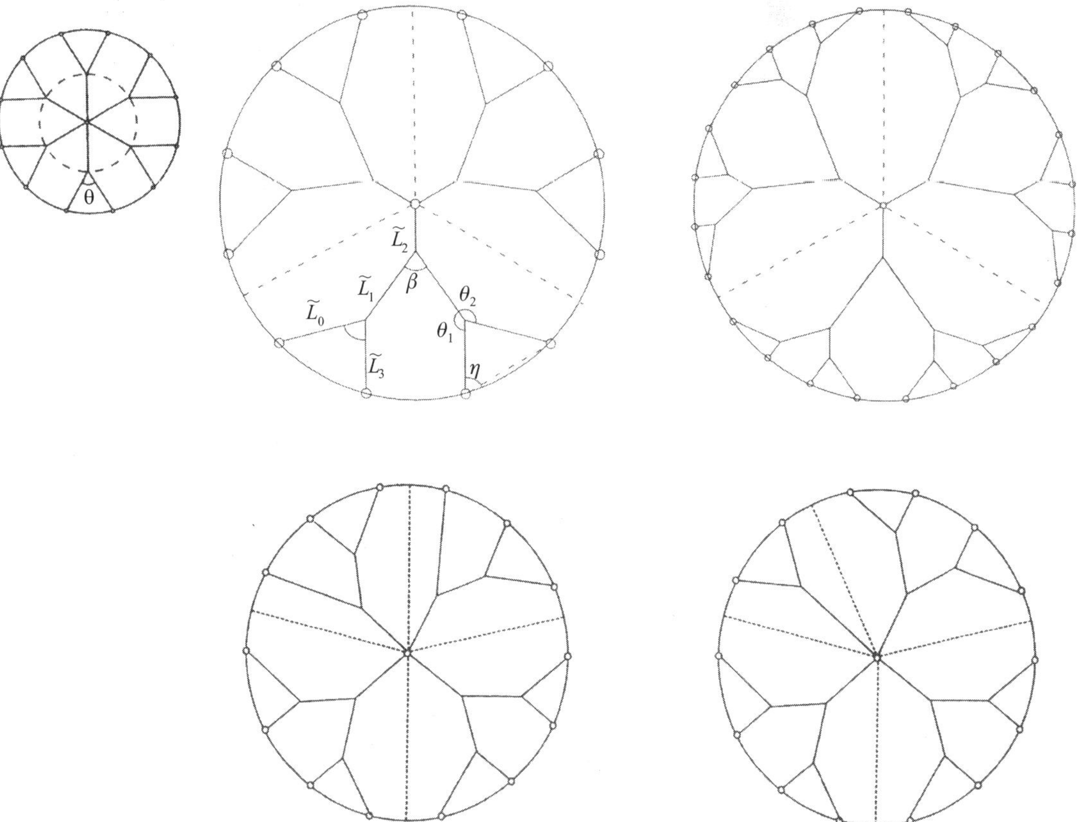

Figure 4.23 More freedom to morph leads to asymmetry and higher performance in circle-point tree networks [9].

this higher level of performance is made possible by increasing the freedom of the morphing architecture. How this is done is explained in Fig. 4.23.

The upper-left frame of Fig. 4.23 illustrates the type of circle-point trees that have been studied in the literature until now. It is based on the reasonable assumption that at each level of pairing or bifurcation, the mother tube splits into two daughter tubes that have the same length. This assumption is so "reasonable" and so popular (no doubt, because of fractal geometry) that no one questioned it. In truth, however, *any simplifying assumption is a straitjacket*—it curtails the freedom of the morphing architecture.

The other frames of Fig. 4.23 show what happens to the tree architecture when the assumption of daughter tubes of equal length is not made. We have discovered in this way the emergence of several types of asymmetry in equilibrium tree architectures: different tube lengths at the same level of branching, different mass flow rates at junctions, and different main branches [9]. The emergence of asymmetry in the

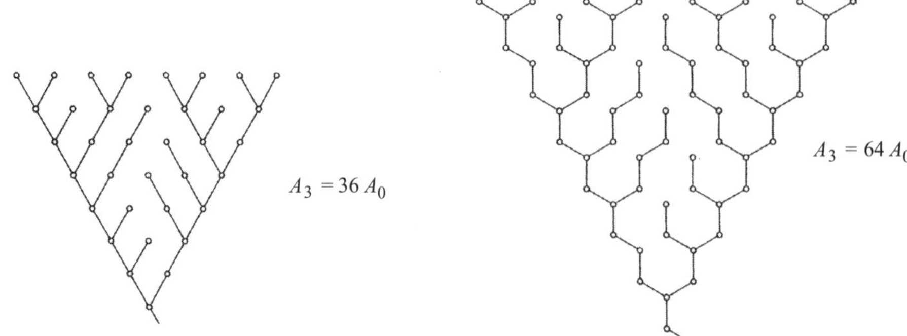

Figure 4.24 Emergence of asymmetry in the optimal layout of area-point tree networks in Figs. 4.21 and 4.22 [24].

best tree networks (human-made or natural) is the fingerprint of the constructal law in action.

We saw tree asymmetry in section 4.7, during the completely free search for a tree layout (A_3) that consists of hexagonal and triangular area elements (A_0), Figs. 4.21 and 4.22. The layouts of channels are isolated in Fig. 4.24. The fact that some branches appear unlike the others (e.g., single branch attached laterally to a stem) is a statement of perfection, not imperfection.

Freedom is good for the performance and survival of a flow structure, be it natural or engineered. Design improvement without freedom to change the structure is nonsense. Rigid flow structures are brittle.

4.9 THREE-DIMENSIONAL TREES

The selection of flow problems in this chapter followed an exploratory course, from the simple toward the complex. Most of the work referred to tree-shaped connections in a plane because flat trees are relatively simpler and easier to illustrate on paper. Their importance in practice remains undiminished, in view of the many natural trees found in river basins, vascularized materials (Chapter 8), traffic patterns, and power distribution networks.

One direction for future work is to extend the ideas of this chapter to trees in three dimensions. Such trees would connect one point to many points distributed uniformly through a volume. We made a step in this direction in section 4.7, where the connection was between the many points of a plane and one point situated off the plane (Fig. 4.14). The next challenge is to consider the case where the entire space around the single point (S) is populated uniformly by points that must be connected optimally to S.

A basic question is whether the best Y-shaped assemblies of two links joined into one link are flat or three-dimensional. The example of Fig. 4.14 is one where the Y construct lies in two planes that intersect at an angle. Two branches of type

PQ form a plane, while the stem QR pierces that plane at an angle. Is a perfectly flat Y construct better? This question is relevant because careful observations of three-dimensional branchings in the cardiovascular tissue have indicated conclusively that the flat Y construct is the prevailing form in three-dimensional trees [26].

Consider the two bent-Y constructs obtained in the first two stages of Fig. 4.14. The volume occupied by one such construct is the same as that of the smallest shaded rectangle in the upper-left corner of Fig. 4.14. We consider this volume fixed, $V = 2d^2(c + e)$ or $V = 2^{1/2}d^3$, because $c = e = 2^{-3/2}d$. The bent Y structure has two lengths, $\overline{PQ} = L_0 = (c^2 + d^2/4)^{1/2} = (3/2)^{1/2}d/2$, and $\overline{QR} = L_1 = (e^2 + d^2/4)^{1/2} = L_0$. In order to evaluate the performance of this structure as a path of least flow resistance, we assume that the Y is the confluence of two round tubes of diameter D_0 and length L_0, which form one tube of diameter D_1 and length L_1. We assume that the total tube volume is constrained, $V_t = 2(\pi D_0^2/4)L_0 + (\pi D_1{}^2/4)L_1$, that the flow is in the Poiseuille regime, and that the ratio of tube diameters has been optimized for minimal resistance, $D_1/D_0 = 2^{1/3}$. Using this ratio and $L_0 = L_1$, we rewrite the tube volume constraint as $V_t = (\pi/2)(1 + 2^{-1/3})D_0{}^2L_0$. The pressure drop across the Y structure is $\Delta P = \Delta P_1 + \Delta P_0$, where ΔP_1 and ΔP_0 are the pressure drops across the L_1 tube (flow rate $\dot{m}$) and, respectively, across the L_0 tube (flow rate $\dot{m}/2$):

$$\Delta P = C'\dot{m}\frac{L_1}{D_1^4} + C'\frac{\dot{m}}{2}\frac{L_0}{D_0^4} \tag{4.78}$$

where $C' = 128\nu/\pi$. Equation (4.78) can be nondimensionalized into an overall flow resistance, after using the preceding results for L_0, L_1 and D_1/D_0, and by eliminating D_0 and d between Eq. (4.78) and the V and V_t constraints:

$$\tilde{P}_{\text{bentY}} = \frac{\Delta P V_t^2}{\dot{m}C'V} = \frac{3^{3/2}\pi^2}{2^8}\left(1 + 2^{-1/3}\right)^3 = 1.156. \tag{4.79}$$

The analysis and optimization of the flat Y structure is performed in relation to the drawing shown in Fig. 4.25. The Y structure lies in the plane of size $f \times 2d$. The volume inhabited by the construct has the same size as before, $V = 2d^2g$. The height of the rectangle in which the Y structure resides is $f = (g^2 + d^2/4)^{1/2}$. The inclination of this plane is free to change, and so is the length L_1, or the location of the Q junction.

In summary, the Y-shaped construct has two degrees of freedom, which in the following analysis are represented by the aspect ratios a/d (or L_0/L_1) and f/d, where $a = f - L_1$. The overall pressure drop, Eq. (4.78), can be combined with the V_t constraint and $D_1/D_0 = 2^{1/3}$, and the result is

$$\Delta P = \frac{\dot{m}C'\pi^2}{8V_t^2}\left(L_0 + 2^{-1/3}L_1\right)^3 \tag{4.80}$$

The group in the round brackets can be minimized by varying the geometry. We accomplish this in two steps. First, we assume that the rectangular area $f \times (2d)$

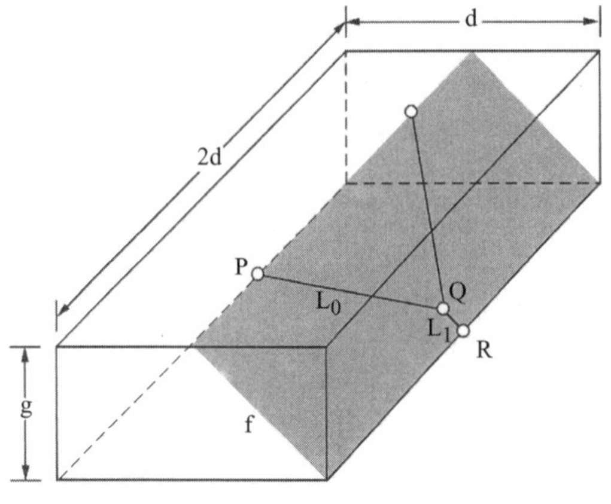

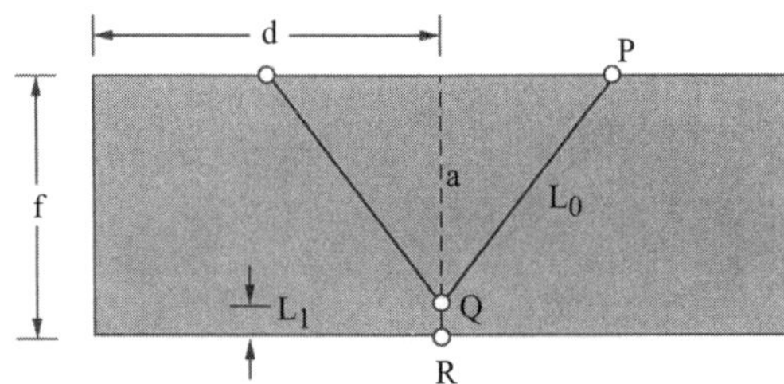

Figure 4.25 Flat Y-shaped flow construct in the elemental volume of a three-dimensional tree network [24].

does not change. On it, the only variable is a, or the position of point Q. The group $(L_0 + 2^{-1/3}L_1)$ can be expressed as a function of a, f, and d, and can be minimized with respect to a. The result is the optimal ratio $a/d = c_a$, where $c_a = 2^{-1}(2^{2/3}-1)^{-1/2}$. Related results are $L_0/d = c_0$, where $c_0 = (1 - 2^{-2/3})^{-1/2}/2$, and $L_1/d = f/d - c_a$. These values can be substituted into Eq. (4.80), and after eliminating d^3 based on the V constraint, we obtain the dimensionless flow resistance of the flat Y construct:

$$\tilde{P}_{\text{flat}} = \frac{\pi^2\left[c_0 + 2^{-1/3}(f/d - c_a)\right]^3}{16\left[(f/d)^2 - 1/4\right]^{1/2}} \tag{4.81}$$

We found numerically that this function reaches its minimum at $f/d = 0.716$, where its value is $\tilde{P}_{\text{flat}Y} = 0.8$. Other relevant geometric characteristics of the optimized structure are $a/d = 0.652$, $g/d = 0.512$, $L_0/L_1 = 13.01$ and $L_1/f = 0.088$.

Equations (4.79) and (4.81) show that the global resistance of the flat Y structure is 30 percent smaller than the global resistance of the bent Y structure. This lends support to the view [13] that natural Y structures, which are flat [26], are results of an evolutionary process of flow access increase.

4.10 LOOPS, JUNCTION LOSSES AND FRACTAL-LIKE TREES

The start made on tree flow architectures in this chapter continues in Chapters 7 and 8 with dendritic designs that combine fluid-flow performance with other objectives. Here we treated the essentials of the fundamental problem of access in area-point and volume-point flows. This, by the way, was the problem that got constructal theory started [25]. The steps taken beyond Chapter 4 are visible in the current literature. They are also visible and in Chapter 5, where we used pure conduction with two conductivities (instead of fluid channels and interstices) to discover analogous tree architectures. For example, in section 5.2 we study the performance and configuration of trees with loops. We proposed this class of architectures originally for fluid flow [22], but, because the lessons of section 5.2 are qualitatively the same as for fluid flow, we do not cover fluid trees with loops here. It suffices to say that the concept of fluid trees with loops [22] is attracting serious interest from the electronics packaging community (e.g., Ref. [27]) because loops endow the tree architecture with greater robustness and resilience and the ability to bathe (to cool) the interstices better.

Another direction is the effect of junction losses on the performance of fluid trees for distribution and collection. In this chapter, we dealt with this important issue in terms of global property of svelteness [20], for example, Eqs. (4.26) and (4.27). We showed analytically and numerically [28, 29] that in agreement with section 1.1, junction losses are negligible when Sv exceeds the range 10 to 100. In this limit the direction of the flow does not affect the outlook of the optimal dendritic architecture: the same trees are recommended for flow from trunk to canopy and for flow from canopy to trunk.

The constructal literature is growing at a fast pace along these and other lines, as shown by the pre-2006 publications reviewed in Refs. [30] through [32]. More recent advances are made in hydrology, for example, the optimal shaping of groundwater flow [33] and the design of the entire river bed [34]. The emergence of asymmetric Y-shaped bifurcations [8, 24] was pursued further in Ref. [35]. Links are made between constructal theory and more established fields of application, such as optimal transport networks [36], biologically inspired tree designs for lab on a chip [37], and the cooling of electronics [38]. The biggest connection is with topology optimization, which until recently was thought to apply only to solid structures. Constructal theory [13] and current topology optimization [34, 39] are showing that topology optimization extends over fluid flow architectures as well. Even better, in Chapter 10 we show that solid structures and fluid flows develop their architectures hand in glove.

Because constructal trees come from principle (the constructal law), and not from an assumed algorithm suggested by nature, the comparative study of constructal theory and fractal geometry continues. Queiros-Conde [40] et al. investigated the properties of the multiscale structure of a constructal tree network, and proposed a new geometric framework to quantify the deviations from scale invariance observed in many fields of physics and life sciences. The constructal tree network was compared with one based on an assumed fractal algorithm. The fractal-like network offers lower performance than the constructal object (in agreement with our discussion of Fig. 4.12), and the constructal object exhibits a parabolic scaling explained in the context of entropic skin geometry based on an equation of scale diffusion in the scale space [41].

REFERENCES

1. A. Bejan and S. Lorente, Thermodynamic optimization of flow geometry in mechanical and civil engineering. *J Non-Equilib Thermodyn*, Vol. 26, 2001, pp. 305–354.
2. A. Bejan, I. Dincer, S. Lorente, A. F. Miguel, and A. H. Reis, *Porous and Complex Flow Structures in Modern Technologies*. New York: Springer, 2004.
3. A. Bejan, L. A. O. Rocha and S. Lorente, Thermodynamic optimization of geometry: T- and Y-shaped constructs of fluid streams. *Int J Thermal Sciences*, Vol. 39, 2000, pp. 949–960.
4. W. R. Hess, Das Prinzip des kleinsten Kraftverbrauches im Dienste hämodynamischer Forschung. *Archiv für Anatomie und Physiologie*, 1914, pp. 1–62.
5. E. R. Weibel, *Symmorphosis: On Form and Function in Shaping Life*. Cambridge, MA: Harvard University Press, 2000.
6. C. D. Murray, The physiological principle of minimal work, in the vascular system, and the cost of blood-volume. *Proc Acad Nat Sci*, Vol. 12, 1926: 207–214.
7. W. Wechsatol, S. Lorente, and A. Bejan, Optimal tree-shaped networks for fluid flow in a disc-shaped body. *Int J Heat Mass Transfer*, Vol. 45, 2002, pp. 4911–4924.
8. L. Gosselin and A. Bejan, Tree networks for minimal pumping power. *Int J Thermal Sciences*, Vol. 44, 2005, pp. 53–63.
9. L. Gosselin and A. Bejan, Emergence of asymmetry in constructal tree flow networks. *J Applied Physics*, Vol. 98, 2005, 104903.
10. W. Wechsatol, A. Bejan, and S. Lorente, Tree-shaped flow architectures: Strategies for increasing optimization speed and accuracy. *Numerical Heat Transfer, Part A*, Vol. 48, 2005, pp. 731–744.
11. A. Bejan and S. Lorente, Thermodynamic optimization of the flow architecture: Dendritic structures and optimal sizes of components. ASME IMECE-Vol. 3, Int. Mech. Eng. Congress and Exposition, New Orleans, November 17–22, 2002.
12. J. C. Ordonez and A. Bejan, System-level optimization of the sizes of organs for heat and fluid flow systems. *Int J Thermal Sciences*, Vol. 42, 2003, pp. 335–342.
13. A. Bejan, *Shape and Structure, from Engineering to Nature*. Cambridge, UK: Cambridge University Press, 2000.
14. A. Bejan and J. H. Marden, Unifying constructal theory for scale effects in running, swimming and flying. *J Exp Biol*, Vol. 209, 2006, pp. 238–248.
15. A. Bejan and J. H. Marden, Constructing animal locomotion from new thermodynamics theory. *American Scientist*, Vol. 49, July–August 2006, pp. 342–349.

16. A. Bejan and J. H. Marden, Locomotion: Une même loi pour tous. *Pour la Science*, No. 346, August 2006, pp. 68–73.
17. A. Bejan, The constructal law of organization in nature: Tree-shaped flows and body size. *J Exp Biol*, Vol. 208, 2005, pp. 1677–1686.
18. A. Bejan and S. Lorente, The constructal law and the thermodynamics of flow systems with configuration. *Int J Heat Mass Transfer*, Vol. 47, 2004, pp. 3203–3214.
19. A. Bejan and S. Lorente, *La Loi Constructale*. Paris: L'Harmattan, 2005.
20. S. Lorente and A. Bejan, Svelteness, freedom to morph, and constructal multi-scale flow structures. *Int J Thermal Sciences*, Vol. 44, 2005, pp. 1123–1130.
21. D. V. Pence, Reduced pumping power and wall temperature in microchannel heat sinks with fractal-like branching channel networks. *Microscale Thermophys Eng*, Vol. 6, 2002, pp. 319–330.
22. W. Wechsatol, S. Lorente, and A. Bejan, Tree-shaped networks with loops. *Int J Heat Mass Transfer*, Vol. 48, 2005, pp. 573–583.
23. A. Bejan, S. Lorente, A. F. Miguel, and A. H. Reis, Constructal theory of distribution of river sizes, Section 13.5, in A. Bejan, *Advanced Engineering Thermodynamics*, 3rd ed., Hoboken, NJ: Wiley, 2006.
24. S. Lorente, W. Wechsatol, and A. Bejan, Tree-shaped flow structures designed by minimizing path lengths. *Int J Heat Mass Transfer*, Vol. 45, 2002, pp. 3299–3312.
25. A. Bejan, Constructal-theory network of conducting paths for cooling a heat generating volume. *Int J Heat Mass Transfer*, Vol. 40, 1997, pp. 799–816 (published November, 1, 1996).
26. M. Zamir, Arterial bifurcations in the cardiovascular system of a rat. *J Gen Physiol*, Vol. 81, 1983, pp. 325–335.
27. X.-Q. Wang, C. Yap, and A. S. Mujumdar, Laminar heat transfer in constructal microchannel networks with loops. *J Electronic Packaging*, Vol. 128, 2006, pp. 273–280.
28. W. Wechsatol, S. Lorente, and A. Bejan, Tree-shaped flow structures with local junction losses. *Int J Heat Mass Transfer*, Vol. 49, 2006, pp. 2957–2964.
29. H. Zhang, S. Lorente, and A. Bejan, Vascularization with trees that alternate with upside-down trees. *J Appl Phys*, Vol. 101, 2007, 094904.
30. A. Bejan and S. Lorente, Constructal theory of configuration generation in nature and engineering. *J Appl Phys*, Vol. 100, 2006, 041301.
31. A. H. Reis, Constructal theory: From engineering to physics, and how flow systems develop shape and structure. *Appl Mech Rev*, Vol. 59, 2006, pp. 269–282.
32. A. Bejan, *Advanced Engineering Thermodynamics*, 3rd ed. (Ch. 13). Hoboken, NJ: Wiley, 2006.
33. A. R. Kacimov, Analytical solution and shape optimization for groundwater flow through leaky porous trough subjacent to an aquifer. *Proc Roy Soc A*, Vol. 462, 2006, pp. 1409–1423.
34. N. Wiker, Topology optimization in flow problems. Thesis No. 1278, Linköping University, Sweden, 2006.
35. W. Wechsatol, J. C. Ordonez, and S. Kosaraju, Constructal dendritic geometry and the existence of asymmetric bifurcation. *J Appl Phys*, Vol. 100, 2006, 113514.
36. M. Durand, Architecture of optimal transport networks. *Phys Rev E*, Vol. 73, 2006, 016116.

37. R. W. Barber and D. R. Emerson, Optimal design of microfluidics networks using biologically inspired principles. *Microfluidics and Nanofluidics*, Vol. 4, 2008, pp. 179–191.

38. X.-Q. Wang, A. S. Mujumdar, and C. Yap, Numerical analysis of blockage and optimization of heat transfer performance of fractal-like microchannel nets. *J Electronic Packaging*, Vol. 128, 2006, pp. 38–45.

39. B. Mohammadi and O. Pironneau, Shape optimization in fluid mechanics. *Annu Rev Fluid Mech*, Vol. 36, 2004, pp. 255–279.

40. D. Queiros-Conde, J. Bonjour, W. Wechsatol, and A. Bejan, Parabolic scaling of tree-shaped constructal network. *Physica A*, Vol. 384, 2007, pp. 719–724.

41. D. Queiros-Conde, A diffusion equation to describe scale- and time-dependent dimensions of turbulent interfaces. *Proc Roy Soc A*, Vol. 459, 2003, pp. 3043–3059.

PROBLEMS

4.1. Lorente et al. [24] showed that near-optimal fluid-flow trees can be constructed by minimizing the length of every tube on its allocated area. The rectangular area A shown in Fig. P4.1 is crossed by a round tube of diameter D. The tube connects one corner with the middle of the opposing side. The tube volume V is fixed. The area A is constant, but its shape (x/y) may vary.

(a) Assume fully developed laminar flow and show that the flow resistance of the tube is governed by the geometric group $R = L/D^4$, where L is the tube length. Find the rectangle shape for which R is minimal.

(b) Show that if the flow is turbulent in the fully rough regime ($f =$ constant), the flow resistance is proportional to $R = L/D^5$. Find the rectangle shape for which R is minimal.

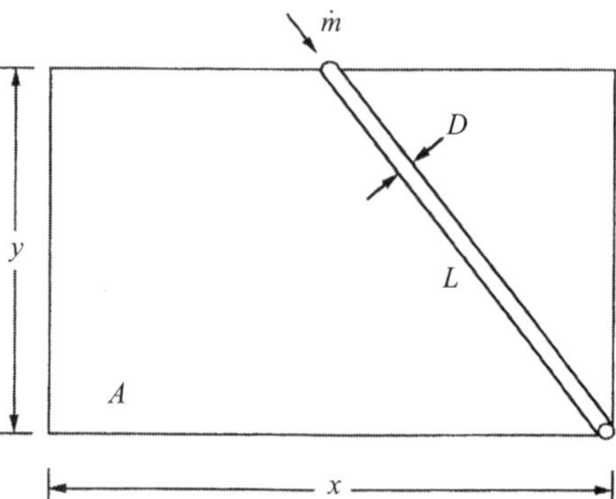

Figure P4.1

4.2. The rectangular territory shown in Fig. P4.2 has the size fixed ($S = AB$), but its shape A/B may vary. The rectangular area S houses a Y-shaped construct of three round ducts, one of length L_1 and diameter D_1 and two of length L_2 and diameter D_2. The Y connects one corner of the rectangle with two points on the side B. The distance between the two points is $B/2$, and the distance between one of these points and the nearest corner is $B/4$. The junction

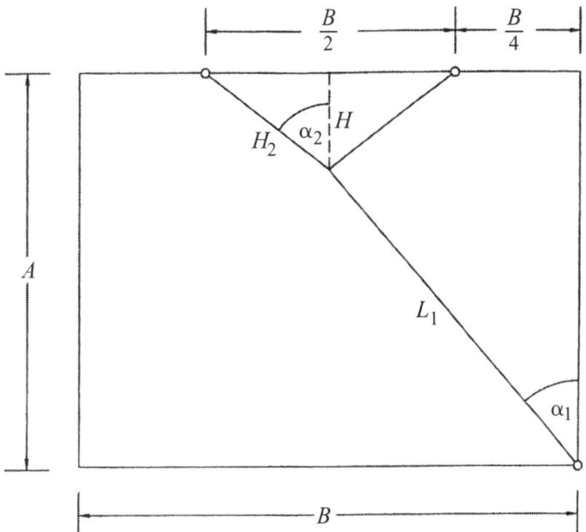

Figure P4.2

between the L_2 ducts and the L_1 duct is located at the distance H from the side B. The total tube volume is fixed.

The objective is to determine the Y configuration for minimum flow resistance. The flow through each pipe is laminar and fully developed. Junction pressure drops are assumed negligible. Use results from the text and earlier problems. For example, the optimal ratio of diameters is $D_2/D_1 = 2^{-1/3}$, and the minimization of global flow resistance is equivalent to minimizing the expression $(L_1 + 2^{1/3}L_2)$. Use $S^{1/2}$ as the length scale in the nondimensionalization of all the lengths. Determine the optimal Y geometry, and report L_1, L_2, H, A/B, and angles α_1 and α_2. Draw the optimal configuration.

4.3. In this problem we learn that by optimizing a simpler configuration we can come close to the performance of the equilibrium flow structure. The equilibrium structure is the one determined in Problem 4.2. A simple version of the original morphing Y is the assumption that L_1 and L_2 are always collinear ($\alpha_1 = \alpha_2 = \alpha$) [24]. In this case the only degree of freedom is the aspect ratio of the area S or the angle α. As in Problem 4.2, begin with the observation that for Poiseuille flow through round tubes the optimal ratio of successive tube diameters is $2^{-1/3}$, and the global flow resistance of the Y-shaped construct is proportional to $R = L_1 + 2^{1/3}L_2$. Show that the optimal geometry is such that $\alpha = 45$ degrees, and the minimized global flow resistance is $R = 1.331S^{1/2}$. The performance of this nonequilibrium flow structure is only 0.5% inferior relative to the performance of the optimized structure that was the most free to morph.

4.4. In this problem we generalize the Hess-Murray rule for bifurcated tubes that do not have identical branches. A round tube of diameter D_1 and length L_1 splits into two tubes of diameters D_2 and D_3 and lengths L_2 and L_3. The tube

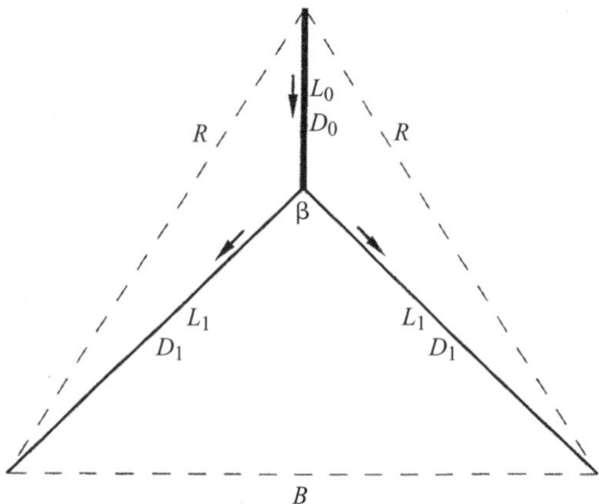

Figure P4.5

length L_2 is not equal to the tube length L_3. The three tubes form a Y-shaped construct: The layout of this Y on a plane or in the three-dimensional space is arbitrary. The tube lengths are given. The tube diameters may vary, but the total volume occupied by the tubes is fixed. The flow is in the Poiseuille regime in every duct. Each tube is *svelte* enough [Eq. (1.1)] so that the pressure drop across the entire Y construct, from the free end of L_1 to the free ends of L_2 and L_3, is dominated by fluid friction along the straight portions of the ducts. In other words, the pressure loss at the Y junction is negligible. Minimize the overall flow resistance and show that the optimal distribution of tube diameters is

$$\frac{D_1}{D_2} = \left[1 + \left(\frac{L_3}{L_2}\right)^3\right]^{1/3}, \qquad \frac{D_2}{D_3} = \frac{L_2}{L_3}$$

Show that when $L_2 = L_3$, these results confirm the Hess-Murray rule [Eq. (4.34)]. In other words, the results obtained in this problem represent a generalization of the Hess-Murray rule for the entire range where $L_2 \neq L_3$.

4.5. One tube of length L_0 and diameter D_0 splits symmetrically into two tubes of length L_1 and diameter D_1. The free ends of the three tubes define an isosceles triangle with the base B and side R (Fig. P4.5). The aspect ratio B/R is a known parameter. An incompressible newtonian fluid flows from the L_0 tube into the L_1 tubes, or in the opposite direction. The flow is laminar and fully developed, and the pressure drop due to the Y junction is negligible. According to the Hess-Murray rule [Eq. (4.34)] the global flow resistance is minimum when $D_0/D_1 = 2^{1/3}$. Minimize the global flow resistance further by selecting the tube lengths. Report the optimal L_0/R and L_1/R as functions of B/R. Show that the optimal bifurcation angle β formed between the L_1 tubes is 74.94 degrees, which is independent of the triangle aspect ratio B/R.

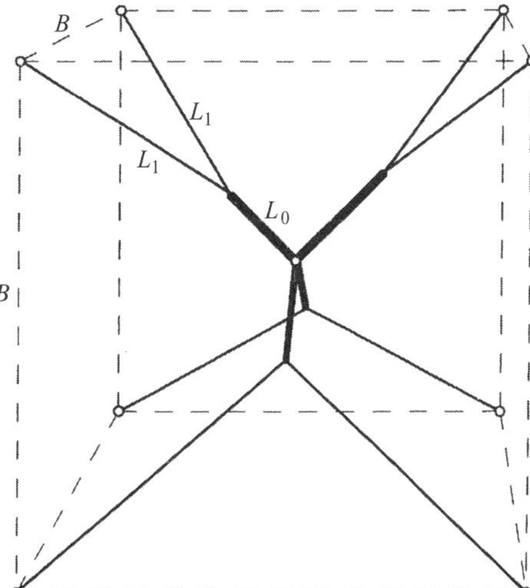

Figure P4.6

4.6. The corners of a cube are the outlets of a tree flow that originates from the center of the cube. There are four central tubes connected to the center, and each undergoes a symmetric bifurcation (Fig. P4.6). The flow is laminar and fully developed with negligible junction losses. Determine the optimal tube lengths L_0 and L_1 in relation to the cube side B, so that you may draw to scale the optimal tree network. Note that this is a special case of the more general solution described in Problem 4.5.

4.7. The minimization of global flow resistance in Problem 4.6 leads to the conclusion that $L_0/B = 0.06319$ and $L_1/B = 0.9491$. Calculate the overall flow resistance of this design. Next, note that the optimal L_0 is so small that perhaps this complication may be omitted. The simpler design is to have eight straight tubes of length R connecting the center of the cube with its corners. Calculate the flow resistance of this simpler design, and make sure that the total tube volume is the same as in the design with bifurcations. How much better is the design with bifurcations relative to the design with eight radial tubes?

4.8. Two identical open-channel streams of width D_2, length L_2, and flow rate $\dot{m}/2$ continue downhill as a single stream (D_1, L_1, $\dot{m}$), Fig. P4.8. The total drop in elevation between the inlets and the outlet of the three-channel assembly is given (and is analogous to the pressure difference ΔP shown in the figure). Also fixed are the size of the territory occupied by the assembly (the area $2L_1L_2$) and the total volume of the three channels. Assume that the shape of each channel cross-section has been optimized for minimal flow resistance. In addition, assume that each channel flow is in the fully turbulent and fully rough regime, so that the friction factor is constant (flow-rate independent) and has the same value for all the channels. Minimize the

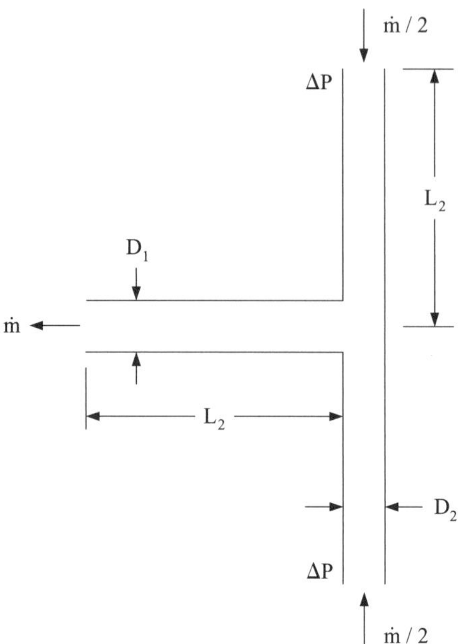

Figure P4.8

overall flow resistance of the assembly and show that the optimal architecture is characterized by $(D_1/D_2)_{opt} = 2^{3/7}$, $(L_1/L_2)_{opt} = 2^{1/7}$, and a junction elevation situated halfway between the elevations of the inlets and the outlet.

4.9. The optimal layout of a bifurcating tube in a two-dimensional space was described in Ref. [3]. Here we consider the analogous problem in three dimensions, for which we showed that the best way to branch one tube into n smaller tubes is to use $n = 3$ [7]. The tube (L_1) and three branches (L_2) are shown in Fig. P4.9. The free ends of the flow structure form the vertices of a pyramid with equilateral triangular base and height $(h + L_1)$, where h is the height of the tube junction above the base. The pyramid volume is fixed. The total volume occupied by the tubes is also fixed. The optimal ratio of tube diameters is $D_1/D_2 = 3^{1/3}$. Determine the optimal ratio of tube lengths L_1/L_2 with h as a parameter, and show that the best design corresponds to $h = 0$.

4.10. One straight tube of length L_1 and diameter D_1 splits into n identical tubes of length L_2 and diameter D_2. The flow is in the Poiseuille regime in all the tubes, and junction pressure losses can be neglected. This "n-pronged fork" is symmetric enough so that the mass flow rate through the mother tube ($\dot{m}_1$) is distributed equally to the n daughter tubes, $\dot{m}_1 = n\dot{m}_2$. The volume occupied by all the tubes is fixed. The tube lengths are fixed. Determine the optimal ratio D_1/D_2 for which the overall pressure drop across the fork is minimum.

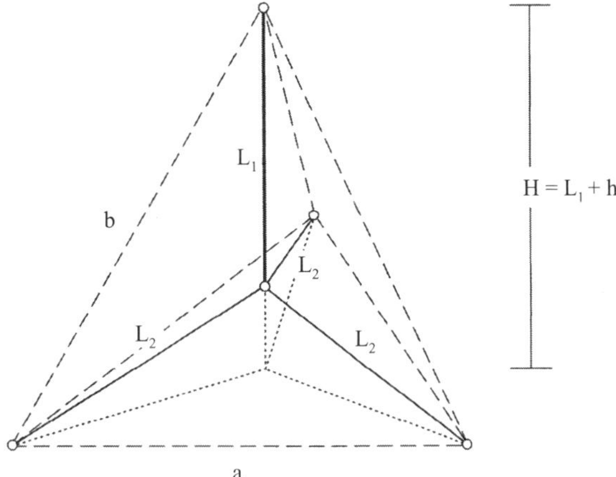

Figure P4.9

4.11. Here we see several implications of the Hess-Murray design of the bifurcation of one mother tube (D_1, L_1) into two identical daughter tubes (D_2, L_2) in Poiseuille flow, Eq. (4.11).

(a) Show that the wall shear stress (τ_1) along the D_1 tube is the same as the shear stress (τ_2) along the D_2 tube. In other words, show that the Hess-Murray design is a rule for how to distribute stresses uniformly in a load-bearing solid, in accordance with the other uniform-stresses configurations of Chapter 10.

(b) Rely on Eq. (4.18) and show that the longitudinal force experienced by the larger tube ($F_1 \sim \tau_1 D_1 L_1$) is $2^{2/3}$ times greater than the longitudinal force experienced by one of the smaller tubes ($F_2 \sim \tau_2 D_2 L_2$).

(c) Finally, show that the pressure drop along the large tube ($\Delta P \sim F_1/D_1^2$) is the same as the pressure drop along the small tube ($\Delta P \sim F_2/D_2^2$). In conclusion, the constructal design represented by Eqs. (4.11) and (4.18) achieves the equipartition of pressure drop between the mother and daughter tubes.

4.12. The stream with the mass flow rate $\dot{m}$ flows in the Poiseuille regime through a single tube of diameter D_1 and length L. The tube volume is specified (V). Determine the pressure drop along the tube (ΔP_1) as a function of $\dot{m}$, L and V. It is proposed to replace the single tube with two tubes in parallel. The mass flow rate through each tube is $\dot{m}/2$, and the tube diameter is D_2. The tube length is the same as before (L). The total volume of the two tubes is equal to the specified volume V. The flow is in the Poiseuille regime. Express the pressure drop along the two-tube assembly (ΔP_2) in terms of $\dot{m}$, L and V. Compare ΔP_2 with ΔP_1, and show that the single-tube design offers greater access to the stream $\dot{m}$ than the two-tube design.

5

CONFIGURATIONS FOR HEAT CONDUCTION

Tree-shaped architectures exist only if they are "alive," that is, if something flows through them to connect one point (source or sink) with the area or volume inhabited by the tree canopy. This "something" does not have to be a fluid, as in Chapter 4. It can also be energy and chemical species; goods; people; and the entire atmosphere, hydrosphere, and biosphere. The principle of how the flow architecture is generated is the same.

In this chapter we focus on configurations through which heat flows by conduction. We begin with tree-shaped designs of high-conductivity blades and fibers (fins, fingers) embedded in heat-generating packages. Solid dendrites become a necessity as the cooling and sensing of smart materials marches toward smaller scales and higher densities of functionality. In this time direction (the future) space becomes more and more expressive. During the current miniaturization revolution there will come a point where the little space that is available must be used by solid that "works" [1], not by void spaces that house channels and pumps for the fluid tree networks that we have developed until now.

5.1 TREES FOR COOLING A DISC-SHAPED BODY

Consider the problem [2] of cooling with one central heat sink (T_0) a disc-shaped body with uniform distribution of heat generation rate (Fig. 5.1). In this two-dimensional geometry the heat generation rates per unit volume and per unit area are, respectively, q''' and $q'' = q''' t$, where t is the thickness of the disc in the direction perpendicular to the plane of Fig. 5.1. The thermal conductivity of the body is k_0. The body temperature is above T_0 such that the generated heat current

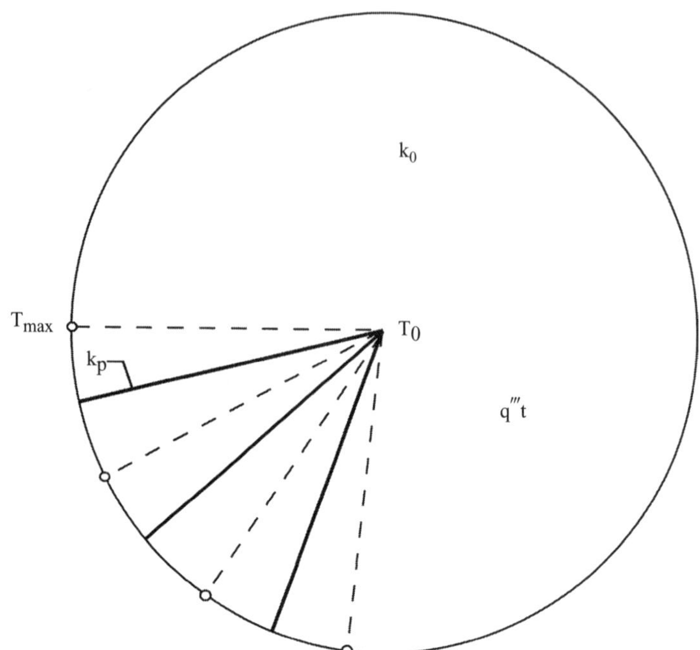

Figure 5.1 Disc-shaped body with uniform heat generation, central heat sink, and radial pattern of high-conductivity paths [2].

flows into the center. We seek ways to minimize the global thermal resistance, that is, the hot-spot temperature $T_{\max}$, which is likely to occur on the rim.

One way to decrease the global resistance is to insert through the k_0 medium an amount of material with considerably higher thermal conductivity, k_p. The composition of the two-material composite is fixed, and is accounted for by the volume fraction

$$\phi = \frac{\text{volume of } k_p \text{ material}}{\text{total volume}} \tag{5.1}$$

The challenge that comes with designing the composite is to optimize the geometry of the paths for heat transfer. The question is how to distribute the k_p paths on the k_0 background, how to shape the k_0 material that is allocated to one blade of k_p material, and how to connect k_p blades to each other and to the central heat sink.

The simplest architectural feature that can be anticipated intuitively is that the thin blades of k_p material (the "veins" of the heat-flow structure) must be arranged radially and equidistantly with one end touching the heat sink. In the simplest design, the k_p blades do not have branches and stretch radially all the way to the rim. As shown in Fig. 5.1, to each k_p blade corresponds a circular sector with adiabatic radial sides (dashed lines). More complex constructs with branches form the subject of sections 5.1.3 and 5.2.

5.1.1 Elemental Volume

A fundamental feature of constructal designs for volume-point and area-point paths is that the smallest volume (or area) scale of the flow structure is known and fixed. The smallest scale is invariant, that is, independent of the size and type of system into which the more and more complex flow structure may grow. In applications such as the cooling of small-scale electronics, the smallest volume scale is often dictated by manufacturing constraints, electromagnetic interference between neighboring components, and the space competition between electrical and thermal design functions.

The elemental volume of the flow structures constructed in this work is the sector of circle shown in Fig. 5.2. We assume that there are many radial k_p blades so that one sector is sufficiently slender to be approximated by an isosceles triangle of base *2H* and height *R*. The area of the elemental sector is fixed,

$$A_s = HR \tag{5.2}$$

while both *H* and *R* may vary. The chief unknown of the architecture is the shape of the sector: the aspect ratio of the element, *H*/*R*; the tip angle of the circular sector; or the number of elements (or blades) that should be assembled into a complete disc (Fig. 5.1).

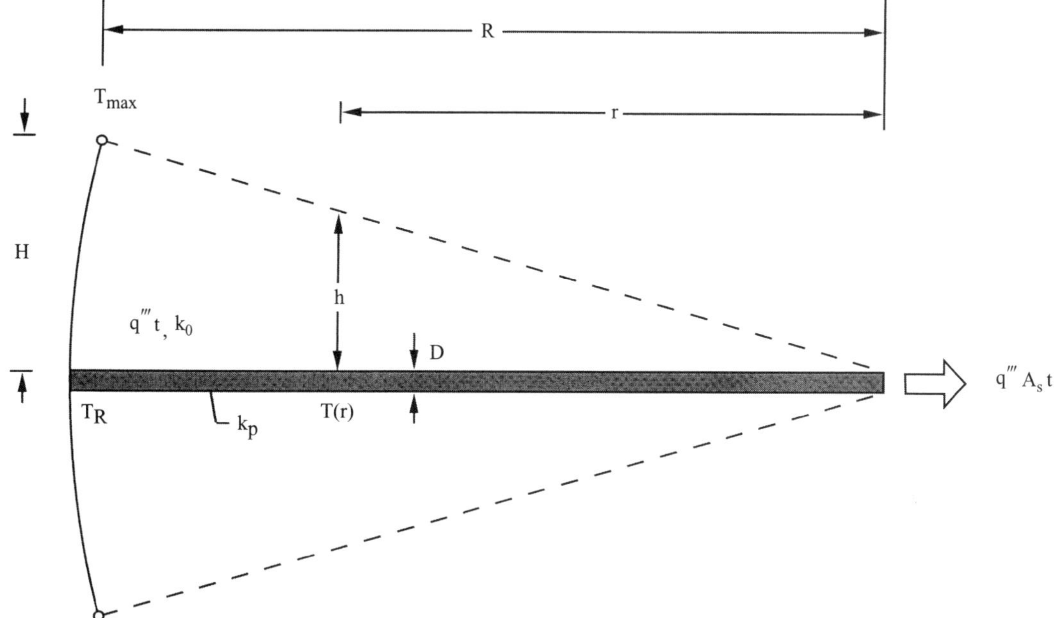

Figure 5.2 Elemental system: circular sector with high-conductivity blade on its center line, and the optimal blade shape [2].

The second constraint of the elemental architecture is the volume fraction of k_p material

$$\phi = \frac{DR}{HR} = \frac{D}{H} \tag{5.3}$$

where D is the thickness of the k_p blade. For the sake of simplicity in analysis, we assume that:

1. The thickness D is constant.
2. The k_p volume fraction is fixed and small, $\phi \ll 1$.
3. The ratio of thermal conductivities is fixed and large.

$$\tilde{k} = \frac{k_p}{k_0} \gg 1 \tag{5.4}$$

It is shown later in Eq. (5.18) that assumption 3 is consistent with the slenderness assumption $H/R \ll 1$ when the design of the element is optimized for minimum overall thermal resistance. Assumption 1 is relaxed in section 5.1.2.

Assumption 2 means that the k_p blade is sufficiently thin to be represented by the axis of symmetry in Fig. 5.2 (the T_R–T_0 line). Review the concept of svelteness, Eq. (1.1). In the drawing of Fig. 5.2, the total size (A_s) is fixed, and so is the "internal size" of the high-conductivity channel (ϕA_s). This means that in this two-dimensional configuration the external length scale is $A_s^{1/2}$, and the internal length scale is $(\phi A_s)^{1/2}$. These lengths are the same for all possible configurations of the type shown in Fig. 5.2. The svelteness of this class of configurations is Sv = (external length scale)/(internal length scale) = $\phi^{-1/2}$, where ϕ is the volume fraction occupied by the high-conductivity channel. The Sv value increases as the channel volume fraction ϕ decreases, that is, as the channels of the architecture become thinner and thinner, as in the direction sketched in Fig. 1.1.

The objective in the search for the geometry of the elemental sector is to minimize the global thermal resistance of the sector,

$$R_s = \frac{T_{\max} - T_0}{q''' \, A_s} \tag{5.5}$$

where $q'''A_s$ is the heat current generated over the entire sector. We evaluate this in a two-part analysis, by calculating $(T_{\max}-T_R)$ and (T_R-T_0). As shown in Ref. [1], this decoupling is possible because under assumptions (2) and (3) the thermal diffusion through the k_0 material is perpendicular to the k_p blade, and the conduction through the k_p material is oriented along the blade.

In the analysis of vertical k_0 conduction along the line $T_{\max}$–T_R, we note that there is heat generation at every point, the $T_{\max}$ end is insulated, and the T_R end (the contact with the k_p blade) is the heat sink. This means that the temperature distribution between the points T_R and $T_{\max}$ is parabolic and has zero slope at

T_{max}. Because of the parabolic distribution, the vertical temperature gradient felt at the T_R point is $2(T_{max} - T_R)/H$. We account for the conservation of energy in the string-shaped fiber T_R–T_{max} by equating the heat current generated in the fiber ($q'''H$) with the heat flux leaving the fiber through the T_R end, namely, $q'''H = 2k_0 (T_{max} - T_R)/H$. From this we deduce the temperature difference between the hot spot and the tip of the high-conductivity blade,

$$T_{max} - T_R = \frac{q''' H^2}{2\,k_0} \tag{5.6}$$

Alternatively, Eq. (5.6) can be derived by solving the equation for heat conduction in the k_0 material:

$$\frac{d^2T}{dy^2} + \frac{q'''}{k_0} = 0 \tag{5.7}$$

where y is oriented in vertical direction in Fig. 5.2. When $H \ll R$, the circular sector approaches an isosceles triangle with base $2H$ and height R. We focus on the left end of the k_p blade, and see the vertical line of height H, from $T = T_R$ where $y = 0$, to $T = T_{max}$ where $y = H$. We integrate Eq. (5.7) along this line,

$$\frac{dT}{dy} + \frac{q'''}{k_0} y + c_1 = 0 \tag{5.8}$$

and determine the constant of integration c_1 by using the top boundary condition:

$$\frac{dT}{dy} = 0 \quad \text{at } y = H \tag{5.9}$$

The result is $c_1 = -q'''y/k$, which substituted in Eq. (5.8) yields

$$\frac{dT}{dy} = \frac{q'''}{k_0} (H - y) \tag{5.10}$$

Next, we integrate Eq. (5.10),

$$T = \frac{q'''}{k_0} \left(Hy - \frac{y^2}{2} \right) + c_2 \tag{5.11}$$

and determine c_2 from the bottom boundary condition

$$T = T_R \text{ at } y = 0 \tag{5.12}$$

This yields $c_2 = T_R$. Finally, by applying Eq. (5.11) to the top end of the H-tall line we obtain the same result as in Eq. (5.6):

$$T_{max} = \frac{q'''}{k_0} \frac{H^2}{2} + T_R \tag{5.13}$$

In the analysis of conduction along the k_p blade, we note that the heat current that flows toward the center increases from $q = 0$ at $r = R$ to the total current $q = q'''tA$ at $r = 0$. The increase experienced by q at an intermediate position (r) is

$$-dq = 2hq''' \, t \, dr \tag{5.14}$$

where $hq'''t$ is the heat current collected over the vertical surface ht, where $h = (H/R)r$. The heat current is proportional to the local temperature gradient:

$$q = k_p \, Dt \, \frac{dT}{dr} \tag{5.15}$$

We determine $(T_{\max} - T_R)$ by eliminating q between Eqs. (5.14) and (5.15), integrating twice in r, and invoking the boundary conditions $dT/dr = 0$ at $r = R$, and $T = T_0$ at $r = 0$. The analytical steps are the same as in Eqs. (5.7) through (5.13). The result is

$$T_R - T_0 = \frac{2q''' R^2}{3 \, k_p \phi} \tag{5.16}$$

The global resistance of the elemental sector is obtained by adding Eqs. (5.6) and (5.16), and nondimensionalizing R_s by using the background conductivity k_0,

$$\tilde{R}_s = \frac{T_{\max} - T_0}{q''' A_s / k_0} = \frac{1}{2} \frac{H}{R} + \frac{2}{3 \, \tilde{k} \phi} \frac{R}{H} \tag{5.17}$$

This expression can be minimized with respect to the aspect ratio of the element, H/R, and the results are

$$\left(\frac{H}{R} \right)_{\text{opt}} = \frac{2}{(3 \tilde{k} \phi)^{1/2}} \ll 1 \tag{5.18}$$

$$\frac{(T_{\max} - T_0)_{\min}}{q''' A_s / k_0} = \frac{2}{(3 \tilde{k} \phi)^{1/2}} \tag{5.19}$$

Equation (5.18) shows that the assumed slenderness of the optimized sector is consistent with assumption (3): the ratio of conductivities $\tilde{k}$ must exceed $1/\phi$ in an order of magnitude sense.

The aspect ratio (5.18) fixes the tip angle of the sector. It also fixes the number of such sectors that fit in a complete disc arrangement, $N = 2\pi R/(2H)$, namely,

$$N = \frac{\pi}{2} \, (3 \tilde{k} \phi)^{1/2} \gg 1 \tag{5.20}$$

From this follows the size of the disc, that is, the size of the construct of N elements:

$$\pi R^2 = N A_s = \frac{\pi}{2} \, (3 \tilde{k} \phi)^{1/2} \, A_s \tag{5.21}$$

Larger discs (R) emerge when $\tilde{k}\phi$ increases. The corresponding thermal resistance of the entire disc is obtained by using NA_s instead of A_s in Eq. (5.19):

$$B_1 = \frac{(T_{\max} - T_0)_{\min}}{q''' N\, A_s / k_0} = \frac{4}{3\pi \tilde{k} \phi} \tag{5.22}$$

The radius of the disc-shaped construct is, from Eq. (5.21):

$$R = (A_s/2)^{1/2}\, (3\tilde{k}\phi)^{1/4} \tag{5.23}$$

The size of the construct (R) is not known a priori. It is the result of construction, that is, the assembly and the constraints that govern the smallest-scale element (A_s, ϕ, $\tilde{k}$). Aggregation, organization, growth and complexity are the result of geometric constraints—trying to fit together a number of smaller optimized parts. A key role is played by the product $\tilde{k}\phi$: larger values of this parameter mean more high-conductivity blades, more slender blades, and a disc-shaped construct with larger radius R. We will examine these properties in the discussion of Fig. 5.9.

5.1.2 Optimally Shaped Inserts

Further improvements in the performance of the construct can be made by relaxing some of the simplifying assumptions, increasing the number of degrees of freedom of the design, and optimizing the design with respect to the new degrees of freedom. One example is the constant-D assumption (1) on which the results of section 5.1.1 are based. Consider instead the general function $D(r)$ that is subjected to the same volume fraction constraint:

$$\phi = \frac{1}{HR} \int_0^R D\, dr \tag{5.24}$$

The choice of k_p blade profile $D(r)$ affects the global resistance through the part $T_R - T_0$. This relationship is obtained by eliminating q between Eqs. (5.14) and (5.15), integrating the resulting equation once in r, invoking $dT/dr = 0$ at $r = R$, and finally integrating from $r = 0$ to $r = R$:

$$T_R - T_0 = \frac{q''' H}{k_p R} \int_0^R \frac{R^2 - r^2}{D}\, dr \tag{5.25}$$

Variational calculus delivers the function $D(r)$ for which the integral (5.25) is minimum subject to the constraint (5.24) (cf. Appendix C and Problem 5.1):

$$D_{\text{opt}} = c\, (R^2 - r^2)^{1/2} \tag{5.26}$$

The constant factor c is calculated by substituting Eq. (5.26) into constraint (5.24):

$$D_{\text{opt}} = \frac{4}{\pi}\, \phi H \left[1 - \left(\frac{r}{R}\right)^2\right]^{1/2} \tag{5.27}$$

The optimal shape of the high-conductivity blade is such that the root is thicker and the tip is blunt (note $dD/dr = -\infty$ at $r = R$), as shown in Problem 5.1. These characteristics match those of other optimized shapes of elemental inserts, nerves and needles. They also agree with the features of natural dendrites [3].

The analytical steps of section 5.1.1 can be repeated with $D_{opt}(r)$ in place of $D = $ constant. Equations (5.18) through (5.23) are replaced, in order, by

$$\left(\frac{H}{R}\right)_{opt} = \frac{\pi/2}{(2\tilde{k}\phi)^{1/2}} \tag{5.28}$$

$$\frac{(T_{max} - T_0)_{min}}{q''' A_s/k_0} = \frac{\pi/2}{(2\tilde{k}\phi)^{1/2}} \tag{5.29}$$

$$N = 2\,(2\tilde{k}\phi)^{1/2} \gg 1 \tag{5.30}$$

$$\frac{(T_{max} - T_0)_{min}}{q''' N A_s/k_0} = \frac{\pi}{8\tilde{k}\phi} \tag{5.31}$$

$$R = (2\,A_s/\pi)^{1/2}\,(2\tilde{k}\phi)^{1/4} \tag{5.32}$$

The decrease in the thermal resistance of the sector is evaluated by dividing Eq. (5.29) by Eq. (5.19): the result is $(\pi/4)(3/2)^{1/2} = 0.96$, or a 4 percent decrease. The resistance decrease registered by the entire disc assembly is estimated by dividing Eq. (5.31) by Eq. (5.22): the result is $3\pi^2/32 = 0.925$, which indicates a 7.5 percent reduction in global resistance.

5.1.3 One Branching Level

Natural and engineered flow structures with fixed smallest scales become more complex as they fill larger spaces. In this section we consider the case of a network of high-conductivity blades that cools a disc-shaped system with uniform heat generation. We begin with a structure with just one level of complexity above that of the radial pattern of Fig. 5.1: one k_p blade stretches radially to the distance L_0 away from the central heat sink, and continues with a number (n) of branches (or, better, tributaries) that reach the rim. In place of the elemental sector of Fig. 5.2, we now analyze the sector with one stem (L_0, D_0) and n tributaries (L_1, D_1) shown in Fig. 5.3. The goal is to assemble with minimum flow resistance a number (N) of branched sectors into a complete disc, as shown in Fig. 5.4.

The following analysis is based on the main result of section 5.1.1: one isolated sector with a single central blade has an optimal aspect ratio when its size and amount of k_p material are fixed, and when its global resistance to heat transfer must be minimized. To use this property, in Fig. 5.3 we view the construct as a combination of n small sectors of aspect ratio H_1/L_1, plus a central sector of aspect ratio H_0/L_0. The length L_1 is the distance from the hot spot (T_{max}) to the point of confluence (T_c).

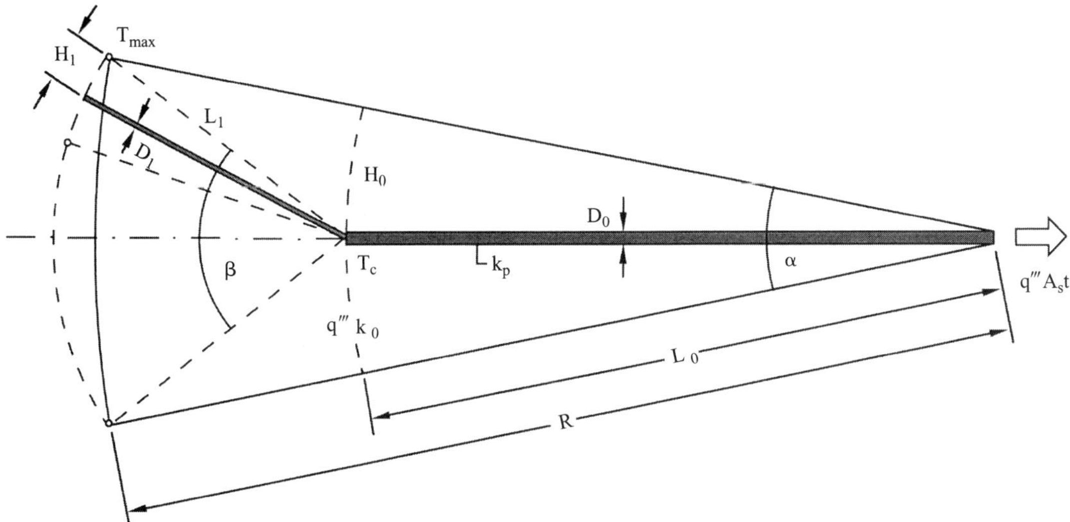

Figure 5.3 One level of branching: one central high-conductivity path (L_0, D_0) with n_1 smaller paths (L_1, D_1) as tributaries [2].

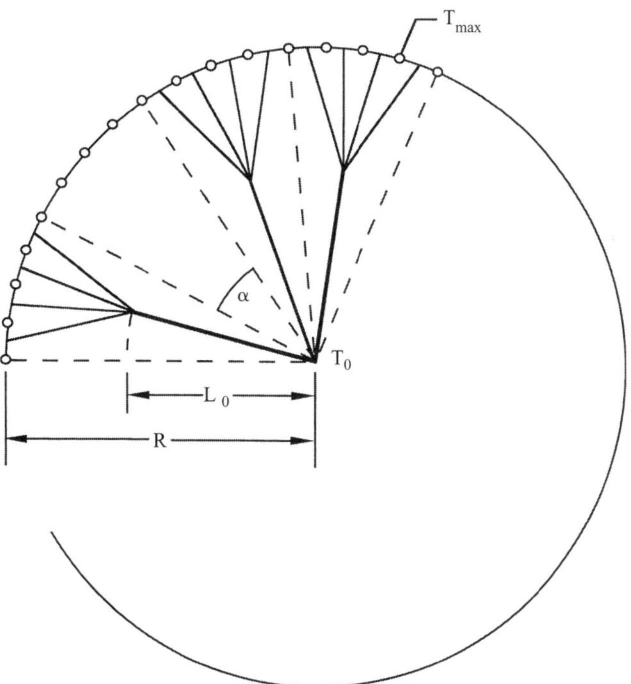

Figure 5.4 Disc-shaped body cooled by a structure consisting of several of the branched sectors shown in Fig. 5.3 [2].

The number of tributaries (n) is an important feature on which the complexity and manufacturability of the structure rests. This feature is particularly important in view of earlier results that show pairing ($n_1 = 2$) is the best option for laminar fully developed flow. Is pairing or bifurcation also recommended here?

We assume that each peripheral sector of radius L_1 is slender enough such that the optimized shape of Eq. (5.18) is correct:

$$\frac{H_1}{L_1} = \frac{2}{(3\tilde{k}\,\phi_1)^{1/2}} \qquad \phi_1 = \frac{D_1}{H_1} \qquad A_1 = H_1 L_1 \tag{5.33}$$

The same cannot be said about the central sector of radius L_0, because, unlike in section 5.1.1, the T_c end of its high-conductivity blade is not insulated. For this reason the aspect ratio of the (H_0, L_0) sector is free to vary, and so is the aspect ratio of the entire sector (radius R) of Fig. 5.3. For the (H_0, L_0) sector we write only

$$\frac{H_0}{L_0} \cong \frac{\alpha}{2} \qquad A_0 = H_0 L_0 \tag{5.34}$$

where the tip angle α is a function of the assumed n and R values. The number of A_1 elements that fit along the perimeter of the R disc is $N = 2\pi R/(2H_1)$. The number of branched sectors of angle α is N/n. The angle α is then

$$\alpha = \frac{2\pi\, n}{N} = \frac{2^{3/2} n}{\tilde{R}\,(3\tilde{k}\phi_1)^{1/4}} \tag{5.35}$$

where the dimensionless radius $\tilde{R}$ is based on $A_1^{1/2}$ as the specified (fixed) elemental length scale of the structure:

$$\tilde{R} = \frac{R}{A_1^{1/2}} \tag{5.36}$$

The area A_0 that is allocated to the stem (L_0, D_0) is $(\alpha/2)L_0^2$, where $L_0 \cong R - L_1$. After some algebra, we obtain

$$A_0 \cong \frac{2^{1/2}\, n\, \tilde{R}\, A_1}{(3\tilde{k}\phi_1)^{1/4}} \left[1 - \frac{(3\tilde{k}\phi_1)^{1/4}}{2^{1/2}\,\tilde{R}}\right]^2 \tag{5.37}$$

Note that Eq. (5.37) is approximate because not all the A_1 elements have the length L_1. This approximation becomes more accurate as the angles β and α decrease.

Next, we estimate the overall thermal resistance $(T_{\max} - T_0)/(q'''A)$. We do this in two parts. For the overall temperature difference sustained by the corner sector of radius L_1 we deduce from Eq. (5.19) that

$$T_{\max} - T_c = \frac{2q'''A_1}{k_0\,(3\tilde{k}\phi_1)^{1/2}} \tag{5.38}$$

The temperature drop from the junction (T_c) to the central link (T_0) requires a new analysis, because unlike Eqs. (5.14) through (5.16) the T_c tip of the D_0 blade receives the heat current collected by the n peripheral sectors of size A_1,

$$q''' \, t n A_1 = k_p \, D_0 \, t \left(\frac{dT}{dt} \right)_{r = L_0} \tag{5.39}$$

As in Fig. 5.2, the radial position in the central sector of Fig. 5.3 is measured from the center ($r = 0$) to the T_c junction ($r = L_0$). The temperature distribution along the D_0 blade is obtained by starting with the equivalent of Eqs. (5.14) and (5.15),

$$-\,dq = 2 \left(\frac{H_0}{L_0} \, r \right) q''' \, t \, dr \tag{5.40}$$

$$q = k_p \, D_0 t \, \frac{dT}{dr} \tag{5.41}$$

eliminating q, integrating twice in r, and invoking the tip condition (5.39) and $T = T_0$ at $r = 0$. In the resulting $T(r)$ expression we set $T(L_0) = T_c$, and obtain

$$T_c - T_0 = \frac{q''' \, L_0}{k_p \, D_0} \left(\frac{2}{3} \, A_0 + n \, A_1 \right) \tag{5.42}$$

Adding Eqs. (5.38) and (5.42), and noting again that $L_0 \cong R - L_1$, we obtain the temperature difference

$$T_{\max} - T_0 \;\cong\; \frac{2 q''' A_1}{k_0 \, (3 \tilde{k} \phi_1)^{1/2}} + \frac{q''' \, (R - L_1)}{k_p \, D_0} \left(\frac{2}{3} \, A_0 + n \, A_1 \right) \tag{5.43}$$

This quantity can be nondimensionalized as

$$\begin{aligned} \tilde{T} = \frac{T_{\max} - T_0}{q''' \, A_1 / k_0} &= \frac{2}{(3 \tilde{k} \phi_1)^{1/2}} + \frac{(3 \tilde{k} \phi_1)^{1/4}}{2^{1/2} \tilde{k} \phi_1 \tilde{D}} \\ &\times \left[\tilde{R} - \frac{(3 \tilde{k} \phi_1)^{1/4}}{2^{1/2}} \right] \times \left\{ \frac{2^{3/2} n \tilde{R}}{3 \, (3 \tilde{k} \phi_1)^{1/4}} \left[1 - \frac{(3 \tilde{k} \phi_1)^{1/4}}{2^{1/2} \, \tilde{R}} \right]^2 + n \right\} \end{aligned} \tag{5.44}$$

where $\tilde{D}$ is the ratio of k_p blade thicknesses,

$$\tilde{D} = \frac{D_0}{D_1} \tag{5.45}$$

The temperature difference $\tilde{T}$ depends on geometry (n, $\tilde{D}$, $\tilde{R}$), and on the presence of k_p material ($\tilde{k}$, ϕ_1). The total amount of k_p material in the R disc is represented by the cross-sectional area

$$A_p = N D_1 L_1 + \frac{N}{n} \, D_0 L_0 \tag{5.46}$$

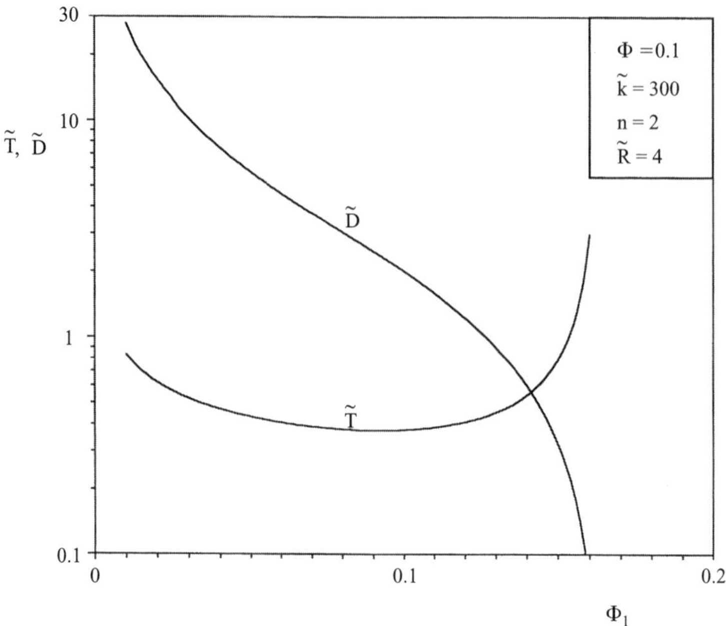

Figure 5.5 The minimization of the overall temperature difference in the disc-shaped construct of Fig. 5.4 [2].

or by the fraction that A_p occupies in the entire disc (πR^2):

$$\phi = \frac{A_p}{\pi R^2} = \frac{(3\tilde{k}\phi_1)^{1/4}\phi_1}{2^{1/2}\tilde{R}} + \frac{\tilde{D}\phi_1}{n\tilde{R}}\left[\tilde{R} - \frac{(3\tilde{k}\phi_1)^{1/4}}{2^{1/2}}\right] \tag{5.47}$$

The ϕ fraction is fixed (e.g., $\phi = 0.1$), and provides a relation between $\tilde{k}$, ϕ_1, $\tilde{D}$, n and $\tilde{R}$. The $\tilde{k}$ ratio is fixed by the choice of materials (e.g., $\tilde{k} = 300$). We expect a tradeoff between ϕ_1 and $\tilde{D}$, which will represent the optimal allocation of k_p material to the D_0 and D_1 blades.

To start with, we set $n = 2$ and $\tilde{R} = 4$, and minimized $\tilde{T}$ by varying ϕ_1 and $\tilde{D}$, where ϕ_1 and $\tilde{D}$ are related by Eq. (5.47). Figure 5.5 confirms that $\tilde{T}$ has a minimum with respect to how the k_p material is allocated. The resulting features of the optimal configuration ($\phi_{1,\text{opt}}$, $\tilde{D}_{\text{opt}}$, $\tilde{T}_{\min}$) are reported in Fig. 5.6. This figure also shows how the optimum responds to changes in the size of the construct, $\tilde{R}$. The optimal allocation of high-conductivity material is almost insensitive to changes in $\tilde{R}$. The temperature difference $\tilde{T}_{\min}$ is almost proportional to $\tilde{R}$.

The numerical work summarized in Fig. 5.6 was repeated for other numbers of elemental branches, $n = 4, 6, \ldots$ The key feature of these results is that the $\phi_{1,\text{opt}}$ and $\tilde{T}_{\min}$ curves, which in Fig. 5.6 were plotted for $n = 2$, do not shift as n increases. Figure 5.7 shows that the $\tilde{D}_{\text{opt}}$ curve rises as n increases. A larger $\tilde{D}_{\text{opt}}$ means an elemental insert (D_1) that is thinner relative to the stem (D_0). The

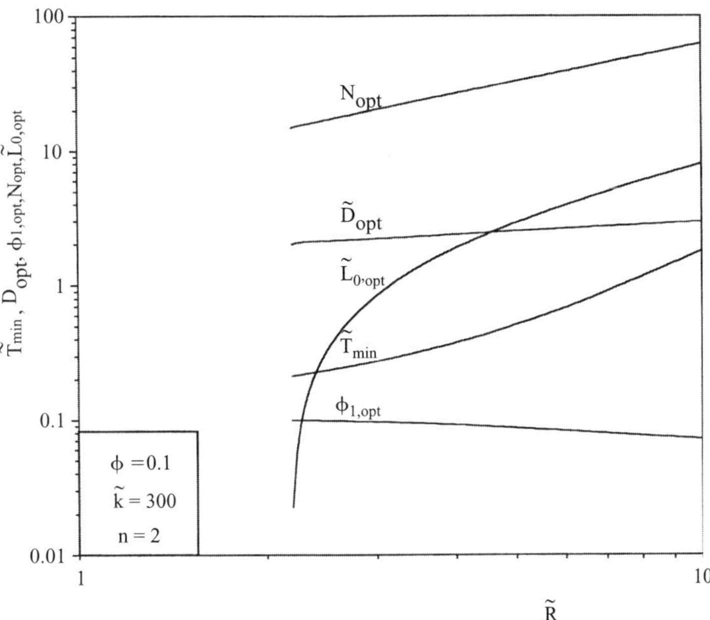

Figure 5.6 The effect of the disc size ($\tilde{R}$) on the optimal configuration determined in Fig. 5.5 [2].

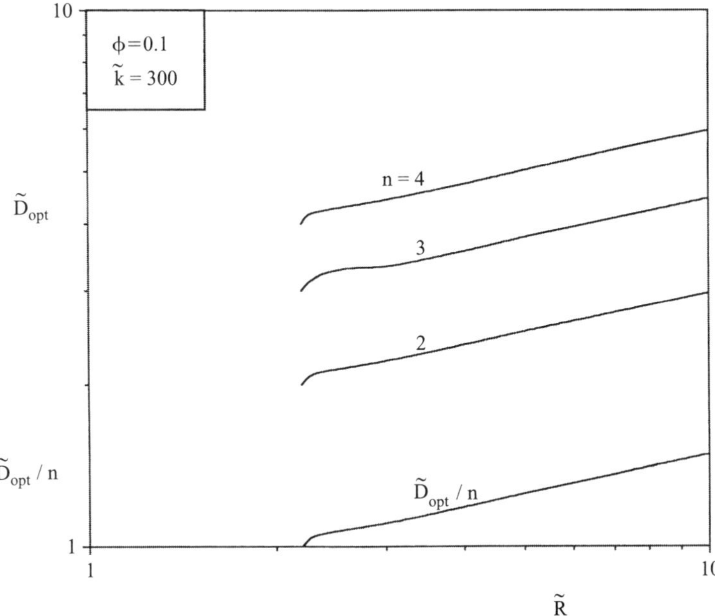

Figure 5.7 The effect of the number of elemental branches (n) on the optimized ratio of the k_p insert thicknesses ($\tilde{D}_{opt}$) [2].

numerical values plotted in Fig. 5.7 indicate that $\tilde{D}_{opt}/n$ is almost constant, which means that the optimum is characterized by $D_0 \approx nD_1$. This approximation is more exact when $\tilde{R}$ is smaller. In that limit the cross-sectional area of the k_p inserts is conserved at the junction between the stem and the branches. When $\tilde{R}$ is larger, the stem cross-section is larger than the combined cross-section of the branches, $D_0 > nD_1$.

Figure 5.6 also shows the required number of peripheral elements,

$$N_{opt} = 2^{-1/2} \pi \tilde{R} \, (3 \tilde{k} \, \phi_{1,opt})^{1/4} \tag{5.48}$$

This number increases as $\tilde{R}$ increases, and is independent of n. The corresponding central length scale of the optimized branched pattern, $\tilde{L}_{opt} = L_{0,opt}/A_1^{1/2}$, is obtained by writing $L_0 = R - L_1$ and using the first of Eqs. (5.33):

$$\tilde{L}_{opt} = \tilde{R} - 2^{-1/2} \, (3 \, \tilde{k} \, \phi_{1,opt})^{1/4} \tag{5.49}$$

This length increases with $\tilde{R}$, and is independent of n (Fig. 5.6). Note that L_0 shrinks to zero (hence $\phi_{1,opt} = \phi$) when $\tilde{R}$ drops to 2.18: this critical $\tilde{R}$ value corresponds to the optimized radial pattern without branches (Fig. 5.1), as we will see again in Fig. 5.8. The disappearance of the branched pattern at $\tilde{R} = 2.18$ is why all the curves in Fig. 5.6 vanish below $\tilde{R} = 2.18$.

In each of the cases optimized in Figs. 5.5 through 5.7 the elemental area A_1 and the disc size R were fixed. This means that the minimization of $\tilde{T}$ is equivalent to the minimization of the thermal resistance of the entire disc:

$$B_2 = \frac{T_{max} - T_0}{q''' \pi R^2 / k_0} = \frac{\tilde{T}}{\pi \tilde{R}^2} \tag{5.50}$$

Because $\tilde{T}_{min}$ is almost proportional to $\tilde{R}$, the minimized resistance $[B_{2,min} = \tilde{T}_{min}/(\pi \tilde{R}^2)]$ decreases almost as $\tilde{R}^{-1}$ as $\tilde{R}$ increases.

The minimized resistance $B_{2,min}$ depends on $\tilde{R}$, $\tilde{k}$ and ϕ. It does not depend on n because $\tilde{T}_{min}$ does not depend on n. If we compare the resistance $B_{2,min}$ with the corresponding resistance (B_1) of the disc with radial inserts (Fig. 5.1), we can determine the recommended "transition" from radial patterns to branched patterns. We make this comparison based on the same elemental size ($A_s = A_1$) and the same amounts and properties of conducting materials ($\tilde{k}$, ϕ). The dimensionless radius of the radial design of section 5.1.1 is [cf. Eq. (5.23)],

$$\tilde{R} = 2^{-1/2} \, (3 \tilde{k} \phi)^{1/4} \tag{5.51}$$

In Fig. 5.1 this radius is fixed when $\tilde{k}$ and ϕ are fixed. In the branched design $\tilde{R}$ can be increased freely. This is why in Fig. 5.8 one case ($\tilde{k}$, ϕ) is represented by one radial-design point $B_1(\tilde{R})$ and one branched-design curve $B_{2,min}(\tilde{R})$. The figure shows that when $\tilde{R}$ exceeds 2.18 (in the case $\phi = 0.1$, $\tilde{k} = 300$), the global

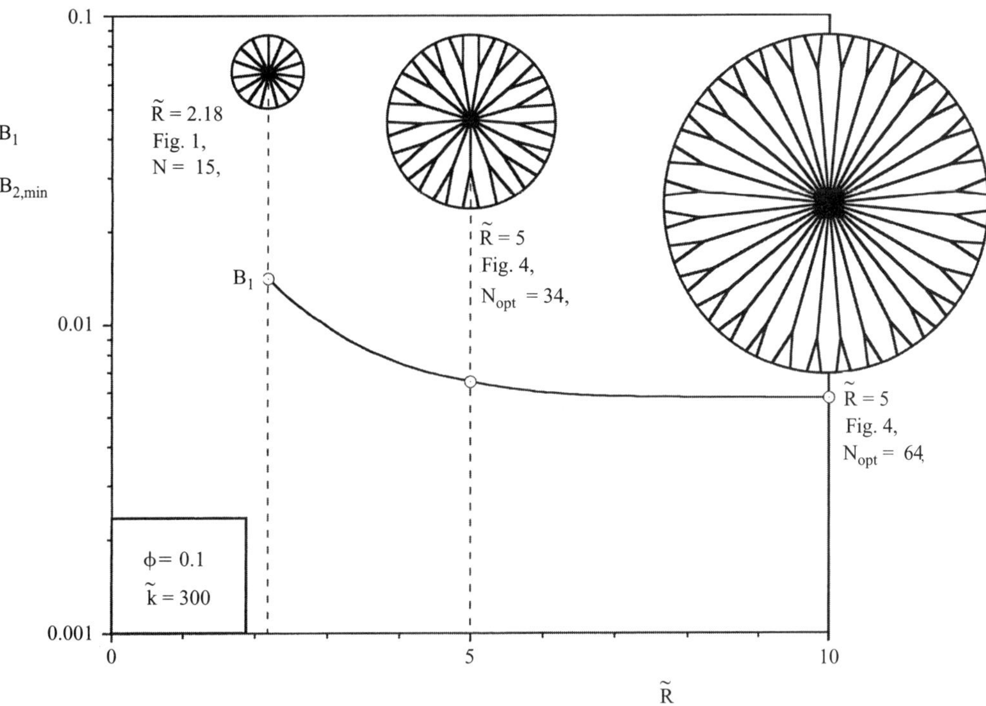

Figure 5.8 The overall thermal resistances of the optimized radial pattern (B_1) and the optimized branched patterns ($B_{2,\min}$) [2].

resistance is smaller when the high-conductivity material is distributed according to the optimized branched pattern.

We repeated the calculations of Fig. 5.8 for other values of $\tilde{k}$ and ϕ, in the range $30 \le \tilde{k} \le 1000$ and $0.01 \le \phi \le 0.1$. We found the same qualitative behavior as in Fig. 5.8, and the additional feature that the effect of $\tilde{k}$ and ϕ on $B_{2,\min}$ is exercised through the product $\tilde{k}\phi$. Figure 5.9 is a condensation of all these results. Lower global resistances are achieved by decreasing $\tilde{k}\phi$, and by switching from the optimized radial pattern (Fig. 5.1) to the optimized branched pattern (Fig. 5.4) when $\tilde{R}$ can be made greater than the $\tilde{R}$ value of the optimized radial pattern. Figure 5.8 illustrates this transition in the case of $\tilde{k} = 300$ and $\phi = 0.1$, for which the radial pattern has $\tilde{R} = 2.18$ and 15 elemental sectors, and the branched pattern has 34 peripheral elements for $\tilde{R} = 5$ and 64 peripheral elements for $\tilde{R} = 10$.

In section 5.1 we illustrated a hierarchical strategy for developing the optimal flow structure for cooling with high-conductivity inserts a disc-shaped body that generates heat at every point. The strategy consists of optimization of heat-flow performance at every scale, followed by the assembly of optimized systems into larger systems. Every geometric detail of the heat-flow structure is derived from

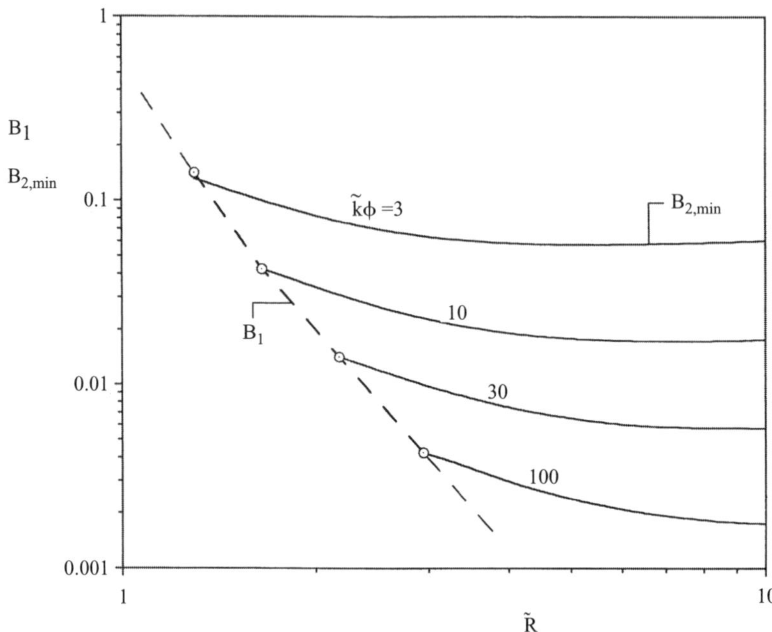

Figure 5.9 The effect of $\tilde{k}$ and ϕ on the overall thermal resistances of the optimized radial and branched patterns [2].

principle. The flow structure is the construction (configuration) of the two-material conductive body.

The analytical work in this section was simple enough to allow the construction to proceed to two levels of assembly, that is, disc-shaped flow structures with one level of branching (Fig. 5.4). Beyond this stage, en route to cooling larger and larger heat-generating bodies, the method calls for optimizing structures with two levels of branching. This step would begin with optimizing the architecture of a sector-shaped construct such as Fig. 5.3, in which each of the sectors that make up the "fan" of angle β has the one-branching structure optimized based on Fig. 5.3. Earlier work on the development of growing tree-shaped conductive trees in rectangular coordinates [3] suggests that, when plotted on Fig. 5.9, the global resistance curve of the two-branchings structure would intersect the existing $B_{2,\min}(\tilde{R})$ curve for one-branching designs. This would mean that there should be a second transition, at a higher $\tilde{R}$ value, where the choice of optimized patterns switches from one-branching designs to two-branching designs. The first transition, or the "onset" of the first branching occurred at $\tilde{R} = 2.18$ for $\phi = 0.1$ and $\tilde{k} = 300$. There is an analogy between this pattern of successive transitions and the discrete transitions toward more complex flows in Bénard convection and turbulence, which is why the occurrence of the latter has also be reasoned on the basis of the constructal law [3, 4].

Example 5.1 A cylindrical surface of low temperature is connected to a thin pipe of high temperature, which is located on the axis of the cylinder. Two-dimensional blades of high thermal conductivity connect the hot center with N spots on the cold perimeter of the cross-section. The rest of the space is evacuated, and radiation heat transfer can be neglected. The heat transfer is purely by conduction along the blades.

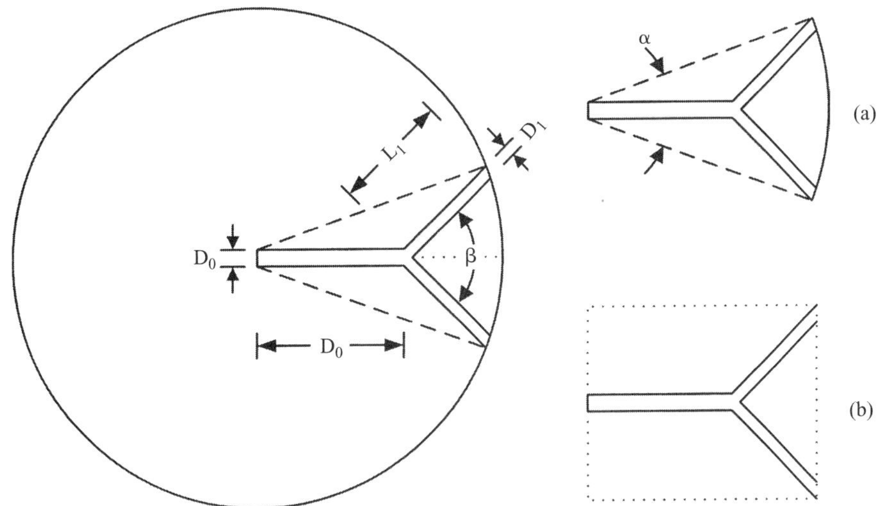

Figure E5.1

It is proposed to arrange the blades into Y-shaped constructs in order to minimize the thermal resistance between the center and the perimeter (see Fig. E5.1). The cylinder radius (R) and the total cross-sectional area of the high-conductivity blades (A_c) are fixed. The dimensions L_0, D_0, L_1 and D_1 vary.

(i) Determine the optimal ratios D_0/D_1 and L_0/L_1 such that the global thermal resistance has the lowest possible value. In other words, identify the best of all possible Y-shaped drawings that can be fitted inside one sector of angle $2\pi/N$.

(ii) Consider next the Y-shaped structure inside a rectangular domain, such that the locations of the three extremities of the Y are fixed. The left end of the L_0 trunk is at high temperature, and the right ends of the L_1 branches are at low temperature. Show that the solution obtained in part (i) is also valid for part (ii), that is, valid for domains of any shape, provided that the ends of the Y construct are fixed.

We begin with the observation that the volume fraction (ϕ) occupied by high-conductivity solid is fixed. Assume that there is only one level of pairing

($n = 1$). Each Y-shaped construct fits inside a sector of angle:

$$\alpha = \frac{2\pi}{N}, \quad \text{fixed} \tag{a}$$

The mother blade has the length L_0 and thickness D_0, while the two daughter blades have the same features, L_1 and D_1. Because the sector is fixed, we consider only one Y, which connects the center with two peripheral points. The thermal conductivity is constant. The thermal resistance of one Y construct is proportional to the geometric group:

$$R_t = \frac{L_0}{D_0} + \frac{1}{2}\frac{L_1}{D_1} \tag{b}$$

in which the second term is the resistance of two L_1/D_1 resistances in parallel. The dimensions L_0, D_0, L_1 and D_1 may vary.

There are several constraints. The first is that the total amount of high-conductivity material is fixed. We express this as the total cross-sectional area of the three blades:

$$A_c = L_0 D_0 + 2L_1 D_1 \tag{c}$$

The next two constraints are the geometric relations

$$L_0 + L_1 \cos\beta = R\cos\frac{\pi}{N} \tag{d}$$

$$L_1 \sin\beta = R\sin\frac{\pi}{N} \tag{e}$$

where R is the disc radius, and β is the sharp angle between the L_0 and L_1 directions. The angle β varies.

In sum, we have one function to minimize [Eq. (a)], five variables, and three constraints [Eqs. (b) through (d)]. This means that the flow configuration has two degrees of freedom, that is, we can change the drawing in two ways. One is by varying the angle β, which determines the length L_0 relative to its maximum (R). The other is by varying the ratio of thicknesses D_0/D_1, which we optimize first.

According to the method of Lagrange multipliers (see Appendix B), to find the extremum of R_{th} subject to the A_c constraint is equivalent to finding the extremum of the aggregate function $\Phi = R_{th} + \lambda A_c$, where λ is the Lagrange multiplier:

$$\Phi = \frac{L_0}{D_0} + \frac{1}{2}\frac{L_1}{D_1} + \lambda(L_0 D_0 + 2L_1 D_1) \tag{f}$$

Both D_0 and D_1 vary. We solve the system

$$\frac{\partial\Phi}{\partial D_0} = 0 \qquad \frac{\partial\Phi}{\partial D_1} = 0 \tag{g, h}$$

and by eliminating λ between Eqs. (g) and (h), we find the optimal ratio (which is independent of L_0, L_1 and β):

$$\left(\frac{D_0}{D_1}\right)_{\text{opt}} = 2 \qquad \text{(i)}$$

Substituting this ratio into Eqs. (a) and (b) we obtain

$$R_{t,\,\min} = \frac{L_0 + L_1}{2D_1} \qquad \text{(j)}$$

$$A_c = 2D_1\,(L_0 + L_1) \qquad \text{(k)}$$

Multiplying Eqs. (j) and (k) we conclude the minimization of R_{th} subject to A_c:

$$R_{t,\,\min} = \frac{1}{A_c}\ (L_0 + L_1)^2 \qquad \text{(l)}$$

The second R_t minimization opportunity is with respect to β. According to Eq. (l), this means to minimize $(L_0 + L_1)$ by selecting β. The shortest $L_0 + L_1$ linkage is when L_0 and L_1 are collinear (radial), which means that $L_0 = 0$. The best of the Y constructs is the one shaped as a V, with just two radial blades L_1.

We can reach the same conclusion formally by minimizing $(L_0 + L_1)$ with respect to β. To do this, we use Eqs. (d) and (e) to obtain $L_0\,(\beta\,)$ and $L_1\,(\beta\,)$, which substituted into $(L_0 + L_1)$ yield

$$\frac{L_0 + L_1}{R} = \cos\frac{\pi}{N} + \sin\frac{\pi}{N}\cdot\frac{1-\cos\beta}{\sin\beta} \qquad \text{(m)}$$

To minimize this expression with respect to β means to minimize the function

$$f = \frac{1-\cos\beta}{\sin\beta} \qquad \text{(n)}$$

This function decreases monotonically as β decreases from $\pi/2$ to 0. In conclusion, the best β is the smallest β that the drawing allows, and that β value is π/N (again, the drawing with $L_0 = 0$).

Note that the whole design ($D_0/D_1 = 2$, $L_0 = 0$) was obtained before we had Eq. (m), that is, without having to take into account the shape of the domain on which the Y construct is installed. The reason is that the positions of the three ends of the Y were fixed. They can be fixed on any plane domain, for example, in the rectangle circumscribed to the Y. Inside any two-dimensional domain with proposed bifurcations of conducting blades, the best Y is the V, that is, the fork with no stem, just two branches.

5.2 CONDUCTION TREES WITH LOOPS

In this section, we examine a special class of tree-shaped structures for conduction cooling: trees with loops [5]. This direction of inquiry is bio-inspired, because "trees with loops" is the basic architecture for leaf venation. We examine how loops and complexity affect the optimized architecture and its maximized global performance and robustness.

5.2.1 One Loop Size, One Branching Level

Consider the two-dimensional structure with loops of one size shown in Fig. 5.10. The structure consists of blades of high conductivity (k_p) embedded in a heat-generating disc of radius R_0 and lower conductivity k_0. There are two blade sizes, (D_0, L_0) and (D_1, L_1), where the L's are lengths and the D's are thicknesses.

There are N points in which the structure touches the rim. The structure is an assembly of trees with only one level of bifurcation, where each L_0 blade is continued by two L_1 blades. The assembly is such that each of the N points on the rim is touched by two of the branched trees. This last feature generates the loops, which are quadrilateral. In this configuration the number of blades that touch the center (n_0) is equal to N.

Heat is generated steadily in the k_0 material, and the volumetric heat generation rate is uniform, q'''. The heat current integrated over the entire disc ($q''' \pi R_0^2$) is collected by the k_p blades and channeled to the center of the disc, which serves as heat sink of temperature $T_{\min}$. The rim is insulated. The highest temperature $T_{\max}$ occurs on the rim, in the spots where the radius that is collinear with L_0 intersects the rim. The global thermal resistance of this heat-flow structure is expressed in dimensionless form by

$$\tilde{T}_{\max} = \frac{T_{\max} - T_{\min}}{q''' R_0^2 / k_0} \tag{5.52}$$

The objective is to morph the flow structure in every possible way so that its global resistance $\tilde{T}_{\max}$ is minimized. There are several constraints to consider. One

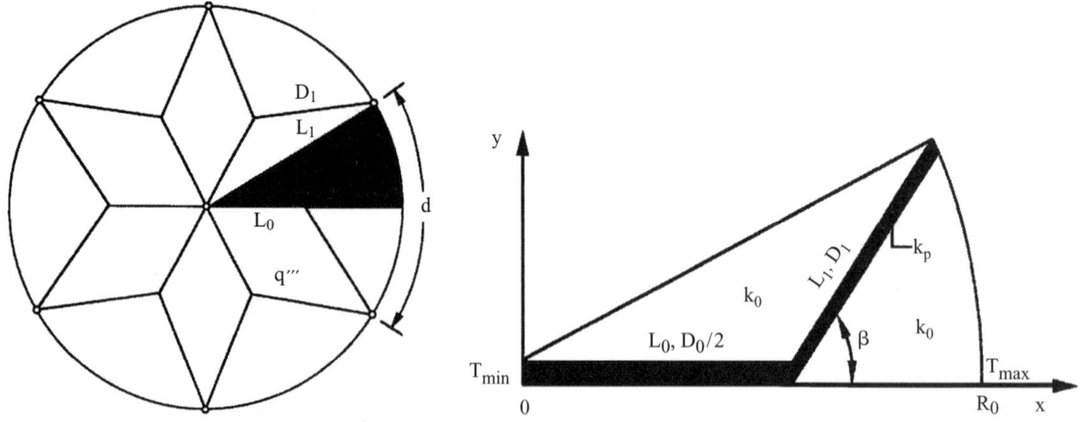

Figure 5.10 Disc-shaped body with uniform heat generation, central heat sink, high-conductivity blades, loops of one size, and one level of bifurcation or pairing [5].

is the disc radius R_0. If, in addition, we fix the distance d between two consecutive k_p points on the rim, then the number of peripheral points is fixed ($N = 2\pi R_0/d$), and so is the area of the sector shown in the lower part of Fig. 5.10:

$$A_0 = \frac{\pi R_0^2}{2N} \tag{5.53}$$

Symmetry allows us to focus on heat conduction in the sector of area A_0. The perimeter of this area is insulated, with the exception of the heat sink (T_{min}). A second important constraint is the total amount of k_p material, which is represented by the total cross-sectional area of the blades present in A_0:

$$A_p = L_0 \frac{D_0}{2} + L_1 D_1 \tag{5.54}$$

Alternatively, the A_p and A_0 constraints can be combined to define the fixed volume fraction occupied by k_0 material, $\phi = A_p/A_0$.

The structure of Fig. 5.10 was optimized numerically by simulating the temperature field in a large number of configurations, calculating the global resistance $\tilde{T}_{max}$, and seeking morphological changes that induce decreases in $\tilde{T}_{max}$. The details are given in Ref. [5]. The numerical results are nondimensionalized in accordance with Ref. [2], namely, $\tilde{R} = R_0/A_1^{1/2}$ and $\tilde{T}_{max} = (T_{max} - T_{min})/(q''' A_1^{1/2}/k_0)$, where the elemental area A_1 is shown later in the upper part of Fig. 5.17.

Figure 5.11 shows the first stage of geometric optimization. The global resistance has a minimum with respect to the bifurcation radius $\tilde{L}_0$, when the other degree of freedom (D_1/D_0) and constraints (N, k_p/k_0, ϕ) are specified. The lower part of Fig. 5.11 shows how $\tilde{L}_1$ and β vary as $\tilde{L}_0$ changes.

The morphing illustrated in Fig. 5.11 was performed for many values of D_1/D_0, and the results are summarized in Fig. 5.12. There are two results, the optimized radial blade length $\tilde{L}_0$, which is practically independent of D_1/D_0, and the minimized global resistance $(\tilde{T}_{max})_{min}$, which has a shallow minimum with respect to D_1/D_0. We denote this smallest of all $(\tilde{T}_{max})_{min}$ values by $(\tilde{T}_{max})_{mm}$. The configuration that corresponds to $(\tilde{T}_{max})_{mm}$ has been drawn to scale in the inset of Fig. 5.12.

To summarize, the conductive flow structure can be optimized with respect to two free geometric features, $\tilde{L}_0$ and D_1/D_0. In Figs. 5.13 through 5.15 we explored the sensitivity of the optimized configuration to changes in the constraints. Figures 5.13 and 5.14 show that the radius of bifurcation $(\tilde{L}_0)_{opt}$ is relatively insensitive to changes in both ϕ and k_p/k_0. We may take $\tilde{L}_0 \cong 0.5$ as a representative value for all the optimized configurations that have $N = 6$ points on the rim. Figures 5.13 and 5.14 also show that the ratio of blade thicknesses $(D_1/D_0)_{opt}$ increases weakly as either ϕ or k_p/k_0 increase. This trend is reasonable because in this direction the peripheral branches become more effective as conduits for heat flow, hence their increasing thickness. In the same direction, the minimized global thermal resistance

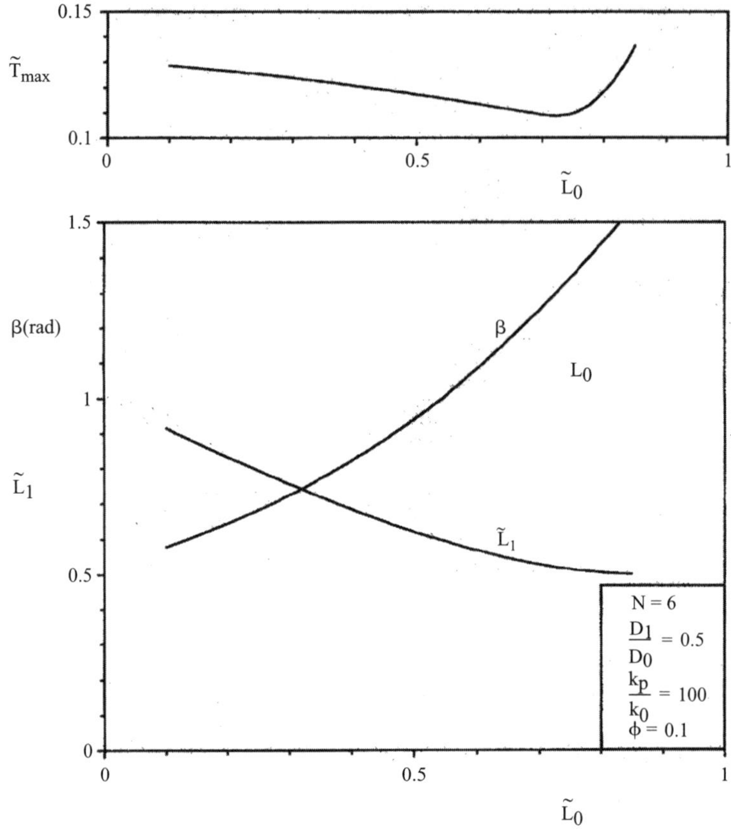

Figure 5.11 The minimization of the overall thermal resistance with respect to the central blade length $\tilde{L}_0$ [5].

decreases. In the ϕ and k_p/k_0 ranges covered by Figs. 5.13 and 5.14, the minimized global conductance varies roughly as

$$(T_{\max})_{\rm mm} = 0.25\phi^{-0.65}(k_p/k_0)^{-0.56} \tag{5.55}$$

In Figure 5.15 we show how the optimized structure and its global performance respond to changes in global size (R_0), when the spacing between k_p points on the rim (d) is fixed. This is equivalent to changing the number of k_p points on the rim, or the number of radial blades N. The disc radius increases in proportion with N, namely, $R_0 = Nd/(2\pi)$. One important conclusion is that the dimensionless minimized global thermal resistance decreases significantly with the size of the system until N is approximately equal to 20, beyond which $(\tilde{T}_{\max})_{\min}$ is less sensitive to changes in the value of N.

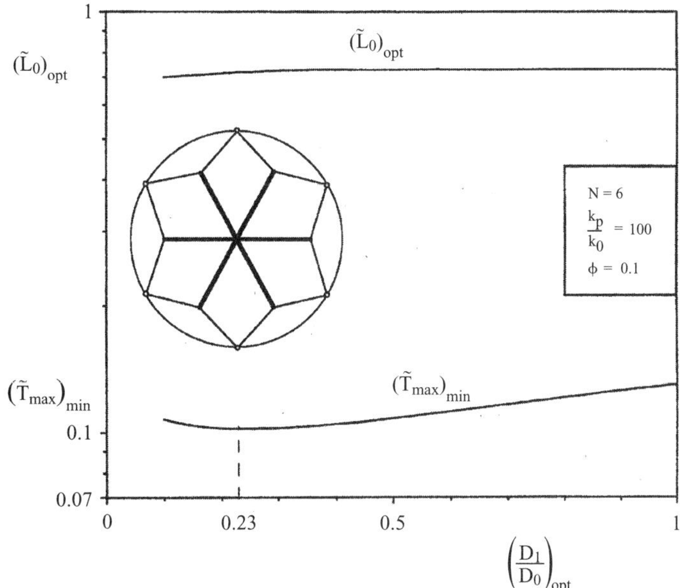

Figure 5.12 The minimization with respect to D_1/D_0 of the global resistance already minimized with respect to $\tilde{L}_0$ [5].

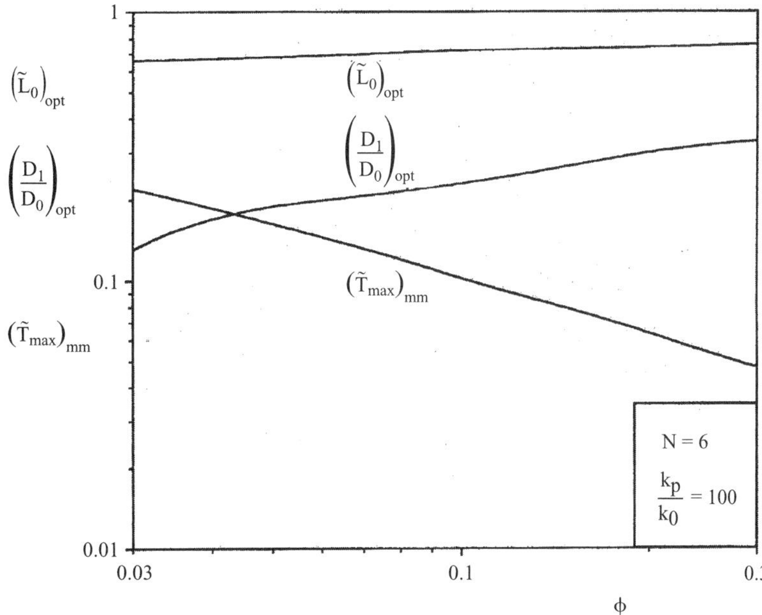

Figure 5.13 The effect of the high-conductivity volume fraction ϕ on the optimized structure of Fig. 5.10 [5].

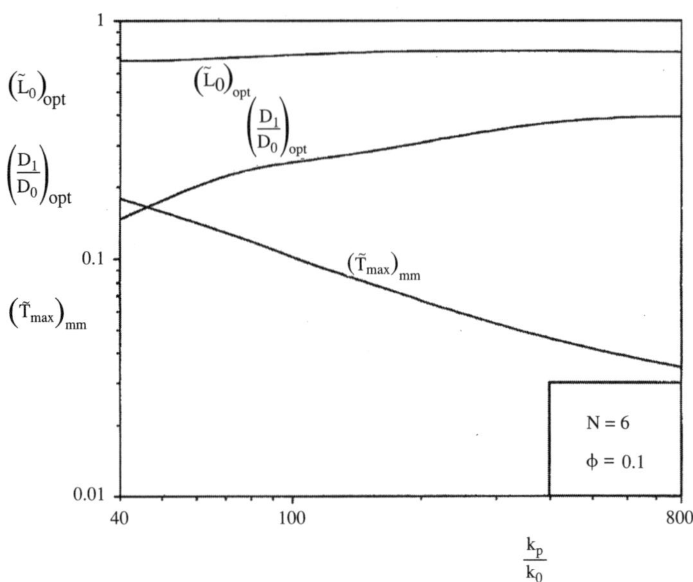

Figure 5.14 The effect of the ratio of conductivities k_p/k_0 on the optimized structure of Fig. 5.10 [5].

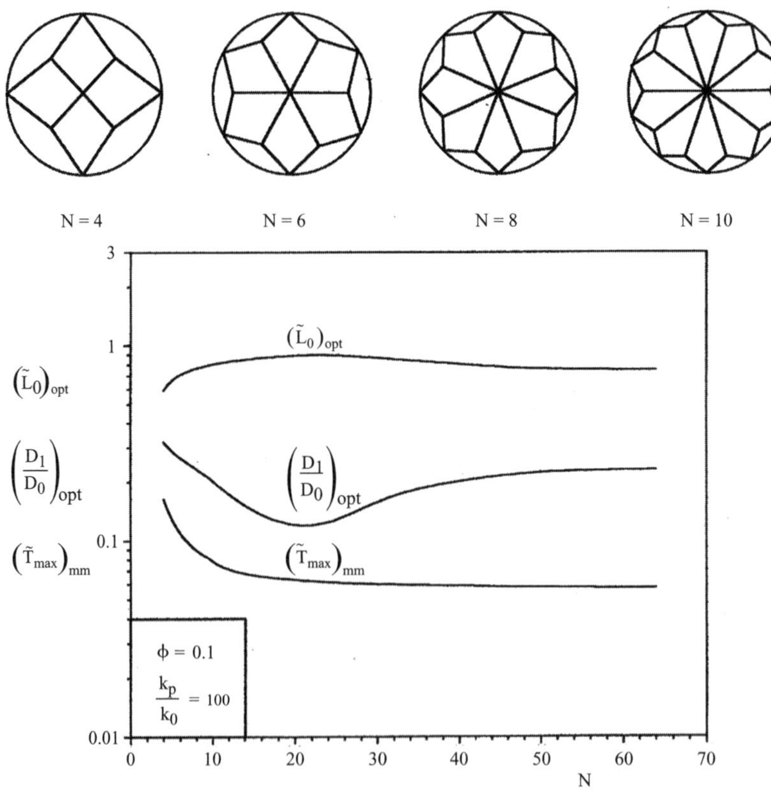

Figure 5.15 The effect of the size of the disc on the optimized structure of Fig. 5.10 [5].

5.2.2 Radial, One-Bifurcation and One-Loop Designs

The radial design of section 5.1.1 is repeated here in Fig. 5.16. The structure consists of N blades of high conductivity (k_p) embedded in a heat generating disc of radius R_0 and lower conductivity k_0. The blades have the length R_0 and thickness D. The global thermal resistance of this flow structure is expressed in dimensionless form by Eq. (5.52). The area of the sector shown in Fig. 5.16 is given by Eq. (5.53). The area occupied by the k_p material is

$$A_p = R_0 D \tag{5.56}$$

The fixed volume fraction occupied by the k_p material is $\phi = A_p/A_0$.

The design with one level of bifurcation is shown in Fig. 5.17: one k_p blade stretches radially to the distance L_0 from the central heat sink, and continues with two tributary branches that reach the rim. Each tributary branch reaches the rim in the middle of the sector shown in Fig. 5.17.

Figure 5.18 shows that the global thermal resistance decreases when the number of radial blades increases. The design with one level of bifurcation performs better than the design with one loop size. Both designs perform better than the radial design when they are compared based on the same number of central blades N. However, if they are compared with the radial structure with $2N$ central blades, the radial design performs better than the bifurcated design when $N \leq 10$. Said another way, in the bifurcated design $L_{0,\text{opt}}$ vanishes when $N \leq 10$ (Fig. 5.19), and the optimal bifurcated structure becomes a radial structure with $2N$ blades. This is why in Fig. 5.18 the curve for bifurcated structures merges with the curve for radial structures with $2N$ blades at $N \sim 10$.

Figure 5.19 shows that the optimal length of the central blade $(\tilde{L}_0)_{\text{opt}}$ in the design with one level of bifurcation is smaller than $(\tilde{L}_0)_{\text{opt}}$ of the design with one

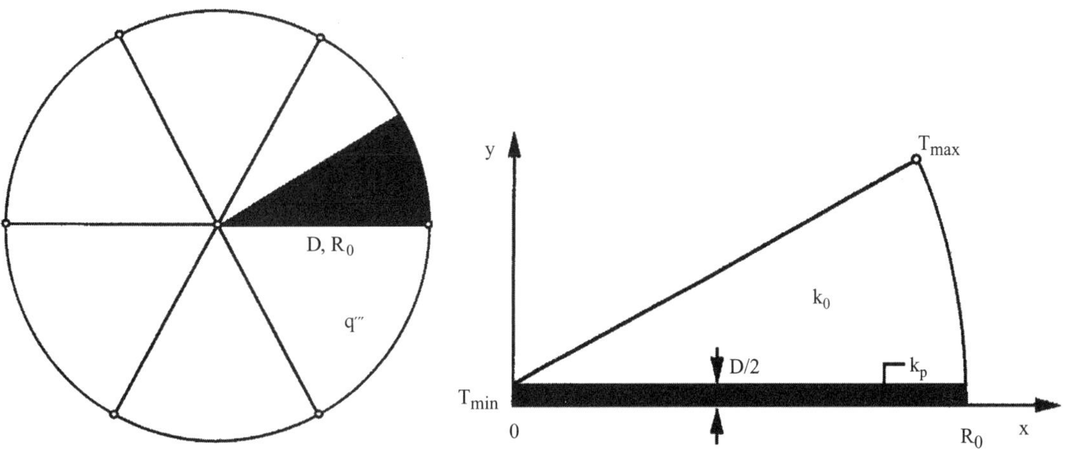

Figure 5.16 Design with radial blades [5].

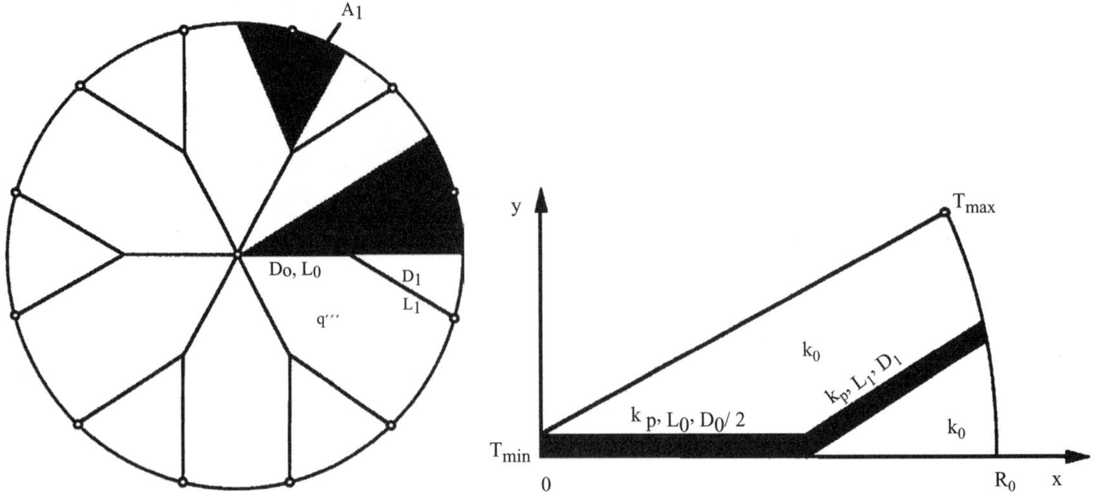

Figure 5.17 Design with one level of bifurcation [5].

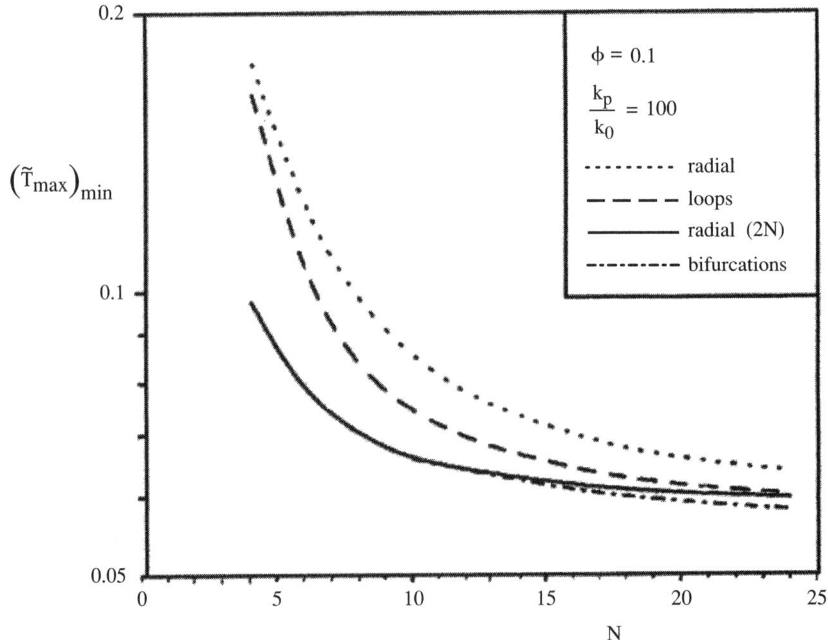

Figure 5.18 The minimized overall thermal resistance as function of the number of central blades [5].

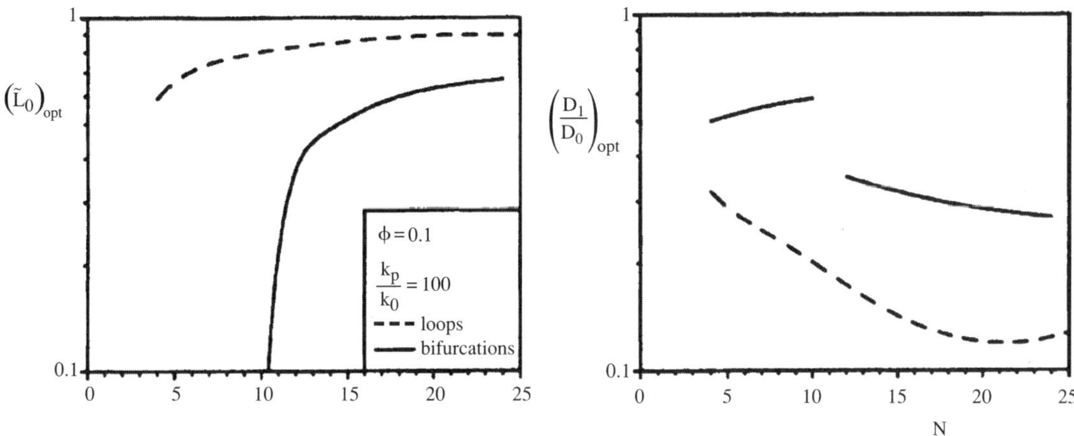

Figure 5.19 The effect of the number of central blades N on the optimal central blade length $(\tilde{L}_0)_{opt}$ and optimal ratio $(D_1/D_0)_{opt}$. [5].

loop size. In the design with one level of bifurcation, the length $(\tilde{L}_0)_{opt}$ tends to zero. This means that when $N \leq 10$ the best design with one level of bifurcation is a radial design with $2N$ central blades.

The right side of Fig. 5.19 shows the effect of N on the ratio $(D_1/D_0)_{opt}$ in the design with one level of bifurcation and one loop size. The $(D_1/D_0)_{opt}$ ratio of the design with loops is smaller. See again the discontinuity at $N \cong 10$, which is due to the disappearance of $L_{0,opt}$ when $N \leq 10$. Figure 5.20 shows examples of the optimized structures documented in Figs. 5.18 and 5.19. The smallest length scale (d) is the same in every design, while N (or R) varies. When N (or R) is fixed, the global performance increases from left to right.

5.2.3 Two Loop Sizes, Two Branching Levels

Considerably more complicated than the configuration optimized until now is the configuration of Fig. 5.21, where there are two characteristic loops, both quadrilateral. One loop has two radial blades L_0 and two branches L_1, while the other has two branches L_1 and two branches L_2. The respective thicknesses of the blades are D_0, D_1 and D_2. The number of k_p points on the rim (N) is the same as the number of radial blades. The disc radius is R_0, and the dimensionless lengths of the design are $(\tilde{L}_i, \tilde{D}_i) = (L_i, D_i)/R_0$. The relative amount of high-conductivity material is

$$\phi = \frac{N}{\pi R_0^2}(D_0 L_0 + 2D_1 L_1 + 2D_2 L_2) \tag{5.57}$$

When ϕ, N and k_p/k_0 are fixed, the design has four degrees of freedom, two length ratios, and two thickness ratios, for example, $\tilde{L}_0$, $\tilde{L}_2$, D_1/D_0 and D_2/D_0. In

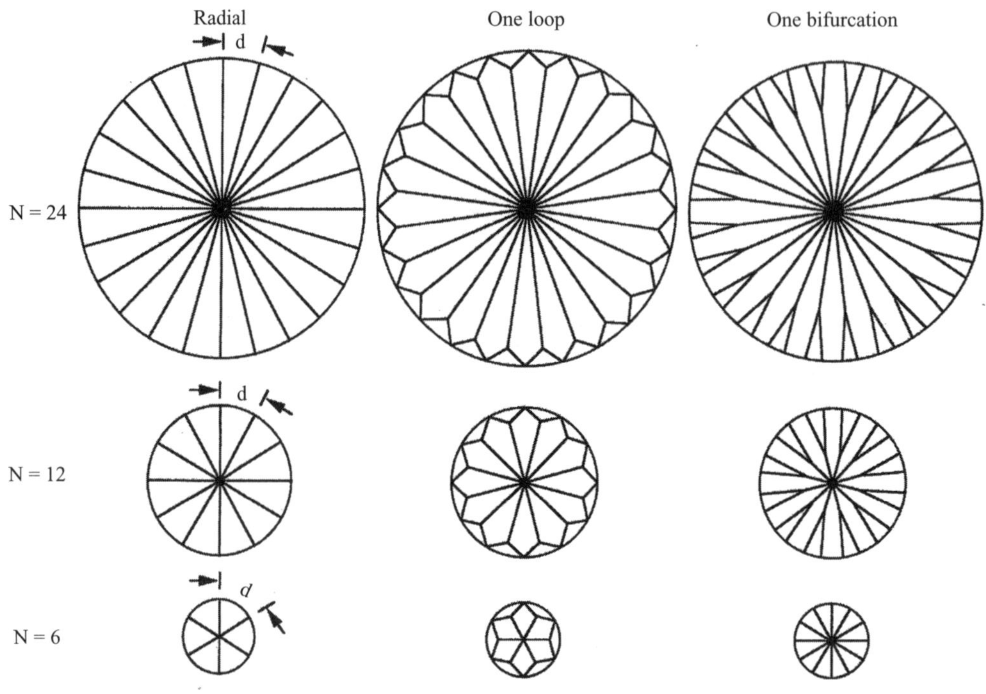

Figure 5.20 Optimized structures with the same smallest scale (d) and increasing size (N or R).

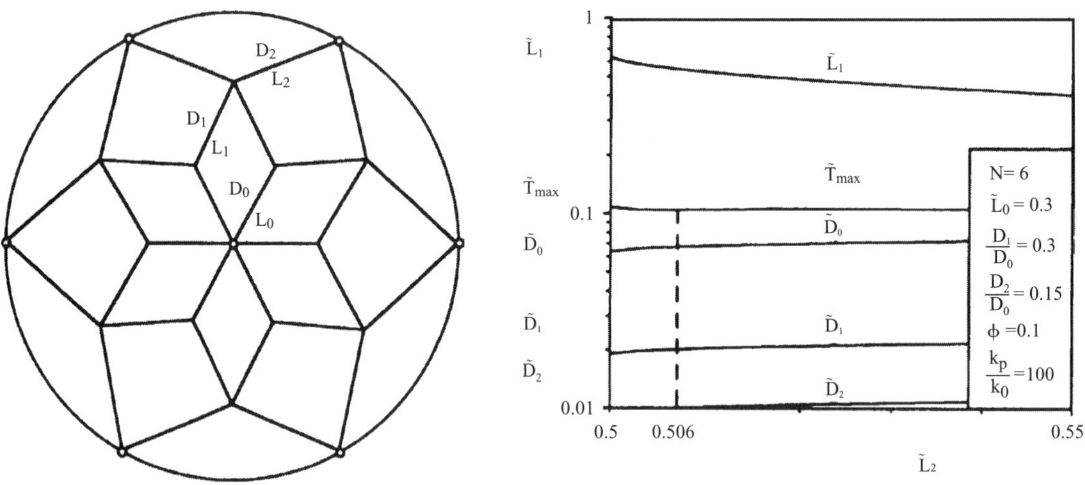

Figure 5.21 Design with loops of two sizes and two levels of bifurcation, and optimization by varying the peripheral length $\tilde{L}_2$ [5].

Rocha et al. [5] we simulated steady conduction with uniform heat generation in a very large number of configurations, and we identified the changes in geometry that cause decreases in the hot-spot temperature $\tilde{T}_{max}$, regardless of its location. We conducted this search hierarchically, in four nested levels.

At the innermost level, we fixed $\tilde{L}_0$, D_1/D_0 and D_2/D_0, and varied $\tilde{L}_2$ until $\tilde{T}_{max}$ reached its lowest value. An example of a search at this level is shown in Fig. 5.21. Noteworthy is the shallowness of the minimum exhibited by $\tilde{T}_{max}$. This means that the accurate selection of the most peripheral length $\tilde{L}_2$ is not a critical design decision. This quality (robustness) of complex flow structures was encountered before, for example in the performance of conduction trees with two or more levels of pairing or bifurcation [3]. It was found that a complex flow structure in which not every feature is optimized performs at nearly the same level as the fully optimized structure.

At the next level, the optimum identified at the first level $[(\tilde{T}_{max})_{min}]$ was generated for a new class of configurations in which D_2/D_0 varied (in addition to $\tilde{L}_2$), while $\tilde{L}_0$ and D_1/D_0 remained fixed. In this way, we found how $(\tilde{T}_{max})_{min}$ varies with D_1/D_0, and we were able to determine the design with the smallest of all the $(\tilde{T}_{max})_{min}$ values, which is labeled $(\tilde{T}_{max})_{mm}$. The optimal ratio D_2/D_0 determined in Fig. 5.22 is called $(D_2/D_0)_o$.

The procedure used in Figs. 5.21 and 5.22 was repeated by keeping $\tilde{L}_0$ constant and varying the ratio D_1/D_0. The results are summarized in Fig. 5.23, which shows that there is an optimal ratio of blade thicknesses, $(D_1/D_0)_o$. The corresponding

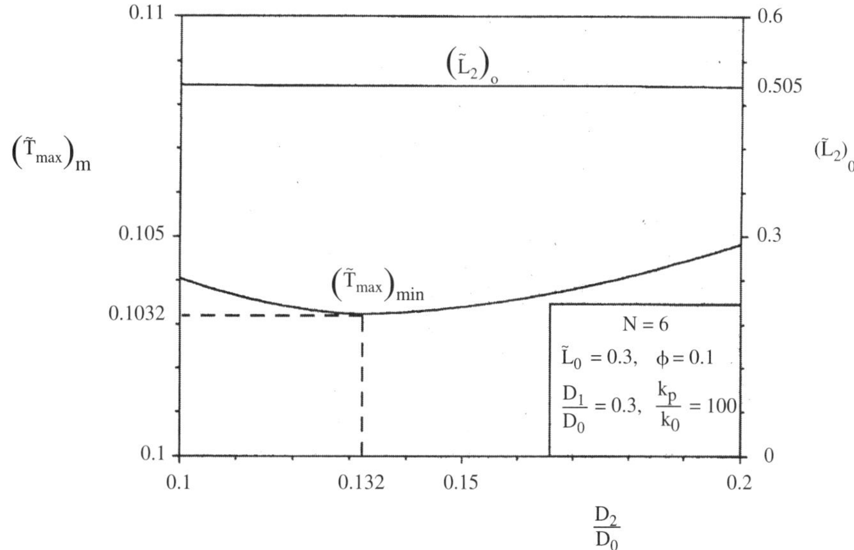

Figure 5.22 The minimization with respect to D_2/D_0 of the global thermal resistance already minimized with respect to the peripheral length $\tilde{L}_2$[5].

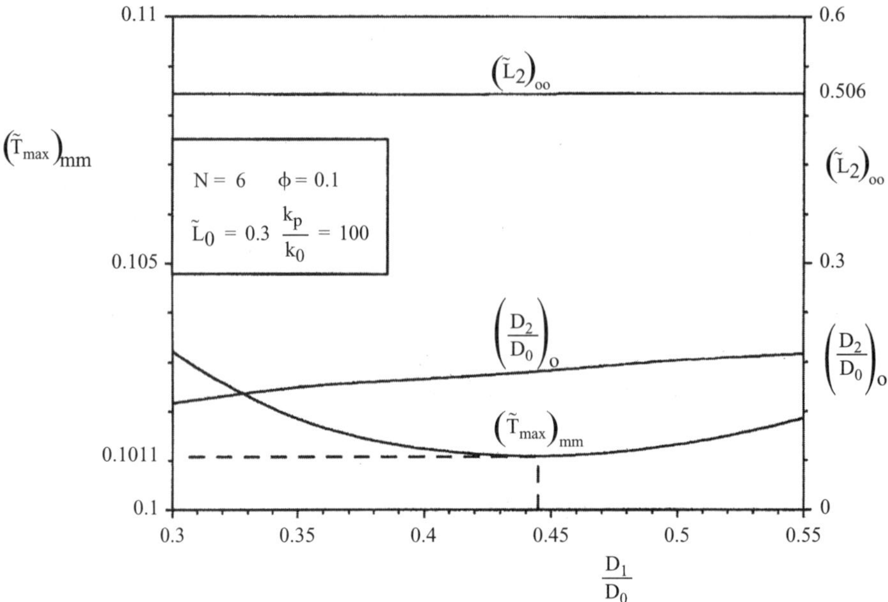

Figure 5.23 The minimization with respect to D_1/D_0 of the global thermal resistance already minimized with respect to $\tilde{L}_2$ and D_2/D_0 [5].

values at this optimum are the minimized global resistance $(\tilde{T}_{max})_{mmm}$, the optimal ratio $(D_2/D_0)_{oo}$, and the optimal length $(\tilde{L}_2)_{ooo}$.

In the final optimization stage (the outermost loop), we exploited the fourth degree of freedom, $\tilde{L}_0$. Figure 5.24 shows that $(\tilde{T}_{max})_{mmm}$ decreases monotonically as $\tilde{L}_0$ vanishes. In this limit, the optimal two-loop structure becomes the same as the optimal one-loop structure studied in section 5.2.1. Figure 5.25 shows four examples of the structures optimized in Fig. 5.24.

We repeated the procedure of Figs. 5.21 through 5.24 by assuming N values other than 6 for the number of central blades, namely $N = 9, 12, 24, 32, 48$, and 64. Figure 5.26 is another example ($N = 9$) of the structures optimized according to the procedure used in Figs. 5.21 through 5.24. The results show that the optimized one-loop design always performs better than the optimized two-loop design. In other words, the message of Fig. 5.24 is general: the optimal L_0 is zero, and the optimized two-loop size structure becomes an optimized one-loop size structure.

In conclusion, if we complicate the design by using two loop sizes (Fig. 5.21), the optimized structure evolves toward one with loops of only one size (Fig. 5.24). This evolution, however, is very gentle in that the global performance and optimized geometric features do not change much from one configuration to the next. There are many configurations with two loop sizes that perform nearly as well as the best configuration. This finding is another illustration of how complex flow structures exhibit robustness, such that in the vicinity of the *equilibrium flow structure* [6, 7] reside many configurations that have nearly the same global performance level.

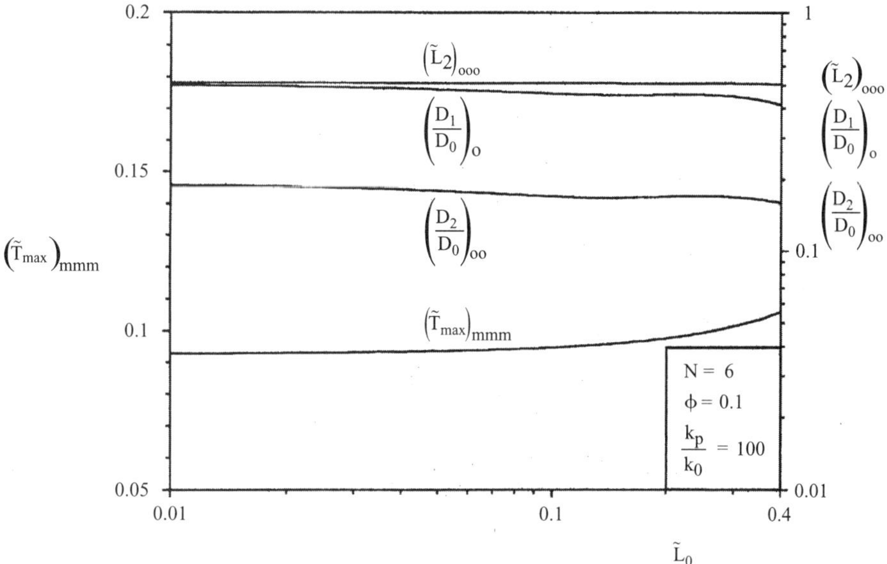

Figure 5.24 The effect of the central blade length $\tilde{L}_0$ on the global thermal resistance already minimized with respect to $\tilde{L}_2$, D_2/D_0, and D_1/D_0 [5].

By comparing Fig. 5.21 with Fig. 5.22 we see that conduction trees with two loop sizes are marginally inferior to trees with only one loop size: then why should designers be interested in tree structures with multiscale loops? Robustness is one reason, because more scales mean more degrees of freedom en route to the optimized multiloop structure. Robustness also means that the global performance of the cooling structure does not change much if the structure is damaged locally. Another advantage of the structure with more loop sizes is that, locally, it spreads the temperature more uniformly even though its peak temperature is slightly higher than in the corresponding structure with only one loop size.

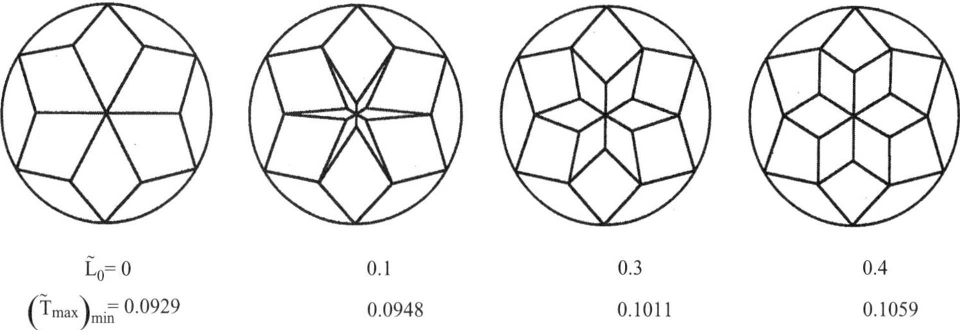

Figure 5.25 Examples of the optimized structures documented in Fig. 5.24 ($N = 6$) [5].

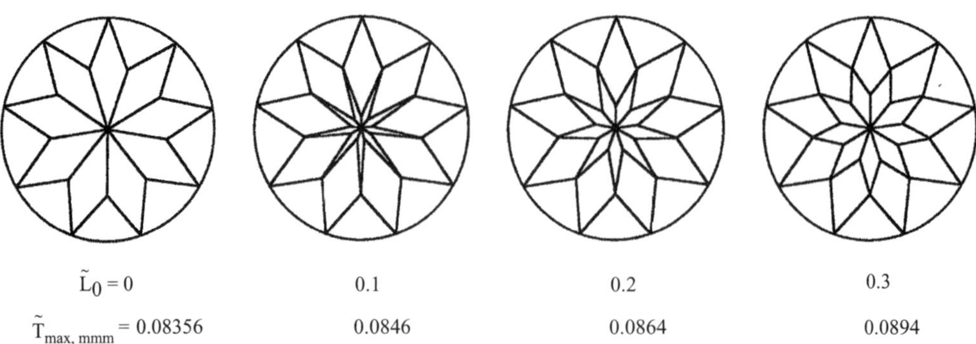

Figure 5.26 Examples of optimized structures with $N = 9$ [5].

5.3 TREES AT MICRO AND NANOSCALES

Earlier in this chapter, thermal conductivities were considered constant during the optimization of geometry. A new aspect of the present problem is that the thermal conductivity at very small length scales is no longer a constant, but varies with the shape and dimensions of the system [8]. The modeling of the transport properties of nanostructures is, in itself, an active field of research, which is motivated in part by thermoelectric properties and applications of small-scale structures. Modeling small-scale transport is a complex task that requires quantum mechanics and statistical physics. For example, in order to calculate the thermal conductivity, it is necessary to consider the different types of energy carriers (electrons and phonons), the effect of temperature, the scattering of the energy carriers, the presence of impurities, the lattice structure, the surface effects, and so on. For details on the modeling of transport properties, See Cahill et al. [9].

Consider the thin slab shown in Fig. 5.27. This could represent a three-dimensional view of the layer of high-thermal-conductivity material shown in Fig. 5.28. Such a layer could be embedded in or deposited on a heat-generating plate. We assume that the x dimension (L) is much larger than the y and z dimensions

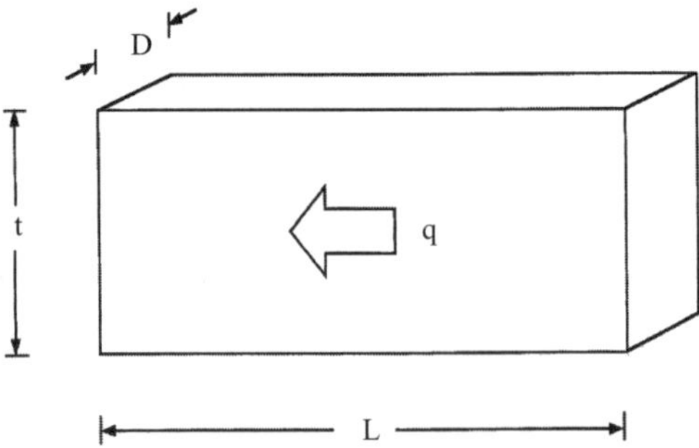

Figure 5.27 Elemental volume [8].

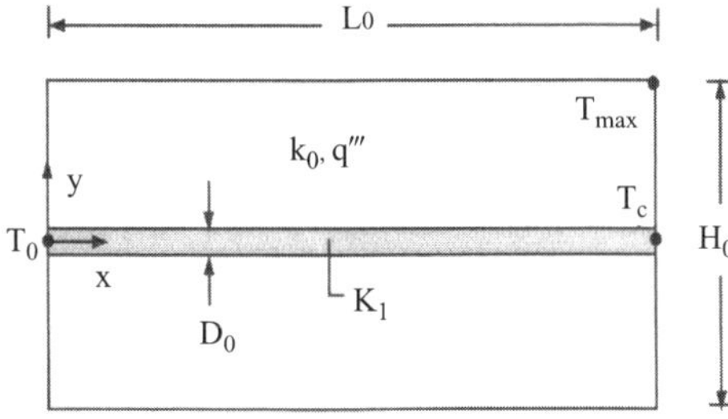

Figure 5.28 The geometry of the high-conductivity blade [8].

(D and t). We also assume that the slab depth t is greater than or equal to D. In this case, the limiting dimension is the smallest one, namely, the width D of the high-conductivity blade. This is the dimension that controls the effect that size has on thermal conductivity.

Thin-film structures such as Fig. 5.27 have been studied experimentally and theoretically. Even though the explicit calculation of the dependence of thermal conductivity on D is complex, it is observed that in semiconductors at very small scales the thermal conductivity decreases as D decreases. When D is sufficiently large, the conductivity does not depend on D, and is equal to the bulk conductivity. These observations recommend the following approximate model for the thermal conductivity of the material that fills the D blade:

$$\frac{k_x}{k_b} = \begin{cases} \dfrac{D}{\lambda} & (D \le \lambda) \\ 1 & (D > \lambda) \end{cases} \tag{5.58}$$

From kinetic theory [10, 11], one estimates the thermal conductivity as

$$k = \frac{1}{3} C\, v l \tag{5.59}$$

where C, v, and l are the heat capacity, the average velocity, and mean free path of the energy carriers. The idea of replacing the mean free path of the phonons in Eq. (5.59) with the smallest dimension of the system (in the present case, D) below a critical length scale (λ) has been used in the literature [12, 13] to describe the phonon contribution to the thermal conductivity. This idea is equivalent to the model assumed in Eq. (5.58). The size effect becomes significant around 10 to 100 nm, which indicates the scale of λ.

The conductivity model of Eq. (5.58) is consistent with models presented in literature in the two extremes [14, 15]: $D \gg \lambda$ and $D \ll \lambda$. Discrepancies between Eq. (5.58) and the actual conductivity can be expected where the two limits intersect, $D \sim \lambda$. In any case, the model of Eq. (5.58) reflects the correct scale of the thermal conductivity even in the intermediate range where $D \sim \lambda$.

In the elemental system of Fig. 5.28, the smaller length scale is D_0. Therefore, it is in the high-conductivity blade that the size effect is most likely to occur. The heat flowing on that path (in the x direction) encounters a conductivity described by Eq. (5.58), or in dimensionless form by

$$\frac{k_{1,x}}{k_{1,b}} \sim \frac{\phi_0}{\tilde{\lambda}}\left(\frac{H_0}{L_0}\right)^{1/2} \tag{5.60}$$

The dimensionless number $\tilde{\lambda} = \lambda / A_0^{1/2}$ indicates how close the system size ($A_0^{1/2}$) is relative to the size effect threshold (λ).

A simple analysis can be performed in order to describe how the optimal configuration changes as dimensions decrease. Note that the hot-spot temperature is located in the upper right corner of $A_0^{1/2}$. The temperature difference ($T_{max} - T_0$) can be expressed as the sum of two contributions $(T_{max} - T_c) + (T_c - T_0)$, where T_c is the temperature of the warm end of the high-conductivity blade. The temperature difference ($T_{max} - T_c$) is sustained by thermal diffusion in the y direction over the distance $H_0/2$ through a low-conductivity material, the bulk conductivity of which is constant (k_0). The temperature difference ($T_c - T_0$) drives the heat current in the x direction, along the D_0 thin blade that has the thermal conductivity modeled according to Eq. (5.49), with $k_{1,b}$ as the bulk thermal conductivity. The analysis follows the steps outlined in section 5.1.1. The result is the global thermal resistance of the elemental system,

$$\tilde{T}_0 = \frac{T_{max} - T_0}{q''' A_0 / k_0} = \frac{H_0}{8 L_0} + \frac{\tilde{\lambda}}{2 \phi_0^2 \tilde{k}} \left(\frac{H_0}{L_0}\right)^{-3/2} \tag{5.61}$$

where $\tilde{k}$ is the ratio of the bulk thermal conductivities, $\tilde{k} = k_{1,b}/k_0 \gg 1$. The global resistance $\tilde{T}_0$ can be minimized with respect to the aspect ratio of the element (H_0/L_0). The optimal aspect ratio and minimal thermal resistance are (Figs. 5.29 and 5.30)

$$\left(\frac{H_0}{L_0}\right)^n_{opt} = \left(\frac{6\tilde{\lambda}}{\phi_0^2 \tilde{k}}\right)^{2/5} \tag{5.62}$$

$$\tilde{T}^n_{0,m} = \frac{5}{3^{3/5} 2^{13/5}} \left(\frac{\tilde{\lambda}}{\phi_0^2 \tilde{k}}\right)^{2/5} \tag{5.63}$$

The superscript n indicates that these results are based on the nano-size conductivity model, Eq. (5.58).

Equations (5.62) and (5.63) are valid for $D_0 \le \lambda$, or in dimensionless form, for $\tilde{\lambda} \ge \tilde{\lambda}_{c,0}$, where $\tilde{\lambda}_{c,0} = \phi_0 (H_0/L_0)^{1/2}$. The conductivity model that has been used has a discontinuity in the derivative dk/dD_0 at $D_0 = \lambda$. Consequently, depending on the expression used for the optimal aspect ratio, Eq. (5.62) or the corresponding result for bulk conductivity (large D_0), one would obtain slightly different results for the value of $\tilde{\lambda}_{c,0}$. The convenience offered by the simple k model of Eq. (5.58) is certainly worth this approximation, and in the present analytical context (an order of magnitude analysis) is fully justified. For the optimal aspect ratio of Eq. (5.62),

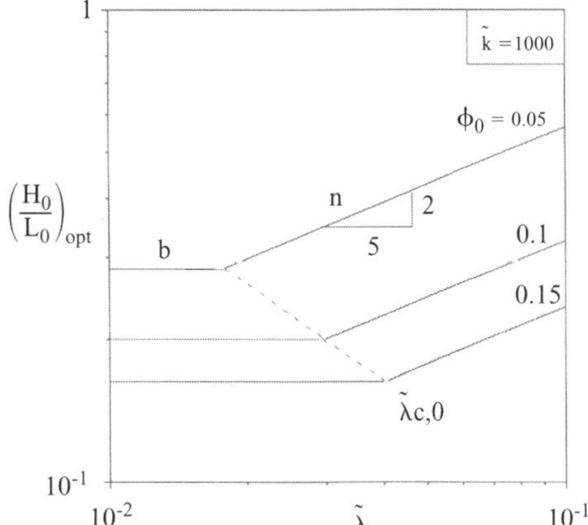

Figure 5.29 The optimal aspect ratio of the elemental volume [8].

we find that the critical $\tilde{\lambda}_{c,0}$ is

$$\tilde{\lambda}_{c,0} = 6^{1/4} \frac{\phi_0^{3/4}}{\tilde{k}^{1/4}} \tag{5.64}$$

When $\tilde{\lambda} \leq \tilde{\lambda}_{c,0}$ the size effect is not present, and the optimal elemental area developed in Ref. [1] and section 5.1.1 applies. However, when $\tilde{\lambda} \geq \tilde{\lambda}_{c,0}$ the present results apply. Equation (5.64) can also be written as $\tilde{\lambda}_{c,0}\tilde{k}^{1/4}\phi_0^{-3/4} \sim 1$.

It is interesting to compare the above results with the result obtained in the absence of size effect (superscript "*b*"), that is, with constant *bulk* thermal conductivities,

$$\left(\frac{H_0}{L_0}\right)_{opt}^{b} = \frac{2}{\left(\tilde{k}\,\phi_0\right)^{1/2}} \qquad \tilde{T}_{0,m}^{b} = \frac{1}{2\left(\tilde{k}\,\phi_0\right)^{1/2}} \tag{5.65}$$

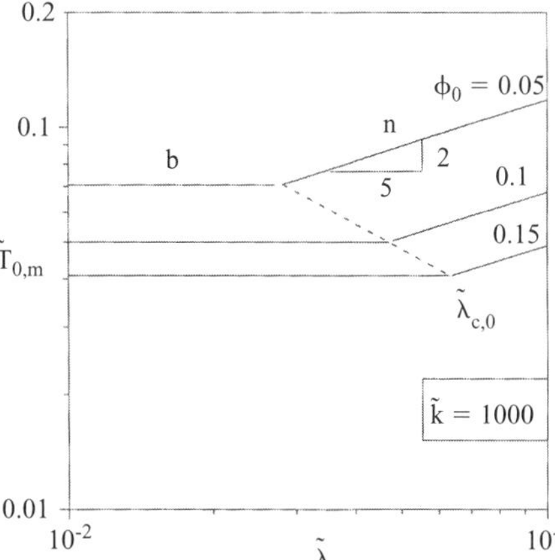

Figure 5.30 The minimized thermal resistance of the elemental volume [8].

The differences between the "n" and "b" regimes are described by the ratios

$$\frac{(H_0/L_0)^n_{\text{opt}}}{(H_0/L_0)^b_{\text{opt}}} = \frac{3^{2/5}\,\tilde{\lambda}^{2/5}\,\tilde{k}^{1/10}}{2^{3/5}\,\phi_0^{3/10}} \qquad \frac{\tilde{T}^n_{0,m}}{\tilde{T}^b_{0,m}} = \frac{5\,\tilde{\lambda}^{2/5}\,\tilde{k}^{1/10}}{2^{8/5}\,3^{3/5}\,\phi_0^{3/10}} \tag{5.66}$$

In view of Eq. (5.64), it follows that the thermal resistance of the nano-element is slightly larger than for the bulk element. The aspect ratio of the nano-element is also slightly larger than for the bulk element: the nano element is less slender. These results are reported in graphical form in Figs. 5.29 and 5.30.

Contrary to the result obtained for bulk materials [1, 3], there is no "equipartition" of thermal resistance between the two terms on the right hand side of Eq. (5.61). Instead, there is "optimal allocation" of thermal resistance, approximate partitioning of thermal resistance, or *optimal distribution of imperfection*: the first thermal resistance ($T_{\max} - T_c$) accounts for 3/5 of the total resistance, while the second ($T_c - T_0$) represents only 2/5 of the total resistance. The assembling and optimizing of larger constructs consisting of such elements is continued further in Ref. [8].

5.4 EVOLUTION OF TECHNOLOGY: FROM FORCED CONVECTION TO SOLID-BODY CONDUCTION

Here, we continue the questioning that started in section 3.7: how must cooling technology change in the pursuit of higher heat-transfer densities? The winner in section 3.7 was forced convection, which promised to displace natural convection when length scales dropped a certain level [cf. Eq. (3.47)]. The reason for this change in technology is that forced-convection heat transfer densities increase as L^{-1} when the swept length L decreases, whereas in natural convection heat-transfer densities increase more slowly (as $H^{-1/2}$, where H is the length swept by natural convection). Is there a cooling mechanism more effective than forced convection when length scales become even smaller?

Consider the main results of section 5.1.1 and Problem 5.5, which show that the maximum heat-transfer density of an elemental two-dimensional volume ($A_0 = H_0 L_0$) cooled by a high-conductivity blade is of order

$$q'''_{\text{C}} \sim \frac{k_0}{A_0}\left(\tilde{k}\phi_0\right)^{1/2}\Delta T \tag{5.67}$$

where $\Delta T = T_{\max} - T_0$, and the C subscript stands for conduction. The optimal shape of the rectangular element A_0 is $H_0/L_0 \sim (\tilde{k}\phi_0)^{-1/2}$ (cf. Problem 5.5), which means that

$$A_0 = L_0^2 \frac{H_0}{L_0} \sim \frac{L_0^2}{\left(\tilde{k}\phi_0\right)^{1/2}} \tag{5.68}$$

Eliminating A_0 between Eqs. (5.67) and (5.68), and using $\tilde{k} = k_p/k_0$, we discover that the heat transfer density removable by pure conduction increases as L_0^{-2} when the k_p blade length L decreases:

$$q_C''' \sim k_p \phi_0 \frac{\Delta T}{L_0^2} \tag{5.69}$$

This decrease is faster than the decrease of q_{FC}''' in the direction of decreasing length scales. In conclusion, there comes a length scale L_0 below which solid-body conduction with high-conductivity inserts is more promising than even the best configuration (with optimal spacings) cooled by forced convection. This threshold length scale is obtained by requiring $q_C''' > q_{FC}'''$, and comparing Eqs. (5.69) and (3.37). The resulting critical length scale for conduction cooling is

$$L_0 < \left(k_p \phi_0\right)^{1/2} \left(\frac{\mu \alpha L^2}{k^2 \Delta P}\right)^{1/4} \tag{5.70}$$

where $\left(k_p \phi_0\right)$ refers to properties of the conduction design, and $\left(\mu \alpha L^2 / k^2 \Delta P\right)$ to properties of the forced-convection design. If at transition, L is the same as L_0, then the length scale below which pure conduction is superior is

$$L_0 < k_p \phi_0 \left(\frac{\mu \alpha}{k^2 \Delta P}\right)^{1/2} \tag{5.71}$$

We arrived in this way at the prediction made in the opening to the second (1996) paper on constructal theory [1], where conduction was foreseen as a replacement for forced convection, and where the point-volume flow problem of constructal conduction trees was first formulated:

> This paper is about one of those fundamental problems that suddenly appear "obvious" but only after considerable technological progress has been made on pushing the frontier. The technology in this case is the cooling of electronics (components and packages), where the objective is to install a maximum amount of electronics (heat generation) in a fixed volume in such a way that the maximum temperature does not exceed a certain level. The work that has been done to devise cooling techniques to meet this objective is enormous. In brief, most of the cooling techniques that are in use today rely on convection or conjugate convection and conduction, where the coolant is either a single phase fluid or one that boils.
>
> The frontier is being pushed in the direction of smaller and smaller package dimensions. There comes a point where miniaturization makes convection cooling impractical, because the ducts through which the coolant must flow take up too much space. The only way to channel the generated heat out of the electronic package is by conduction. This conduction path will have to be very effective (of high thermal conductivity, k_p), so that the temperature difference between the hot spot (the heart of the package) and the heat sink (on the side of the package) will not exceed a certain value.

> Conduction paths also take up space. Designs with fewer and smaller paths are better suited for the miniaturization evolution. The fundamental problem addressed in this paper is this:
>
> Consider a finite-size volume in which heat is being generated at every point and which is cooled through a small patch (heat sink) located on its boundary. *A* finite amount of high conductivity (k_p) material is available. Determine the optimal distribution of k_p material through the given volume such that the highest temperature is minimized [1].

Constructal theory is now a fast growing field, the pre-2006 status of which was reviewed in Refs. [16] through [18]. Here, we draw attention to even newer developments. The optimization of conduction configuration at the elemental level (fibers, particles) formed the subject of Refs. [19] and [20]. Heat conduction in trees was generated by means of an evolutionary algorithm in Ref. [21], which distributed, better and better, a finite amount of high-conductivity material through a low-conductivity domain with area-point heat flow. This evolution of tree configuration is conceptually equivalent to the one described in Ref. [22] for area-point fluid flow in an erodable river basin, where instead of high-k channels on low-k domains, the river basin grew out of the distribution of high-permeability flow channels on a background with low-permeability Darcy flow. The flow configurations evolved in Ref. [21] are similar to those evolved in Ref. [22].

Completely analogous to these tree designs is the evolution of photovoltaic cells [23]. The distribution of high-conductivity material on the collector acquires a dendritic shape if enough morphing freedom is invested in its development. This electric tree development, coupled with that of trees for fluid flow (Chapter 4) and heat trees, makes obvious the future of constructal trees in the architectures of novel packages of fuel cells, as indicated in Ref. [24].

Conduction paths (blades) assembled into trees were analyzed for global thermal resistance in Ref. [25]. Unlike in this chapter, however, no consideration was given to the layout of the high-k paths on the available territory. Nothing was said about how the tree fills its allotted space, and for this reason the global conclusions reached in Ref. [25] do not have a meaning in relation to the properties of the conduction trees developed throughout constructal theory [1–2, 16–18]. The neglect of tree layout [25] is analogous to the choice made by West et al. [26] in their model of "space filling" trees for fluid flow: the way in which the tree fills its space is the unknown (the design).

Industry is beginning to put constructal theory concepts to good use. Witness the development at IBM Zürich of hierarchically sized and nested channels for cooling high-density electronics by direct thermal contact across multiscale interfaces [27]. Multiscale interfaces were developed earlier for maximal thermal conductance across rough interfaces [28, 29]. Cavities shaped as an H and embedded in a conducting medium were optimized in Ref. [30]. Novel Y-shaped fins with features similar to those of Fig. E5.1 were optimized in Ref. [31].

REFERENCES

1. A. Bejan, Constructal-theory network of conducting paths for cooling a heat generating volume. *Int J Heat Mass Transfer*, Vol. 40, 1997, pp. 799–816 (issue published on November 1, 1996).
2. L. A. O. Rocha, S. Lorente, and A. Bejan, Constructal design for cooling a disc-shaped area by conduction. *Int J Heat Mass Transfer*, Vol. 45, 2002, pp. 1643–1652.
3. A. Bejan, *Shape and Structure, from Engineering to Nature*. Cambridge, UK: Cambridge University Press, 2000.
4. R. A. Nelson Jr. and A. Bejan, Constructal optimization of internal flow geometry in convection. *J Heat Transfer*, Vol. 120, 1998, pp. 355–364.
5. L. A. O. Rocha, S. Lorente and A. Bejan, Conduction tree networks with loops for cooling a heat generating volume. *Int J Heat Mass Transfer*, Vol. 49, 2006, pp. 2626–2635.
6. A. Bejan and S. Lorente, The constructal law and the thermodynamics of flow systems with configuration. *Int J Heat Mass Transfer*, Vol. 45, 2004, pp. 3203–3214.
7. A. Bejan and S. Lorente, *La Loi Constructale*. Paris: L'Harmattan, 2005.
8. L. Gosselin and A. Bejan, Constructal trees at micro and nanoscales. *J Appl Phys*, Vol. 96, No. 10, 2004, pp. 5852–5859.
9. D. G. Cahill, W. K. Ford, K. E. Goodson, *et al.*, Nanoscale thermal transport. *J Appl Phys*, Vol. 93, 2003, pp. 593–818.
10. F. Reif, *Fundamentals of Statistical and Thermal Physics*. New York: McGraw-Hill, 1965.
11. C. Kittel, *Introduction to Solid Sate Physics*, 7th ed., New York: Wiley, 1996.
12. L. D. Hicks and M. S. Dresselhaus, Effect of quantum-well structures on the thermoelectric figure of merit. *Phys Rev B*, Vol. 47, 1993, pp. 12727–12731.
13. L. D. Hicks and M. S. Dresselhaus, Thermoelectric figure of merit of a one-dimensional conductor. *Phys Rev B*, Vol. 47, 1993, pp. 16631–16634.
14. A. I. Chervanyov, Effects of boundary scattering and optic phonon drag on thermal conductivity of a slab of rectangular cross-section. *Phys Rev B*, Vol. 66, 2002, 214302.
15. A. Majumdar, On thermal conductivity of a slab of rectangular cross-section. *J Heat Transfer*, Vol. 115, 1993, pp. 5–11.
16. A. Bejan and S. Lorente, Constructal theory of configuration generation in nature and engineering. *J Appl Phys*, Vol. 100, 041301.
17. A. H. Reis, Constructal theory: from engineering to physics, and how flow systems deelop shape and structure. *Appl Mech Rev*, Vol. 59, 2006, pp. 269–282.
18. A. Bejan, *Advanced Engineering Thermodynamics*, 3rd ed. (Ch. 13). Hoboken, NJ: Wiley, 2006 .
19. E. G. Youngs and A. R. Kacimov, Conduction through an assembly of spherical particles at low liquid contents. *Int J Heat Mass Transfer*, Vol. 50, 2007, pp. 292–302.
20. A. R. Kacimov, Optimal design of fibers subject to steady conduction. *Heat Mass Transfer*, Vol. 43, 2007, pp. 319–324.
21. F. Mathieu-Potvin and L. Gosselin, Optimal conduction pathways for cooling a heat-generating body. *Int J Heat Mass Transfer*, Vol. 50, 2007, pp. 2996–3006.
22. M. R. Errera and A. Bejan, Deterministic tree networks for river drainage basins. *Fractals*, Vol. 6, 1998, pp. 246–261.

23. A. M. Morega and A. Bejan, A constructal approach to the optimal design of photovoltaic cells. *Int J Green Energy*, Vol. 2, 2005, pp. 233–242.
24. J. V. C. Vargas, J. Ordonez, and A. Bejan, Constructal flow structure for a PEM fuel cell. *Int J Heat Mass Transfer*, Vol. 47, 2004, pp. 4177–4193.
25. P. Xu, B. Yu, M. Yun, and M. Zou, Heat conduction in fractal tree-like branched networks. *Int J Heat Mass Transfer*, Vol. 49, 2006, pp. 3746–3751.
26. G. B. West, J. H. Brown, and B. J. Enquist, A general model for the origin of allometric scaling laws in biology. *Science*, Vol. 276, 1997, pp. 122–126.
27. T. Runschwiler, U. Kloter, R. J. Linderman, H. Rothuizen, and B. Michel, Hierarchically nested channels for fast squeezing interfaces with reduced thermal resistance. *IEEE Transactions on Components and Packaging Technologies*, Vol. 30, 2007, pp. 226–234.
28. M. Neagu and A. Bejan, Constructal placement of high-conductivity inserts in a slab: optimal design of "roughness." *J Heat Transfer*, Vol. 123, 2001, pp. 1184-1189.
29. J. V. C. Vargas and A. Bejan, The optimal shape of the interface between two conductive bodies with minimum thermal resistance. *J Heat Transfer*, Vol. 124, 2002, pp. 1218–1221.
30. C. Biserni, L. A. O. Rocha, G. Stanescu, and E. Lorenzini, Constructal H-shaped cavities according to Bejan's theory. *Int J Heat Mass Transfer*, Vol. 50, 2007, pp. 2132–2138.
31. M. Lorenzini and S. Moretti, Numerical analysis on heat removal from Y-shaped fins: efficiency and volume occupied for a new approach to performance optimization. *Int J Thermal Sciences*, Vol. 46, 2007, pp. 573–579.

PROBLEMS

5.1. The elemental volume optimized in section 5.1.1 contained a high-conductivity blade with uniform thickness. Relax the assumption that the k_p blade thickness D_0 is uniform. Assume instead that D_0 has an arbitrary power-law shape, $D_0\ (x) = bx^n$, where b and n are two constants. Instead of a circular sector, use the rectangular element shown in Fig. P5.1. The

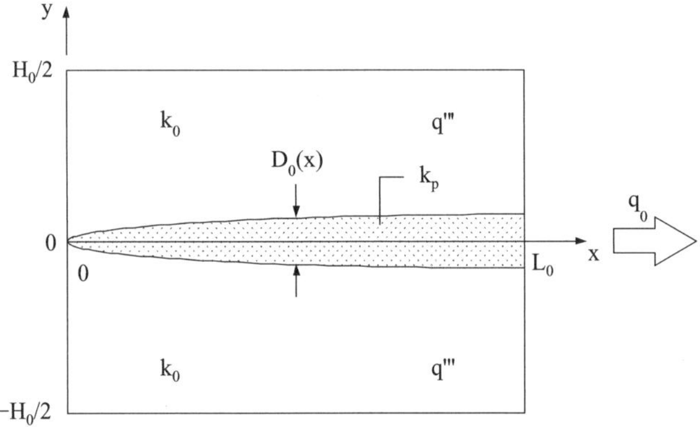

Figure P5.1

k_p material constraint is $\phi_0 = A_{p,0}/A_0$, constant. Derive the global thermal resistance and minimize this expression with respect to the exponent n and the external shape H_0/L_0. Show that the global resistance minimized in this fashion is

$$\left(\frac{\Delta T_0 k_0}{q''' A_0}\right)_{\min} = \frac{2^{1/2}}{3(\tilde{k}\phi_0)^{1/2}}$$

and that it is 6% smaller than the minimized resistance of the corresponding system with constant D_0 blade.

5.2. It is necessary to heat a wall uniformly. One proposal is to use a stream of hot fluid but to arrange it in a hairpin counterflow shape such that the stream returns along itself. Let x be the coordinate along the hot portion of the stream, from the inlet ($x = 0$) to the hairpin turn ($x = L$). The stream capacity rate is $\dot{m}c_P$. The inlet temperature (T_{in}) and wall temperature (T_w) are given. All the thermal resistances between the stream and wall are dominated by the convective resistance $(hD)^{-1}$, where h is the heat transfer coefficient (constant) for fully developed flow inside the tube and D is the tube diameter. The perimeter of thermal contact is approximately D between stream and wall, and between stream and stream. Determine analytically the temperature distributions along the hot and cold portions of the counterflow. Show that the rate at which the counterflow deposits heat into the wall is independent of x.

5.3. The side of a container is heated by a ring-shaped electrical resistance of radius R_m. The ring is embedded in a concentric disc-shaped "heat spreader" of radius R and conductivity k. The disc is attached with very good thermal contact to the container wall. The objective is to determine the ring radius R_m so that the temperature distribution over the disc is the least nonuniform distribution possible. This is a constructal problem of deriving geometry from the optimal distribution of imperfection. To solve the problem analytically, analyze the two-dimensional model shown in Fig. P5.3, which represents a disc of unit thickness. The heat generation effect is represented by a ring that generates the heat flux q''[W/m^2]. The heat transmission from the disk to the container wall is represented by the heat sink effect q'''[W/m^3] that is distributed uniformly over the disk. Determine the temperature distribution over the disk, where R_m may vary. If T_m, T_c, and T_p are, in order, the temperature of the hot spot (the ring), the temperature in the disc center, and the temperature on the disc perimeter, determine the optimal ring radius such that the larger of $T_m - T_c$ and $T_m - T_p$ is the smallest possible. Show that in this configuration T_c is equal to T_p.

5.4. Related to Example 5.1 is the three-dimensional configuration in which each Y-shaped structure is made of three bars (struts) of cross-sectional areas A_0 and A_1, and lengths L_0 and L_1. The Y structures conduct heat from the center

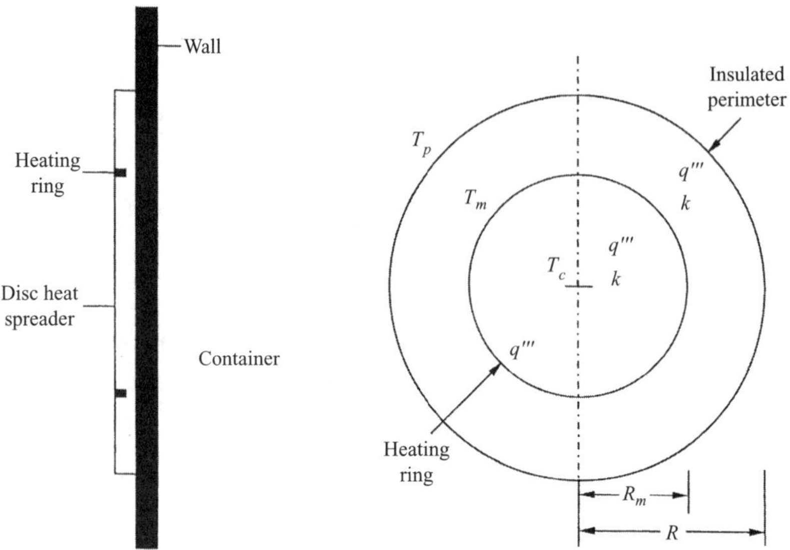

Figure P5.3

of a ring (the heat source) to the ring itself, which is the heat sink. The total volume of conductive material is fixed. This configuration is visualized by replacing (D_0, D_1) with (A_0, A_1) in Fig. E5.1 in the text. The ends of each Y structure are fixed. Show that the global thermal resistance between the center and the ring is minimum when $A_0/A_1 = 2$ and $L_0/L_1 = 0$. In other words, show that of all the possible Y drawings, the best is the special case represented by V.

5.5. Consider again the minimization of the thermal resistance of a volume element with uniform heat generation, but instead of a circular sector (Fig. 5.2), use the rectangular element $H_0 \times L_0$ sketched in Fig. 5.28. Show that Eqs. (5.18) through (5.19) are replaced by

$$\left(\frac{H_0}{L_0}\right)_{\text{opt}} = \frac{2}{(\tilde{k}\phi_0)^{1/2}} \ll 1 \qquad \frac{(T_{\max} - T_0)_{\min}}{q''' A_0/k_0} = \frac{1}{2(\tilde{k}\phi)^{1/2}}$$

where $\tilde{k} = k_1/k_0$ and $A_0 = H_0 L_0$. Comment on the effect of the element shape (sector vs. rectangle) on the essential features of the optimal elemental design.

5.6. In this problem you are asked to derive the heat conduction equivalent of the Hess-Murray rule. Consider the following two-dimensional configuration. One blade of length L_1 and thickness D_1 is continued by two identical blades of length L_2 and thickness D_2. The three blades form a Y-shaped construct in the plane of the paper. The dimension of the construct in the direction perpendicular to the paper is larger than the Y. Heat flows by Fourier conduction along the Y, from stem to branches, or vice versa. All the blades have the constant thermal conductivity k. The total volume of conducting material is fixed. The lengths L_1 and L_2 are given. The only variables are D_1 and D_2. Determine the optimal ratio D_1/D_2 for which the global thermal resistance of the Y construct is minimum.

6

MULTISCALE CONFIGURATIONS

The generation of tree architectures in Chapters 4 and 5 introduced us to the benefits of multiple length scales and optimized complexity. In this chapter we exploit this idea to generate classes of flow configurations that offer greater compactness (e.g., maximum heat-transfer density): stacks of parallel plates with multiple lengths and spacings, surfaces with discrete heat sources (e.g., electronic components in a package), bundles of cylinders with multiple spacings and diameters, and optimal layout and sequencing of droplets for maximum mass transfer density. The direction toward more compactness is the same as the direction toward more svelteness (cf. Figs. 11.10 and 11.11).

In thermofluid design there is a growing body of work that illustrates the constructal method. In all this work, the global objective is the maximization of heat-transfer density or the minimization of the hot-spot temperatures when the total heat generation rate, volume, and other constraints are specified. Compactness, miniaturization, and multiscale flow structures are generated along this route.

In this chapter we start with a fundamental problem in the making of packages of components connected by flows: the optimal positioning of discrete heat sources in a fixed volume with natural convection. The performance of packages with equidistant discrete heat sources has been studied extensively, numerically and experimentally, on a case-by-case basis. Here, we adopt a constructal attitude and allow the configuration to morph freely, to be unknown. In the next two sections we search for the optimal configuration in two settings: a large number of small heat sources distributed freely on a wall, and a small number of finite-size heat sources mounted freely inside one wall of a two-dimensional enclosure.

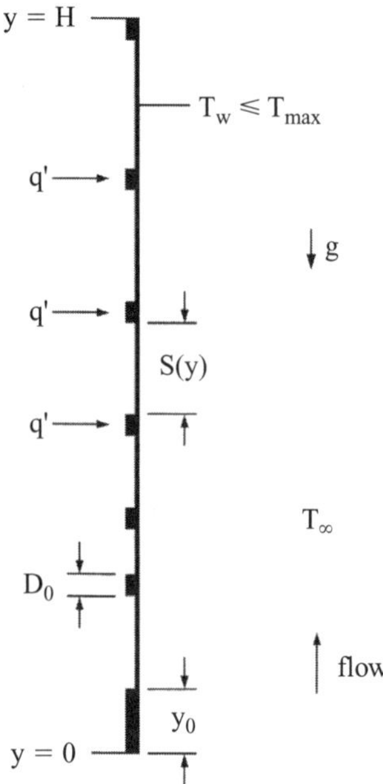

Figure 6.1 The multiple length scales of the nonuniform distribution of finite-size heat sources on a vertical wall [1].

6.1 DISTRIBUTION OF HEAT SOURCES COOLED BY NATURAL CONVECTION

Consider a vertical wall of height H, which is in contact with a fluid reservoir of temperature T_∞ (Fig. 6.1). The wall is heated by horizontal line heat sources. Each source has the strength q'[W/m]. To start, assume that the heat sources appear as points on the wall sketched in Fig. 6.1. Each line heat source extends in the direction perpendicular to the figure. The flow is two-dimensional and by natural convection in the boundary layer regime. The number of sources per unit of wall height is unknown, $N'(y)$:

$$N' = \frac{\text{number of sources}}{\text{unit length}} \tag{6.1}$$

According to the method of constructal design (the optimal distribution of imperfection), the global system (the wall and the fluid that bathes it) will perform best when all its elements work as hard as the hardest working element [2]. This means that if $T_{\max}$ is the maximal temperature that must not be exceeded at the hot spots that occur on the wall, then the entire wall should operate at $T_{\max}$. The problem is to determine the distribution of heat sources on the wall, $N'(y)$, such that the wall temperature is near the allowed limit:

$$T_w(y) \cong T_{\max}, \quad \text{constant} \tag{6.2}$$

Assume that the density of line sources is sufficiently high so that we may regard the distribution of discrete q' sources as a nearly continuous distribution of nonuniform heat flux:

$$q''(y) = q' N' \tag{6.3}$$

The heat flux distribution that corresponds to Eqs. (6.2) and (6.3) and $\mathrm{Pr} \gtrsim 1$ is [3]:

$$\mathrm{Nu} \cong 0.5\mathrm{Ra}_y^{1/4} \tag{6.4}$$

or, more explicitly,

$$\frac{q''}{T_{\max} - T_\infty} \frac{y}{k} \cong 0.5 \left[\frac{g\beta \left(T_{\max} - T_\infty\right) y^3}{\alpha\nu} \right]^{1/4} \tag{6.5}$$

By eliminating $q''(y)$ between Eqs. (6.3) and (6.5) we obtain the required distribution of heat sources:

$$N'(y) \cong 0.5 \frac{k}{q'} \left(T_{\max} - T_\infty\right)^{5/4} \left(\frac{g\beta}{\alpha\nu} \right)^{1/4} y^{-1/4} \tag{6.6}$$

This function shows that the heat sources must be positioned closer when they are near the start of the boundary layer. They must be farther apart on sections of the wall near $y = H$. The total number of q' sources that must be installed according to Eq. (6.6) on the wall of height H is

$$N = \int_0^H N' dy \cong \frac{2}{3} \frac{k}{q'} \left(T_{\max} - T_\infty\right) \mathrm{Ra}^{1/4} \tag{6.7}$$

where

$$\mathrm{Ra} = \frac{g\beta H^3 \left(T_{\max} - T_\infty\right)}{\alpha\nu} \tag{6.8}$$

The Rayleigh number is a dimensionless parameter that accounts for two global constraints, the wall height H and the maximal allowable excess temperature at the hot spots. The total heat-transfer rate from the q' sources to the T_∞ fluid is

$$Q' = q'N \cong \frac{2}{3} k \left(T_{\max} - T_\infty\right) \mathrm{Ra}^{1/4} \tag{6.9}$$

This represents the global performance level to which any of the optimized nonuniform distributions of concentrated heat sources will aspire.

The physical implementation of the preceding results begins with the observation that the smallest scale that can be manufactured in the heating scheme of Fig. 6.1 is the D_0 height of the line heat source. The local spacing between two adjacent lines is $S(y)$. This spacing varies with altitude in accordance with the N' distribution function (6.6). The wall height interval that corresponds to a single line heat source is $D_0 + S(y)$. This means that the local number of heat sources per unit of wall height is

$$N'(y) = \frac{1}{D_0 + S(y)} \tag{6.10}$$

The heat strength of one source (q') is spread uniformly over the finite height of the source, $q_0'' = q'/D_0$. The heat flux q_0'' is a known constant, unlike the function $q''(y)$ of Eq. (6.5), which will be the result of design. By eliminating $N'(y)$ between Eqs. (6.6) and (6.10), we obtain the rule for how the wall heating configuration should be constructed:

$$\frac{S(y)}{H} \cong \frac{2q'\mathrm{Ra}^{-1/4}}{k\,(T_{\max} - T_\infty)} \left(\frac{y}{H}\right)^{1/4} - \frac{D_0}{H} \tag{6.11}$$

The function $S(y)$ of Eq. (6.11) has negative values in the vicinity of the start of the boundary layer. The smallest value that S can have is 0. This means that there is a starting wall section ($0 < y < y_0$) over which the line sources should be mounted flush against each other. The height of this section (y_0) is obtained by setting $S = 0$ and $y = y_0$ in Eq. (6.11):

$$\frac{y_0}{H} \cong \mathrm{Ra}\left(\frac{D_0}{H}\right)^4 \left[\frac{k\,(T_{\max} - T_\infty)}{2q'}\right]^4 \tag{6.12}$$

From $y = 0$ to $y = y_0$ the wall is heated with uniform flux of strength $q_0'' = q'/D_0$. The number of sources that cover the height y_0 is $N_0 = y_0/D_0$. Above $y = y_0$ the wall is heated on discrete patches of height D_0, and the spacing between patches increases with height.

These basic features of the optimal design are illustrated in Fig. 6.1. The design has multiple length scales: H, D_0, y_0 and $S(y)$. The first two are constraints. The last two are interrelated, and are results of global maximization of performance, subject to the constraints. Taken together, the lengths represent the constructal design—the flow architecture that out of an infinity of possible architectures brings the entire wall to the highest performance level possible.

The global heat-transfer performance of the constructal design can be estimated in the limit where the number of heat-source strips D_0 is sufficiently large. In this limit, the integral (6.7) applies only in the upper region of the wall ($y_0 < y < H$), where the concentrated sources are spaced according to Eq. (6.6). In the lower region of height y_0, the D_0 strips are mounted without spacings between them, and their number is

$$N_0 = \frac{y_0}{D_0} = \mathrm{Ra}\left(\frac{D_0}{H}\right)^3 \left(\frac{k\,(T_{\max} - T_\infty)}{2q'}\right)^4 \tag{6.13}$$

The total number of D_0 strips on the H wall is

$$N = N_0 + \int_{y_0}^{H} N'dy = \frac{2}{3}\frac{k}{q'}(T_{\max} - T_\infty)\,\mathrm{Ra}^{1/4} \times \left[1 - \frac{3}{16}\left(\frac{k\,(T_{\max} - T_\infty)}{2q'}\right)^3 \left(\frac{D_0}{H}\right)^3 \mathrm{Ra}^{3/4}\right] \tag{6.14}$$

This expression shows that in the limit $D_0 \to 0$, where the D_0 strips become line heat sources, Eq. (6.14) reduces to the simpler form shown in Eq. (6.7).

The total rate of heat transfer from the wall to the fluid is $Q' = q'N$, which yields

$$Q' \cong \frac{2}{3}k\,(T_{\max} - T_\infty)\,\mathrm{Ra}^{1/4}\left[1 - \frac{3}{16}\left(\frac{k\,(T_{\max} - T_\infty)}{2q'}\right)^3 \left(\frac{D_0}{H}\right)^3 \mathrm{Ra}^{3/4}\right] \tag{6.15}$$

By comparing Eq. (6.15) with Eq. (6.9), we see that when D_0 is finite the total heat transfer rate is less than in the limit of line heat sources ($D_0 = 0$).

In summary, Eq. (6.15) is the more general, and represents the maximal performance of all the designs that are possible in the limit of large N. The optimization has already been performed, and is expressed by the optimal distribution shown in Eq. (6.6). This is why the only parameters that may still affect the global performance are visible in Eq. (6.15): they are the constraints, $(T_{\max} - T_\infty), H, D_0$ and q'.

These analytical results were tested based on numerical simulations of natural convection in a three-dimensional enclosure [1, 4] that has a discretely heated vertical wall of height H (Fig. 6.2). The opposing wall serves as heat sink. The

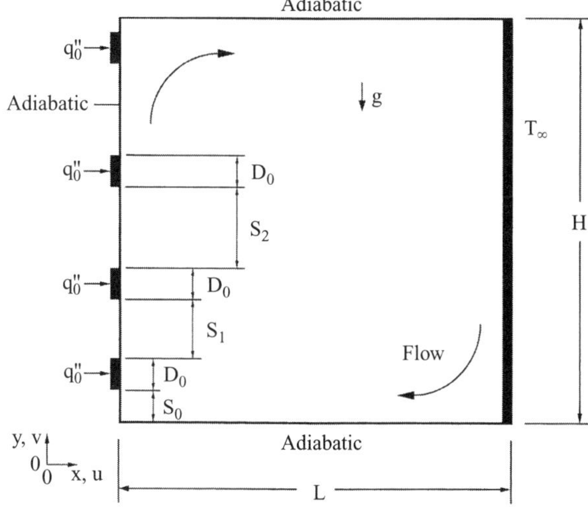

Figure 6.2 Enclosure with discrete heat sources on one of the vertical walls [1].

governing equations and boundary conditions were nondimensionalized by using $(\tilde{D}_0, \tilde{S}_0, \tilde{S}_1, \ldots) = (D_0, S_0, S_1, \ldots)/H$ and the dimensionless temperature $\tilde{T} = (T - T_\infty)/(q_0''H/k)$. The Rayleigh number ($\text{Ra}_*$) is based on the heat source strength, $\text{Ra}_* = g\beta q_0'' H^4/(\alpha\nu k)$. Note the difference between Ra_* and the Rayleigh number based on temperature difference, Eq. (6.8). The flow boundary conditions are no slip and no penetration along all internal surfaces. The right wall is isothermal at $\tilde{T} = 0$. The left wall has one or more heat strips of height $\tilde{D}_0$. Each strip dissipates uniformly a dimensionless heat flux of $\tilde{q} = 1$, which is defined as $\tilde{q} = k(\partial T/\partial x)_{x=0}/q_0''$. The top and bottom walls, and the wall sections $(\tilde{S}_0, \tilde{S}, \ldots)$ that separate the $\tilde{D}_0$ strips are adiabatic. The Prandtl number was fixed at $\text{Pr} = 0.7$. The numerical method is detailed in Ref. [1].

In the numerical optimization of the distribution of heat sources, we search not only for accurate optimal designs but also for the patterns in which the designs organize themselves. Behind this search was the question of whether the numerically optimized distributions approach in some ways the theoretical distribution anticipated in Eqs. (6.6) and (6.11). If the answer is "yes", then the theoretical distribution represents a very useful strategy for all future design.

The parameters that are fixed for each group of simulations in which the spacings vary are $\tilde{D}_0$ and Ra_*. The parameters $H/L = 1$ and $\text{Pr} = 0.7$ are fixed. The objective in every case is to maximize the global thermal conductance of the enclosure, namely,

$$C = \frac{Q'}{k(T_{\max} - T_\infty)} \tag{6.16}$$

When three sources are to be installed on the wall ($N = 3$) there are three degrees of freedom: the spacings S_0, S_1, and S_2. The optimal spacings depend on the Rayleigh number and the size (D_0) of the individual heat source, cf. Eq. (6.11). A sample of optimized configurations with three degrees of freedom is presented in Fig. 6.3. The spacings decrease as Ra_* increases, and the sources migrate toward the starting end of the wall boundary layer. The corresponding (maximized) global conductance of the arrangement increases as more heat sources are added (i.e., as N increases, Fig. 6.4), but the rate of increase slows down. Diminishing returns is an important characteristic in the evolution of complex constructal structures. It is also an important consideration in practice: how much optimized complexity is enough?

Another interesting effect becomes visible if we use larger heat sources and repeat the three-source optimization. Figure 6.5 shows the results for the optimal spacings when $\tilde{D}_0 = 0.2$. Unlike in Fig. 6.3, optimal spacings for $\tilde{S}_0$ and $\tilde{S}_1$ are found only if Ra_* is less than 3×10^3 and 5×10^2, respectively. At greater Ra_* values, the largest C values belong to designs with $\tilde{S}_0 = \tilde{S}_1 = 0$, which represent a continuous heating strip of size $2\tilde{D}_0$ placed in the starting corner of the hot wall. The coalescence of discrete heat sources at the start of the boundary layer is consistent with the theory developed earlier in this section. Coalescence strengthens as Ra_* increases.

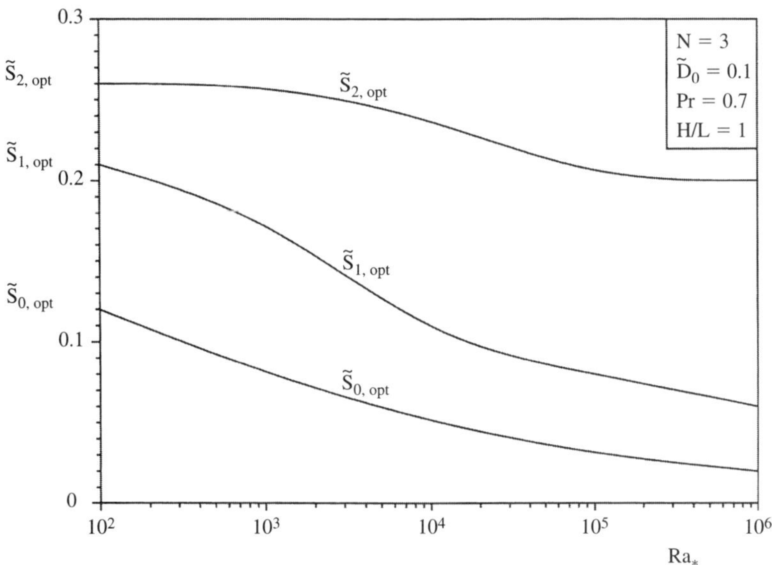

Figure 6.3 The optimal locations of three heat sources on the left wall of the enclosure of Fig. 6.2 [1].

A simpler application in which to use the strategy provided in this section is the design of a chimney for cooling a box of electronics by convection. A two-dimensional formulation of this problem is shown in Fig. 6.6. The vertical walls have heated sections (labeled $\tilde{T} = 1$) and adiabatic sections (labeled $\partial\tilde{T}/\partial\tilde{x} = 0$). The wall height is H. The transition from one type of section to the other occurs at $\tilde{y} = \tilde{H}_0$, where $\tilde{y} = y/H$ and $\tilde{H}_0 = H_0/H$. There are two choices: (a) heated entrance, and (b) heated exit. Which is better, when, and by how much?

From the preceding theory, the message of which is sketched qualitatively in Fig. 6.1, we know to expect that the better configuration must be (a). The exact answer is provided in Fig. 6.7, where $Ra_H = g\beta H^3(T_w - T_0)/(\alpha\nu)$, and $T_w - T_0$ is the temperature difference between the heated sections (isothermal) and the inflowing fluid. The angle $\gamma = 0$ degrees means that the two walls are parallel (vertical). The dimensionless global thermal conductance of the channel is

$$\tilde{q} = \frac{q''' H^2}{k(T_W - T_0)} \tag{6.17}$$

The heat-transfer density is $q''' = q'/(HD)$, where D is the wall to wall spacing, and q' is the total heat-transfer rate from one wall into the channel.

The configuration has two degrees of freedom, D/H and H_0/H, and $\tilde{q}_{max}$ is the value of $\tilde{q}$ maximized with respect to the spacing D. This preliminary optimization is illustrated in Fig. 6.8, where $\tilde{D} = D/H$. This preliminary step confirms the validity of the theory of section 3.3: there exists an optimal spacing for maximum

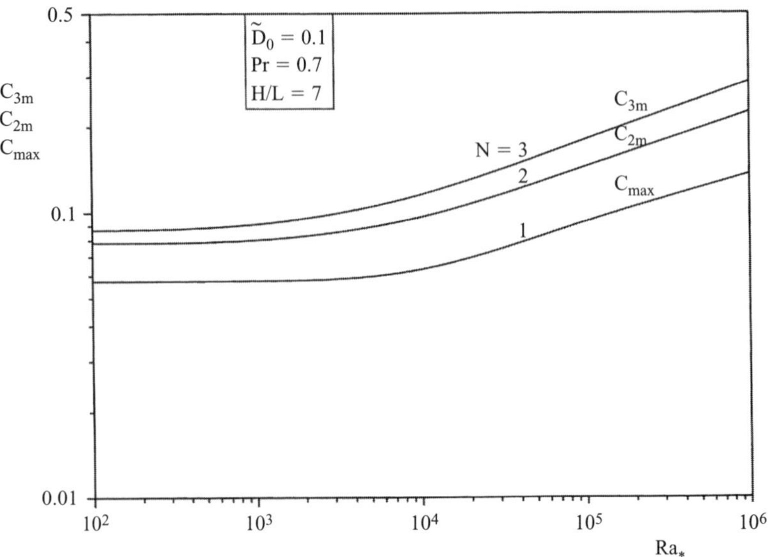

Figure 6.4 The effect of the number of discrete heat sources on the maximized global conductance of the enclosure of Fig. 6.2 [1].

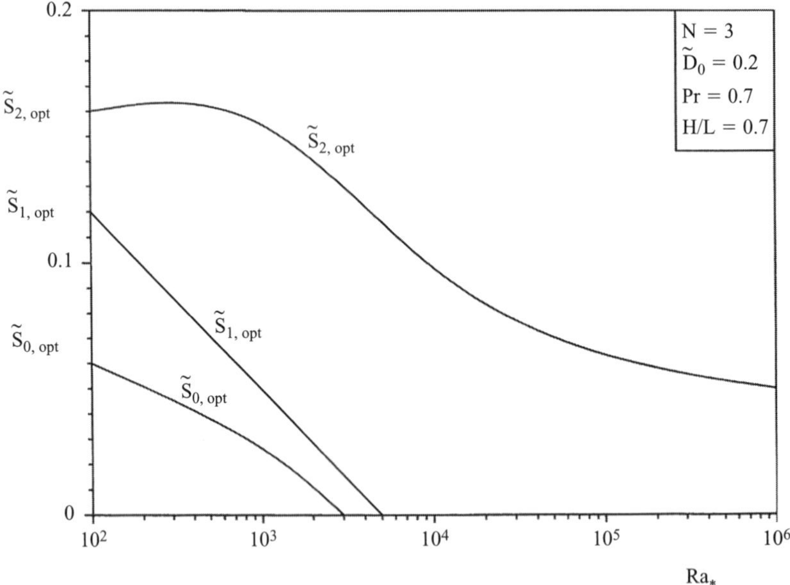

Figure 6.5 The optimal locations of three heat sources the sizes of which are larger than in Fig. 6.3 [1].

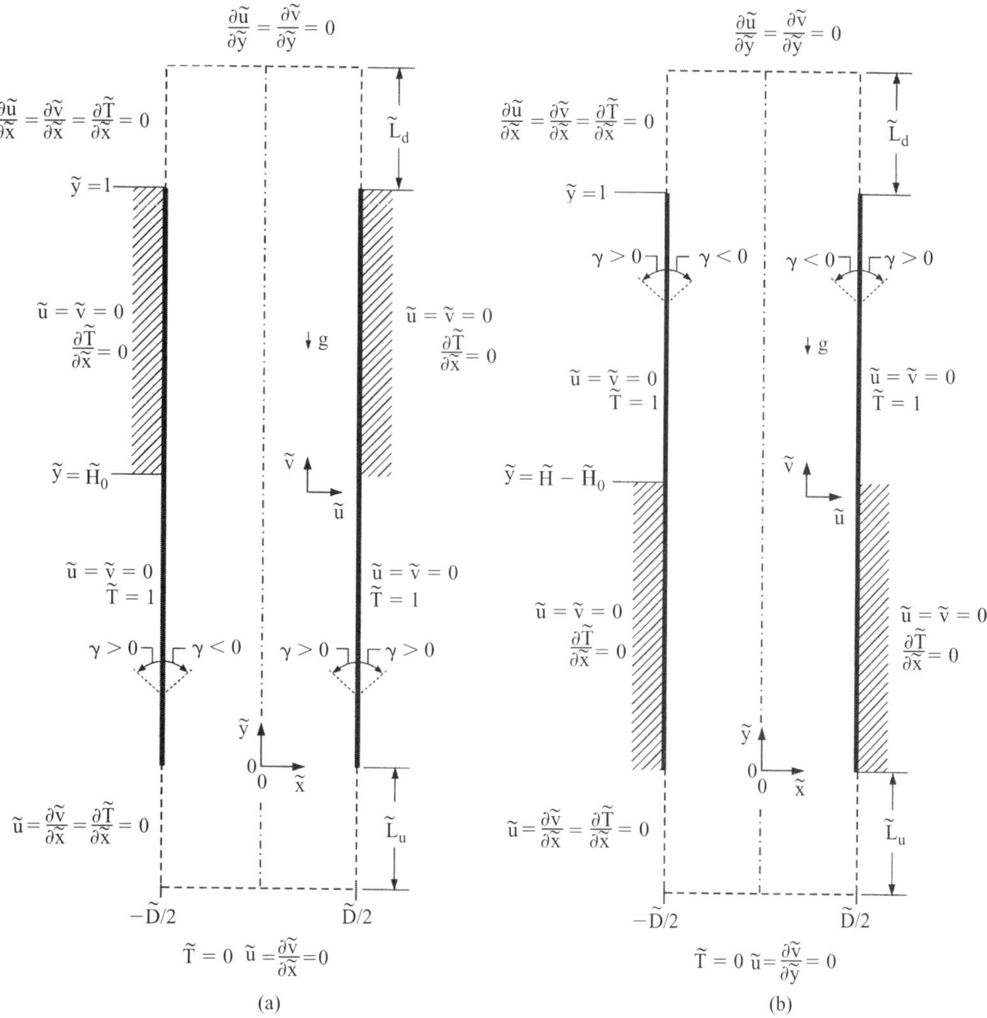

Figure 6.6 Vertical channel with nonuniform heating: (a) heated entrance, and (b) heated exit [5].

heat-transfer density by natural convection, and the "hard" numerical results confirm the estimate based on the intersection of asymptotes. The optimal spacing is such that $(D/H)\mathrm{Ra}_H^{1/4}$ is a number comparable with 3, and the maximum heat-transfer density $\tilde{q}_{\max}\mathrm{Ra}_H^{1/2}$ is a number comparable to but smaller than 0.5.

The heat-transfer densities maximized for configurations *a* and *b* in Fig. 6.7 show that configuration *a* is superior. The difference between *a* and *b* decreases as H_0 approaches H, that is, as the unheated section disappears. Additional recent work on the performance and maximization of natural convection cooling of discrete heat sources is available in Refs. [6] through [13].

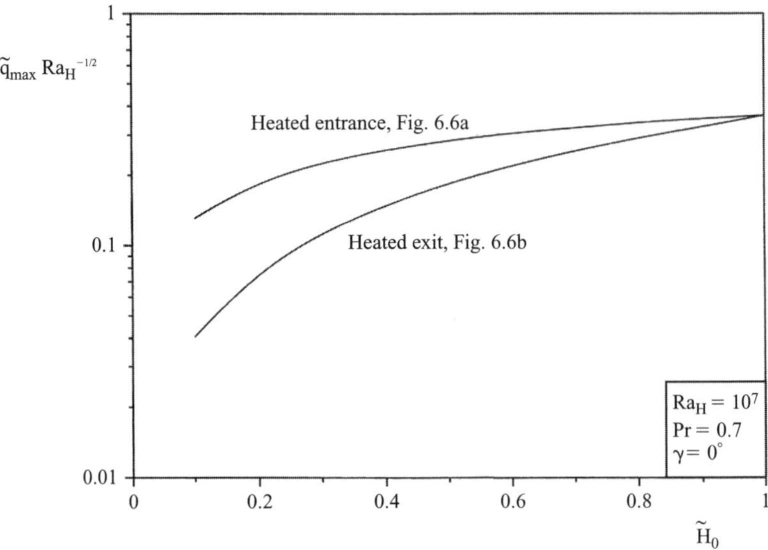

Figure 6.7 The effect of the size and location of the heated zones on the maximum heat-transfer rate density in the configurations of Fig. 6.6 [5].

6.2 DISTRIBUTION OF HEAT SOURCES COOLED BY FORCED CONVECTION

Internal forced convection continues to be one of the most common modes of heat-transfer in thermal design today. The reason is the strong emphasis on the miniaturization of cooling and heating configurations. Forced convection promises significantly greater compactness (greater heat-transfer density) than natural convection when dimensions decrease: review section 3.7.

In this section we consider the forced convection counterpart of the fundamental problem of distributing discrete heat sources on a wall [14]. Consider a horizontal plate of length L, which is in contact with a free stream of velocity U_∞ and temperature T_∞. The plate is heated by line heat sources of fixed strength q'[W/m]. The heat sources appear as points on the plate sketched in Fig. 6.9. Each line heat source extends in the direction perpendicular to the figure. The flow is two-dimensional and in the laminar boundary layer regime. The number of heat sources per unit of plate length defined in Eq. (6.1) is unknown.

In accordance with the method of constructal design, if T_{max} is the maximal temperature that must not be exceeded at the hot spots that occur on the plate, then the entire plate should operate at T_{max}. The problem is to determine the distribution of heat sources on the plate, $N(y)$, such that the wall temperature is near the allowed limit $T_w(x) = T_{max}$, constant. The analysis follows the same steps as in section 6.1. First, we assume that the density of line sources is sufficiently high so that we may express the distribution of discrete q' sources as a nearly continuous distribution of

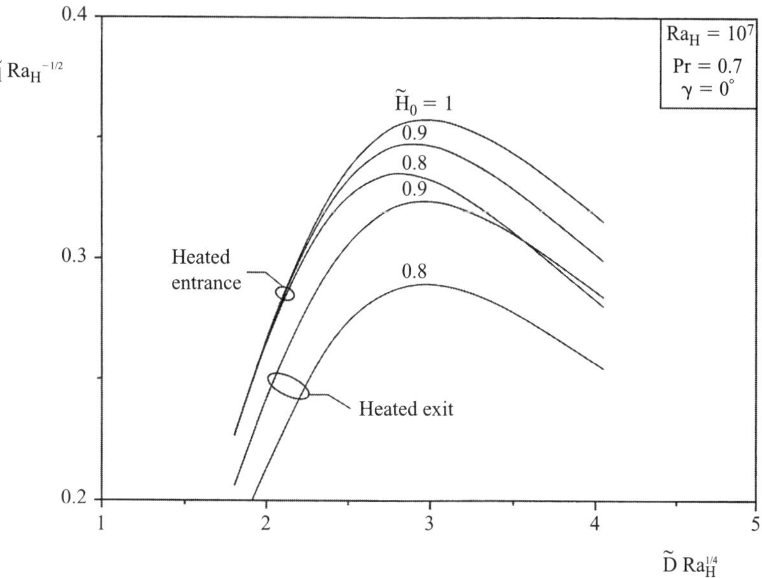

Figure 6.8 The effect of channel spacing and distribution of wall heating with the heat-transfer density in the configurations of Fig 6.6 [5].

heat flux:

$$q''(x) = q'N' \tag{6.18}$$

The heat flux distribution that corresponds to Eq. (6.18) and Pr $\gtrsim$ 1 is part of the solution for the laminar boundary layer on an isothermal wall [3], $\mathrm{Nu} = 0.332\,\mathrm{Pr}^{1/3}\,\mathrm{Re}_x^{1/2}$, or

$$\frac{q''x}{k\,(T_{\max} - T_\infty)} = 0.332\,\mathrm{Pr}^{1/3}\left(\frac{U_\infty x}{\nu}\right)^{1/2} \tag{6.19}$$

By eliminating q'' between Eqs. (6.18) and (6.19) we obtain the required distribution of heat sources:

$$N'(x) = 0.332\frac{k}{q'}\,(T_{\max} - T_\infty)\,\mathrm{Pr}^{1/3}\left(\frac{U_\infty}{\nu}\right)^{1/2} x^{-1/2} \tag{6.20}$$

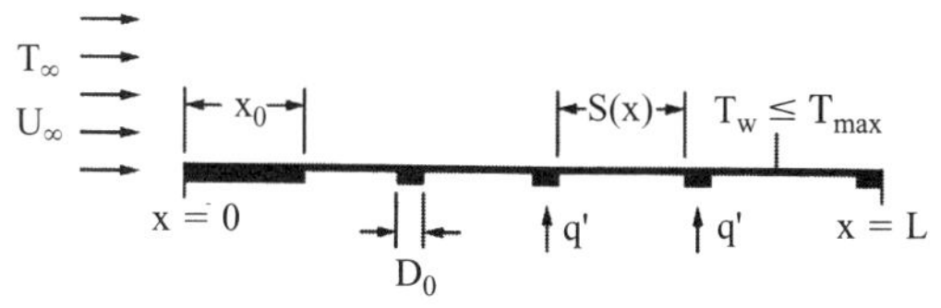

Figure 6.9 The multiple length scales of the distribution of finite-size heat sources on a horizontal plate [14].

The function $N'(x)$ represents the optimal configuration of heat sources. It shows that the sources must be positioned closer when they are near the start of the boundary layer. The total number of heat sources is

$$N = \int_0^L N' dx = 0.664 \frac{k}{q'} (T_{\max} - T_\infty) \, \mathrm{Pr}^{1/3} \mathrm{Re}^{1/2} \tag{6.21}$$

where $\mathrm{Re} = U_\infty L/\nu$. The rate of heat transfer from all the heat sources to the T_∞ fluid is

$$Q'_{\max} = q' N = 0.664 k \, (T_{\max} - T_\infty) \, \mathrm{Pr}^{1/3} \mathrm{Re}^{1/2} \tag{6.22}$$

This $Q'_{\max}$ expression is the same as the total heat transfer rate from an isothermal wall at $T_{\max}$. Equation (6.22) represents the maximized global performance of the wall with discretely distributed heat transfer.

The physical implementation of the optimal distribution is limited by an important manufacturing constraint: there exists a smallest scale in the design—the D_0 thickness of the line heat source. Features smaller than D_0 cannot be made. This constraint endows the design with structure, graininess (coarseness), and visibility.

The local spacing between two adjacent heat lines is $S(x)$. This spacing varies with x in accordance with the optimal N' distribution function, Eq. (6.20). The plate length interval that corresponds to a single line heat source q' is $D_0 + S(x)$. This means that the local number of heat sources per unit of wall height is

$$N'(x) = \frac{1}{D_0 + S(x)} \tag{6.23}$$

The strength of one source (q') is spread uniformly over the finite thickness of the source ($q''_0 = q'/D_0$). The heat flux q''_0 is a known constant, unlike the function $q''(x)$ of Eq. (6.19), which is the result of design. By eliminating $N'(x)$ between Eqs. (6.20) and (6.23), we obtain the rule for how the wall heating scheme should be constructed:

$$\frac{S(x)}{L} \cong \frac{3q' \, \mathrm{Pr}^{-1/3} \, \mathrm{Re}^{-1/2}}{k \, (T_{\max} - T_\infty)} \left(\frac{x}{L}\right)^{1/2} - \frac{D_0}{L} \tag{6.24}$$

The spacing S increases as x increases. Near the start of the boundary layer, the $S(x)$ function of Eq. (6.24) has negative values. This means that the above description breaks down in a region ($0 \le x \le x_0$) near the start of the boundary layer. Because D_0 is the smallest length scale of the structure, the spacings S cannot be smaller than D_0. We define x_0 as the longitudinal scale where S is as small as D_0 in an order of magnitude sense,

$$S \sim D_0 \quad \text{when } x \sim x_0 \tag{6.25}$$

By substituting this into Eq. (6.24) we determine the starting length scale over which Eq. (6.24) is not valid

$$\left(\frac{x_0}{L}\right)^{1/2} \sim 0.664 \frac{D_0}{L} \frac{k}{q'} (T_{max} - T_0) \, \mathrm{Pr}^{1/3} \mathrm{Re}^{1/2} \tag{6.26}$$

In summary, the wall structure has two distinct sections. Downstream of $x \sim x_0$, the wall is heated on discrete patches of length D_0, which are spaced according to Eq. (6.24). Upstream of $x \sim x_0$, the heat sources are mounted flush against each other. We model this starting section as one with uniform heat flux, in such a way that at the end of this section (at $x \sim x_0$) the wall temperature reaches the same maximum level (T_{max}) that the optimized spacings (6.24) are designed to maintain downstream of $x \sim x_0$. The wall temperature is T_0 at $x = 0$. It reaches T_{max} at the transition distance x_0, and continues undulating at T_{max} (and slightly under) from x_0 until L.

These basic features of the constructal design are illustrated in Fig. 6.9. The design has multiple length scales: L, D_0, x_0 and $S(x)$. The first two length scales are constraints. The last two are results of global maximization of performance in a morphing architecture subjected to the constraints. Taken together, the lengths represent multiscale constructal design—the flow architecture that brings the entire wall to the highest performance level possible.

The global heat-transfer performance of the optimal design can be estimated analytically in the limit where the number of heat-source strips D_0 is sufficiently large. In this limit, the integral (6.21) applies only in the downstream region of the plate ($x_0 < x < L$), where the concentrated sources are spaced according to Eq. (6.20). The heat transfer rate collected from $x \sim x_0$ to $x = L$ is

$$Q'_{x_0-L} \cong 0.664k(T_{max} - T_0)\mathrm{Pr}^{1/3}\mathrm{Re}^{1/2}\left[1 - \left(\frac{x_0}{L}\right)^{1/2}\right] \tag{6.27}$$

For the starting section of length x_0, we use the classical result for the wall with uniform heat flux, $\mathrm{Pr} \gtrsim 1$, and temperature T_{max} at $x \sim x_0$ (cf. Ref. [3]):

$$Q'_0 \cong 0.453k(T_{max} - T_0)\left(\frac{U_\infty x_0}{\nu}\right)^{1/2} \mathrm{Pr}^{1/3} \tag{6.28}$$

The total heat transfer rate can be expressed as

$$Q' \cong Q'_0 + Q'_{x_0-L} \cong Q'_{max}\left[1 - 0.318\left(\frac{x_0}{L}\right)^{1/2}\right] \tag{6.29}$$

where the approximate sign stems from the order of magnitude estimation of x_0, Eq. (6.26). In the limit $x_0/L \to 0$, the total heat transfer rate Q' approaches Q'_{max}, because in this limit the wall temperature rises uniformly to T_{max}. The right side of Eq. (6.26) shows that this limit is approached as D_0/L decreases, and as $q'/[k(T_{max} - T_0)]$ increases, that is, when the heat sources are concentrated, numerous, and strong.

The geometric features unveiled by the preceding analysis were verified based on numerical simulations of forced convection through a two-dimensional channel formed between two parallel plates [14]. On one of the surfaces facing the channel, several heat sources of size D_0 and strength q' were allowed to move freely until the heat transfer density of the entire channel was maximized. The other surface facing the channel was adiabatic. An additional degree of freedom of the flow structure (in addition to the spacings between the heat sources) was the transverse spacing between the two parallel plates. The agreement between the theoretically expected and the numerically generated architecture is similar to what we illustrated in section 6.1 for discrete sources with natural convection. The conclusion to both sections 6.1 and 6.2 is that the generation of constructal multiscale flow structures is useful, and that the analyses presented here constitute good and dependable strategy.

Even better flow architectures for high heat-transfer density can be generated by endowing the flow system with additional and more realistic features that can be changed freely [15]. An example is shown in Fig. 6.10. Here, we relaxed two of the classical assumptions that modelers of parallel-plate packages make. First, we did not assume that the thickness of a heat-generating plate is negligible. The numerical results showed that the effect of finite plate thickness is to generate secondary flow features such as stagnation, separation, and recirculation. These flow features have an effect on the optimized plate-to-plate spacings; however, this effect does not change the order of magnitude of the results.

The second classical assumption that we discarded is that the package contains a very large number of plates. The elemental symmetry (i.e., the repeatability of the channel flow field) that was the centerpiece of past numerical studies was replaced by the need to simulate the entire (asymmetric) flow field through the package and its immediate vicinity. The results show that when there are very few plates in the package, and when some of the plate surfaces are adiabatic, there is more than one optimal internal length scale. The package emerges as a structure with multiple scales, the sizes and positions of which are discovered. Flow

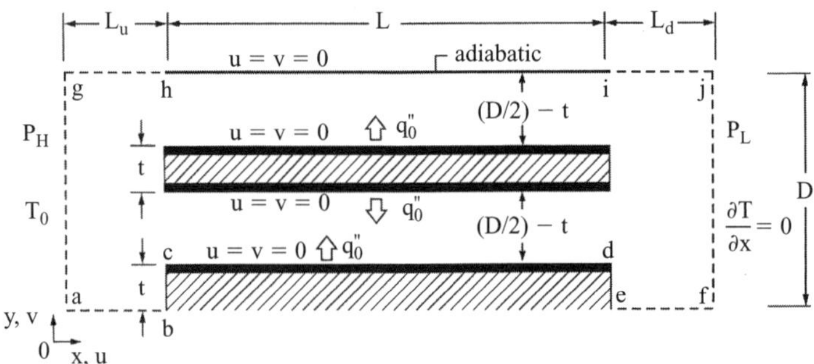

Figure 6.10 Multiscale configuration with forced convection through two unequal channels formed between plates with finite thickness [15].

structures with multiple scales that are distributed optimally (hence nonuniformly) are encountered elsewhere in constructal design and throughout this book.

The global performance of the package increases as the number of optimized dimensions (degrees of freedom) increases. Diminishing returns occur in the direction of higher optimized complexity.

Additional current work on forced convection in architectures with high heat-transfer density can be found in Refs. [16] through [22]. Future work may extend the strategy of this section to heat-transfer package models with even higher levels of complexity and realism. Good candidates are packages with larger numbers of components and asymmetric surface heating conditions, three-dimensional flow simulations that account for all the walls that confine the package, and the finite thermal conductivity of the parts that generate heat.

Another aspect that deserves future study is the robustness of this class of constructal structures with multiple scales, and the performance of corresponding structures built based on existing results for a single spacing. If the difference is not great, then the constructal complex flow structure based on a single constructal spacing is robust, and it promises to perform close to the highest level even if not every single degree of freedom is exploited during the development of the flow architecture. Robustness is good for performance: it places many "winners" in the vicinity of the equilibrium flow structure in the design space (e.g., Figs. 3.2 and 11.10).

6.3 MULTISCALE PLATES FOR FORCED CONVECTION

A key result of constructal theory is the prediction of spacings for the internal flow structure of volumes that must transfer heat and mass to the maximum. This discovery holds for both forced and natural convection, and is outlined in sections 3.3 and 3.4. Constructal spacings have been determined for several configurations, depending on the shape of the heat-transfer surface that is distributed through the volume: stacks of parallel plates, bundles of cylinders in crossflow, and arrays of staggered plates (e.g., Fig. 6.11). In each configuration, the reported optimal spacing is a single value, that is, a *single length scale* that is distributed uniformly through the available volume.

Is the stack of Fig. 6.11 the best way to pack heat transfer into a fixed volume? It is, but only when a single length scale is to be used, that is, if the structure is to be *uniform*. The structure of Fig. 6.11 is uniform because it does not change from $x = 0$ to $x = L_0$. At the most, the geometries of single-spacing structures vary periodically, as in the case of arrays of cylinders and staggered plates.

6.3.1 Forcing the Entire Flow Volume to Work

The structure of Fig. 6.11 can be improved if more length scales (D_0, D_1, D_2, ...) are available [23]. The technique consists of installing more heat transfer

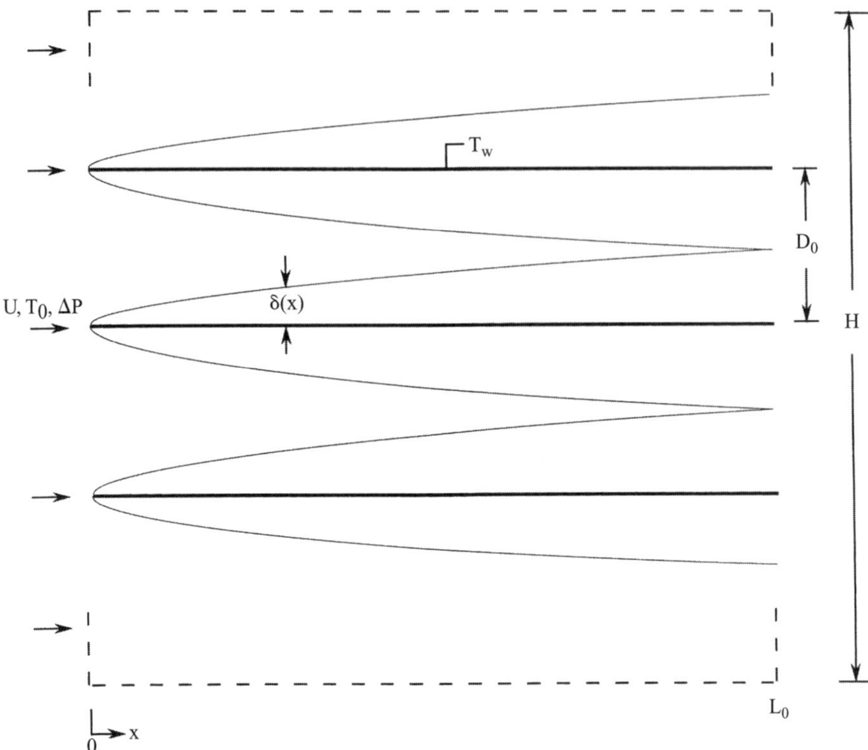

Figure 6.11 Constructal package of parallel plates with one spacing and forced convection [23].

in regions of the volume HL_0 where the boundary layers are thinner. Those regions are situated immediately downstream of the entrance plane $x = 0$. Regions that do not work for the global enterprise must be either put to work or eliminated. In Fig. 6.11 the wedges of fluid contained between the tips of opposing boundary layers are not involved in transferring heat. They can be involved if heat-generating blades of shorter lengths (L_1) are installed on their planes of symmetry. This new design is shown in Fig. 6.12.

Each new L_1 blade is coated by laminar boundary layers [3] with the thickness

$$\delta(x) \cong 5x(Ux/\nu)^{-1/2} \tag{6.30}$$

Because δ increases as $x^{1/2}$, the boundary layers of the L_1 blade merge with the boundary layers of the L_0 blades at a downstream position that is approximately equal to $L_0/4$. The approximation is due to the assumption that the presence of the L_1 boundary layers does not affect significantly the downstream development ($x > L_0/4$) of the L_0 boundary layers. This assumption is made for the sake of simplicity. The order-of-magnitude correctness of this assumption is clear, and it comes from geometry: the edges of the L_1 and L_0 boundary layers must intersect at a distance

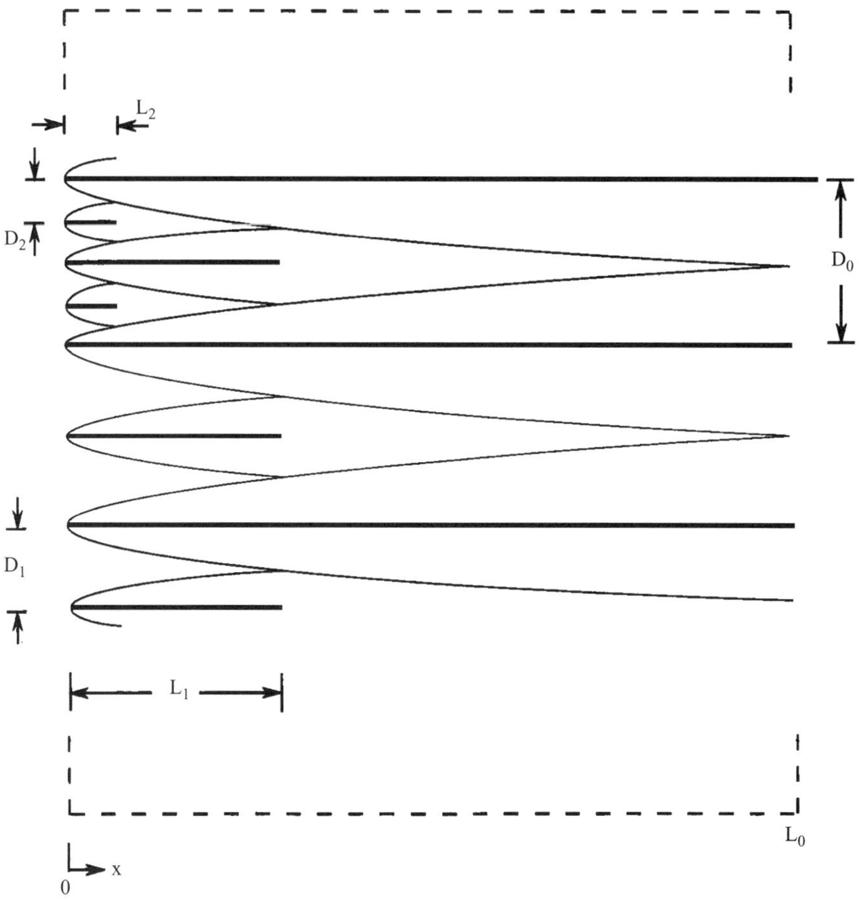

Figure 6.12 Constructal multiscale package of parallel plates with forced convection [23].

of order

$$L_1 \cong \frac{1}{4} L_0. \tag{6.31}$$

Note that by choosing L_1 such that the boundary layers that coat the L_1 blade merge with surrounding boundary layers at the downstream end of the L_1 blade, we invoke once more the constructal packing principle of sections 3.3 and 3.4 and Fig. 6.11. We are being *consistent* as constructal designers and, because of this, every structure with merging boundary layers will be constructal, optimal, or near optimal, no matter how complicated.

The wedges of isothermal fluid (T_0) remaining between adjacent L_0 and L_1 blades can be populated with a new generation of even shorter blades, $L_2 \cong L_1/4$. Two such blades are shown in the upper-left corner of Fig. 6.12. The length scales become smaller (L_0, L_1, L_2), but the shape of the boundary layer region is the same for all the blades, because the blades are all swept by the same flow (U). The merging and expiring boundary layers are arranged according to the algorithm

$$L_i \cong \frac{1}{4} L_{i-1}, \quad D_i \cong \frac{1}{2} D_{i-1}, \quad (i = 1, 2, \ldots, m) \tag{6.32}$$

for which we will show that m must be finite, not infinite. In other words, as was stressed when constructal tree structures were discovered [24], the image generated by the algorithm (6.32) is not a fractal object. The sequence of decreasing length scales is finite, and the smallest size (D_m, L_m) is predictable, as we will see in Eq. (6.47).

To complete the description of the sequential construction of the multiscale flow structure we note that the number of blades of a certain size increases as the blade size decreases. Let $n_0 = H/D_0$ be the number of L_0 blades in the uniform structure of Fig. 6.11, where

$$D_0 \cong 2\delta(L_0) \cong 10 \left(\frac{\nu L_0}{U} \right)^{1/2} \tag{6.33}$$

The number of L_1 blades is $n_1 = n_0$, because there are as many L_1 blades as there are D_0 spacings. At scales smaller than L_1, the number of blades of one size doubles with every step:

$$n_i = 2n_{i-1} \quad (i = 2, 3, \ldots, m) \tag{6.34}$$

Two conflicting features emerge as the structure grows in the sequence started in Fig. 6.12. One is attractive: the total surface of temperature T_w installed in the HL_0 volume increases. The other is detrimental: the flow resistance increases, the flow rate driven by the fixed ΔP decreases, and so does the heat-transfer rate associated with a single boundary layer. The meaningful question is how the volume is being used: what happens to the heat-transfer rate *density* as complexity increases?

6.3.2 Heat Transfer

The total heat-transfer rate from the T_w surfaces to the T_0 fluid can be estimated by summing up the contributions made by the individual blades. The heat-transfer rate through one side of the L_0 blade is equal (in an order of magnitude sense) to the heat-transfer rate associated with a laminar boundary layer (in accordance with the Pohlhausen solution for Prandtl numbers of order 1; e.g., Ref. [3]):

$$\frac{\bar{q}_0''}{\Delta T} \frac{L_0}{k} \cong 0.664 \left(\frac{U L_0}{\nu} \right)^{1/2} \tag{6.35}$$

Here, $\bar{q}_0''[W/m^2]$ is the L_0-averaged heat flux, $\Delta T = T_w - T_0$, and k is the fluid thermal conductivity. There are $2n_0$ such boundary layers, and their combined contribution to the total heat-transfer rate of the package of Fig. 6.12 is

$$\bar{q}_0' = 2n_0\bar{q}_0'L_0 \cong 1.328k\Delta T n_0 \left(\frac{UL_0}{\nu}\right)^{1/2} \tag{6.36}$$

The same calculation can be performed for any group of blades of one size, L_i. Their total heat transfer rate q_i'[W/m] is given by a formula similar to Eq. (6.36), in which n_0 and L_0 are replaced by n_i and L_i. The heat-transfer rate of all the blades is the sum

$$q' = \sum_{i=0}^{m} q_i' \cong 1.328k\Delta T n_0 \left(\frac{UL_0}{\nu}\right)^{1/2} S \tag{6.37}$$

where S is a sum that depends solely on geometry:

$$S = 1 + \frac{n_1}{n_0}\left(\frac{L_1}{L_0}\right)^{1/2} + \frac{n_2}{n_0}\left(\frac{L_2}{L_0}\right)^{1/2} + \cdots + \frac{n_m}{n_0}\left(\frac{L_m}{L_0}\right)^{1/2} = 1 + \frac{m}{2} \tag{6.38}$$

This analysis confirms the anticipated trend: the total heat-transfer rate increases monotonically as the complexity of the structure (m) increases.

6.3.3 Fluid Friction

It is necessary to evaluate the flow resistance of the multiscale structure, because the velocity U that appears in Eq. (6.37) is a result, not an assumption. The pressure difference ΔP is specified, and it is related to all the friction forces felt by the blades. We rely on the same approximation as in the case of heat transfer, and estimate the friction force along one face of one blade by using the solution for the laminar boundary layer (e.g., Ref. [3]):

$$\tau_i \cong C_{fi}\frac{1}{2}\rho U^2, \qquad C_{fi} = \frac{1.328}{(UL_0/\nu)^{1/2}} \tag{6.39}$$

Here, τ_i and C_{fi} are the averaged shear stress and skin friction coefficient, respectively. The total force felt by the blades of size L_i is

$$F_i = 2n_i\tau_i L_i \cong 1.328\rho(\nu L_i)^{1/2} n_i U^{3/2} \tag{6.40}$$

where F_i is expressed in N/m. The total force on the multiscale package is

$$F = \sum_{i=0}^{m} F_i \cong 1.328\rho(\nu L_0)^{1/2} n_0 U^{3/2} S \tag{6.41}$$

This force is balanced by the longitudinal force imposed on the control volume $\Delta PH = F$, which combined with Eq. (6.41) and the D_0 formula (6.33), yields the order of magnitude of the average velocity of the fluid that permeates the structure:

$$U \cong 2.7 \left(\frac{\Delta P}{\rho S} \right)^{1/2} \tag{6.42}$$

This result confirms the second expected trend: the flow slows down as the complexity of the structure (S, or m) increases.

6.3.4 Heat Transfer Rate Density: The Smallest Scale

Putting together the results of the heat transfer and fluid flow analyses, we find how the structure performs globally when its constraints are specified (ΔP, $\Delta T, H$, L_0). Eliminating U between Eqs. (6.37) and (6.42) yields the dimensionless global thermal conductance,

$$\frac{q'}{k\Delta T} \cong 0.36 \frac{H}{L_0} \mathrm{Be}^{1/2} S^{1/2} \tag{6.43}$$

where the dimensionless pressure drop is

$$\mathrm{Be} = \frac{\Delta P L_0^2}{\mu \alpha} \tag{6.44}$$

In this expression μ and α are the viscosity and thermal diffusivity of the fluid. The alternative to using the global conductance is to use the heat transfer rate density $q''' = q'/HL_0$. Both quantities increase with the applied pressure difference (Be) and the complexity of the flow structure (S, or m). In conclusion, in spite of the conflicting effects of S in Eqs. (6.37) and (6.42), the effect of increasing S is beneficial from the point of view of packing more heat transfer in a given volume.

How large can the factor S be? The answer follows from the observation that the geometry of Fig. 6.12 is valid when boundary layers exist, that is, when the boundary layers are *distinct*. To be distinct, boundary layers must be slender. Figure 6.12 makes it clear that boundary layers are less slender when their longitudinal scales (L_i) are shorter. The shortest blade length L_m below which the boundary layer convection mechanism breaks down is

$$L_m \sim D_m \tag{6.45}$$

In view of Eqs. (6.32), this means that

$$L_0 \sim 2^m D_0 \tag{6.46}$$

Finally, by using Eqs. (6.33) and (6.42) we find the smallest scale, which occurs at the level m given implicitly by

$$2^m \left(1 + \frac{m}{2}\right)^{1/4} \sim 0.17 \mathrm{Be}^{1/4} \tag{6.47}$$

In view of the order-of-magnitude character of the analysis based on Eq. (6.45), the right side of Eq. (6.47) is essentially $(\text{Be}/10^3)^{1/4}$. Equation (6.47) establishes m as a slowly varying monotonic function of $\text{Be}^{1/4}$. This function can be substituted in Eq. (6.43) to see the complete effect of Be on the global heat-transfer performance:

$$\frac{q'}{k\Delta T} \cong 0.36 \frac{H}{L_0} \text{Be}^{1/2} \left(1 + \frac{1}{2}m\right)^{1/2}. \tag{6.48}$$

In conclusion, the required complexity (m) is *finite*, and it increases monotonically with the imposed pressure difference (Be). More flow means more length scales, and smaller smallest length scales. The structure becomes not only more complex but also finer.

The monotonic effect of m is accompanied by diminishing returns. Each new length scale (m) contributes to global performance less than the preceding length scale ($m - 1$).

The hierarchical multiscale flow architecture constructed in this section is one more comment on fractal geometry. Fractal structures are generated by assuming (postulating) certain algorithms. In the fractal literature, the algorithms are selected such that the resulting structures resemble flow structures observed in nature. For this reason, fractal geometry is a descriptive method, not a predictive theory [24,25].

Contrary to the fractal approach, the constructal method used in this section generated the construction algorithms [Eqs. (6.32) and (6.34)], including the smallest-scale cutoff, Eq. (6.47). The algorithms were generated by the constructal law of increase of flow access: more heat transfer rate from a fixed volume. This principle was invoked every time the optimal spacing between two blades was used. Optimal spacings were assigned to all the length scales, and were distributed throughout the volume. With regard to fractal geometry and why (empirically) some fractal structures happen to resemble natural flow structures, the missing link has been the *origin* of the algorithm [26]. Constructal theory delivers the algorithm as a result of the constructal law.

The multiscale flow architectures developed in this section were validated based on numerical simulations and optimization of geometry [27]. The regime was laminar forced convection in the range $10^5 \leq \text{Be} \leq 10^8$ and $\text{Pr} = 1$. Structures with one, two, and three length scales confirmed the theoretical features, including the diminishing returns associated with increasing m.

6.4 MULTISCALE PLATES AND SPACINGS FOR NATURAL CONVECTION

Forced convection was used in the preceding section only for illustration, that is, as a flow mechanism for which to build the multiscale structure. A completely analogous multiscale structure can be deduced for laminar natural convection. The complete analogy that exists between optimal spacings in forced and natural convection was described by Petrescu [28]. In brief, if the structure of Fig. 6.11 is rotated by 90 degrees counterclockwise, and if the flow is driven upward by the buoyancy effect, then the role of the overall pressure difference ΔP is played by the difference between two hydrostatic pressure heads, one for the fluid column of height L_0 and temperature T_0, and the other for the L_0 fluid column of temperature T_w. If the Boussinesq approximation applies, then the effective pressure difference that drives the flow is [3],

$$\Delta P = \rho g \beta \Delta T L_0 \tag{6.49}$$

where $\Delta T = T_w - T_0$, β is the coefficient of volumetric thermal expansion, and g is the gravitational acceleration aligned vertically downward (against y in Fig. 6.13). By substituting the ΔP expression (6.49) into the Be definition (6.44) we find that the dimensionless group that replaces Be in natural convection is the Rayleigh number $\mathrm{Ra} = g\beta \Delta T L_0^3/(\alpha\nu)$. Other than the Be $\rightarrow$ Ra transformation, all the features that are due to the generation of multiscale blade structure for natural convection should mirror the features described for forced convection in section 6.3.

The upper portions of Fig. 6.13 show another way to arrive at the optimal single-spacing conclusion of Chapter 3. This other way is the idea that a convective flow has a "body", which must be fitted "just right" in the space between the solid walls of the channel [29] (cf. section 3.6). The designer must first "see" the body and then fit the walls around the body. The foot comes before the shoe.

In Fig. 6.13 (top), thermal boundary layers develop as the fluid sweeps the heat generating surfaces, and a volume of unheated fluid appears between the boundary layers. The amount of unheated fluid depends on the shape of the total volume V. Two extreme shapes of V are shown in Figs. 6.13a and 6.13c. If V is too square, Fig. 6.13a, the unheated region is large, and the volume occupied by working fluid (the boundary layers) is small. If V is too slender, Fig. 6.13c, the boundary layers merge early, and the stream warms up. It becomes "overworked", that is, poor as a coolant. These two extremes suggest the existence of an optimal volume shape: V is occupied most fully by working fluid when its boundary layers merge at the end of the channel, Fig. 6.13b. This principle is employed in the lower part of Fig. 6.13, where a fixed two-dimensional volume $L_0 \times W$ is filled optimally with vertical plates cooled by laminar natural convection. The spacing D_0 is such that the thermal boundary layers merge at the top of the structure. This is the geometric meaning of the intersection of asymptotes executed in section 3.3.

The generation of vertical multiscale plates with natural convection was performed [30] on configurations such as Fig. 6.12 turned 90 degrees counterclockwise. The parametric domain in which these architectures have been generated was laminar natural convection with $10 \leq \mathrm{Ra} \leq 10$ and $\mathrm{Pr} = 0.7$. The plates had one, two, and three length scales. The numerically generated features confirmed the theoretical features. Figure 6.14 shows the numerical temperature distribution inside a section of the multiscale package, for three optimized structures: $m = 0$, 1,

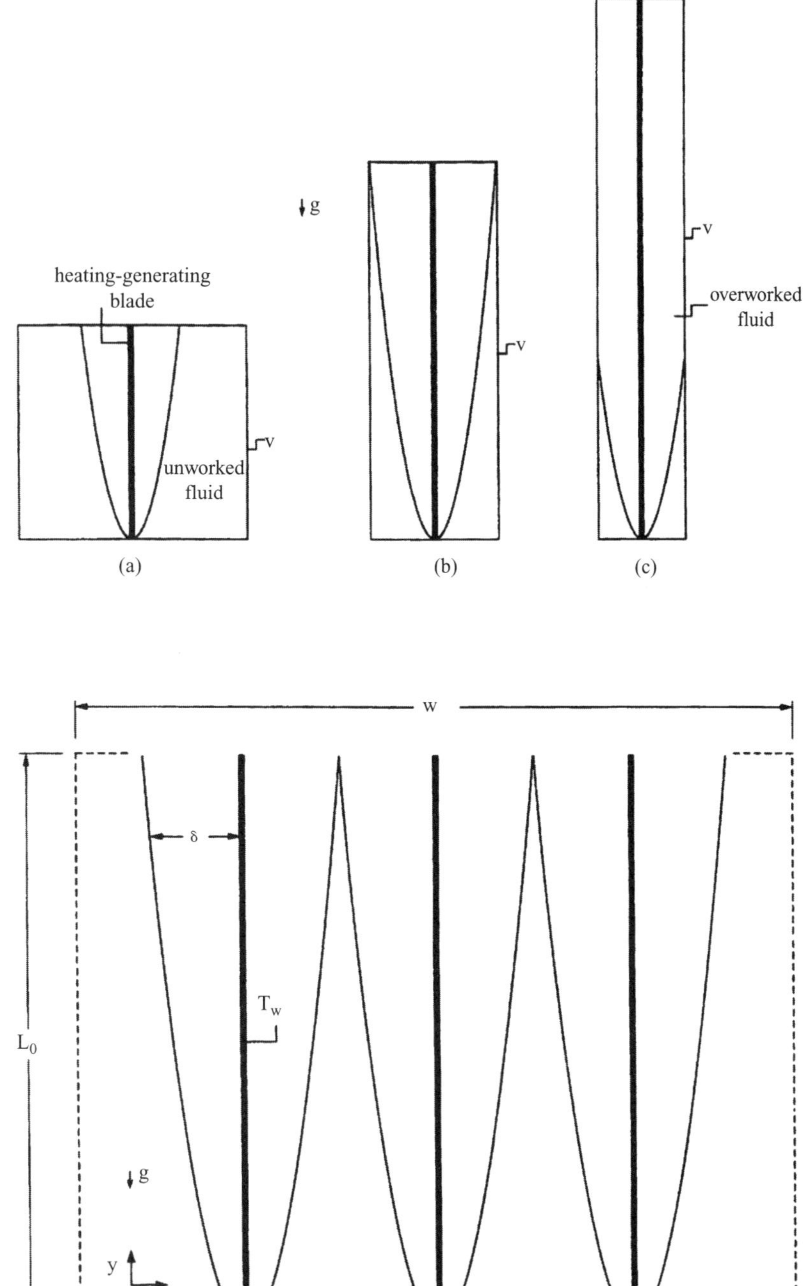

Figure 6.13 Constructal package of vertical plates with one spacing [30]. The upper drawing shows the optimization of the shape of one elemental volume.

Figure 6.14 Optimal distribution of imperfection: the temperature distribution in multiscale natural convection packages with one, two, and three length scales [30]. See also Color Plate 9.

and 2 at Ra $= 10^6$ and Pr $= 0.7$. The section consists of only four L_0-long channels. The temperature ranges between two main colors, red ($\tilde{T} = 1$) and blue ($\tilde{T} = 0$). As the number of length scales increases, the color red is distributed more uniformly, illustrating the progress toward maximal heat-transfer rate density, or optimal distribution of imperfection.

6.5 MULTISCALE CYLINDERS IN CROSSFLOW

What we demonstrated for packages of multiscale plates in sections 6.3 and 6.4 holds for packages of more complicated components that generate heat. Multiple scales that are distributed wisely (hence nonuniformly) are good. It pays to search for and find the best place and best neighbors for every component. This yields the best arrangement of seemingly "diverse" components: this is the origin of *diversity*, that is, multiscale components that give the uneducated eye the impression that the arrangement is random. It is not; in fact, it is the opposite of random. It is deterministic.

We see this if we consider volumes filled with components that are shaped so that they are more difficult to pack than the parallel plates considered until now. Parallel cylinders are less well behaved because they leave open spaces between them even when they are pressed tightly against each other. They do not fit. In spite of this, uniform distribution is the norm: throughout heat exchanger design, the philosophy is to assemble cylinders in crossflow by using one cylinder size and one cylinder-to-cylinder spacing throughout the volume (e.g., Fig. 3.9b). One size fits all.

The constructal approach is to free the solid to change size and location so that it fills the available flow volume best. This leads to configurations such as Fig. 6.15,

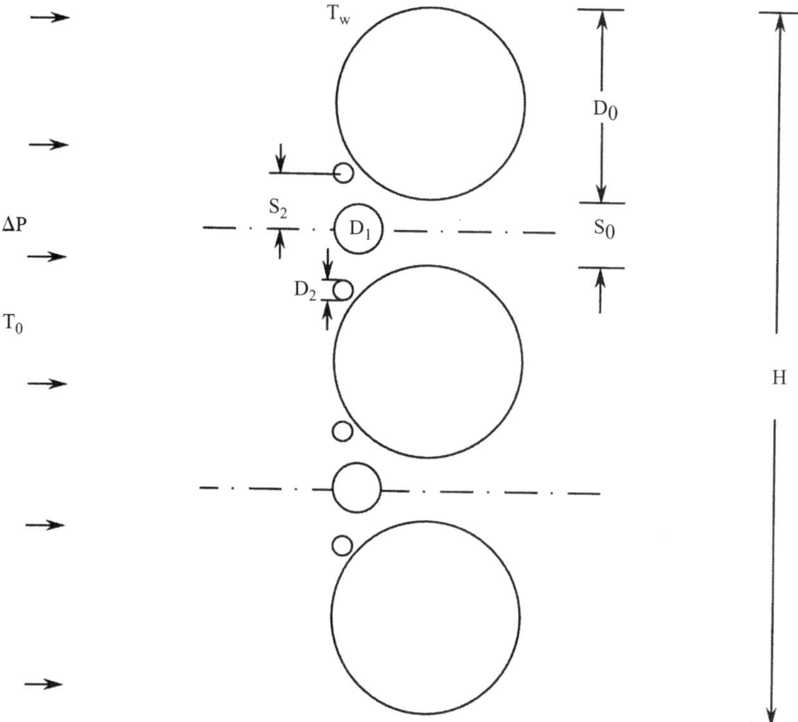

Figure 6.15 Column of cylinders with three diameters (D_0, D_1, D_2) cooled by forced convection [31].

where the available flow space $H \times D_0$ is occupied by cylinders of three sizes (D_0, D_1, D_2) [31]. According to the old philosophy, only D_0-cylinders would inhabit the flow space. The smaller cylinders (D_1, D_2) could be placed anywhere in the $H \times D_0$ rectangle, but, after a search of the kind illustrated in sections 6.3 and 6.4, the best places for them are such that they are tangent to the inlet plane. In other words, the configuration shown in Fig. 6.15 has already been optimized with respect to the horizontal positions of the D_1 and D_2 cylinders.

The degrees of freedom that are left in Fig. 6.15 are four: the diameters D_1 and D_2 and the spacings S_0 and S_2. These dimensions can be optimized such that the total heat-transfer rate from the cylinders (T_w) to the inflowing coolant (T_0) is maximum. The resulting structures have multiple scales that are distributed nonuniformly. Bello-Ochende and Bejan [31] did this work for columns of cylinders with one, two, and three length scales. Figure 6.16 shows how these optimized multiscale structures change when the specified pressure difference (ΔP) across the column changes. The pressure difference number is defined as $\text{Be} = \Delta P D_0^2/(\alpha\mu)$.

Read from left to right, Fig. 6.16 shows how the spacing between older cylinders increases when new cylinders are placed in the existing gaps. The montage shows that when the flow becomes faster (larger Be) the spacings become noticeably smaller, while the cylinder diameters do not change much. This observation is

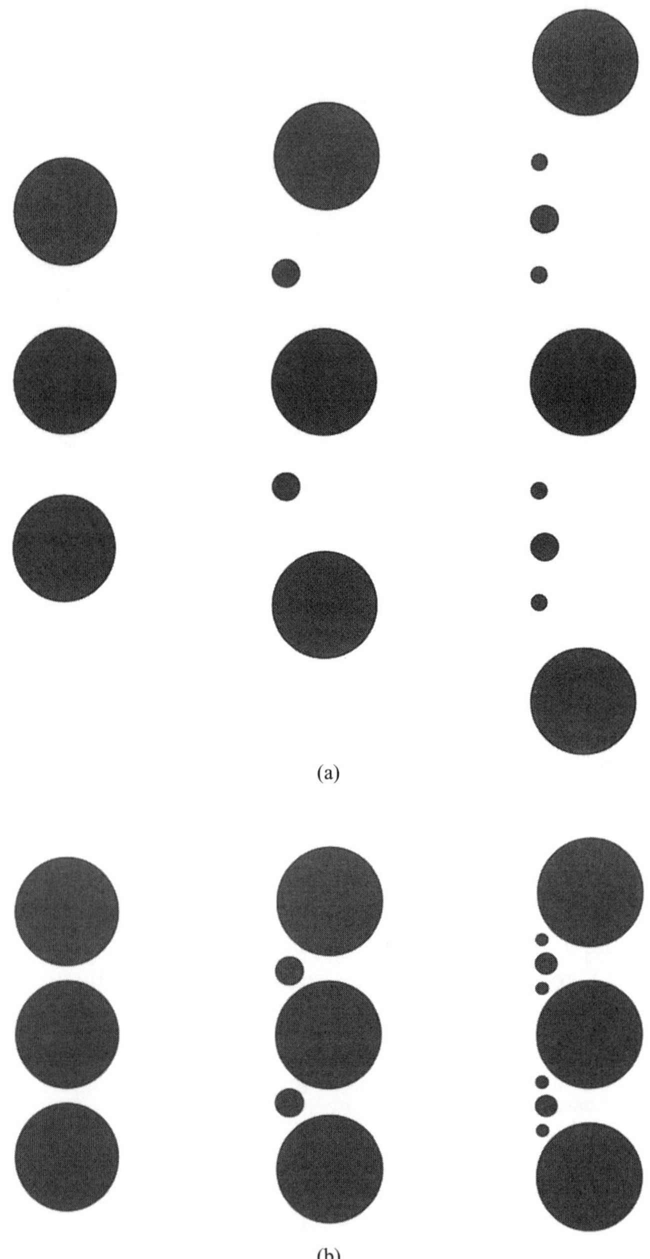

Figure 6.16 Columns of cylinders with one, two, and three diameters: (a) Be = 10^3; (b) Be = 10^6. See also Color Plate 10.

relevant not only in heat exchanger design but also in animal design. A morphing multiscale structure is robust when it can perform optimally under different flow conditions (slow, fast) by using the same solid parts. It is a lot easier for the structure to adapt itself by adjusting its fluid spacings, as opposed to redesigning its solid components. One example from the animal realm is the multichannel organization of a swarm of bees [32]. The solid components (the bees) are permanent features, while the airways formed between bees change with the inlet temperature of the ambient air that cools the swarm.

Similar multiscale structures and performance qualities have been demonstrated for cylinders in crossflow natural convection [33]. The reason for the analogy is the correspondence between Be and the Rayleigh number [28], as we explained at the start of section 6.4.

Another line of new work is the multiscale distributing of discrete heat sources cooled by natural convection in a fixed space. The proposal made in Ref. [1] and section 6.1 has been validated and extended numerically and by means of genetic algorithms and neural networks [34–36].

The fundamental value of the design strategy taught in this chapter is that multiscale flow structures are applicable to every sector of heat exchanger design. The novelty is the increase in heat-transfer density and the *nonuniform distribution of length scales* through the available space. This approach promises leads to the discovery of new and *unconventional* internal flow structures for heat exchangers and cooled electronic packages.

6.6 MULTISCALE DROPLETS FOR MAXIMUM MASS TRANSFER DENSITY

Figures 6.12 through 6.16 showed how to construct the multiscale hierarchical structure for maximum heat-transfer density in a given volume with forced or natural convection. The following scale analysis outlines the structure of a multiscale concept for a mass exchanger with maximum mass transfer density. Spherical drops of one liquid are injected into another liquid, and both liquids move along a pipe. For simplicity, assume no relative motion between drops and surrounding liquid: both move in one direction with the speed V, and travel the pipe length L_0. The travel time associated with V and L_0 is

$$t_0 = \frac{L_0}{V} \tag{6.50}$$

Mass diffuses into the drops with the mass diffusivity D_i, and into the surrounding liquid with the mass diffusivity D_e.

We seek the configuration of the flow system—the hierarchy of the population of drops—that packs maximum mass transfer in the liquid flow space of length L_0. The answer, as we saw in Fig. 6.12, is to fill the L_0 long space with liquid (drops and interstices) that are just penetrated by mass diffusion as they leave the flow space.

We begin with the largest drops (radius R_0), which require the entire time t_0 (or the longest flow length L_0) to be fully penetrated by mass diffusion. This requirement, the full penetration of R_0 at $t \sim t_0$, yields the scale of the largest drops:

$$R_0 \sim (D_i t_0)^{1/2} \tag{6.51}$$

On the outside of R_0, diffusion is controlled by D_e and time. If $D_e \gg D_i$, the diffusion thickness into the D_e liquid is $(D_e t_0)^{1/2}$, and so must be the spacing between the centers of adjacent R_0 drops:

$$S_0 \sim (D_e t_0)^{1/2} \quad \text{if} \quad D_e \gg D_i \tag{6.52}$$

If $D_e \ll D_i$, the external diffusion thickness is again $(D_e t_0)^{1/2}$, but the center-to-center spacing between R_0 drops is

$$\begin{aligned} S_0 &\sim R_0 + (D_e t_0)^{1/2} \\ &\sim t_0^{1/2}\left(D_i^{1/2} + D_e^{1/2}\right) \end{aligned} \tag{6.53}$$

We retain Eq. (6.53) because it is more general than Eq. (6.52). The way in which D_i and D_0 diffusion collaborate to determine R_0 and S_0 during t_0 is sketched in Fig. 6.17.

Because of the similarities between Figs. 6.17 and 6.12, the rest of the constructal packing sequence of the L_0 space is evident. In the space between three (or four) R_0 drops, we inject one R_1 drop at the same time (at $t = 0$). The R_1 drop is fully

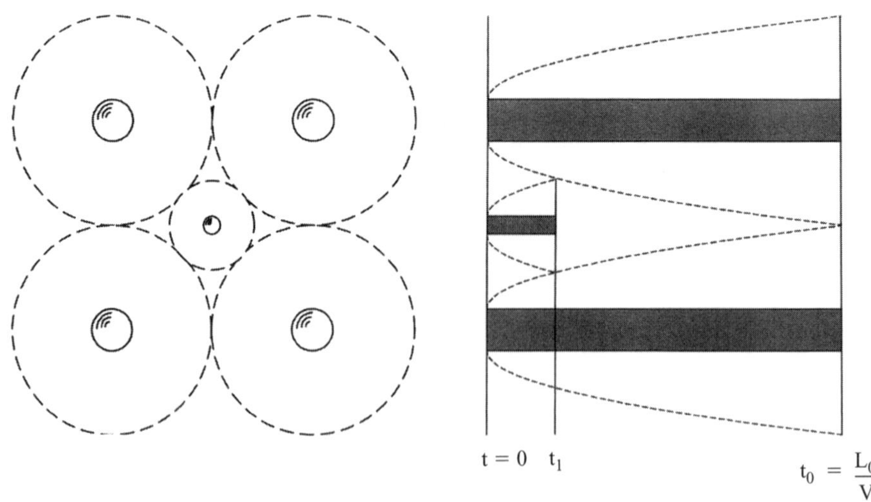

Figure 6.17 Square pattern for the injection of multiscale droplets in a liquid flow system.

penetrated by D_i diffusion during the time t_1, hence

$$R_1 \sim (D_i t_1)^{1/2} \tag{6.54}$$

On the outside of R_1, diffusion penetrates a liquid shell of thickness $(D_e t_1)^{1/2}$. For maximum packing (Fig. 6.12), this external thickness and the external thickness of the R_0 drops (during the same t_1) must add up to S_0 (recall that in scale analysis we neglect factors of order one [3]):

$$(D_e t_1)^{1/2} + (D_e t_1)^{1/2} \sim S_0 \tag{6.55}$$

After using Eq. (6.53), we find that

$$\left(\frac{t_1}{t_0}\right)^{1/2} \sim \frac{1}{1 + (D_i/D_e)^{1/2}} \leq 1 \tag{6.56}$$

or that

$$\frac{t_1}{t_0} \sim f < 1, \quad \text{constant} \tag{6.57}$$

In conclusion,

$$\frac{R_1}{R_0} \sim f^{1/2} < 1 \tag{6.58}$$

How many R_1 drops should be injected? Assume a square pattern of injecting R_0 drops in the entrance plane (Fig. 6.17). If there are n_0 drops of size R_0 in the entrance plane (perpendicular to L_0), then in the same plane, and at the same time, there should be $n_1 = n_0$ drops of size R_1 injected. In this way, we have deduced not only sizes and spacings, but also numbers, that is, hierarchy (n_0, n_1). The equality $n_1 = n_0$ also holds if the large-drop injection pattern is triangular (Fig. 6.18). The rest of the R_i scales are determined in the same manner as R_1. For R_2-size drops we find

$$\frac{t_2}{t_1} \sim f \qquad \frac{R_2}{R_1} \sim f^{1/2} \qquad n_2 = 4n_1 \tag{6.59}$$

For the next, finer drops, we find

$$\frac{t_3}{t_2} \sim f \qquad \frac{R_3}{R_2} \sim f^{1/2} \qquad n_3 = 3n_2 \tag{6.60}$$

which is a pattern that holds for $i \geq 1$:

$$\frac{t_{i+1}}{t_i} \sim f \qquad \frac{R_{i+1}}{R_i} \sim f^{1/2} \qquad n_{i+1} = 3n_i \tag{6.61}$$

This multiscale population of drops is injected at the same time ($t = 0$) through appropriately spaced and sized orifices. The next "puff" of such multiscale drops

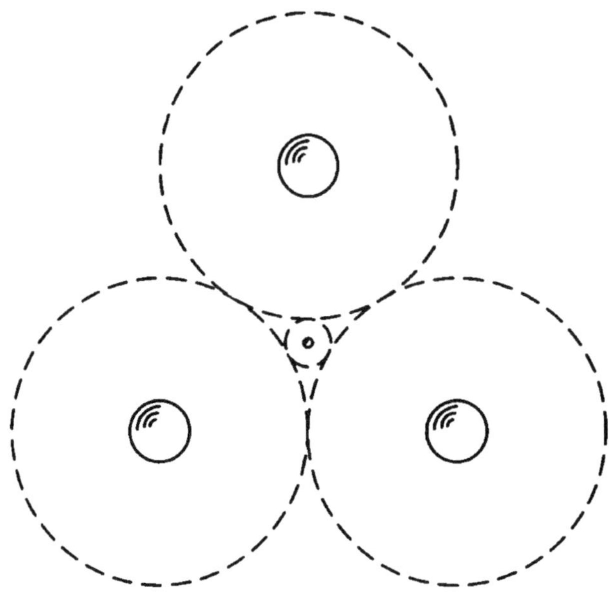

Figure 6.18 Triangular pattern for the injection of multiscale droplets in a liquid flow system.

comes at

$$t_s \sim \frac{R_0 + S_0}{V} \tag{6.62}$$

that is, after enough time such that the liquid column has enough liquid in which the R_i population can diffuse *without* interference from older and younger populations.

How well is this architecture using the available space? The total volume of D_i liquid penetrated by mass transfer during one puff (t_s) is

$$V_\Sigma \sim n_0 R_0^3 + n_1 R_1^3 + n_2 R_2^3 + \cdots \tag{6.63}$$

This volume is greater than the volume of optimally spaced drops of one size,

$$V_0 \sim n_0 R_0^3 \tag{6.64}$$

One can describe the merit of this constructal packing in terms of a *packing efficiency* defined as,

$$\eta \sim \frac{V_\Sigma}{V_0} > 1 \tag{6.65}$$

which depends weakly on the large-scale injection pattern (e.g., square versus triangular), and on the number of scales s ($i = 0, 1, \ldots,$ or $R_0, R_1, \ldots$). With a more exact analysis one could show that η is such that diminishing returns are reached as the number of scales becomes greater than a moderate number.

This number of scales in droplet hierarchy should be determined because it is essential.

REFERENCES

1. A. K. da Silva, S. Lorente, and A. Bejan, Optimal distribution of discrete heat sources on a wall with natural convection. *Int J Heat Mass Transfer*, Vol. 47, 2004, pp. 203–214.
2. A. Bejan, *Shape and Structure, from Engineering to Nature*. Cambridge, UK: Cambridge University Press, 2000.
3. A. Bejan, *Convection Heat Transfer*, 3rd ed. Hoboken, NJ: Wiley, 2004.
4. A. K. da Silva, S. Lorente, and A. Bejan, Constructal multi-scale structures for maximal heat transfer density. *Energy*, Vol. 31, 2006, pp. 620–635.
5. A. K. da Silva, A. Bejan, and S. Lorente, Maximal heat transfer density in vertical morphing channels with natural convection. *Numerical Heat Transfer, Part A*, Vol. 45, 2004, pp. 135–152.
6. H. Y. Wang, F. Penot, and J. B. Saulnier, Numerical study of a buoyancy-induced flow along a vertical plate with discretely heated integrated circuit packages. *Int J Heat Mass Transfer*, Vol. 40, 1997, pp. 1509–1520.
7. S. K. W. Tou, C. P. Tso, and X. Zhang, 3-D numerical analysis of natural convective liquid cooling of a 3×3 heater array in rectangular enclosures. *Int J Heat Mass Transfer*, Vol. 42, 1999, pp. 3231–3244.
8. I. Sezai and A. A. Mohamad, Natural convection from a discrete heat source on the bottom of a horizontal enclosure. *Int J Heat Mass Transfer*, Vol. 43, 2000, pp. 2257–2266.
9. Y. Liu and N. Phan-Thien, An optimum spacing problem for three chips mounted on a vertical substrate in an enclosure. *Numerical Heat Transfer, Part A*, Vol. 37, 2000, pp. 613–630.
10. Q. H. Deng, G.-F. Tang, and Y. Li, A combined temperature scale for analyzing natural convection in rectangular enclosures with discrete wall heat sources. *Int J Heat Mass Transfer*, Vol. 45, 2002, pp. 3437–3446.
11. Q. H. Deng, G.-F. Tang, Y. G. Li, and M. Y. Ha, Interaction between discrete heat sources in horizontal natural convection enclosures. *Int J Heat Mass Transfer*, Vol. 45, 2002, pp. 5117–5132.
12. E. Yu and Y. Joshi, Heat transfer enhancement from enclosure discrete components using pin-fin heat sinks. *Int J Heat Mass Transfer*, Vol. 45, 2002, pp. 4957–4966.
13. S. K. W. Tou and X. F. Zhang, Three-dimensional numerical simulation of natural convection in an inclined liquid-filled enclosure with an array of discrete heaters. *Int J Heat Mass Transfer*, Vol. 46, 2003, pp. 127–138.
14. A. K. da Silva, S. Lorente, and A. Bejan, Optimal distribution of discrete heat sources on a plate with laminar forced convection. *Int J Heat Mass Transfer*, Vol. 47, 2004, pp. 2139–2148.
15. A. K. da Silva, S. Lorente, and A. Bejan, Constructal multi-scale structures with asymmetric heat sources of finite thickness. *Int J Heat Mass Transfer*, Vol. 48, 2005, pp. 2662–2672.
16. H. J. Shaw and W. L. Chen, Laminar forced-convection in a channel with arrays of thermal sources. *Wärme Stoffübertrag*, Vol. 26, 1991, pp. 195–201.

17. C. Y. Wang, Optimum placement of heat-sources in forced-convection. *J Heat Transfer*, Vol. 114, 1992, pp. 508–510.
18. A. Gupta and Y. Jaluria, Forced convective liquid cooling of arrays of protruding heated elements mounted in a rectangular duct. *J Electron Packaging*, Vol. 120, 1998, pp. 243–252.
19. C. P. Tso, G. P. Xu, and K. W. Tou, An experimental study on forced convection heat transfer from flush-mounted discrete heat sources. *J Heat Transfer*, Vol. 121, 1999, pp. 326–332.
20. M. M. Rahman and J. Raghavan, Transient response of protruding electronic modules exposed to horizontal cross flow. *Int J Heat Fluid Flow*, Vol. 20, 1999, pp. 48–59.
21. G. I. Sultan, Enhancing forced convection heat transfer from multiple protruding heat sources simulating electronics components in a horizontal channel by passive cooling. *Microelectron J*, Vol. 31, 2000, pp. 773–779.
22. S. Chen, Y. Liu, S. F. Chan, C. W. Leung, and T. L. Chan, Experimental study of optimum spacing problem in the cooling of simulated electronics package. *Heat Mass Transfer*, Vol. 37, 2001, pp. 251–257.
23. A. Bejan and Y. Fautrelle, Constructal multi-scale structure for maximal heat transfer density. *Acta Mechanica*, Vol. 163, 2003, pp. 39–49.
24. A. Bejan, *Advanced Engineering Thermodynamics*, 2nd ed. (p. 765). New York: Wiley, 1997.
25. P. Bradshaw, Shape and structure, from engineering to nature. *AIAA J*, Vol. 39, 2001, p. 983.
26. L. Nottale, *Fractal Space-Time and Microphysics*. Singapore: World Scientific, 1993.
27. T. Bello-Ochende and A. Bejan, Maximal heat transfer density: Plates with multiple lengths in forced convection. *Int J Thermal Sciences*, Vol. 43, 2004, pp. 1181–1186.
28. S. Petrescu, Comments on the optimal spacing of parallel plates cooled by forced convection. *Int J Heat Mass Transfer*, Vol. 37, 1994, p. 1283.
29. T. Bello-Ochende and A. Bejan, Fitting the duct to the "body" of the convective flow. *Int J Heat Mass Transfer*, Vol. 46, 2003, pp. 1693–1701.
30. A. K. da Silva and A. Bejan, Constructal multi-scale structure for maximal heat transfer density in natural convection. *Int J Heat Fluid Flow*, Vol. 26, 2006, pp. 34–44.
31. T. Bello-Ochende and A. Bejan, Constructal multi-scale cylinders in cross-flow. *Int J Heat Mass Transfer*, Vol. 48, 2005, pp. 1373–1383.
32. B. Heinrich, The mechanisms of energetics of honeybee swarm temperature regulation. *J Exp Biol*, Vol. 91, 1981, pp. 25–55.
33. T. Bello-Ochende and A. Bejan, Constructal multi-scale cylinders with natural convection. *Int J Heat Mass Transfer*, Vol. 48, 2005, pp. 4300–4306.
34. T. Dias Jr. and L. F. Milanez, Optimal location of heat sources on a vertical wall with natural convection and genetic algorithm. *Int J Heat Mass Transfer*, Vol. 49, 2006, pp. 2090–2096.
35. A. Muftuoglu and E. Bilgen, Natural convection in an open square cavity with discrete heaters at their optimized positions. *Int J Thermal Sciences*, Vol. 47, 2008, pp. 369–377.
36. R. R. Madadi and C. Balaji, Optimization of the location of multiple discrete heat sources in a ventilated cavity using artificial neural networks and micro genetic algorithm. *Int J Heat Mass Transfer*, Vol. 51, 2008, pp. 2299–2312.

PROBLEMS

6.1. The steam generator of a modern steam power plant uses the siphon circulation scheme shown in Fig. P6.1. There are two vertical columns of height H. The downcomer (D_1) contains dense fluid (liquid) with density ρ_1. The riser (D_2) contains a mixture of liquid and vapor with density ρ_2. The flow along each column is fully developed turbulent in the fully rough limit, meaning that the friction coefficients f_1 and f_2 are two constants, independent of flow rate. The total volume of the two columns is specified. Determine the optimal allocation of volume (D_1/D_2) so that the global flow resistance encountered by the circulation is minimal, or that the flow rate $\dot{m}$ is maximal. Show that the result is $(D_1/D_2)_{opt} = (f_1/f_2)^{1/7}\,(\rho_2/\rho_1)^{1/7}$.

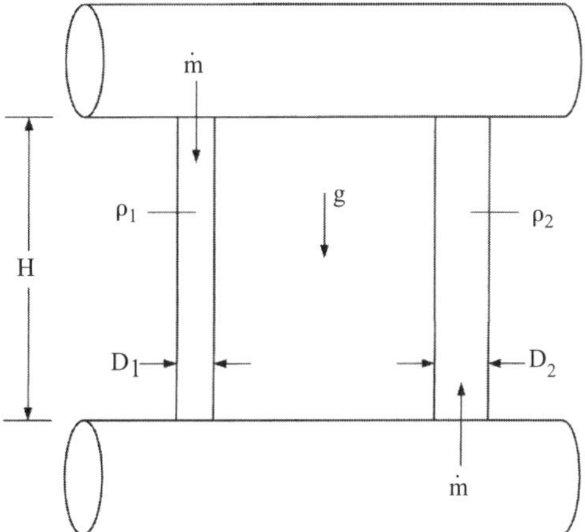

Figure P6.1

6.2. Generalize the preceding volume allocation design by taking account of the fact that the friction factors (f_1, f_2) depend on the Reynolds numbers in each volume, and that the density in the riser (ρ_2) depends on the quality of the liquid vapor mixture (x_2). The density ρ_1 and the total volume V are known parameters. Compare this solution with the solution to the preceding problem.

7

MULTIOBJECTIVE CONFIGURATIONS

7.1 THERMAL RESISTANCE VERSUS PUMPING POWER

The convective flow structures developed in the preceding chapter introduced the student to the idea that one flow configuration serves two functions: low fluid-flow resistance and low thermal resistance. The two functions compete, and from this clash emerges the configuration. In this chapter we pursue this idea further and focus on how to meet these two objectives at the same time. One topic is how to generate tree-shaped flow structures for cooling electronics. Another is the generation of tree-shaped networks for distributing hot or cold water, that is, trees for maximum point-area flow access and minimum heat loss from all the pipes to the ambient. This approach to flow configurations that meet two flow objectives puts us in a good position to tackle some of the most promising applications today: high-density heat exchangers and fuel cells that are configured as tree-shaped flows.

We begin with Fig. 7.1, which serves as an icon for the idea that we teach in this chapter. The convection cooling of a heat-generating or heat-absorbing body is a design challenge with at least two objectives: minimal thermal resistance and minimal pumping power. Unlike in thermodynamic optimization (Chapter 2), where the search is for a unique balance (for minimal irreversibility) between heat-transfer performance and fluid-mechanics performance, in constructal design the search is for flow architectures. On a graph with thermal resistance versus fluid-flow resistance or pumping power, there is one curve for each flow architecture subject to specified global constraints. In forced convection configurations the thermal resistance decreases as the pumping power increases.

The lesson is to "morph" the flow configuration, that is, to change the flow architecture so that it is represented by the curve that is situated as close as possible

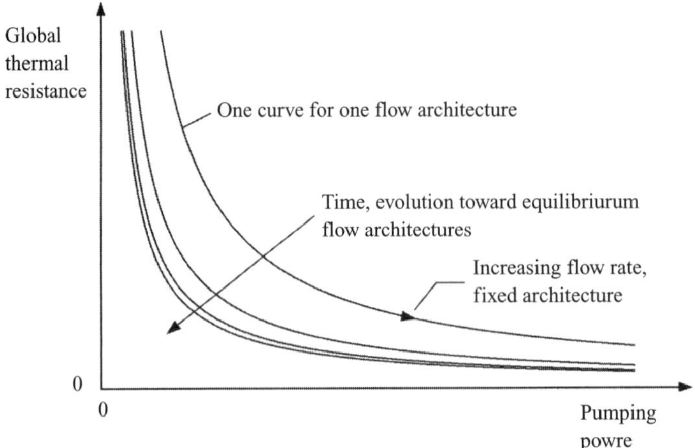

Figure 7.1 The generation of flow architectures in the pursuit of low thermal resistance and low pumping power [1].

to the origin. The push toward lower and lower curves, or better and better architectures, is a general trend in all flow systems, engineered and natural. This tendency is summarized in the constructal law, as a self-standing principle in the thermodynamics of nonequilibrium (flow) systems [2]. Flow architectures evolve in time, in the direction of *equilibrium configurations* where geometry is most free to change, while global performance does not change anymore. To illustrate the diminishing returns, in Fig. 7.1 we show curves that become denser as "low thermal resistance" and "low pumping power" are achieved at the same time.

7.2 ELEMENTAL VOLUME WITH CONVECTION

Two-objective design maps such as Fig. 7.1 can and should be drawn for every class of convective flow architectures, from the simplest (Fig. 7.2) to more complex (sections 7.3 and 7.4). Here, we illustrate the method for a two-dimensional rectangular volume $H \times L$ with uniform volumetric heat generating rate $q'''[\mathrm{W/m^3}]$. This flow system is the smallest-scale element of much larger and complex structures such as tree-shaped vascularized systems.

The coolant that sweeps the volume is single phase and with constant properties. The mass flow rate through the entire volume is $\dot{m}'\ [\mathrm{kg/s\,m}]$. The flow rate $\dot{m}'$ is expressed per unit length in the direction perpendicular to the plane of Fig. 7.2. This problem was first proposed in Ref. [3], where the perfusion rate $\dot{m}'/H$ was assumed given. This assumption is not made in the following analysis. Assume that the $H \times L$ volume is a package of electronics that generates heat. The coolant inlet temperature is T_0. The temperature at any point in the package cannot exceed a prescribed level ($T_{\max}$).

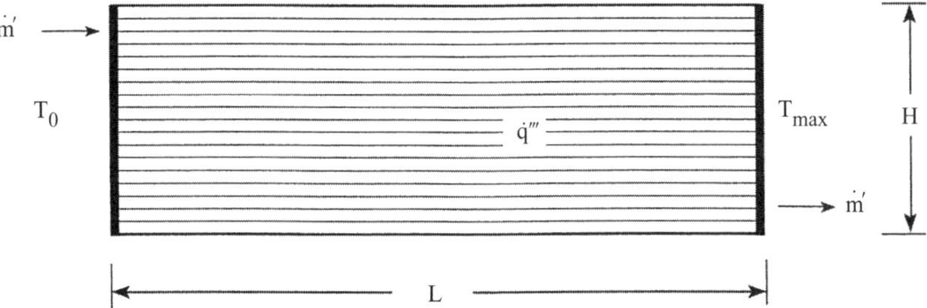

Figure 7.2 Two-dimensional volume with volumetric heat generation and unidirectional permeating flow.

For simplicity, assume that $\dot{m}'$ is distributed uniformly over H, and flows through a large number (n) of parallel-plate channels. The arrangement requires the use of two headers (the larger fluid volumes indicated with thick black lines in Fig. 7.2), one upstream, to spread $\dot{m}'$, and the other downstream, to reconstitute the $\dot{m}'$ stream. The uniform unidirectional flow sweeps the $H \times L$ space and pushes the hot spots ($T_{\max}$) to the right extremity of the space. When the multichannel structure is sufficiently fine [i.e., when n is sufficiently large, cf. Eq. (7.29) later in this section], the peak temperature of the heat-generating solid is essentially the same as the peak bulk temperature of the permeating fluid. Because the highest fluid temperature occurs in the exit plane, the first law of thermodynamics applied to the HL control volume requires

$$\dot{m}' c_P (T_{\max} - T_0) = q''' HL \tag{7.1}$$

When the size of the elemental volume is fixed (HL), the total heat current generated by the volume is fixed ($q''' HL$). Equation (7.1) shows that the global thermal conductance of the system varies inversely with the flow capacity rate:

$$\frac{T_{\max} - T_0}{q''' HL} = \frac{1}{\dot{m}' c_P} \tag{7.2}$$

This behavior accounts for the downward trend along each of the curves plotted in Fig. 7.1. The flow configuration sketched in Fig. 7.2 is represented by a single curve on the more general design map suggested by Fig. 7.1.

The pumping power required to push the flow through the n two-dimensional channels of Fig. 7.2 is estimated as follows. First, the mass flow rate through a single channel is

$$\dot{m}'_0 = \frac{\dot{m}'}{n} = \rho D U \tag{7.3}$$

Here, D is the channel spacing (the size of the slit through which the fluid flows), and U is the fluid velocity averaged over D. The end-to-end pressure drop is [cf.

Eq. (1.22)],

$$\Delta P = f \frac{4L}{D_h} \frac{1}{2} \rho U^2 \tag{7.4}$$

where the hydraulic diameter of the slit is $D_h = 2D$. If we assume Poiseuille flow through each channel, then in accordance with section 1.2.1 we write

$$f = \frac{24}{\mathrm{Re}_{D_h}} \tag{7.5}$$

where $\mathrm{Re}_{D_h} = UD_h/\nu$ and 24 is the value of the Poiseuille constant for parallel plates. Combining Eqs. (7.4) and (7.5) we obtain a proportionality between ΔP and $\dot{m}'_0$, namely $\Delta P = 12\nu L\dot{m}'_0/D^3$. Finally, to estimate the pumping power we assume an incompressible fluid and a reversible pump (cf. section 2.5),

$$\dot{W}' = \frac{\dot{m}'\Delta P}{\rho} = \frac{12\nu L(\dot{m}')^2}{\rho n D^3} \tag{7.6}$$

which reveals a proportionality between $\dot{W}'$ and $\left(\dot{m}'\right)^2$. This trend corresponds to the direction of the arrow plotted on the top curve in Fig. 7.1. Taken together, Eqs. (7.2) and (7.6) mean that in the configuration of Fig. 7.2 the global thermal conductance is proportional to the pumping power raised to the power $-1/2$.

Another way to express the pumping power is to note that the total volume fraction ϕ occupied by all the channels is fixed,

$$\phi = \frac{nDL}{HL} < 1 \tag{7.7}$$

In this notation, Eq. (7.6) becomes

$$\dot{W}' = 12 \frac{\nu}{\rho} \frac{n^2}{\phi^3} \frac{\left(\dot{m}'\right)^2}{HL} \left(\frac{L}{H}\right)^2 \tag{7.8}$$

and indicates that the pumping power increases rapidly as one configuration is replaced by one with more channels.

The attractive configurations are those with fewer channels: this brings us to the $n = 1$ configuration sketched in Fig. 7.3, for which the pumping power is obtained by setting $n = 1$ in Eq. (7.8),

$$\dot{W}' = \frac{12\nu \left(\dot{m}'\right)^2}{\rho\phi^3 HL} \left(\frac{L}{H}\right)^2 \tag{7.9}$$

The global thermal resistance of the one-channel configuration is estimated as follows. All the heat generated inside the elemental volume $\left(q'''HL\right)$ is convected away by the $\dot{m}'$ stream that flows through the D channel. In Fig. 7.3 the rectangular boundary $H \times L$ is adiabatic, with the exception of the coldest spot (T_0) that occurs

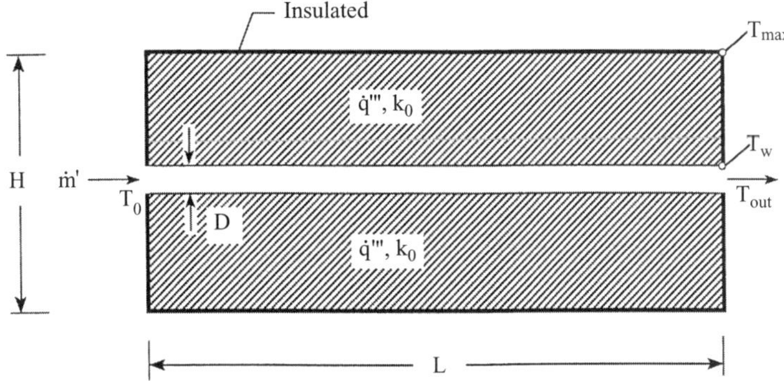

Figure 7.3 Two-dimensional volume with heat generation, vertical conduction in the solid, and horizontal single-stream convection.

in the vicinity of the inlet to the D channel. The hot spots (T_{max}) occur in the two corners that are situated the farthest from the inlet.

An analytical expression for the peak excess temperature ($T_{max} - T_0$) can be developed when the solid thermal conductivity k_0 is small and the aspect ratio H/L is sufficiently smaller than 1 such that the conduction through the heat-generating material (k_0) is oriented perpendicularly to the fluid channel. If we also assume that $D \ll H$ (i.e., $\phi \ll 1$), the temperature drop between the hot-spot corner (T_{max}) and the wall spot near the channel outlet is (Problem 5.5)

$$T_{max} - T_w = \frac{q''' H^2}{8k_0} \tag{7.10}$$

in accordance with the solution for steady unidirectional heat conduction in a $H/2$-thick slab with one side insulated and the other at T_w. The increase experienced by the bulk temperature of the $\dot{m}'$ stream from the inlet to the outlet is [cf. Eq. (7.1)],

$$T_{out} - T_0 = \frac{q''' HL}{\dot{m}' c_P} \tag{7.11}$$

There is also a temperature difference between the bulk temperature T_{out} and the duct wall temperature (T_w) in the plane of the outlet. Temperature differences of this kind are neglected in this section based on the assumption that the flow is fully developed and the channel spacing is sufficiently small. The validity condition is discussed later, in the derivation of Eq. (7.31).

Summing up, the peak excess temperature is given by a two-term expression

$$T_{max} - T_0 = (T_{max} - T_w) + (T_w - T_0) = \frac{q''' H^2}{8k_0} + \frac{q''' HL}{\dot{m}' c_P} \tag{7.12}$$

The global thermal resistance can be nondimensionalized as

$$R_1 = \frac{T_{\max} - T_0}{q'''HL/k_0} = \frac{1}{8}\frac{H}{L} + \frac{1}{M} \tag{7.13}$$

where M is the dimensionless mass flow rate

$$M = \frac{\dot{m}' c_P}{k_0} \tag{7.14}$$

In the same notation, the pumping power of Eq. (7.9) is nondimensionalized as

$$W_1 = \dot{W}' \frac{\rho}{\nu} \left(\frac{c_P}{k_0} \right)^2 HL\phi^3 = \left(\frac{L}{H} \right)^2 M^2 \tag{7.15}$$

Subscript 1 is a reminder that Eqs. (7.13) and (7.15) refer to Fig. 7.3, in which all the flow resides in a single channel. The corresponding dimensionless expressions for n channels (Fig. 7.2) are obtained from Eqs. (7.2) and

$$R_n = \frac{1}{M} \tag{7.16}$$

$$W_n = 12\, n^2 \left(\frac{L}{H} \right)^2 M^2 \tag{7.17}$$

where subscript n indicates that these expressions are valid for n channels, where $n \gg 1$.

The corresponding performance functions for the rectangular volume with two cooling channels ($n = 2$) are obtained by substituting $H/2$ in place of H in Eq. (7.12), and $n = 2$ in Eq. (7.8). The dimensionless expressions for the global thermal resistance and pumping power are

$$R_2 = \frac{1}{32}\frac{H}{L} + \frac{1}{M} \tag{7.18}$$

$$W_2 = 48 \left(\frac{L}{H} \right)^2 M^2 \tag{7.19}$$

Similarly, for the rectangular element cooled by three channels separated by the distance $H/3$, we obtain

$$R_3 = \frac{1}{72}\frac{H}{L} + \frac{1}{M} \tag{7.20}$$

$$W_3 = 108 \left(\frac{L}{H} \right)^2 M^2 \tag{7.21}$$

Equations (7.13) through (7.21) express the two objectives (low R, low W) as functions of two parameters: the shape H/L and the flow rate M. When one parameter is specified, then the other parameter relates R to W in such a way that the flow architecture is represented by a single curve in the R–W plane

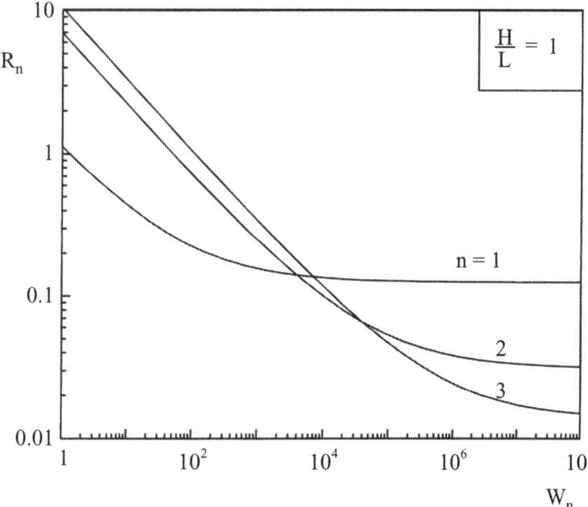

Figure 7.4 Two objectives: low thermal resistance and low pumping power in the configurations of Figs. 7.2 and 7.3.

(cf. Fig. 7.1). For example, if we fix the rectangular shape ($H/L = 1$) we obtain the family of two-objective curves shown in Fig. 7.4. Each flow configuration ($n = 1, 2, 3$) has its own curve.

If we fix the mass flow-rate parameter ($M = 1$, Fig. 7.5) we obtain another family of curves that show again that the simplest ($n = 1$) is the best. This is in accord with the beginnings of constructal theory [4], where the smallest ("elemental") subvolume of a more complex flow architecture had only one channel. One longitudinal stream in balance with diffusion perpendicular to the stream is the recipe for design at the smallest (finite) scale. We see this everywhere in natural design, from the alveolus of the lung to the eddy of turbulence and the smallest rivulet of a river basin.

There are other ways in which to use the lesson taught in this section, but the important thing to keep in mind is that the two objectives are pursued with equal vigor, and from the start. A more traditional view is to see the performance of one configuration as the trade-off between two penalties,

$$R = a\frac{H}{L} + \frac{1}{M} \tag{7.22}$$

$$W = b\left(\frac{L}{H}\right)^2 M^2 \tag{7.23}$$

where (a, b) is a pair of constants—one pair for one configuration. If the pumping power is specified (because of the pumping apparatus that is available), then W becomes a constraint, instead of an objective. With $W =$ constant, the system (7.22) through (7.23) expresses R as a function of H/L. This function can be minimized with respect to H/L:

$$\left(\frac{H}{L}\right)_{\text{opt}} = \frac{(b/W)^{1/4}}{a^{1/2}} \qquad R_{\min} = 2a^{1/2}\left(\frac{b}{W}\right)^{1/4} \tag{7.24}$$

Another traditional view is to treat R and W as two contributions to the "cost" associated with the imperfection of the elemental rectangle as a convection instrument. The combined cost is

$$C = R + rW \tag{7.25}$$

in which r represents the "importance" of W relative to R in the cost formula. After using Eqs. (7.22) and (7.23), we find that C can be minimized independently with respect to H/L and M:

$$\left(\frac{H}{L}\right)_{\text{opt}} = (2rb)^{1/5}\, a^{-3/5} \qquad M_{\text{opt}} = \left(2rba^2\right)^{-1/5} \tag{7.26}$$

An interesting aspect of this economic optimization of the elemental convection volume is that the optimal shape varies inversely with the optimal flow rate,

$$a\left(\frac{H}{L}\right)_{\text{opt}} = M_{\text{opt}}^{-1} \tag{7.27}$$

and that their product is independent of the assumed economics weighting factor r.

In this section we were able to express our ideas analytically because we made useful assumptions of simplification before we started. This technique of simplifying (but not simplifying too much) is important to learn. It is a dying art in a market taken over by industrial-scale numerical simulations that look impressive

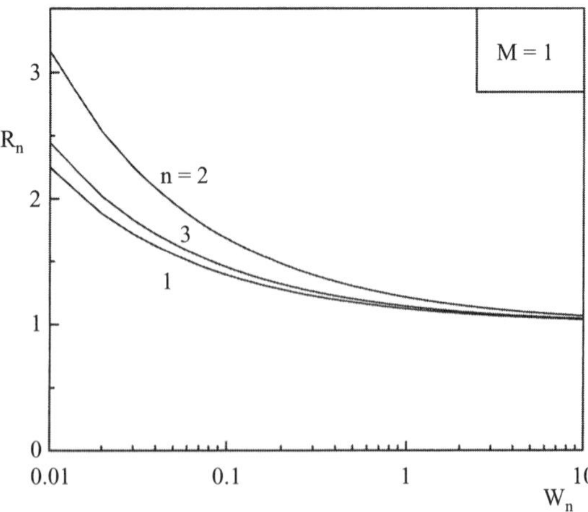

Figure 7.5 Thermal resistance versus pumping power when the mass flow rate is fixed.

(complicated) but explain nothing. The value of the above board (pencil and paper) demonstration cannot be overstated.

By the same token, simplifications must always be *justified*. For example, above Eq. (7.1) we assumed that the parallel channels are numerous enough such that the highest temperature is the outlet temperature of the fluid. How many channels are "enough" so that this assumption is valid? It is valid when the vertical temperature difference across the solid ($T_{max} - T_{out}$) is small in comparison with the temperature rise along the channel ($T_{out} - T_0$). This is equivalent to saying that if we write Eq. (7.13) for n equidistant channels separated by H/n,

$$\frac{T_{max} - T_0}{q'''HL/k_0} = \frac{1}{8n^2}\frac{H}{L} + \frac{1}{M} \tag{7.28}$$

then the first term on the right side must be much smaller than the second term. This delivers the condition that must be satisfied by n so that Eq. (7.1) is valid.

$$n \gg \left(\frac{M}{8}\frac{H}{L}\right)^{1/2} \tag{7.29}$$

Another assumption [made above Eq. (7.12)] was that the temperature difference between the wall and the exiting fluid must be smaller than the longitudinal temperature rise experienced by the fluid in the channel,

$$T_w - T_{out} \ll T_{out} - T_0 \tag{7.30}$$

For the left side, we write that the convective heat flux (in the plane of the outlet) from the wall to the fluid is $q'' = h\,(T_w - T_{out})$. The same heat flux is the heat generation rate integrated over the cut of height H, namely $q'' = q'''H$. Furthermore, in laminar fully developed flow the heat transfer coefficient is $h = \mathrm{Nu}\, k_f/D$, where k_f is the fluid thermal conductivity, and the Nusselt number Nu is a constant in the 1–10 range [5]. For the right side of Eq. (7.30), we write $T_{out} - T_0 = q'''HL/\left(\dot{m}'c_P\right)$. In view of all these scales, Eq. (7.30) becomes

$$\frac{H}{L}M \ll \frac{k_f}{k_0}\,\mathrm{Nu}\,\frac{n}{\phi} \tag{7.31}$$

This places a constraint on the parameters H/L and M that one must assume in order to plot on the R–W field the curves based on Eqs. (7.13) through (7.21).

7.3 DENDRITIC HEAT CONVECTION ON A DISC

In this section we turn our attention to the generation of more complex (tree-shaped) flow configurations based on the two-objective method presented so far. The selected flow system is a disc-shaped body that generates heat volumetrically. This flow system has been optimized extensively from the fluid mechanics point of view [6, 7] (cf. Chapter 4). The cooling effect is provided by a single stream of single-phase fluid, which enters the disc through its center, and flows toward the periphery. The objective is to find the flow patterns that minimize *at the same time* the solid hot-spot temperature, and the pumping power requirement [1].

For simplicity, we assume that the disc is sufficiently thin so that it can be modeled as a disc with uniform rate of heat generation per unit area (q''). The disc radius is R, and the fluid mass flow rate is $\dot{m}$. The heat is generated at every point in the solid material of thermal conductivity k. It is then conducted through the solid to the nearest flow channel, and, later, it is convected by the stream that flows along the channel.

The following treatment has three parts. In the first, we analyze the heat-transfer performance of tree-shaped flows that have been optimized for minimal flow resistance or minimal pumping power. The second part is devoted to the pure heat-transfer problem of determining the flow patterns that minimize the global thermal resistance of the system. We also determine the corresponding fluid mechanics performance (pressure drop, pumping power) of the heat-transfer-optimized cooling scheme. In the third part we consider the combined fluid-flow and heat-transfer optimization of the flow architecture. An important issue is whether the performance of the designs optimized in the first part is much different than the performance of the structures optimized in the second part. Furthermore, is it necessary to optimize the flow structure for minimum thermal resistance and minimum pumping power at the same time? Are the fluid-flow-optimized tree networks nearly optimal from a heat-transfer standpoint?

7.3.1 Radial Flow Pattern

Consider the simplest design first: the coolant enters through the center of the disc, and flows along n_0 radial ducts (Fig. 7.6). The mass flow rate through each duct is $\dot{m}_0 = \dot{m}/n_0$. When n_0 is of order 10 or greater, a sector that has one duct as centerline can be approximated by an isosceles triangle of base $2\pi R/n_0$ and height R. This triangular model is shown on the right side of Fig. 7.6. The triangular element is analogous to the rectangular element of Fig. 7.3. In order to determine the global thermo-fluid performance of the disc-shaped system, it is sufficient to describe the flow of heat and fluid in one sector. As in Eq. (7.7), we assume that the space occupied by the duct is a small fraction of the entire space:

$$\phi = \frac{n_0 DR}{\pi R^2} \ll 1 \tag{7.32}$$

Here, D is the thickness of the duct image projected on the disc area. The duct geometric aspect ratio is $D/R = \pi\phi/n_0$. The approximation (7.32) permits an analysis in which the ducts are treated as lines drawn on the disc. This limits the validity domain of the analysis but justifies the calculation of pressure drops based on the long and thin duct sections, by neglecting the losses at the junctions [8] (cf. section 1.2.2).

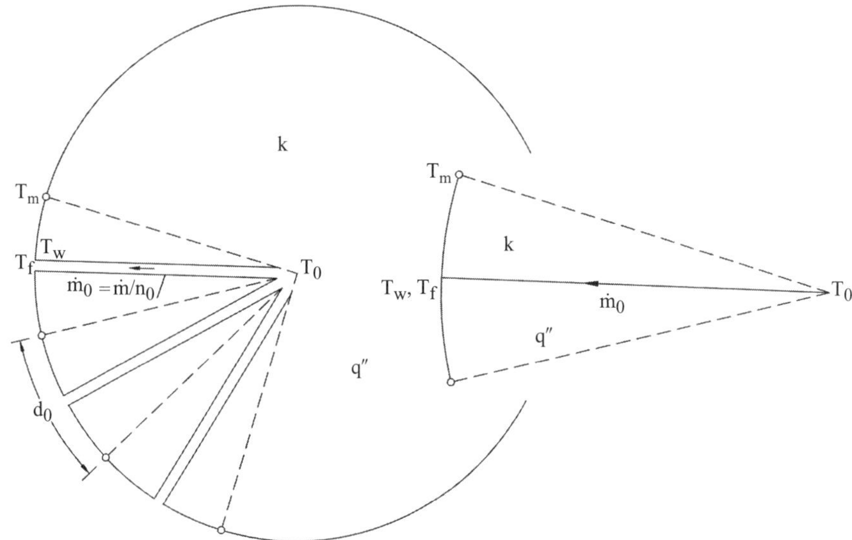

Figure 7.6 Flow pattern with radial ducts, and isosceles triangle approximation of the sector associated with one duct [1].

If q is the total heat current generated by the disc, then the heat generation rate per unit area is $q'' = q/(\pi R^2)$. The effect of q'' is to raise all the temperatures above the lowest temperature (T_0), which belongs to the coolant inlet. The highest temperature (T_m) occurs in the two peripheral corners of the sector. The overall temperature difference that drives the heat-transfer process is $T_m - T_0$. The overall thermal resistance of the disc is $(T_m - T_0)/q$, which will be nondimensionalized later in Eq. (7.36).

We evaluate the thermal resistance by noting that when the sector is slender the conduction heat flux through the solid material of thermal conductivity k proceeds in the direction perpendicular to the duct. The overall temperature difference $T_m - T_0$ is the sum of three contributions,

$$T_f - T_0 = \frac{\pi q'' R^2}{n_0 \dot{m}_0 c_p} \tag{7.33}$$

$$T_w - T_f = \frac{\pi q'' R D}{n_0 p k_f \mathrm{Nu}} \tag{7.34}$$

$$T_m - T_w = \frac{\pi^2 R^2 q''}{2 n_0^2 k t} \tag{7.35}$$

where T_f, T_w, t, k_f and Nu are the fluid bulk temperature at the duct outlet, the duct wall temperature at the outlet, the disc thickness, the fluid thermal conductivity, and the Nusselt number based on D. Equation (7.33) is the first law of thermodynamics applied to one sector. Equation (7.34) is the statement that the heat current that arrives in the angular direction at the wall end of the duct, $q''\pi R/n_0$, is the same as the heat current that flows from the wall to the fluid in the plane of the outlet, $hp(T_w - T_f)$, where h and p are the heat transfer coefficient and the wetted perimeter of the duct cross-section. We also used the Nusselt number definition $\mathrm{Nu} = hD/k_f$, with the observation that for fully developed laminar flow Nu is a constant with a value between 1 and 10 [5].

Equation (7.35) is obtained from the temperature distribution along the perimeter of length $d_0/2 = 2\pi R/(2n_0)$, from T_m to T_w. Because the solid generates heat at every point, the temperature distribution along the perimeter is parabolic, with zero temperature gradient at T_m, and the largest temperature gradient at T_w. The parabolic shape is special in the sense that the temperature gradient at T_w is $4(T_m - T_w)/d_0$, i.e., *twice* the gradient based on the temperature difference separated by $d_0/2$. The heat flux that arrives at T_w, namely, $kt[4(T_m - T_w)/d_0]$, is equal to the heat generation rate integrated along $d_0/2$, namely $(d_0/2)q''$. From this equality follows Eq. (7.35).

By adding Eqs. (7.33) through (7.35), and nondimensionalizing the sum as the overall thermal resistance of the disc,

$$\tilde{T}_m = \frac{T_m - T_0}{q/(kt)} \tag{7.36}$$

we obtain the expression

$$\tilde{T}_m = \frac{1}{M} + \frac{\pi \phi kt}{n_0^2 k_f p \mathrm{Nu}} + \frac{\pi}{2n_0^2} \tag{7.37}$$

As in Eq. (7.14), the number M is a dimensionless mass flow rate,

$$M = \frac{\dot{m} c_P}{kt} \tag{7.38}$$

Alternatively, because of the heat-transfer-units concept popular in heat exchanger design, we propose to call M^{-1} the *number of conduction heat-transfer units*.

The three terms of $\tilde{T}_m$ represent, in order,

(i) The radial temperature increase experienced by the coolant.
(ii) The temperature difference between the duct wall and the fluid.
(iii) The maximum temperature difference across the heat generating solid.

The ratio between terms (ii) and (iii) is $2\phi kt/(\mathrm{Nu}\, k_f p)$, in which 2/Nu is a number of order 1. This means that the second term is negligible in the limit of thin ducts (svelte drawings), so thin that the projected area fraction ϕ is smaller than $k_f p/(kt)$. The following numerical work on radial ducts and more complicated configurations was based on the assumption that $\phi < k_f p/(kt)$, such that the wall-fluid thermal resistance is negligible.

The solid curves in Fig. 7.7 show the behavior of $\tilde{T}_m$ when the second term of Eq. (7.37) is neglected. The thermal resistance drops when the flow rate (M)

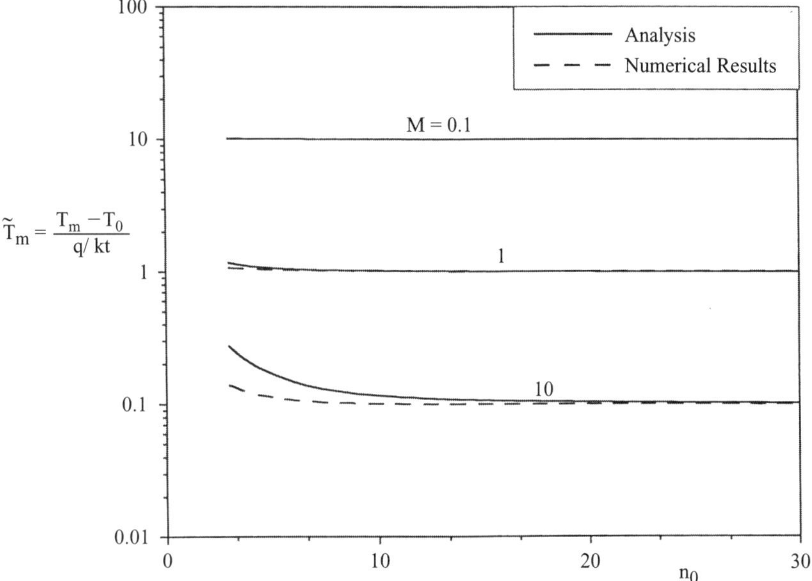

Figure 7.7 The thermal resistance of the radial flow structure shown in Fig. 7.6: comparison between the numerical and analytical rcsults [1].

increases. This trend matches the behavior anticipated in Fig. 7.1, because along with the increase in flow rate comes an increase in pumping power. The thermal resistance also decreases as n_0 increases, that is, as the flow structure becomes more complex. This second trend—the effect of flow architecture—is the direction explored in the remainder of this section.

Here is how to use the preceding analysis and the radial pattern of Fig. 7.6 for the purpose of developing and testing (benchmarking) the numerical simulation of heat flow in the disc-shaped system. Heat conduction can be simulated numerically in a single sector, which has uniform heat generation, and a line-thin stream of coolant on the centerline [1]. The boundaries of the sector are insulated. The equation for steady two-dimensional heat conduction in polar coordinates (r, θ) is

$$\frac{1}{\tilde{r}}\frac{\partial}{\partial \tilde{r}}\left(\tilde{r}\frac{\partial \tilde{T}}{\partial \tilde{r}}\right) + \left(\frac{n_0/\pi}{\tilde{r}}\right)^2 \frac{\partial^2 \tilde{T}}{\partial \tilde{\theta}^2} + \frac{1}{\pi} = 0 \tag{7.39}$$

where

$$\tilde{r} = \frac{r}{R} \qquad \tilde{\theta} = \frac{\theta}{\pi/n_0} \qquad \tilde{T} = \frac{T - T_0}{q/kt} \tag{7.40}$$

The boundary conditions are $\partial\tilde{T}/\partial\tilde{r} = 0$ at $\tilde{r} = 1$, $\partial\tilde{T}/\partial\tilde{\theta} = 0$ at $\tilde{\theta} = 1$, $\tilde{T} = 0$ and $\partial\tilde{T}/\partial\tilde{r} = 0$ at $\tilde{r} = 0$, and $\tilde{T}_f = (T_f - T_0)/(q/kt) = M^{-1}$ at $\tilde{r} = 1$, $\tilde{\theta} = 0$. The

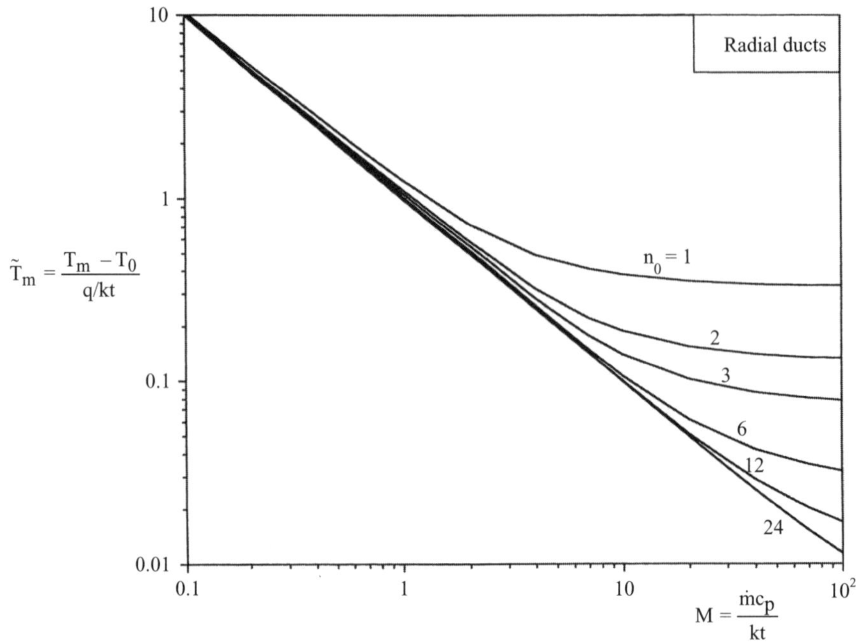

Figure 7.8 Numerical solution for the thermal resistance of the radial design of Fig. 7.6 [1].

continuity of heat flux through the duct wall ($\tilde{\theta} = 0$) requires

$$\frac{1}{2} M \frac{\partial \tilde{T}}{\partial \tilde{r}} = \frac{n_0^2}{\pi \tilde{r}} \frac{\partial \tilde{T}}{\partial \tilde{\theta}} \tag{7.41}$$

Equation (7.39) was solved by the second-order accurate finite differences method by using a nonuniform grid. The highest temperature occurred at $\tilde{r} = 1$ and $\tilde{\theta} = 1$, in accordance with the T_m location indicated in Fig. 7.6.

Figure 7.7 shows that the numerical results agree with Eq. (7.37) in the limit where the analysis is valid: slender sectors (large n_0). The numerical results are used in the remainder of this section because they did not require the use of the slender-sectors approximation. In Fig. 7.8 the numerical analysis of the radial design was extended to the extreme cases $n_0 = 2$ and $n_0 = 1$, which are far from the domain of applicability of Eq. (7.37). The overall thermal resistance decreases monotonically as n_0 increases, however, when n_0 is sufficiently large the thermal resistance is relatively insensitive to further increases in n_0.

The radial design of Fig. 7.6 provides another example of the competition between the overall thermal resistance and the required pumping power. To show this analytically, assume that each radial duct is a round tube of diameter D with Poiseuille flow. This is the same D that in Eq. (7.32) represented the thickness of the duct image projected on the disc area. The round tube assumption is made for the purpose of illustration, because any other duct cross-section with a hydraulic diameter comparable to D has fluid flow characteristics similar to those described in this section (cf. sections 1.2.1 and 3.1.2).

To evaluate the overall flow resistance encountered by the total stream $\dot{m}$, we begin with the mean fluid velocity through one tube,

$$U_0 = \frac{D^2}{32\mu}\frac{\Delta P}{R} \tag{7.42}$$

where ΔP is the overall pressure difference measured between the center and the periphery of the disc. The flow rate through one tube is $\dot{m}_0 = U_0 \rho \pi D^2/4$. Combining Eq. (7.42) and $\dot{m}_0 = \dot{m}/n_0$, where $\dot{m}$ is the total flow rate for the disc, we obtain the overall flow resistance

$$\frac{\Delta P}{\dot{m}} = \frac{128\nu R}{\pi n_0 D^4} \tag{7.43}$$

The flow resistance decreases rapidly as the duct diameter increases. This trend is resisted by requirements such as Eq. (7.32): the fluid-flow network can occupy only a small fraction of the space occupied by the system. Assume that the total volume occupied by the ducts is fixed,

$$V = n_0 \frac{\pi D^2}{4} R \tag{7.44}$$

so that D varies as $n_0^{-1/2}$. The minimal pumping power required by the entire assembly is $\dot{W} = \dot{m}\Delta P/\rho$. This can be expressed in dimensionless form by using Eqs. (7.38) and (7.43),

$$\tilde{W} = 8\pi n_0 M^2 \tag{7.45}$$

where

$$\tilde{W} = \dot{W}\frac{\rho c_p^2 V^2}{\nu k^2 t^2 R^3} \tag{7.46}$$

It can be shown that the general form of Eq. (7.45) is

$$\tilde{W} = 8\pi f M^2 \tag{7.47}$$

where f is the dimensionless flow resistance

$$f = n_0 \left(\sum_{j=0}^{p} 2^{j/3} \frac{L_j}{R} \right)^3 \tag{7.48}$$

Here, p is the construct level, and L_j is the pipe length at the pairing level j. In the case of radial tubes without ramifications (Fig. 7.6), the factor f is equal to n_0.

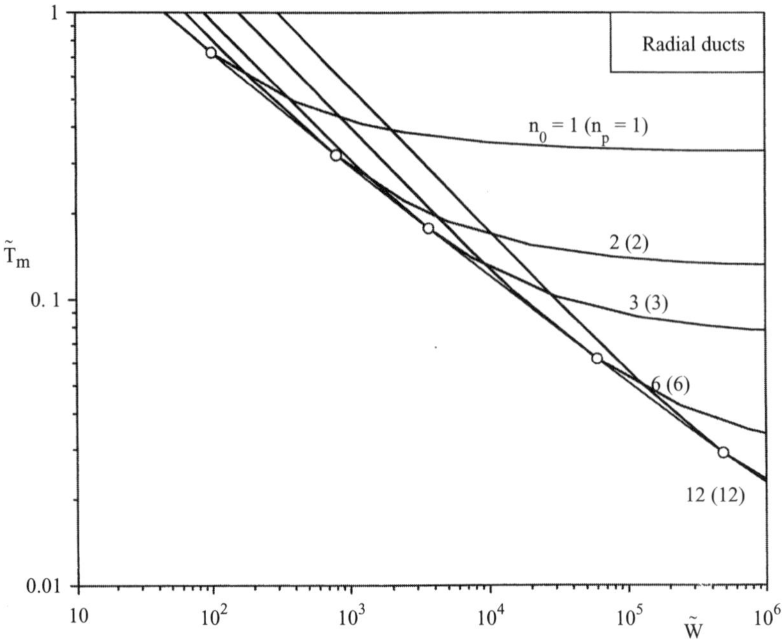

Figure 7.9 The competition between overall thermal resistance and pumping power in the radial flow structure of Fig. 7.6 [1].

In conclusion, Eq. (7.45) shows that the pumping power increases in proportion to n_0. A small number of tubes is desirable from the fluid mechanics point of view. This message is in conflict with that of Fig. 7.8, which showed that a larger n_0 is better from the heat transfer point of view. Preferable are low $\tilde{W}$ and low $\tilde{T}_m$ *at the same time*. This is why in Fig. 7.9 we show the result of combining Fig. 7.8 with Eq. (7.45): M is the parameter that varies along each of the curves.

Which radial design (n_0) is best depends on the application. If the available pumping power is in the range $10^4 < \tilde{W} < 10^5$, then the best performance (the minimal $\tilde{T}_m$) is offered by the $n_0 = 6$ design. The recommended number of radial tubes increases as $\tilde{W}$ increases. The ability to vary n_0, or the freedom to contemplate an entire family of radial designs, is represented by the envelope of the curves shown in Fig. 7.9, which is approximately $\tilde{T}_m \cong 3.75\tilde{W}^{-0.37}$. How this relation brings the performance in the desirable operating domain (cf. Fig. 7.1) is discussed later in Fig. 7.18. This relation is not a true envelope because in Fig. 7.9 we do not have an infinity of constant-n_0 curves. What we have is a power-law expression that plays adequately the role of common tangent for the few constant-n_0 curves. The points of tangency are identified with circles. They are noted (recorded and reproduced later in Fig. 7.18) as indicators of how closely each curve approaches the desirable limit (low $\tilde{T}_m$, low $\tilde{W}$). The number n_p represents the number of outlets on the disc perimeter, namely $n_p = n_0$ in the case of radial ducts.

7.3.2 One Level of Pairing

Figure 7.9 showed the locus of low $\tilde{T}_m$ and low $\tilde{W}$ designs when the competing designs are of the same class (namely, radial structures, Fig. 7.6). We show next that there exist flow structures that are represented by curves lower than those shown in Fig. 7.9. They belong to classes of flow structures that are more complex than the radial designs examined until now. The more complex structures are dendritic.

Figure 7.10 shows the main features of the dendritic flow structure when a single level of pairing is placed at the radial distance L_0. There are $n_0 = 3$ radial tubes connected to the center, and $n_p = 2n_0$ equidistant outlets on the disc perimeter. The design has three degrees of freedom: n_0, L_0, and the ratio of the diameters of the L_0, and L_1 tubes. The optimal diameter ratio subject to total tube volume is $2^{1/3}$ [cf. the Hess-Murray rule, Eq. (4.11)]. In addition to this optimization, the dimensionless flow resistance f can be minimized with respect to L_0 subject to the constraint of total tube volume and fixed disc radius. The total tube volume V is the same as for the radial designs discussed earlier; in the present case, $V = n_0(\pi/4)\, D_0^2$ $(L_0 + 2^{1/3}L_1)$. The optimal dimensions $(\hat{L}_0, \hat{L}_1)$ and performance $(f_{\min})$ are reported in Table 7.1, where the dimensionless lengths are $\left(\hat{L}_0, \hat{L}_1\right) = (L_0,\ L_1)/R$. The drawing shown in Fig. 7.10 is for the case $n_0 = 3$ of Table 7.1, and represents an architecture optimized for minimal flow resistance.

Next, we turn our attention to the thermal performance of the structures optimized for minimal flow resistance, Table 7.1. The global thermal resistance of such structures was calculated and reported in Fig. 7.11. The behavior is qualitatively

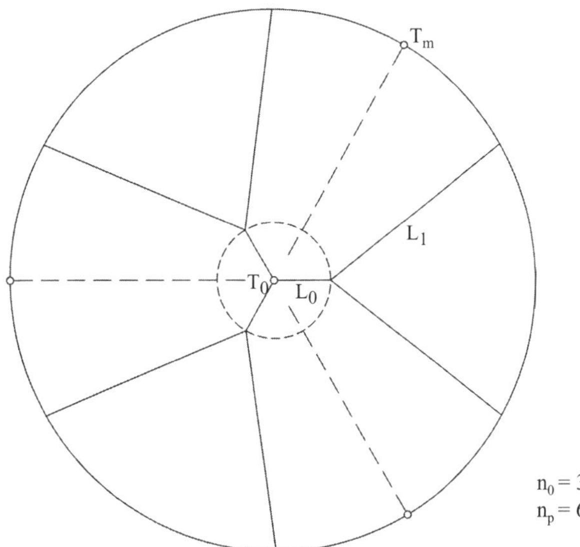

Figure 7.10 The structure with three central channels, one level of pairing, and minimal flow resistance [1].

Table 7.1 Optimized flow structures with one level of pairing and minimal flow resistance (e.g., Fig. 7.10).

n_0	$\hat{L}_0$	$\hat{L}_1$	f_{min}
3	0.214	0.822	5.849
6	0.628	0.426	9.471
12	0.821	0.216	15.605
24	0.913	0.108	27.625

the same as what we saw in Fig. 7.8 for radial designs, where $\tilde{T}_m$ and M continue to be defined as in Eqs. (7.36) and (7.38).

The pumping power that corresponds to the optimal designs of Table 7.1 is reported in Fig. 7.12, where $\tilde{W}$ is defined as in Eq. (7.46). The relationship between thermal performance ($\tilde{T}_m$), fluid mechanics performance ($\tilde{W}$) and complexity (n_0) is similar to what we found for radial designs (Fig. 7.9). The common tangent of the constant-n_0 curves in Fig. 7.12 is approximately $\tilde{T}_m \cong 5.94\ \tilde{W}^{-0.4}$, which is different than the tangent found in Fig. 7.9. The points of tangency are identified with circles, and are reproduced later in Fig. 7.18. Recall that in dendritic structures with one level of pairing, the number of outlets on the perimeter is twice the number of central ducts, $n_p = 2n_0$.

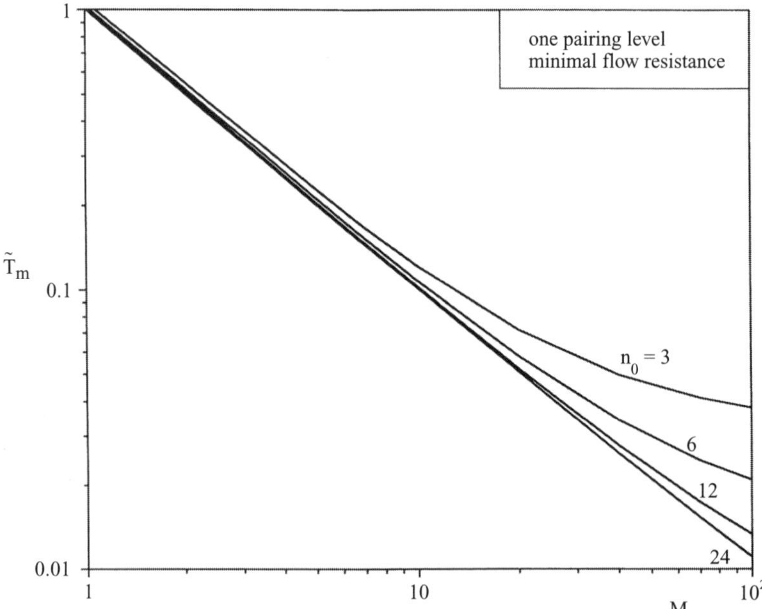

Figure 7.11 The thermal resistance of structures with one level of pairing and minimal flow resistance (Table 7.1) [1].

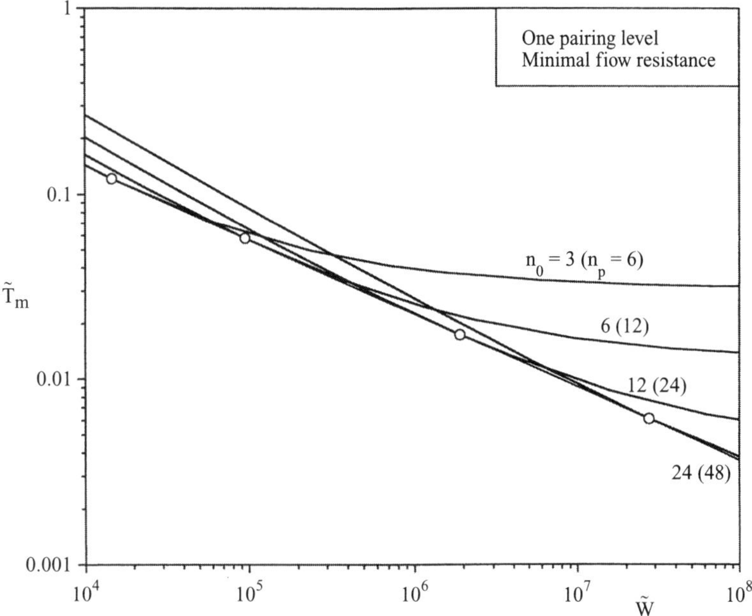

Figure 7.12 The competition between overall thermal resistance and pumping power in structures with one level of pairing and minimal flow resistance (Table 7.1) [1].

The alternative to minimizing the flow resistance is to minimize the thermal resistance of the structure with one level of pairing. The degrees of freedom are n_0 and $\hat{L}_0$, while the flow-rate number M is a specified parameter. Figure 7.13 shows the behavior of $\tilde{T}_m$ when $n_0 = 3$. We found numerically that the best designs—the lowest $\tilde{T}_m$ values—are found in the limit $\hat{L}_0 \rightarrow 0$. In other words, from the point of view of minimizing the thermal resistance, the design with $2n_0$ radial tubes is better than any of the designs with paired tubes.

The thermal performance of the $\hat{L}_0 = 0$ designs recommended by Fig. 7.13 is the same as what is reported in Fig. 7.8 for radial designs, provided that the n_0 values of Fig. 7.8 are replaced by $n_0/2$. For the same reason, the combined thermal resistance and pumping power performance is the same as in Fig. 7.9, provided that in Fig. 7.9 we put $n_0/2$ in place of n_0. With this substitution in mind, the common tangent of the $\tilde{T}_m(\tilde{W})$ curves for the $\hat{L}_0$ designs of Fig. 7.13 is the same as in Fig. 7.9, namely $\tilde{T}_m = 3.75\ \tilde{W}^{-0.37}$. Or, if we think in terms of the number of outlets on the perimeter (n_p) instead of n_0, the common tangent found in Fig. 7.9 can be extended safely to the right, toward larger $\tilde{W}$ and n_p values.

7.3.3 Two Levels of Pairing

The next step toward structures with greater complexity is in Fig. 7.14. The dendritic structure has two levels of pairing and two constraints, the total tube volume and

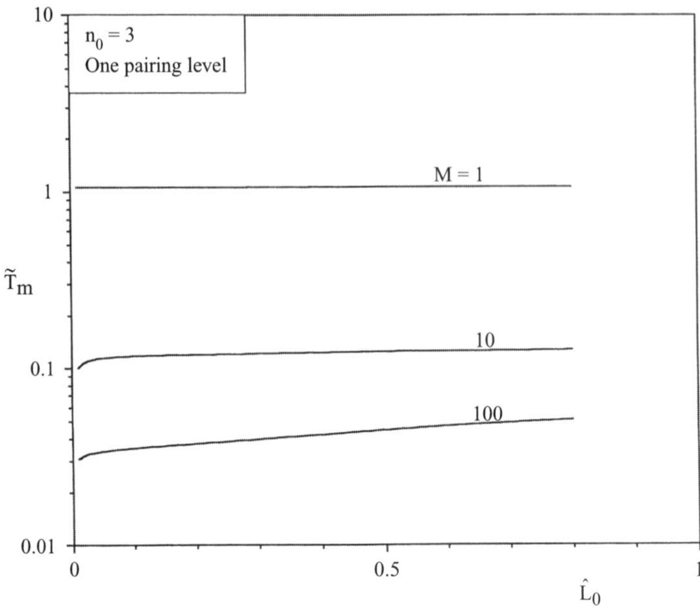

Figure 7.13 The effect of the length of the innermost radial pipes ($\hat{L}_0$) on the thermal resistance of structures with one level of pairing and $n_0 = 3$ [1].

the disc radius. The optimal ratios of successive tube diameters continue to be equal to $2^{1/3}$. The remaining degrees of freedom of the flow architecture are n_0, $\hat{L}_0$ and $\hat{L}_1$. The overall flow resistance f has been minimized with respect to $\hat{L}_0$ and $\hat{L}_1$. Table 7.2 shows a sample of the numerical results. The $n_0 = 3$ design of Table 7.2 has been drawn to scale in Fig. 7.14.

The overall thermal resistance that corresponds to the flow-optimized designs of Table 7.2 is reported in Fig. 7.15. Finally, the minimized pumping power that corresponds to these designs is reported on the abscissa of Fig. 7.16. The common tangent of the three curves is $\tilde{T}_m = 8.5\ \tilde{W}^{-0.43}$.

We also examined the alternative, which is to minimize the overall thermal resistance ($\tilde{T}_m$), instead of the overall resistance to fluid flow (f, or $\tilde{W}$). The degrees of freedom of the flow geometry are n_0, $\hat{L}_0$ and $\hat{L}_1$ (or $\hat{x}_1$), where ($\hat{x}_1 = x_1/R$) is the radial distance between the two levels of pairing (i.e., between the two

Table 7.2 Constructal flow structures with two levels of pairing and minimal flow resistance (e.g., Fig. 7.14).

n_0	$\hat{L}_0$	$\hat{L}_1$	$\hat{L}_2$	f_{min}
3	0.157	0.509	0.432	9.82
6	0.543	0.358	0.192	13.16
12	0.770	0.200	0.090	18.98

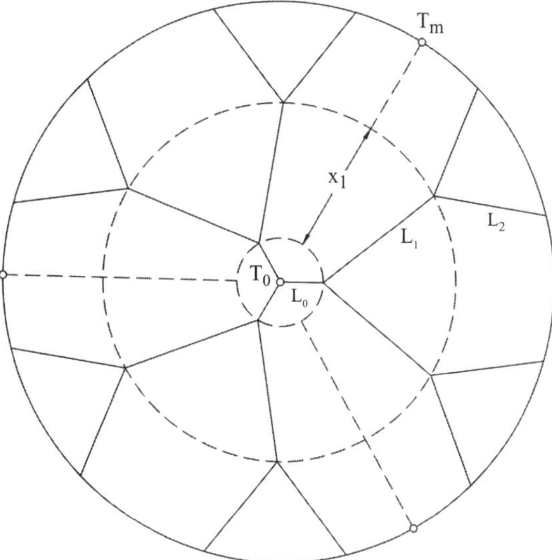

Figure 7.14 Structure with two levels of pairing and minimal flow resistance ($n_0 = 3$) [1].

dashed circles in Fig. 7.14). Figure 7.17 shows the behavior of the overall thermal resistance when $\hat{L}_0$ and $\hat{x}_1$ vary subject to constant n_0 and M. Although there is an optimal $\hat{x}_1$ when $\hat{L}_0$ is greater than 0.2, it is clear that the lowest thermal resistance belongs to designs characterized by $(\hat{L}_0, \hat{x}_1) \rightarrow 0$. These are radial designs, and their performance is again covered by Figs. 7.8 and 7.9, provided that for the present case the n_0 values of Figs. 7.8 and 7.9, are replaced by $n_0/4$, where n_0 is the number of central tubes in Fig. 7.14 and Table 7.2. The common tangent of the curves for the present designs ($\hat{L}_0 \rightarrow 0, \hat{x}_1 \rightarrow 0$) is the same as in Fig. 7.9, namely, $\tilde{T}_m = 3.75\ \tilde{W}^{-0.37}$.

Figure 7.18 is a concise summary of the two classes of flow architectures that were generated in this section: architectures with minimal pumping power vs. architectures with minimal global thermal resistance. The number of levels of pairing makes a difference when the flow geometry is selected based on the minimization of pumping power. When the objective is minimal global thermal resistance, the best flow pattern consists of radial ducts, and the number of pairing levels loses its effect.

An approximate reading of Fig. 7.18 justifies the conclusion that constructal flow structures that are complex are also robust. An optimized complex structure performs nearly as well as the next (stepwise more complex) structure. This characteristic of complex flow structures has been encountered in the design of other dendritic flows, for example, in trees for pure fluid flow and pure heat conduction (cf. Chapters 4 and 5).

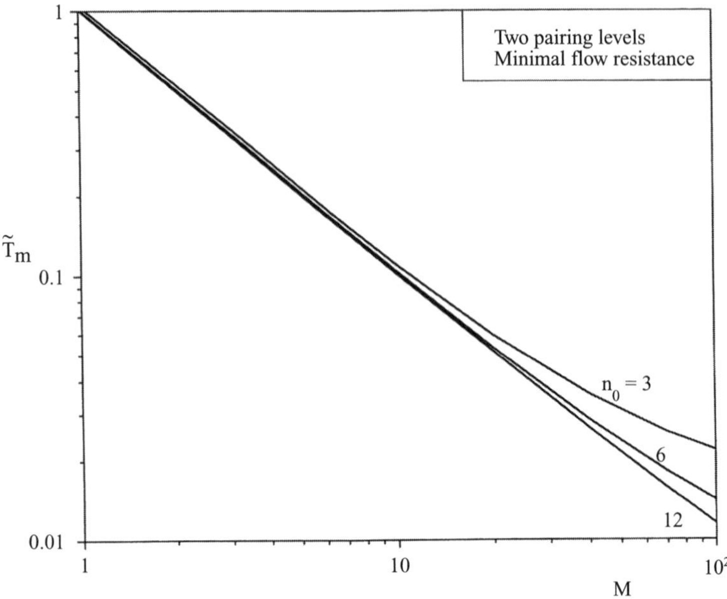

Figure 7.15 The thermal resistance of structures with two levels of pairing and minimal flow resistance (Table 7.2) [1].

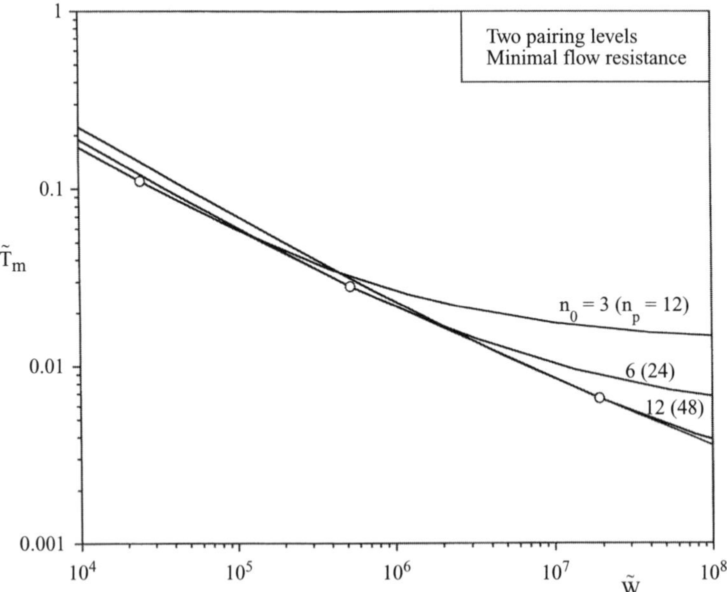

Figure 7.16 The competition between overall thermal resistance and pumping power in structures with two levels of pairing and minimal flow resistance (Table 7.2) [1].

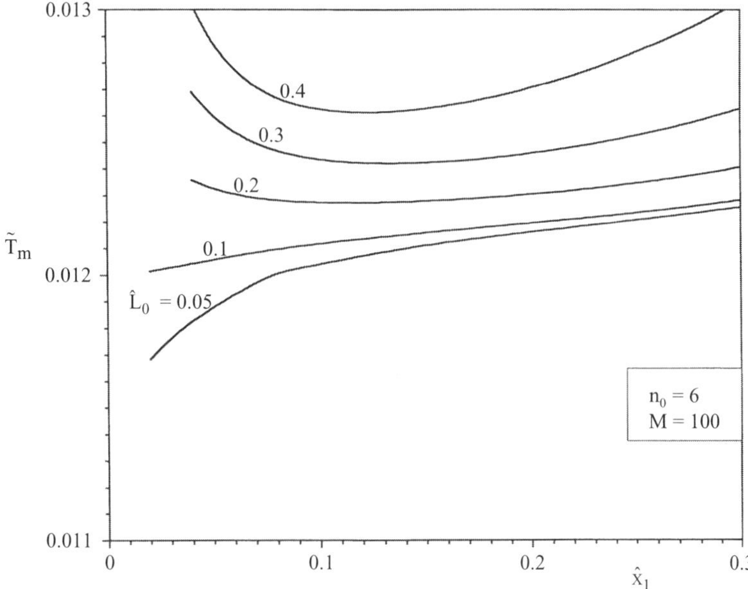

Figure 7.17 The effect of the innermost length ($\hat{L}_0$) and the radial distance between the two pairing levels ($\hat{x}_1$) on the thermal resistance of structures with two levels of pairing (e.g., Fig. 7.14) [1].

A more careful reading of Fig. 7.18, however, leads to a conclusion of great importance in the quest for convective flow structures at smaller and smaller scales. Miniaturization and greater heat generation densities mean lower $\tilde{T}_m$ values and higher $\tilde{W}$ values. We see that in this direction the complexity of the flow structure and the optimization objective matter. In this limit, the idea expressed in Fig. 7.1 (low $\tilde{T}_m$, low $\tilde{W}$) can be achieved by using dendritic structures, and by optimizing the structure for minimal flow resistance.

Another important conclusion is that optimized complexity is a good idea only below a certain $\tilde{T}_m$, that is, above a certain level of miniaturization or above a certain level of volumetric heat-transfer density. The world of competing flow architectures is characterized by a sharp transition between the simplest structure (radial), and progressively more complex structures (dendritic). The simplest structures are preferable when $\tilde{T}_m > 0.03$, or $\tilde{W} < 5 \times 10^5$. The more complex structures prevail in this competition when $\tilde{T}_m < 0.03$, or $\tilde{W} > 5 \times 10^5$. This stepwise change in flow configuration reminds us of natural occurrences in the quest for global performance, for example, the eddy formation in fluid flow (e.g., shear layers, Bénard convection) and the sudden appearance (and disappearance) of morphological changes in natural flow systems, animate or inanimate (e.g., the sudden appearance of thicker links in the time-dependent development of a river basin, while rain falls steadily on an erodable medium [9]). Similar to animal design is the pursuit of more than one

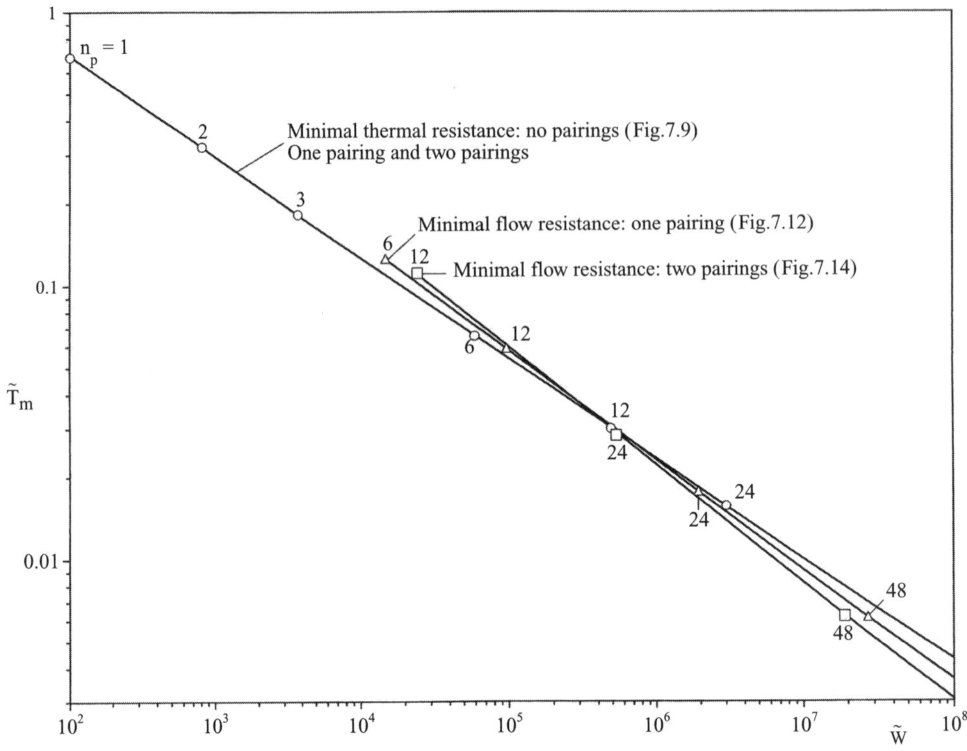

Figure 7.18 The common tangents of the $\tilde{T}_m(\tilde{W})$ curves for the optimized flow architectures: minimal flow resistance vs. minimal thermal resistance [1].

objective at the same time, in the present case, minimal flow resistance and minimal pumping power.

In closing, consider the engineering problem of how to select the flow pattern when the disc size (R) and the smallest length scale (d_0) are specified. These scales determine the number of outlets, n_p. For example, if $n_p = 24$ we may choose between the designs represented by three points in Fig. 7.18. The design with the smallest thermal resistance requires the largest pumping power. How to select one design out of the three possibilities depends on other considerations, such as the available pumping power. As n_p increases, the thermal resistance becomes insensitive to the increasing complexity, while the competing designs perform better and better hydrodynamically. In the limit of decreasing length scales and increasing complexity, the dendritic architectures are preferable.

An alternative practical point of view is to recognize the distance between two outlets as the smallest length scale of the design (d_0, Fig. 7.19) and to regard it as a manufacturing constraint. This distance is held fixed. The disc radius R and area A increase as the number of outlets increases. Figure 7.19 shows several designs

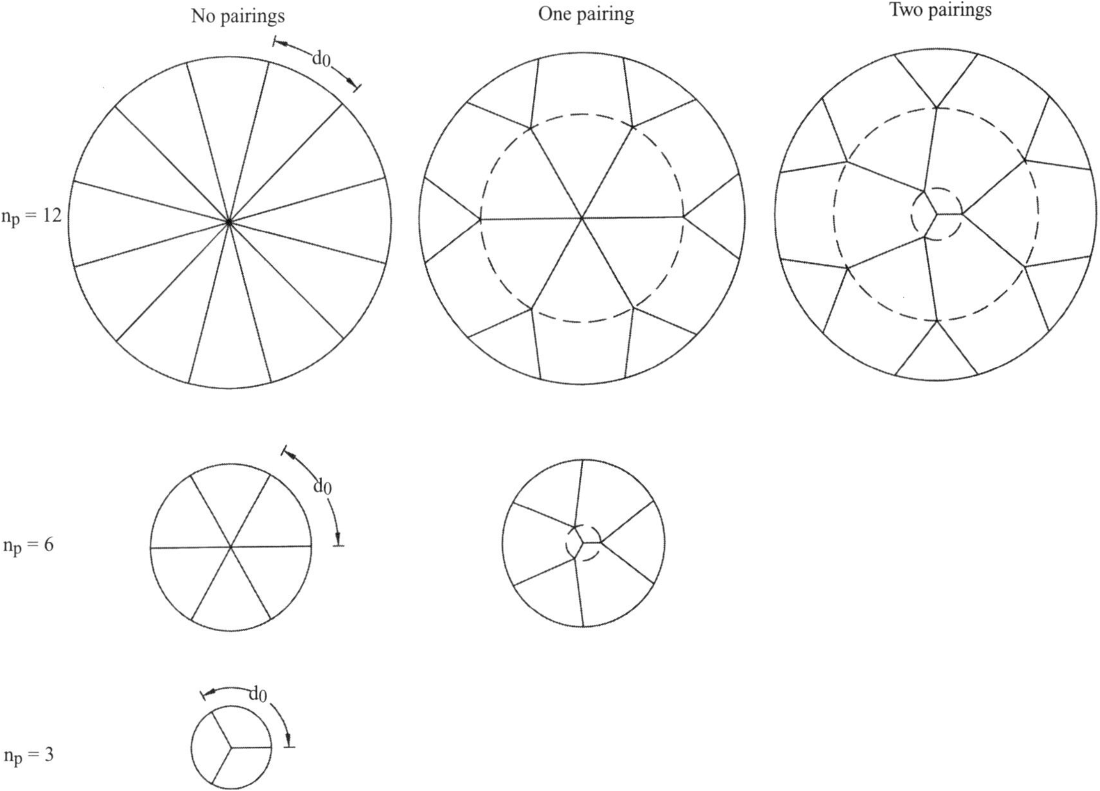

Figure 7.19 Examples of hydrodynamically optimized flow structures with the same smallest scale (d_0) and increasing size (n_p, or R) [1].

for $n_p = 3, 6$ and 12, where the dendritic patterns have been optimized for minimal fluid-flow resistance.

The fluid-flow performance of the hydrodynamically optimized designs with fixed d_0 is presented in Fig. 7.20. The overall pressure drop between the disc center and the periphery (ΔP) is nondimensionalized as

$$\hat{P} = \frac{V^2 \Delta P}{\dot{m}'' \nu d_0^5} \tag{7.49}$$

where the mass flow rate per unit disc area [$\dot{m}'' = \dot{m}/(\pi R^2)$] is assumed independent of the disc radius. It can be shown that for Poiseuille flow and optimized ratios of successive tube diameters (section 7.3.2) the dimensionless pressure drop is

$$\hat{P} = \frac{n_p^5 f}{4\pi^3} \tag{7.50}$$

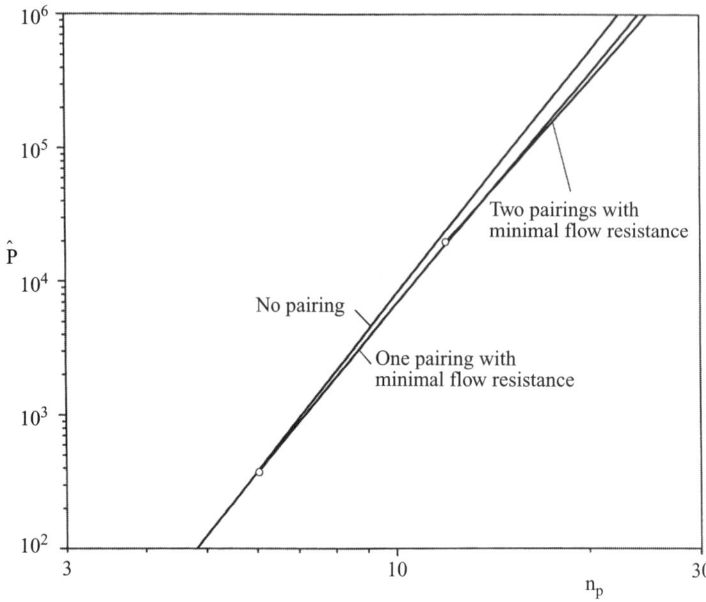

Figure 7.20 The minimized overall flow resistance of structures with the same smallest scale (d_0) and increasing size (n_p, or R) [1].

for which the optimization of the flow architecture yields the minimized f values (e.g., Table 7.1). In particular, the radial design has $f = n_p$, which means that $\hat{P} = n_p^6/(4\pi^3)$.

Figure 7.20 shows that the global flow resistance $\hat{P}$ increases with n_p, i.e., with the size of the flow system. This increase, however, can be slowed down through the use of progressively more complex dendritic structures. The symbols plotted in Fig. 7.20 represent the starting n_p values of the indicated structure. Dendrites with one level of pairing start with $n_p = 6$, while dendrites with two levels of pairing start with $n_p = 12$.

A qualitatively similar picture emerges if, instead of $\hat{P}$, we plot on the ordinate the dimensionless pumping power required by the design. In conclusion, we see that *optimized* complexity can be beneficial, and that the best flow structure becomes more complex as the flow system becomes larger. Again we conclude that optimized complexity is a result of constructal design, not an objective—it must not be confused with the maximization of complexity.

7.4 DENDRITIC HEAT EXCHANGERS

Constructal theory and design focuses attention on the judicious distribution of fluid streams that perform functions over volumes. To maximize the global performance of the macroscopic flow system means that every volume element should function at the same (the highest) level of performance as any other volume element. The flow structures that emerge along this design route are tree shaped, with multiple scales that are arranged hierarchically. Like the bronchial tree of the human lung, such structures achieve maximal density of transport, or maximal compactness.

In this section we bring ideas of sections 7.2 and 7.3 to a class of even more complex flow architectures: dendritic heat exchangers [10, 11]. This is a distinctly new direction for the development of heat exchanger architectures. Current heat exchanger design methods call for the use of "uniform" (one scale) flow structures, for example, banks of parallel tubes in human-scale heat exchangers, and arrays of parallel microchannels for electronics cooling. One tube or channel is designed to perform the same as its neighbor. Uniform architectures are at work in many other applications, for example, nuclear reactor cores, packed beds, volumetrically cooled electric windings, and packages of electronics.

Fluid streams have a characteristic analogous to the economies of scale known in economics: the transportation cost per unit decreases when the goods are transported in larger and larger quantities. Larger streams flow more easily through correspondingly wider ducts. Said another way, two small streams flow more easily when they flow together through a single duct. Geometric features (e.g., coalescence) generated by the search for less resistance endow the larger flow system with organization, structure, geometry, or topology. What is perceived as bad ("mal"distribution) because of the designer's simplifying assumption of flow uniformity is, in fact, good [12] from the point of view of minimizing all the internal flow resistances together, by balancing the streams against each other in such a way that the global resistance of the macroscopic and highly complex system is minimum. The emerging flow structure is tree shaped (multiscale, nonuniform), not parallel and uniform.

7.4.1 Geometry

We begin with the assumption that the heat exchanger has the flow structure shown in Fig. 7.21. This structure was developed based on constructal theory in Ref. [4], and has several advantages: it is well documented, the flow path length is the same for all the source-sink paths, and the source-sink pressure drops and flow rates are the same for all the paths. In brief, the constructal tree of Fig. 7.21 distributes a stream (e.g., $\dot{m}_n = \dot{m}_2$) *uniformly* over a square-shaped territory.

The balanced counterflow heat exchanger has two identical trees—one for the hot fluid, and the other for the cold fluid. The two trees mate perfectly, so that one tube of the hot tree is parallel to and in excellent thermal contact with the corresponding tube of the cold tree. In the lower part of Fig. 7.21, the hot tree distributes the single stream ($\dot{m}_n$) over the square area occupied by the structure. The cold tree collects a large number of mini-streams ($\dot{m}_0$) and forms a single cold stream ($\dot{m}_n$). The same single-phase fluid flows through the two trees.

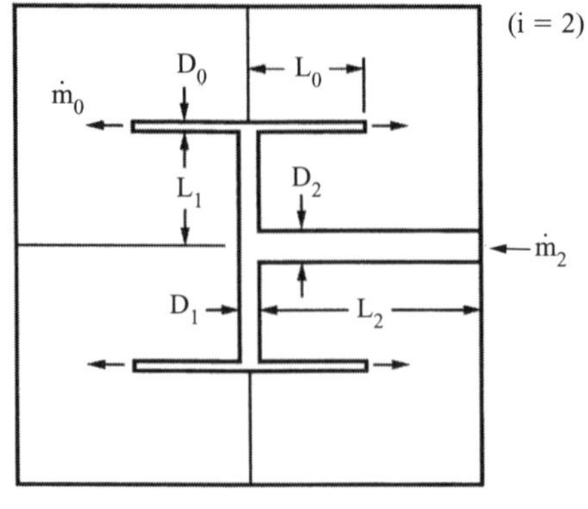

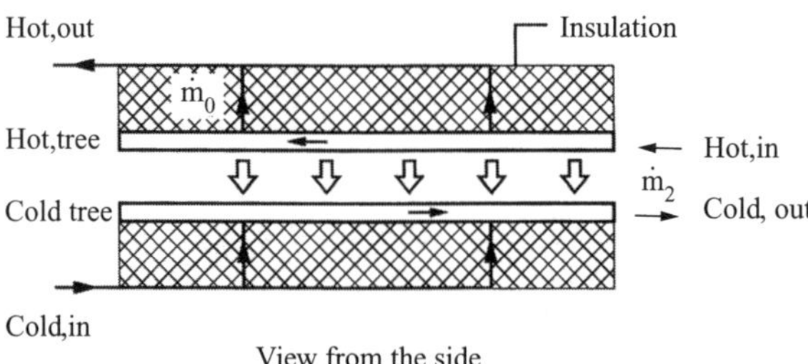

Figure 7.21 Counterflow of tree-shaped streams distributed over a square area [11].

The hot mini-streams are collected by a manifold on the back side of the hot tree, and are led out as a single stream. The manifold is insulated from the hot tree. The mixing of the hot mini-streams inside the manifold is not a source of irreversibility because, as we shall see in section 7.4.3, the $\dot{m}_0$ streams arrive at the manifold not only at the same pressure but also at the same temperature (cf. Ref. [13]). On the back side of the cold tree, the cold stream $\dot{m}_0$ is distributed by a manifold as a large number of mini-streams $\dot{m}_0$, which feed the canopy of the cold tree. The cold manifold is insulated with respect to the cold tree flow structure. The arrows that indicate the flow of fluid (not heat) in Fig. 7.21 could be reversed, but the arrangement itself (the drawing) and its thermo-fluid performance do not change.

One tree is made of many tubes of $(n + 1)$ sizes. One tube has the length L_i and internal diameter D_i, where $i = 0, 1, \ldots, n$. The number of tubes of type i is n_i. In constructal design the smallest scale is primordial: the construction of the entire flow architecture rests on the smallest tube scale (L_0, D_0), which inhabits the smallest square area element (called elemental system). Larger constructs are made by pairing smaller constructs. For example, the first constructs $(i = 1)$ consist of joining two elemental systems, and joining the two elemental streams into a first-construct stem of size (L_1, D_1). Dichotomy (pairing, or bifurcation) is used here as a deduced constructal feature, not as an assumption. In the second construct ($i = 2$), the stem (L_2, D_2) is twice as long as L_1, or L_0. Tube lengths double after two consecutive construction steps. When there are many construction levels, we may express the length-doubling rule by writing approximately

$$L_{i+1} \cong 2^{1/2} L_i \qquad (i = 0, 1, \ldots, n) \tag{7.51}$$

Pairing at every construction level means that the tube numbers and flow rates are ordered as

$$n_i = 2^{n-i} \qquad \dot{m}_i = 2^i \dot{m}_0 \quad (i = 0, 1, \ldots, n) \tag{7.52}$$

There are n construction stages, and in the end a single stream of flow rate $\dot{m}_n$ (= $2^n\ \dot{m}_0$) flows into or out of the largest (n^{th}) construct. The tree structure bathed by $\dot{m}_n$ has $(n + 1)$ length scales, which are organized hierarchically.

7.4.2 Fluid Flow

To determine the fluid mechanics benefits of the dichotomous (pairing) structure, we make the assumption that every tube is slender so that the svelteness Sv is large (cf. section 1.1), and that the flow is in the Poiseuille regime. The pressure drops are due mainly to friction along the straight sections of the network. In other words, we neglect the local pressure drops associated with joining two tubes together [8]. The pressure drop along one tube of type i is

$$\Delta P_i = \frac{128}{\pi} \nu \dot{m}_i \frac{L_i}{D_i^4} \tag{7.53}$$

Because the path between the root of the tree ($\dot{m}_n$) and one point in the canopy ($\dot{m}_0$) is the same for all the canopy points (the path consists of the same numbers and sizes of tubes), the canopy points are all at one pressure. The overall pressure difference (root-canopy, or canopy-root) is

$$\Delta P = \sum_{i=0}^{n} \Delta P_i \tag{7.54}$$

According to the Hess-Murray rule, when the volume occupied by all the tubes (V) is fixed, ΔP is minimal when the tube diameters obey the proportionality, $D_{i+1} = 2^{1/3} D_i$, ($i = 0, 1, 2, \ldots, n$). This optimization result is robust, because it does not depend on the lengths L_i, their sizes and relative positions. By combining Eqs. (7.51) through (7.54), we can write the minimal overall pressure drop as

$$\Delta P = \frac{128}{\pi} \nu \dot{m}_n \frac{L_0}{D_0^4} 2^{-n} S_1 \tag{7.55}$$

where the sum S_1 is a function of n, namely $S_1 = (2^{(n+1)/6} - 1)/(2^{1/6} - 1)$. The total tube volume is

$$V = \sum_{i=0}^{n} n_i \frac{\pi}{4} D_i^2 L_i = \frac{\pi}{4} D_0^2 L_0 2^n S_1 \tag{7.56}$$

Another global dimension of the design is the total area covered by largest construct (the nth construct),

$$A = 2^n (2L_0)^2 \tag{7.57}$$

The optimized tree structure depends on three geometric features: the smallest scales (D_0, L_0) and the complexity (n). By using constraints (7.56) and (7.57) we obtain $L_0 = 2^{-(n+2)/2} A^{1/2}$ and $D_0 = \pi^{-1/2} 2^{3/2-n/4} V^{1/2} A^{-1/4} S_1^{-1/2}$, such that the architecture is described by A, V and n. The ideal-limit pumping power required to force $\dot{m}_n$ to flow through one of the (A, V)-size trees is $\dot{W} = \dot{m}_n \Delta P/\rho$, which after using Eq. (7.55) becomes

$$\tilde{W} = \pi^3 2^{-n/2} S_1^3 M^2 \tag{7.58}$$

The dimensionless power requirement and total mass flow rate are defined as

$$\tilde{W} = \frac{\dot{W} V^2}{(k\,\text{Nu}/c_P)^2 (\text{Nu}/\rho)\, A^{5/2}} \qquad M = \frac{\dot{m}_n\, c_P}{\pi k \text{Nu}\, A^{1/2}} \tag{7.59}$$

Note the similarities between these definitions and Eqs. (7.46) and (7.38).

7.4.3 Heat Transfer

The heat-transfer performance of two streams in counterflow is condensed in the relations between effectiveness and number of heat transfer units. The equivalent relation for two trees in counterflow were developed in Ref. [11]. The two trees are aligned perfectly, like one palm pressed against the other (Fig. 7.21, bottom). One tube of the hot tree (L_i, D_i) is placed right next to its counterpart from the cold tree. This local counterflow is balanced, because the capacity rate $\dot{m}_i c_P$ flows through each of the two tubes. This is why the temperature difference ΔT_i between the two $\dot{m}_i$ streams does not vary in the flow direction (Fig. 7.22). The stream-to-stream heat-transfer rate is

$$q_i = U_i \pi D_i L_i \Delta T_i \tag{7.60}$$

where the overall heat-transfer coefficient U_i is based on the internal area of one tube. We assume that the heat current q_i is impeded primarily by the internal (convective) thermal resistances of the two laminar flows, and not by the thermal diffusion through the material in which the tube pair is embedded. The overall

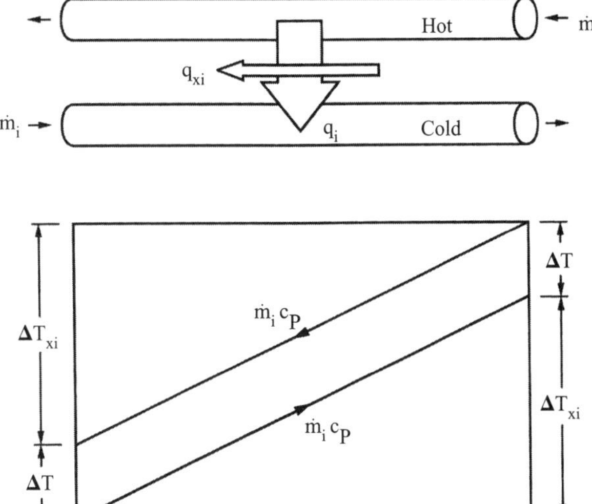

Figure 7.22 Counterflow of two matching ducts of the two trees of Fig. 7.21 [11].

heat-transfer coefficient is

$$\frac{1}{U_i} = \frac{1}{h_i} + \frac{1}{h_i} \tag{7.61}$$

where h_i is the heat transfer coefficient for thermally and hydrodynamically fully developed flow in one tube,

$$h_i = \frac{k}{D_i}\mathrm{Nu} \tag{7.62}$$

and the Nusselt number is a constant on the order of 1, cf. Ref. [5]. The heat current q_i is also equal to the enthalpy decrease experienced by the warm stream,

$$q_i = \dot{m}_i c_P \Delta T_{xi} \tag{7.63}$$

where ΔT_{xi} is the temperature variation along the stream. Dividing Eqs. (7.60) and (7.63), we find

$$\frac{\Delta T_{xi}}{\Delta T_i} = \frac{\pi k \mathrm{Nu} L_i}{2\dot{m}_i c_P} = N_i \tag{7.64}$$

The group N_i is the number of heat transfer units of the counterflow formed by the two $\dot{m}_i$ streams. This counterflow convects energy longitudinally in the direction of the warmer stream [14–16], that is, in the direction of the energy current q_{xi} shown in Fig. 7.22. The hot stream carries enthalpy to the left at the rate $\dot{m}_i c_P\, T_{\text{hot}}$. The cold stream carries enthalpy to the right at the rate $\dot{m}_i c_P T_{\text{cold}}$. The net energy current, which flows to the left, is the difference between the two enthalpy streams. If we write ΔT_i for the stream-to-stream temperture difference $T_{\text{hot}} - T_{\text{cold}}$, the net longitudinal convective energy current is

$$q_{xi} = \dot{m}_i c_P \Delta T_i \tag{7.65}$$

Combining this with Eq. (7.64), we see that the longitudinal convective current is proportional to the longitudinal temperature gradient [14–16]:

$$q_{xi} = \frac{2\,(\dot{m}_i c_P)^2}{\pi k \,\text{Nu}} \frac{\Delta T_{xi}}{L_i} \tag{7.66}$$

Figure 7.22 shows that we use the additional subscript x to indicate new quantities that refer to the longitudinal (streamwise) direction. Perpendicular to the stream-to-stream heat transfer q_i, we use the longitudinal energy stream q_{xi}. Perpendicular to the stream-to-stream temperature difference ΔT_i, we use the longitudinal temperature excursion ΔT_{xi}.

The total current convected by the two trees in the direction of the warmer tree is $n_i q_{xi}$. The continuity of this current requires that

$$n_i q_{xi} = q_{xn} \tag{7.67}$$

where q_{xn} is the "root" convective current due to the pair of largest tubes. By combining Eqs. (7.65) and (7.66) with Eq. (7.52), we find

$$2^n \dot{m}_0 c_P \Delta T_i = q_{xn} \tag{7.68}$$

This result shows that the stream-to-stream temperature difference ΔT_i does not depend on the position (i) of the tube pair in the tree hierarchy. The discovery is that there is only one ΔT, and it is found everywhere along the tree sandwich,

$$\Delta T_i = \Delta T \tag{7.69}$$

The total heat transfer rate from the hot tree to the cold tree is

$$q = \sum_{i=0}^{n} n_i q_i \tag{7.70}$$

The same heat transfer rate can be estimated by summing up the longitudinal enthalpy drops (q_{xi}) experienced by all the streams,

$$q = \sum_{i=0}^{n} n_i \dot{m}_i c_P \Delta T_{xi} = \dot{m}_n c_P \Delta T_x \tag{7.71}$$

where ΔT_x is the overall temperature difference between the root (inlet, or outlet) and the canopy points (outlets, or inlets),

$$\Delta T_x = \sum_{i=0}^{n} \Delta T_{xi} = (T_{\text{out}} - T_{\text{in}})_{\text{hot}} = (T_{\text{out}} - T_{\text{in}})_{\text{cold}} \tag{7.72}$$

Performing the summation shown in Eq. (7.70) and dividing by Eq. (7.71), we obtain the global equivalent of the local expression (7.64):

$$\frac{\Delta T_x}{\Delta T} = \frac{\pi k \mathrm{Nu} L_0}{\dot{m}_n c_P} 2^{n-1} S_2 = N \tag{7.73}$$

Here, N is the global number of heat transfer units of the tree counterflow, and S_2 is a sum that depends very weakly on the complexity of the tree structure (n), namely $S_2 = (1 - 2^{-(n+1)/2})/(1 - 2^{-1/2})$.

The global temperature difference sustained by the double-tree heat exchanger is $\Delta T_x + \Delta T = T_{\text{in,hot}} - T_{\text{in,cold}}$. The global thermal resistance of this new type of heat exchanger is

$$R_t = \frac{\Delta T_x + \Delta T}{q} = \frac{1}{\dot{m}_n c_P \varepsilon} \tag{7.74}$$

where ε is the global effectiveness of the tree counterflow heat exchanger

$$\varepsilon = \frac{\Delta T_x}{\Delta T_x + \Delta T} = \frac{N}{N + 1} \tag{7.75}$$

This relationship between effectiveness (ε) and number of heat transfer units (N) happens to be the same as the relationship for the much simpler configuration where two streams are in counterflow and have equal capacity rates [17]. This suggests that the $\varepsilon = N/(N + 1)$ is inherent to the entire class of balanced counterflow heat exchangers, regardless of the complexity of the flow arrangement. What changes from one complex arrangement to the next is the makeup of N formula, namely, Eq. (7.73) for the trees of Fig. 7.21.

An interesting feature of the tree counterflow is its highly *nonuniform* longitudinal variation of temperature. It is known that in the counterflow formed by two streams of equal capacity rate, the temperature gradient in the flow direction is independent of longitudinal position: the longitudinal temperature distribution is linear (Fig. 7.22). For the tree counterflow, Eqs. (7.64) and (7.69) show that the temperature gradient associated with the tube length L_i decreases fast as i increases, i.e., toward the trunk, $\Delta T_{xi}/L_i = \pi k \mathrm{Nu} \Delta T/(2^{i+1} \dot{m}_0 c_P)$. We can present this behavior by plotting the temperature along the flow, between the tree root ($i = n$) and any of the points of the canopy ($i = 0$). The longitudinal position x_i is measured from the canopy point toward the root,

$$x_i = \sum_{i=0}^{i} L_i = L_0 \frac{2^{(i+1)/2} - 1}{2^{1/2} - 1} \tag{7.76}$$

Because the total flow length is $x_n = x_i$ ($i = n$), the dimensionless longitudinal position in the flow direction is

$$\xi_i = \frac{x_i}{x_n} = \frac{2^{(i+1)/2} - 1}{2^{(n+1)/2} - 1} \tag{7.77}$$

Next, we construct a similar coordinate for the longitudinal variation of temperature. The temperature change from the canopy point to the far end of the ith tube is

$$y_i = \sum_{i=0}^{i} \Delta T_{xi} = \frac{\pi k \mathrm{Nu} \Delta T L_0}{2\dot{m}_0 c_P} \frac{1 - 2^{-(i+1)/2}}{1 - 2^{-1/2}} \tag{7.78}$$

This can be referenced to the total temperature change along the canopy-root flow, $y_n = y_i(i = n) = \Delta T_x$, namely

$$\eta_i = \frac{y_i}{y_n} = \frac{1 - 2^{-(i+1)/2}}{1 - 2^{-(n+1)/2}} \tag{7.79}$$

Figure 7.23a shows the streamwise temperature distribution as a sequence of points (ξ_i, η_i) for a given number of pairing levels (n). Each point (i) represents the junction between tubes L_i and L_{i+1}; that is, the junction is at the end of the L_i tube that is closer to the trunk, not the canopy. Unlike the counterflow between two balanced streams, where the temperature distribution has a single longitudinal temperature gradient (Fig. 7.22), in the counterflow formed between two balanced trees the longitudinal temperature gradient becomes greater as we approach the canopy. The temperature distribution is a fractured line, one segment for one value of n. When the number of pairing or bifurcation levels n is large, the temperature distribution resembles a concave curve with one end pinned in the canopy ($\xi_n = \eta_n = 0$) and the other in the root ($\xi_n = \eta_n = 1$).

An interesting observation is highlighted with dashed lines in Fig. 7.23a. The breaks in the temperature distribution curves fall on straight dashed lines that originate from the canopy and root corners. The dashed lines cut the top and left sides of the square frame at $\xi_n = 2^{-n/2}$ and $\eta_n = 1 - 2^{-n/2}$. This observation leads to a fast method of constructing the temperature distribution for a given n. As shown in Fig. 7.23a, the construction begins with marking the ξ_n and η_n cuts on the top and left sides of the square frame. Next, the root and canopy corners are connected with the cuts by straight dashed lines (n_x, n_y). The dashed lines intersect and form a net with loops. The construction of the nth fractured line begins from one corner (e.g., canopy), on the segment that belongs to the radial line n. This first segment ends at a point of intersection that is the lower-left corner of a quadrilateral loop. The next segment is the long diagonal of the loop. The construction continues with segments that serve as long diagonals for subsequent loops, and ends with the segment of the nth radial line that leads to the root point.

Another way to construct the nth fractured line is based on the observation that the n bends of the nth fractured line are the n points of intersection of the straight dashed lines, such that $n_x + n_y = n + 1$. For example, the fractured line drawn for $n = 4$ has four bends located at the intersections of four pairs of dashed lines (n_x, n_y), namely, (1, 4), (2, 3), (3, 2) and (4, 1).

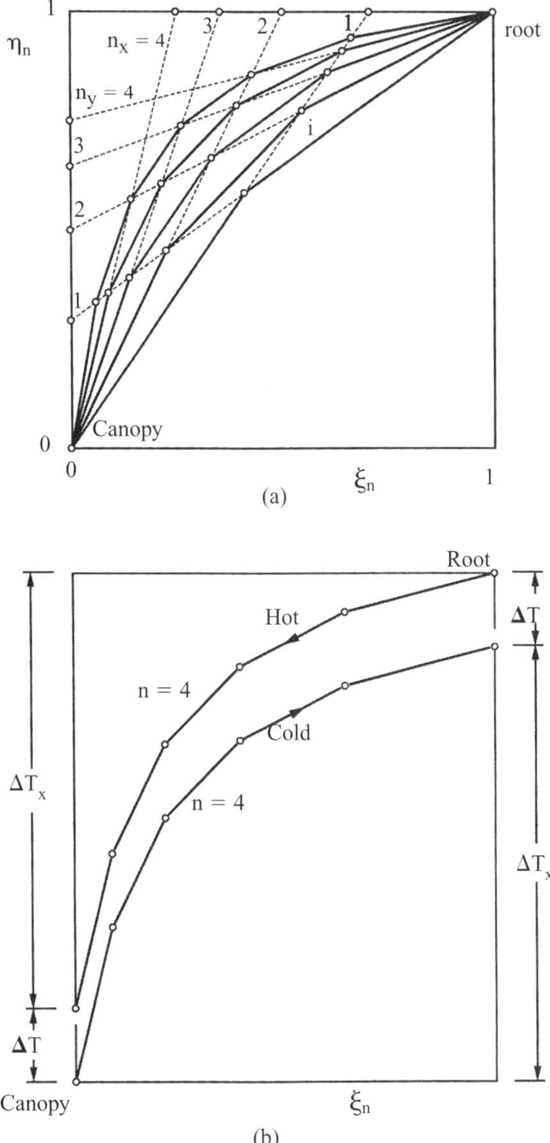

Figure 7.23 The temperature distribution along the constructal tree flow [11].

The temperature distributions of the two tree-shaped streams in counterflow are shown for the case $n = 4$ in Fig. 7.23b. The stream-to-stream temperature difference ΔT is the same at every longitudinal location between the root and the canopy. The hot-tree and cold-tree distributions come together as the global conductance N increases. The differences between Fig. 7.23b and Fig. 7.22 show why tree-shaped heat exchangers are new and depart fundamentally from classical two-stream heat exchangers.

7.4.4 Radial Sheet Counterflow

A simpler and potentially more compact counterflow configuration is shown in Fig. 7.24a. Each stream is shaped as a disc and flows between the center and the rim. The hot stream enters through a central port of radius R_c, and is collected in a peripheral plenum of radius R. The cold stream is supplied by a similar plenum, and flows toward a central outlet of radius R_c. The two discs are placed in close thermal contact, so that they form a counterflow in which each branch connects a circle with its center.

The simplest counterflow of this type consists of fluid sheets that flow through parallel-plates channels with constant spacing (D). The stream flow rate ($\dot{m}$) is the same at every radial position (r). Because the hot and cold capacity rates are balanced, the local stream-to-stream temperature difference is independent of r:

$$T_{\text{hot}}(r) - T_{\text{cold}}(r) = \Delta T \tag{7.80}$$

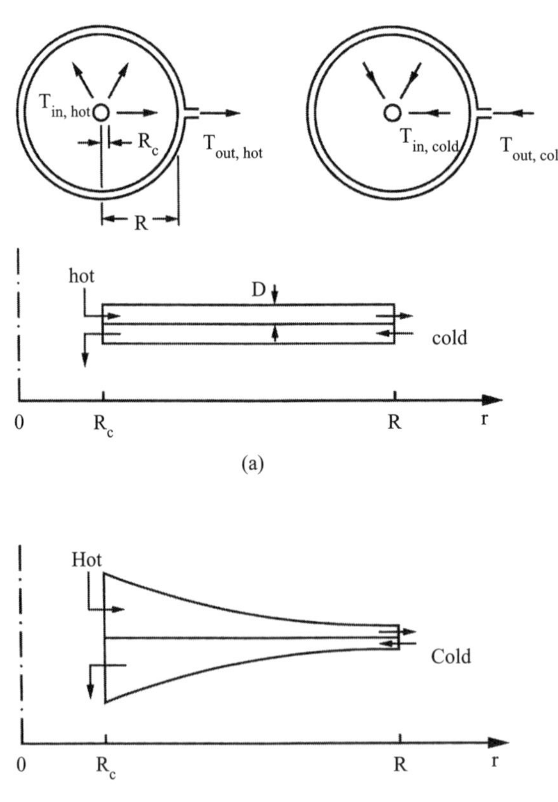

Figure 7.24 Counterflow of two fluid sheets flowing radially: (a) constant-D channels; (b) variable-D channels [11].

An analysis analogous to that of section 7.4.3 and classical heat exchanger treatments leads to the conclusion that the total heat-transfer rate between the two disc-shaped fluid sheets is given by [11]

$$q_a = \dot{m} c_P \varepsilon_a (T_{\text{in,hot}} - T_{\text{in,cold}}) \tag{7.81}$$

$$\varepsilon_a = \frac{N_a}{1 + N_a} \tag{7.82}$$

$$N_a = \frac{\pi R^2 k \text{Nu}}{2 D \dot{m} c_p} \tag{7.83}$$

where the Nusselt number Nu is a dimensionless constant in the range 1–10. This analysis accounts for the fact that the wetted perimeter of thermal contact of each stream is $2 \times 2\pi r$; that is, it depends on radial position.

The fluid mechanics performance of the same arrangement is described by the overall pressure drop, ΔP_a. We assume that the flow channel is sufficiently narrow ($D \ll R$) and the Reynolds number is sufficiently small so that the velocity profile across the channel is parabolic (as in Poiseuille flow) with the pressure P_a a function of radial position. If the flow proceeds from the rim to the center, the pressure gradient is

$$\frac{dP_a}{dr} = \frac{6 \nu \dot{m}}{\pi D^3 r} \tag{7.84}$$

Integrating this we obtain the overall pressure difference $\Delta P_a = P_a(R) - P_a(R_c)$. In this result we replace D in terms of the volume of one disc-shaped channel, $V_a = \pi R^2 D$, and obtain the pressure difference formula $\Delta P_a = 6\pi^2 \nu \dot{m} (R^6 / V_a^3) \ln(R/R_c)$.

An improved radial flow configuration is shown in Fig. 7.24b, where D is larger closer to the center, in order to offset the high velocities and friction that occur near the center when D is constant. There is an optimal channel shape [$D_{\text{opt}}(r)$] so that the overall pressure drop (ΔP_2) is minimal subject to the same channel volume constraint $V_b = \int_{R_c}^{R} 2\pi r D dr = V_a$. The pressure drop ΔP_b is calculated with Eq. (7.82) in which we replace P_a with P_b, and D (constant) with $D(r)$. The function $D_{\text{opt}}(r)$ for which ΔP_b reaches its minimum subject to V_a = constant is determined by variational calculus [11] (cf. Appendix C): $D_{\text{opt}}(r) = 3V_a / [4\pi r^{1/2} (R^{3/2} - R_c^{3/2})]$. The minimal pressure drop that corresponds to this optimal channel shape is $\Delta P_b = (4^4 \pi^2 / 3^3)(\dot{m} \nu / V_a^3)(R^{3/2} - R_c^{3/2})^4$. The improvement in fluid flow performance in going from the constant-D design (ΔP_a) to the optimal-D design (ΔP_b) is indicated by the ratio

$$\frac{\Delta P_b}{\Delta P_a} = \frac{(4/3)^4}{2} \frac{\left[1 - (R_c/R)^{3/2}\right]^4}{\ln(R/R_c)} \tag{7.85}$$

This ratio is plotted in Fig. 7.25: it is approximately equal to $[2 \ln (R/R_c)]^{-1}$, which is less than 1 when $R_c \ll R$.

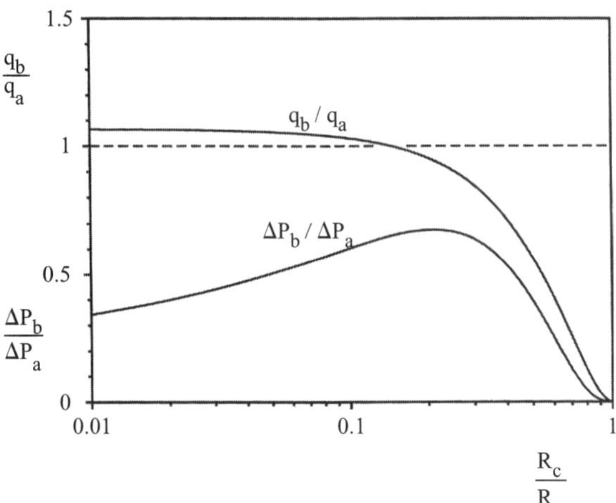

Figure 7.25 Improvement in hydraulic and thermal performance when the constant-D design (Fig. 7.24a) is replaced by the optimal-D design (Fig. 7.24b), when $R_c \ll R$ [11].

The heat-transfer performance of the optimal-D design is described by Eq. (7.80) and $q_b = \dot{m} c_P \varepsilon_b \left(T_{\text{in,hot}} - T_{\text{in,cold}}\right)$, $\varepsilon_b = N_b/(1 + N_b)$, and

$$N_b = \frac{8\pi^2}{15} \frac{k\text{Nu}R^4}{\dot{m} c_p V_a} \left[1 - \left(\frac{R_c}{R}\right)^{5/2}\right]\left[1 - \left(\frac{R_c}{R}\right)^{3/2}\right] \tag{7.86}$$

This can be compared with the heat-transfer performance of the constant-D design, Eqs. (7.81), where we may replace D in terms of V_a by writing $V_a = \pi R^2 D$. When $N_a \gg 1$ and $N_b \gg 1$, the effectivenesses are $\varepsilon_a \simeq 1$ and $\varepsilon_b \simeq 1$, and there is no significant change in thermal performance: $q_a \simeq q_b$. When $N_a \ll 1$ and $N_b \ll 1$, we have $\varepsilon_a \simeq N_a$ and $\varepsilon_b \simeq N_b$, and the relative change in thermal performance is described by the ratio

$$\frac{q_b}{q_a} = \frac{16}{15} \left[1 - \left(\frac{R_c}{R}\right)^{5/2}\right]\left[1 - \left(\frac{R_c}{R}\right)^{3/2}\right] \tag{7.87}$$

This ratio is greater than 1 when $R_c/R \ll 1$, meaning that the optimal shaping of the $D(r)$ channel has two benefits: the intended one (reduced flow resistance), and an unexpected one (reduced thermal resistance).

7.4.5 Tree Counterflow on a Disk

The improvements in performance associated with using a larger D where r is small (Fig. 7.24b) justify the search for alternate configurations that may offer even greater benefits. One approach is to guide each stream through a tree network of radial and quasi-radial ducts (Fig. 7.26). This flow configuration has been optimized

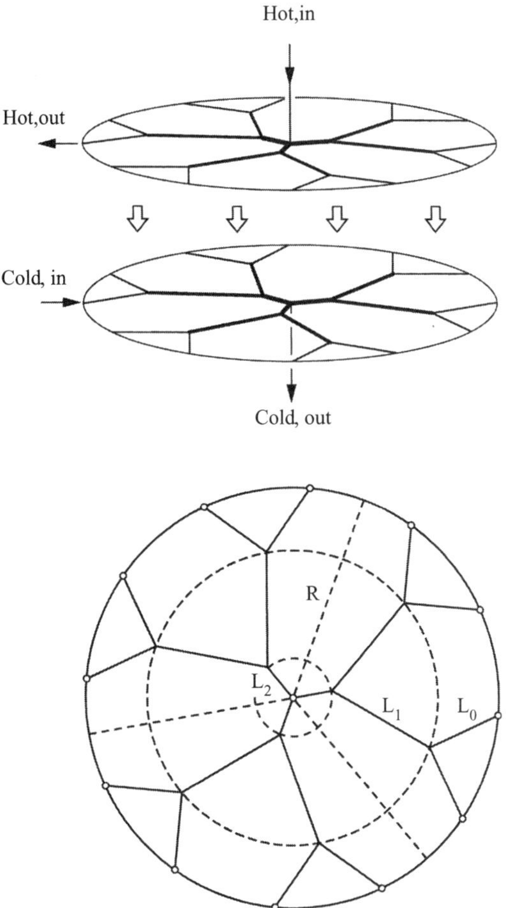

Figure 7.26 Constructal tree network of channels spread on a disk ($n_0 = 3$, $n = 2$) [11].

numerically for minimal global resistance in Chapter 4 and Refs. [6] and [7]. The radial tree configuration resembles the optimal-*D* configuration of Fig. 7.24b in one respect: near the center of the disc the flow cross-section is larger.

The tree counterflow heat exchanger consists of two trees of the type shown in Fig. 7.26, which are sandwiched as shown in Fig. 7.24. The tree layout on the disc follows deterministically from the minimization of the global flow resistance between the disc center and its periphery, subject to fixed disc radius *R*, and fixed total duct volume. The optimal tree architectures reported in Refs. [6] and [7] are based on the svelteness assumption $Sv > 10$, according to which the ducts are slender enough and the Reynolds number is small enough so that the flow in every duct is in the fully developed laminar regime, with negligible junction pressure drop losses. Even though the analysis of Refs. [6] and [7] is reported for round ducts, the optimal duct layout is valid for any other duct cross-sectional shapes, provided that the fully developed laminar flow regime prevails.

The minimization of global flow resistance generates all the geometric features of the tree architecture: the sequence of tube diameters, the number of tributaries (the optimal number is two), and the tube lengths. The example given in Fig. 7.26 is the optimal tube layout for a tree with 12 ports on the rim, one port in the center and two levels of pairing, or bifurcation. The optimal duct architectures for a large class of such flows are documented in Refs. [6] and [7]: 1–400 ports on the rim, 1–16 tubes reaching the center, and 0–7 levels of pairing. Here, we show the heat-transfer performance of a counterflow formed by two of the disc trees optimized for fluid flow.

The optimized layout of ducts has three features: the number of ports on the rim (n_0), the number of pairing levels (n), and the number of ducts that are connected to the center (n_n). Only two of these are degrees of freedom. In place of Eq. (7.51), we have the optimized lengths, which are available in tabulated dimensionless form (cf. Table 4.1),

$$\hat{L}_i = L_i/R \qquad (i = 0, 1, \ldots, n) \tag{7.88}$$

with the observation that L_0 is a duct that touches the rim, and L_n one that touches the center. In place of Eq. (7.52), we have

$$n_i = 2^{-i} n_0 \qquad (i = 0, 1, \ldots, n) \tag{7.89}$$

The number of central tubes is $n_n = 2^{-n}\, n_0$, and indicates that n, n_0 and n_n cannot be specified independently. Finally, the total flow rate through the disc center is

$$\dot{m} = n_n \dot{m}_n = n_0 \dot{m}_0 \tag{7.90}$$

The heat exchanger analysis relies on Eqs. (7.88) through (7.90), and repeats the steps starting with Eq. (7.60). We find that Eq. (7.70) continues to hold, but that Eq. (7.71) is replaced by

$$q = \dot{m} c_P \Delta T_x \tag{7.91}$$

where ΔT_x is the total temperature excursion experienced by one stream as it flows from the rim to the center, or vice versa. Dividing Eq. (7.70) by Eq. (7.91), and using Eqs. (7.60) through (7.63), we obtain $\Delta T_x / \Delta T = N$, where N is the new result—the number of heat transfer units of the disc-tree counterflow:

$$N = \frac{\pi k \, \mathrm{Nu} R}{\dot{m} c_P} S_3 \qquad S_3 = n_n \, 2^{n-1} \sum_{i=0}^{n} 2^{-i} \, \hat{L}_i \tag{7.92}$$

Note the similarities between Eqs. (7.92) and (7.73). The function S_3 (n, n_n) has been calculated based on the $\hat{L}_i$ data available in Ref. [6] and Table 4.1.

The heat-transfer performance of the disc-tree counterflow is described by Eqs. (7.75) and (7.92). The effectiveness increases monotonically as the number of heat transfer units N increases. The N value increases as the complexity (n, n_n) increases. The dendritic pattern took shape based on the minimization of flow resistance between one central port and several ports on the rim. The selection of the ideal pattern is ruled by a combination of n and n_n, which will maximize the performance of the disc-shaped tree counterflow heat exchanger.

Which configuration is ideal depends on what is considered fixed. When the smallest length scale of the flow pattern and the radius of the circle R are fixed, pairing is beneficial for a fixed n_n. This conclusion is clearer when n_n is large. For n_n values smaller than 2, the dendritic pattern is replaced by point-to-point flows, which are not relevant for tree-shaped heat exchangers.

7.4.6 Tree Counterflow on a Square

The disc counterflow of Fig. 7.26 is simpler and more compact than the tree counterflow of Fig. 7.21 for two reasons. One is the benefit of centrosymmetry: streams flow almost radially in Fig. 7.26, and this makes the flow architecture simpler (more reproducible) as one moves in the angular direction. The main source of simplicity, however, is the fact that in Fig. 7.26 the flow is between one point (the center) and many points on a curve (the rim), whereas in Fig. 7.21 the many points are distributed over an area. For this reason in Fig. 7.21 it was necessary to use two insulated plenums to collect the discharged hot stream and to distribute the cold stream. Insulated plenums are not necessary in disc-shaped counterflows: as shown in Fig. 7.24, the hot and cold streams are collected and, respectively, distributed by a rim jacket.

A simpler version of the disc-shaped tree is the square design shown in Fig. 7.27a. The new feature is that the disc is replaced by a square of side L. The hot stream flows through ducts machined on the square, from the center to the square periphery. The cold stream flows in the opposite direction through an identical structure. The duct layout can be optimized numerically; however, a much simpler route is provided by the approximate method proposed in Ref. [7]: the minimization of flow path lengths (section 4.5). The approximate path of least resistance between the center of the square and one of the sides is the multiscale structure shown in the upper quadrant of Fig. 7.27a, where there are $n = 3$ levels of pairing. The number of tubes that reach the center is $n_n = 8$, because when the quadrant dendrite is installed in the remaining areas, two central tubes overlap on each diagonal (hence the approximate character of the method; note that when the same structure is optimized numerically, the two diagonal tubes do not overlap—they are bowed slightly, leaving a space between them, Fig. 7.27b).

The heat-transfer performance of a counterflow sandwich of two square trees of the type shown in Fig. 7.27a can be determined by performing the equivalent of

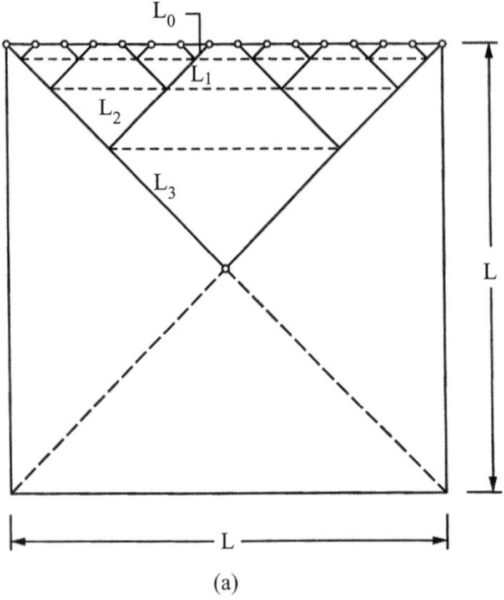

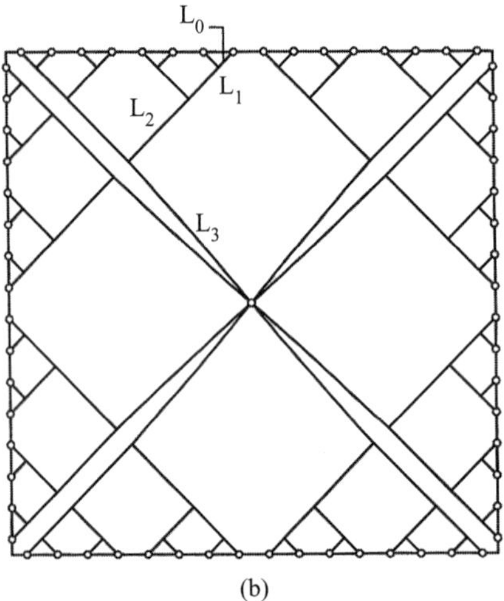

Figure 7.27 (a) Minimal-length tree of channels connecting one point with a square rim ($n_0 = 8, n = 3$) [7]; (b) qualitative presentation of the minimal-resistance tree of channels connecting a square rim with its center ($n = 3, n_0 = 8$) [11].

the analysis of section 7.4.5. The tube lengths are now given by a single formula, $L_i = 2^i L/[2^{1/2}(2^{n+1} - 1)]$, $(i = 0, 1, \ldots, n)$. The only degree of freedom in the selection of the optimized tree architecture is the number of pairing levels (n). The number of channels of one size is $n_i = 2^{n-i+3}$. The analysis concludes with the effectiveness-N relation (7.75), for which the new number of heat transfer units is $N = (\pi k \mathrm{Nu} L/\dot{m} c_P) S_4$, where $S_4 = 2^{n+3/2}(n+1)/(2^{n+1} - 1)$. The function $S_4(n)$ is closely approximated by $2^{1/2}(n + 1)$ when $n > 2$. The number of heat transfer units of the square tree counterflow heat exchanger increases with the number of pairing levels.

7.4.7 Two-Objective Performance

In section 7.4 we documented the thermo-fluid performance of several counterflow heat exchanger configurations in which each stream bathes its allocated space as a tree. How effective are the tree-shaped flow architectures?

Consider first the constructal tree configuration of Fig. 7.21, for which the global size (A) and total tube volume on one side (V) are fixed [cf. Eqs. (7.56) and (7.57)]. The fluid-flow performance is indicated by the pumping power (7.58), where M is the dimensionless total mass flow rate (7.59). The thermal performance of the constructal tree configuration is documented in section 7.4.3. The global thermal resistance $R_t = (T_{\text{in,hot}} - T_{\text{in,cold}})/q$ is summarized by

$$\tilde{R}_t = \frac{1}{M\varepsilon} \tag{7.93}$$

where ε is given by Eq. (7.75), and

$$\tilde{R}_t = \pi k \text{Nu} A^{1/2} R_t, \qquad N = 2^{n/2-2} S_2/M \tag{7.94}$$

The N expression listed above is a rewriting of Eq. (7.73).

Equations (7.93) and (7.58) show how $\tilde{R}_t$ and $\tilde{W}$ compete: when the flow rate increases, $\tilde{R}_t$ decreases and $\tilde{W}$ increases. By eliminating M between these two equations we obtain Fig. 7.28a. Note that when n increases the smallest length scale (L_0) decreases. The effect of complexity (n) is to produce designs that are attractive when more pumping power is available. The apparent envelope of the constant-n curves is the boundary that separates the possible designs from the impossible. This envelope shows how close the architectures of Fig. 7.21 can bring us to the two-objective limit of small $\tilde{R}_t$ and small $\tilde{W}$. If the available $\tilde{W}$ is specified, for example, $\tilde{W} = 10^4$, then there exists an optimally complex constructal tree, which has $n = 5$ pairing levels. This is an important feature: complexity is a *result*. Again, optimized complexity must not be confused with maximized complexity.

Figure 7.28a also reveals a pattern of diminishing returns. Each new increase in the number of pairing levels (n) brings a progressively smaller change in performance. The performance of an optimized flow structure gradually becomes insensitive to the addition of new features. The largest changes in performance occur when the flow structure is the simplest, and the first architectural feature (the "invention") is added to the design, which is later optimized. *The invention has greatest impact when it is first applied*, that is, when it is implemented alone, not as one in a group of competing features (cf. Ref. [4], p. 409).

The thermo-fluid performance of the minimal-length tree on the square (Fig. 7.27a) can be evaluated on the same basis. The total area covered by the tree is $A = L^2$, which means that the smallest length scale is $L_0 = A^{1/2}/[2^{1/2}(2^{n+1} - 1)]$. The total tube volume is $V = 2\pi D_0^2 L_0 2^n S_5$, where $S_5 = (2^{2(n+1)/3} - 1)/(2^{2/3} - 1)$, which yields D_0 as a function of A and V. The dimensionless M, $\tilde{W}$ and $\tilde{R}_t$ continue to be defined as in Eqs. (7.59) and (7.94). The pumping power requirement is

$$\tilde{W} = \pi^3 2^{n+9/2} \frac{S_5^3}{\left(2^{n+1} - 1\right)^3} M^2 \tag{7.95}$$

The thermal resistance is given by Eq. (7.93) with $\varepsilon = N/(1+N)$ and $N = S_4/M$. By eliminating M between the functions $\tilde{W}(M, n)$ and $\tilde{R}_t(M, n)$, we obtain the family of curves plotted in Fig. 7.28b. The $\tilde{R}_t(\tilde{W})$ curves cross at $\tilde{W} \sim 10^3$, indicating that the radial design ($n = 0$) is superior when the available pumping power is small,

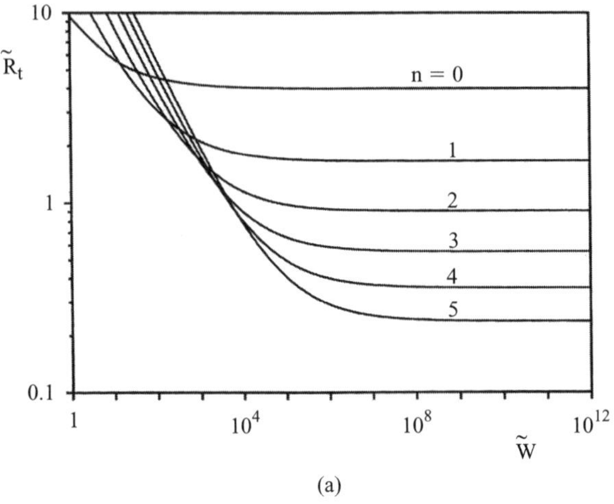

(a)

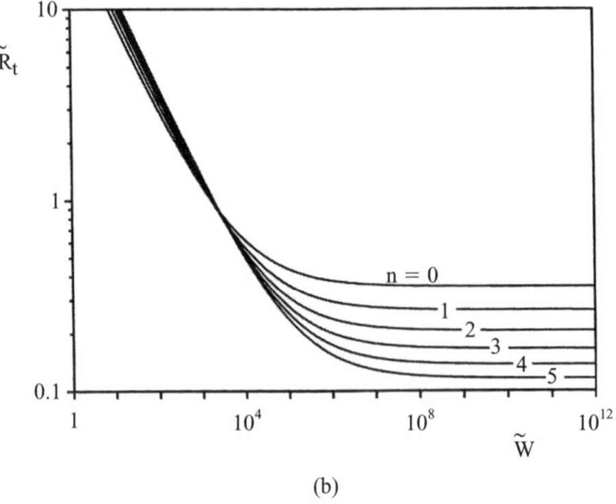

(b)

Figure 7.28 Thermo-fluid performance: (a) the tree architecture of Fig. 7.21; (b) the tree architecture of Fig. 7.27a [11].

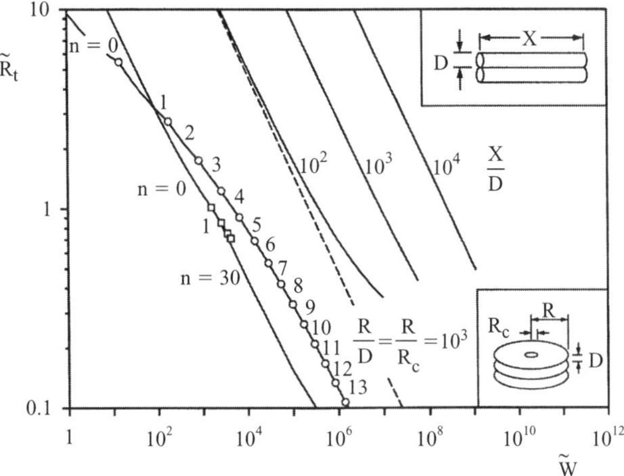

Figure 7.29 The low-$\tilde{R}$ and low-$\tilde{W}$ envelopes of the curves of Figs. 7.28a and 7.28b, and the thermo-fluid performance of designs consisting of two streams in counterflow, and two radial sheets in counterflow [11].

$\tilde{W} < 10^3$. Tree-shaped structures with more levels of pairing (n) become more attractive as the available pumping power increases. When $\tilde{W}$ is specified, there is an optimal n—an optimized complexity—that identifies the design with the lowest thermal resistance $\tilde{R}_t$.

Figure 7.29 summarizes the progress made in the development of flow architectures that offer simultaneously small $\tilde{R}_t$ and small $\tilde{W}$ values. The curve marked by circles is the envelope of the family of curves derived in Fig. 7.28a for the constructal tree family of the type shown in Fig. 7.21. The number printed between two successive circles on this curve represents the n value of the best constructal tree configuration—the tree with the lowest thermal resistance and pumping power. All the constructal tree configurations [the ones optimized in Fig. 7.28a and, especially, the ones that have not been optimized] lie to the right of this curve.

The curve marked by squares in Fig. 7.29 represents the envelope of the performance family of curves shown in Fig. 7.28b for the tree-on-square design class of Fig. 7.27a. To the right of this curve lie all the tree-on-square designs, the optimized ones (Fig. 7.28b) and all the other possible flow connections between the center and perimeter of the square. The n value indicates the number of pairing levels, or the complexity of the tree.

The intersection of the curves marked by circles and square shows that the square designs perform better when larger pumping power levels are possible. This is an important finding, because the square designs of Fig. 7.27a have stream-collecting manifolds on the rim, and are easier to package into larger assemblies, that is, easier than packing the constructs with top and bottom manifolds shown in Fig. 7.21. The intersection of the two curves further refines the demarcation line between the cloud of possible designs (the upper-right domain) and the uncharted design domain (the lower-left domain). The edge of the cloud represents the most promising configurations developed so far.

How good are the best configurations? In Fig. 7.29 we also projected the $\tilde{R}_t$-$\tilde{W}$ performance of the simplest class of designs that serve the same function as in Figs. 7.21 and 7.27a, and have the same constraints (A, V). Again, the circles indicate the envelope of the curves shown in Fig. 7.28a, while the squares refer to the envelope of the curves plotted in Fig. 7.28b. The simplest configuration is a counterflow of two $\dot{m}$ streams flowing through parallel tubes of length X and diameter D. The flow is the Poiseuille regime, and the constraints are $A = XD$ and $V = (\pi/4)D^2X$. The analysis, which is omitted for brevity, shows that there exists a family of designs, one for each slenderness ratio X/D. The family of X/D-curves has the same features as the families of n-curves shown in Figs. 7.28a and 7.28b. The envelope of the X/D-curves falls well to the right of the envelopes of the n-curves, indicating that the two-stream counterflow is never competitive in relation to tree-shaped counterflows.

Figure 7.29 also shows an example of how the radial sheet counter flow of Fig. 7.24a performs in the same $\tilde{R}_t - \tilde{W}$ plane. There is a family of curves, one for each pair (R/D, R/R_c). These curves fall to the right of the n-curves marked with circles and squares.

Figure 7.29 is a clear argument in favor of tree-shaped architectures for transport devices. Current heat exchanger methodology is based on assuming one-scale flow structures such as the single-D structures shown in the two inserts in Fig. 7.29. Tree-shaped architectures have multiple scales (D_i, L_i), which are organized hierarchically, and which can be distributed optimally (and nonuniformly) over the available territory. And their performance is orders-of-magnitude superior.

7.5 CONSTRUCTAL HEAT EXCHANGER TECHNOLOGY

The multiobjective constructal design outlined in this chapter is a new direction of development of concepts for heat exchanger technology. Here, we give a few examples of the emerging field. The first design concept in which dendritic heat exchangers were proposed [10] has been constructed, tested experimentally, and simulated numerically in three dimensions by Raja et al. [18] and Raja [19]. This work demonstrated conclusively the step-change increase in performance offered by the constructal architecture relative to classical architectures [17].

Dendritic configurations for chemical reactors were proposed by Azoumah et al. [20–22]. Here the two objectives are high density of chemical reaction and low pumping power. The comparison between this two-objective approach to dendritic chemical reactors and the method of entropy generation minimization [4, 15] (see also Chapter 2) was made subsequently by Azoumah et al. [21] and Zhou et al. [23].

New papers on dendritic apparatuses for heat and mass transfer appear regularly. The comparison between entropy generation minimization and the dendritic heat exchangers described here in section 7.4 was made by Zimparov et al. [24, 25]. More practical aspects of the implementation of dendritic architectures in the heat exchanger assembly are pursued by Luo et al. [26, 27]. The free morphing of even more classical configurations is advocated by Matos et al. [28], Bello-Ochende et al. [29], and Vargas et al. [30, 31]. The relevance of constructal flow architectures to the design of flow and heat transfer systems for very large and complex vehicles such as aircraft was outlined in Ref. [32]. The march in the opposite direction, toward microscale flow dendrites for vascularized composites (smart materials) is the subject of the next chapter.

7.6 TREE-SHAPED INSULATED DESIGNS FOR DISTRIBUTION OF HOT WATER

We consider here the fundamental problem of distributing a supply of hot water as uniformly as possible over a given territory. This is a classical problem of civil engineering, with related subfields in piping networks, sewage and water runoff, irrigation, steam piping, and so on. Examples of hot water distribution networks are Refs. [13] and [33–37]. Unlike in Ref. [35], where the flow path geometry is assumed and frozen throughout the remaining optimization process, in this section we reserve the freedom to change the configuration, to "morph" it into better patterns en route to higher levels of performance.

The present treatment of the hot water distribution problem is restricted to its two main objectives: thermal and fluid-flow performance [13]. The distribution of hot water to users on a specified territory presents two problems to the thermal designer: the fluid mechanics problem of minimizing the pumping power and the heat-transfer problem of minimizing the loss of heat from the piping network. The water flow is from one point (the source) to an area—the large number of users spread uniformly over the area.

The fluid mechanics problem has been addressed in various forms, not only in engineering [9, 38] but also in physics and biology [39–45]. The addition of the thermal objective is new, especially on the background of the existing work on tree-shaped flows that facilitate the flow of heat from point to area, or point to volume [9]. In the present problem the requirement is to prevent the flow of heat from the branches of the piping network to the area (the "ambient"). This is to be accomplished by using a specified (finite) amount of thermal insulation for all the pipes and by distributing it optimally.

7.6.1 Elemental String of Users

The area A is supplied with hot water by a stream with the mass flow rate $\dot{m}$ and initial temperature T_i. The stream enters the area by crossing one of its boundaries. The area is inhabited by a large number of users, $n = A/A_0$, where A_0 is the area

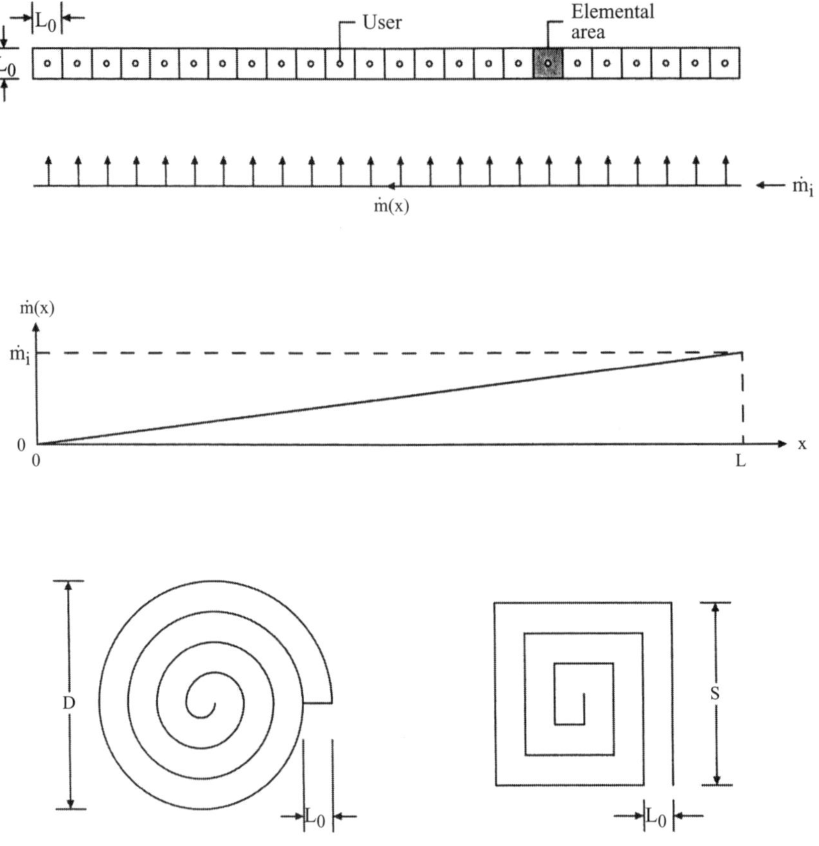

Figure 7.30 String of users supplied by a single stream of hot water, and round and square areas served by the string-shaped stream [13].

element allocated to a single user. Let A_0 be a square with the side length L_0, so that A constitutes a patch work of n squares of size A_0. Each elemental square must receive an equal share of the original stream of hot water, $\dot{m}/n$. As in all the point-area tree flows considered previously [4, 9], the fundamental question is how to connect the elements so that the ensemble (A) performs best.

We begin with the simple option of feeding a large number of elements from the same stream, which is a straight or curved line of length L (Fig. 7.30). The entering flow rate is $\dot{m}_i$. The flow rate $\dot{m}(x)$ decreases linearly to $\dot{m}(0) = 0$, because each user draws the same share of $\dot{m}_i$ per unit length x. It is assumed that the number of elements is sufficiently large so that the variation of $\dot{m}(x)$ may be treated as continuous

$$\dot{m}(x) = \frac{x}{L}\,\dot{m}_i. \tag{7.96}$$

Because of the leakage of heat to the ambient, the temperature of the stream $T(x)$, decreases from its original temperature T_i (at $x = L$). We assume that the dominant thermal resistance between the stream and the ambient is posed by a cylindrical shell of thermal insulation installed on the outside of the pipe that carries the stream. In this limit the rate of heat loss per unit of pipe length is

$$q' = \frac{2\pi k}{\ln(r_0/r_i)}[T(x) - T_\infty], \tag{7.97}$$

where r_0, r_i, k and T_∞ are the outer and inner radii of the insulation, the thermal conductivity of the insulating material, and the ambient temperature. Written for a pipe length dx, the first law of thermodynamics requires

$$\dot{m}c_P dT = \frac{2\pi k}{\ln(r_0/r_i)}(T - T_\infty)dx. \tag{7.98}$$

The temperature distribution along the water stream (or the pipe wall) is obtained by integrating Eq. (7.98) from $x = L$ where $T = T_i$

$$\frac{T(x) - T_\infty}{T_i - T_\infty} = \exp\left(-N\int_x^L \frac{dx}{x\ \ln(r_0/r_i)}\right) \tag{7.99}$$

where, by analogy with the number of heat-transfer units defined for heat exchangers, N is the *number of heat loss units* of the L-long pipe system

$$N = \frac{2\pi kL}{\dot{m}_i c_P} \tag{7.100}$$

The temperature of the hot water delivered to the users is $T(x)$. According to Eq. (7.99), this temperature distribution is known as soon as the distribution of thermal insulation is known, namely, the ratio r_0/r_i as a function of x. The constraint is the total amount of insulation wrapped around the pipe

$$V = \pi\int_0^L \left(r_0^2/r_i^2\right)dx \tag{7.101}$$

7.6.2 Distribution of Pipe Radius

The geometry of the pipe and its insulation is described completely by r_0 and r_i (or r_0/r_i and r_i) as functions of x. The optimal distribution of pipe size $[r_i(x)]$ can be derived from the minimization of the pumping power $\dot{W}$ required to drive $\dot{m}(x)$ through the entire system. The pumping power varies as the product $\dot{m}(dP/dx)$, while in fully rough turbulent flow the pressure gradient is proportional to $\dot{m}^2/r_i^5$. In conclusion, the pumping power per unit length varies as $\dot{m}^3/r_i^5$, or x^3/r_i^5, and the objective is to minimize the integral

$$I_1 = \int_0^L \frac{x^3}{r_i^5}dx \tag{7.102}$$

If the constraint is the pipe wall material, then the pipe size must vary such that the integral

$$I_2 = \int_0^L r_i dx \tag{7.103}$$

remains fixed. Note that $I_2 = V_p/(2\pi t)$, where V_p is the volume of the pipe-wall material, and t is the wall thickness, which is assumed considerably smaller than r_i. The variational calculus solution to the problem of minimizing I_1 subject to $I_2 =$ constant is (cf. Appendix C)

$$r_i = c_i x^{1/2} \tag{7.104}$$

for which the factor c_1 is provided by the actual pipe-wall material constraint (7.103), namely, $c_1 = (3/2)\, I_2/L^{3/2}$. The pipe that supplies the line of users must become narrower in accelerated fashion as the water approaches the last user $(x = 0)$.

7.6.3 Distribution of Insulation

We now turn our attention to the optimal spreading of a specified amount of insulating material, Eq. (7.101). We consider three ways to pursue the optimization by maximizing the temperature of the hot water received by the most disadvantaged user, the "end" user, the farthest from the source, the one who receives the coldest water:

$$T_0 = T(x = 0). \tag{7.105}$$

We obtain an expression for T_0 by setting $x = 0$ in Eq. (7.99). To maximize T_0 means to minimize the integral that appears in the argument of the exponential on the right side of Eq. (7.99),

$$I_3 = \int_0^1 \frac{d\xi}{\xi \ln R} \tag{7.106}$$

where

$$\xi = x/L, \quad R(\xi) = r_0/r_i \tag{7.107}$$

The material constraint (7.101) in combination with the optimized pipe radius (7.105) means that the integral constraint that must be satisfied by $R(\xi)$ is

$$I_3 = \int_0^1 \frac{d\xi}{\xi \ln R} \tag{7.108}$$

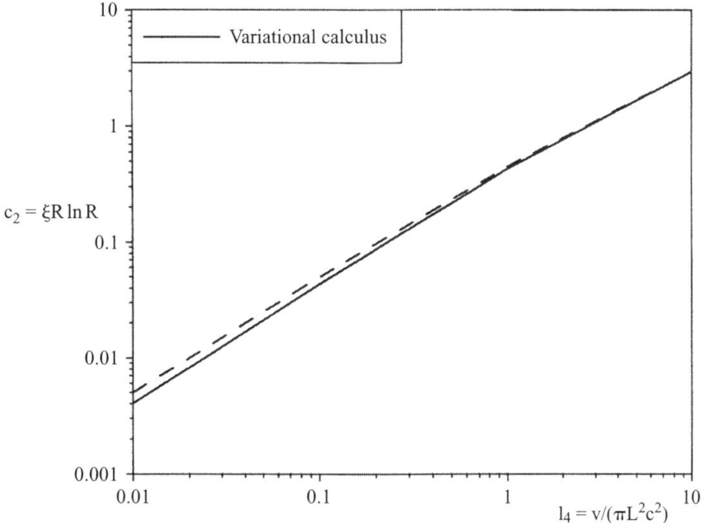

Figure 7.31 The effect of the total amount of insulation material (I_4) on the optimal distribution of insulation (c_2) [13].

where $I_4 = V/(\pi L^2 c_1^2)$. The function R that minimizes I_3 subject to constant I_4 is obtained implicitly by variational calculus,

$$\xi = c_2/(R \ln R) \tag{7.109}$$

The factor c_2 is determined by using this R function into the calculation of the actual (specified) amount of insulating material, I_4. Combining (7.108) and (7.109) we obtain the function $c_2(I_4)$ shown by the solid line in Fig. 7.31. The factor c_2 is almost proportional to the amount of insulation material, I_4. This makes sense, because c_2 is also proportional to $R \ln R$ [cf. Eq. (7.109)]: thicker insulations (larger R values) require more insulation material.

According to Eq. (7.109), the radii ratio $R = r_0/r_i$ increases nearly as $1/\xi$ as the stream approaches the last user ($\xi \to 0$). Since r_{i} decreases as $\xi^{1/2}$ in the same direction [cf. Eq. (7.104)], the outer radius of the insulation varies nearly as $\xi^{-1/2}$. Furthermore, the integrand of the material constraint (7.108) shows that in this design the amount of insulation per unit of pipe length increases toward the $\dot{m} = 0$ end of the pipe. This behavior is confirmed by Fig. 7.32, which shows a plot of Eq. (7.109) for several amounts of insulation material (I_4).

Finally, the maximized end temperature is obtained by substituting Eq. (7.109) into I_3 and, finally, into Eq. (7.99) written for $x = 0$. The chart constructed in Fig. 7.33(a) shows that the maximized end temperature increases monotonically as the number of heat loss units of the entire insulation (N) decreases, and as the total amount of insulation (I_4) increases. Figure 7.33 reports the maximized performance of the entire system as a function of its two global constraints, material and length (or flow residence time).

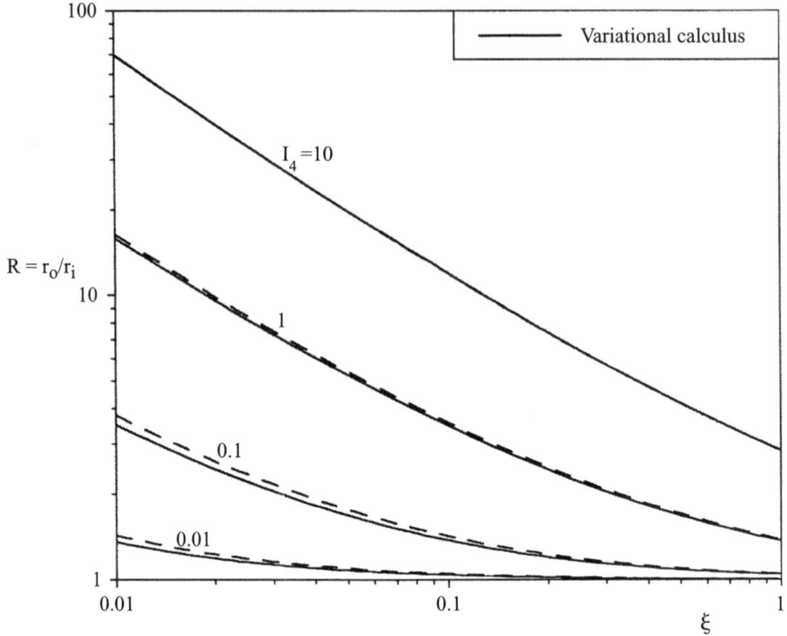

Figure 7.32 The optimal ratio of insulation radii when the temperature of the hot water delivered to the farthest user is maximized [13].

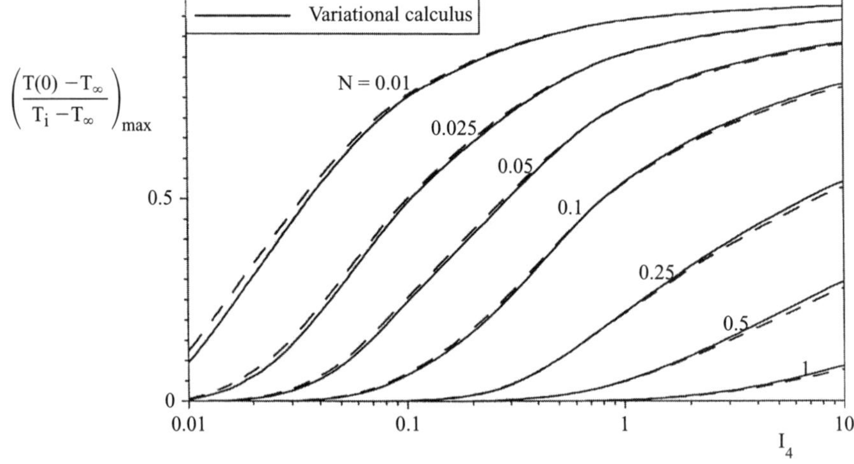

Figure 7.33 The maximized and temperature as a function of the amount of insulation material (I_4) and the number of heat loss units N [13].

7.6.4 Users Distributed Uniformly over an Area

In this section we turn our attention to the more important and practical question of how to distribute a hot water stream to a population of users spread over an area, A. One solution is to coil the optimized string of users in such a way that the area is covered. Two possibilities are sketched in the lower part of Fig. 7.30. The string of length L and width L_0 can cover a disc-shaped area of diameter $D = [(4/\pi)L_0L]^{1/2}$, or a square-shaped area of side $S = (L_0L)^{1/2}$. Are these the best ways to allocate the hot water stream to the entire area?

The alternative is to introduce branches in the path of the stream and to distribute area elements to each branch. We explore this alternative by starting with the smallest (and therefore simplest) area element, and continuing toward larger areas by assembling elements into larger and larger constructs. One simple rule of assembly is to use four constructs into the next, larger assembly (Fig. 7.34). In this case each construct covers a square area, in the sequence $A_0 = L_0{}^2$, $A_1 = 4L_0{}^2$, $A_2 = 4^2L_0{}^2, \ldots, A_i = 4^iL_0{}^2$. This assembly rule is "simple" because the shape (square) of each construct is assumed, not optimized. An alternative construction sequence is described in section 7.5.5.

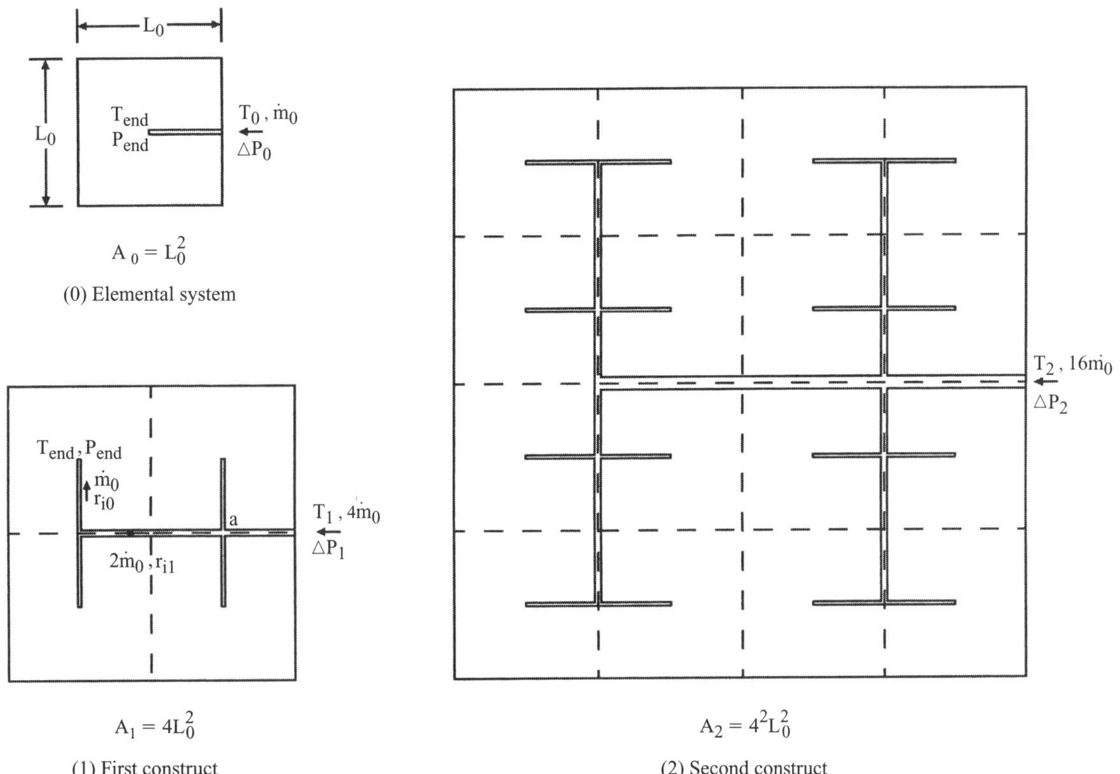

Figure 7.34 Sequence of square-shaped constructs containing tree-shaped streams of hot water [13].

The objective is to supply with hot water the users distributed uniformly over A_i, and to accomplish this task with minimal pumping power and a finite amount of thermal insulation. The geometry of each pipe is described by its length (a fraction or multiple of L_0), inner radius wetted by the flow (r_i), and ratio of insulation radii ($R = r_0/r_i$). The pipe-wall thickness is neglected for the sake of simplicity. The subscripts 0, 1 and 2, indicate the elemental area, first construct, and second construct, in accordance with the notation shown in Fig. 7.34.

To minimize the pumping power requirement at the elemental level ($\dot{W}_0 = \dot{m}_0 \Delta P_0/\rho$) is to minimize the pressure drop along the elemental duct of the length $L_0/2$. Assuming as in the earlier sections that the flow is fully developed turbulent in the fully rough regime (f = constant), we find that the pressure drop derived from the definition of friction factor [cf., Eq. (1.22)] is

$$\Delta P_0 = \frac{f}{\pi^2} \frac{\dot{m}_0^2 L_0/2}{\rho r_{i0}^5} \tag{7.110}$$

The corresponding heat-loss analysis based on Eq. (7.98), with $\dot{m}_0$ constant in place of $\dot{m}$, yields the temperature drop from the inlet to the element (T_0) to the user (T_{end})

$$\frac{T_{\text{end}} - T_\infty}{T_0 - T_\infty} = \exp\left(-\frac{N_0}{\ln R_0}\right) \tag{7.111}$$

where the number of heat loss units is based on elemental quantities

$$N_0 = \frac{\pi k L_0}{\dot{m}_0 c_P} \tag{7.112}$$

At the first-construct level there are two pipe sizes, one central pipe of length $(3/2)L_0$ and radii r_{i1} and $R_1 = (r_0/r_i)_1$, and four elemental branches of length $(1/2)L_0$ and radii r_{i0} and $R_0 = (r_0/r_i)_0$. The flow rate is $4\dot{m}_0$ through the root of the tree, and $\dot{m}_0$ through each small branch. By writing the equivalent of Eq. (7.110) for each segment of pipe without branches, we find that the drop in pressure from the root to the most distant user (the center of the farthest element) is

$$\Delta P_1 = \frac{f}{\pi^2} \frac{\dot{m}_0^2 L_0}{\rho} \left(\frac{12}{r_{i1}^5} + \frac{1}{2r_{i0}^5}\right) \tag{7.113}$$

By analogy with the shaping of the water duct volume in section 7.5.2, the pressure drop ΔP_1 can be minimized by selecting the ratio of pipe sizes r_{i1}/r_{i0} subject to a water volume constraint. If, as in section 7.5.2, we constrain the amount of duct wall material, and if we assume that the duct thickness (t) is a constant independent of duct inner radius, then the constraint means that the total wetted surface is fixed:

$$\frac{3}{2} L_0 r_{i1} + 2 L_0 r_{i0} = \text{constant} \tag{7.114}$$

The same geometric relation applies when the constraint refers to the amount of soil that must be excavated in order to bury the pipe system to a constant depth. The minimization of ΔP_1 subject to constraint (7.114) yields

$$(r_{i1}/r_{i0})_{\text{opt}} = 2^{5/6} \tag{7.115}$$

An alternative is to fix the total volume occupied by the ducts

$$\frac{3}{2}L_0 r_{i1}^2 + 2L_0 r_{i0}^2 = \text{constant} \tag{7.116}$$

This constraint also represents applications where the thickness of each pipe is proportional to the pipe inner radius, and where the total amount of wall material is constrained. In such cases, the minimization of ΔP_1 subject to constraint (7.116) yields a slightly different optimal step in pipe size

$$(r_{i1}/r_{i0})_{\text{opt}} = 2^{5/7} \tag{7.117}$$

The optimization of the geometry of the thermal insulation shells wrapped around each pipe proceeds in the same steps as the pressure-drop minimization. We write a temperature drop expression of type (7.111) for each segment of pipe without branches. Here we omit the algebra and report only the overall temperature drop from the root of the tree (T_1) to the temperature (T_{end}) of the water stream delivered to the most distant user:

$$\theta_1 = \frac{T_{\text{end}} - T_\infty}{T_1 - T_\infty} = \exp\left(-\frac{N_0}{\ln R_0} - \frac{5N_0}{4\ln R_1}\right) \tag{7.118}$$

The dimensionless end temperature θ_1 depends on three parameters, R_0, R_1 and N_0. The geometric parameters R_0 and R_1 are related through the thermal insulation volume constraint

$$V_1 = \pi L_0 r_{i0}^2 \left[\frac{3}{2}\left(\frac{r_{i1}}{r_{i0}}\right)^2 (R_1^2 - 1) + 2\left(R_0^2 - 1\right)\right] \tag{7.119}$$

for which (r_{i1}/r_{i0}) is a number furnished by Eq. (7.115) or Eq. (7.117). Constraint (7.119) may be put into the dimensionless form $\tilde{V}_1 = (R_0,\ R_1)$ by recognizing r_{i0} as the smallest pipe size (section 7.5.2) and defining the dimensionless insulation volume $\tilde{V}_1 = V_1/(\pi r_{i0}^2 L_0)$.

The maximization of expression (7.118) with respect to R_0 and R_1, and subject to constraint (7.119) yields the optimal step change in radii ratio

$$\left(\frac{R_1 \ln R_1}{R_0 \ln R_0}\right)_{\text{opt}} = \left(\frac{5}{6}\right)^{1/2} \left(\frac{r_{i0}}{r_{i1}}\right)_{\text{opt}} \tag{7.120}$$

In view of the $(r_{i1}/r_{i0})_{\text{opt}}$ values listed in (7.115) and (7.117), we conclude that $R_{1,\text{opt}} < R_{0,\text{opt}}$, that is, the shell of the central duct is relatively thin in comparison with the shells of the elemental ducts. "Relatively thin" means that the shell thickness is small in comparison with the radius of the same tube. Combining the

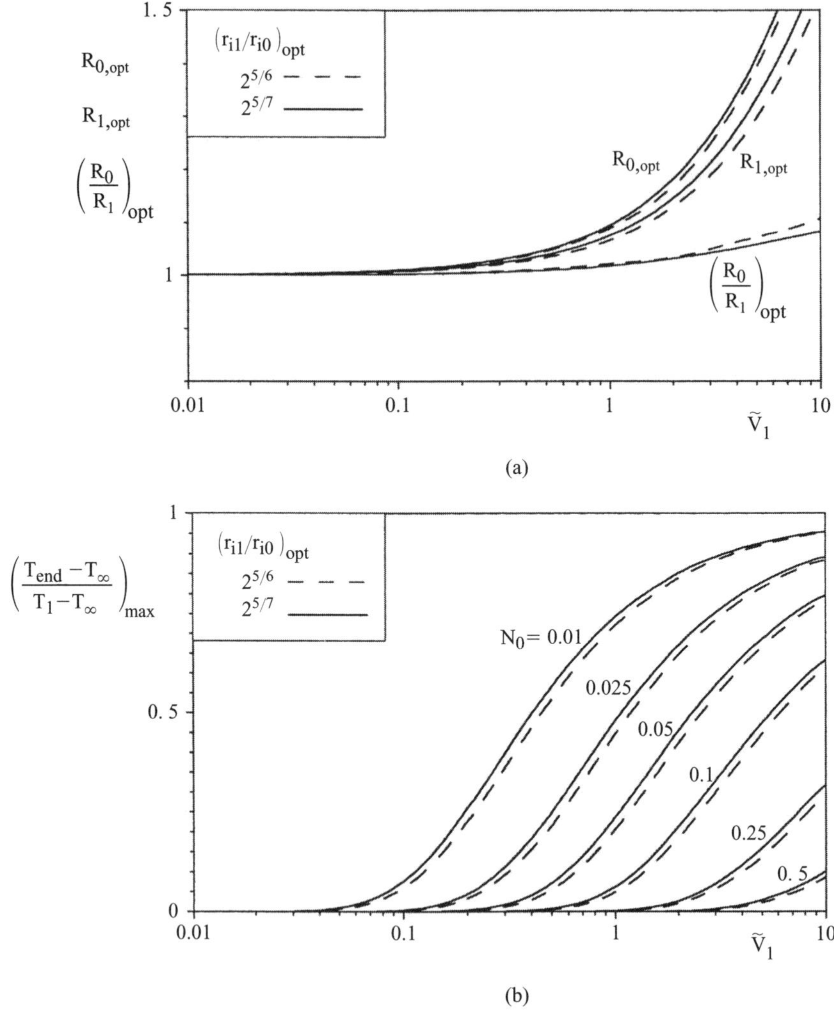

Figure 7.35 The optimal ratios of insulation radii for the first construct (A_1) shown in Fig. 7.34, and the maximized end temperature of the first construct [13].

function $R_{1,\text{opt}}(R_{0,\text{opt}})$ of Eq. (7.120) with the $\tilde{V}_1 = (R_0,\ R_1)$ constraint (7.119) we obtain the radii ratios $R_{0,\text{ opt}}(\tilde{V}_1)$ and $R_{1,\text{ opt}}(\tilde{V}_1)$ displayed in Fig. 7.35a. The same figure shows the evolution of the step change in geometric form, $(R_1/R_0)_{\text{opt}}$, as the amount of insulation $\tilde{V}_1$ increases. Noteworthy is that the optimized architecture of the distributed insulation is independent of N_0. Once again, the value of $(r_{i1}/r_{i0})_{\text{opt}}$, or the choice between constraints (7.114) and (7.116) has little effect on the insulation geometry.

The maximized performance of the first construct is measured by the θ_1 values obtained by substituting into Eq. (7.118) the optimized geometry reported in

Fig. 7.35a. The result is the function $\theta_{1,\max}(N_0, \tilde{V}_1)$ presented in Fig. 7.35b. The temperature of the water stream delivered to the farthest user increases as the amount of insulation increases, and as the thermal conductivity of the insulating material (N_0) decreases. These trends agree with what we saw in Fig. 7.33 following the optimization of the distribution of insulation in the string-shaped system, Fig. 7.30.

The optimization of the internal architecture of the second construct (A_2, Fig. 7.34) is performed by executing the same steps as in the optimization of the first construct. New is the larger size of the construct ($A_2 = 4A_1, \dot{m}_2 = 16\dot{m}_0$) and the new central duct of length $3L_0$, inner radius r_{i2}, and insulation radii ratio $R_2 = r_{o2}/r_{i2}$. The optimized geometric features of the first construct are retained.

In the fluid-flow part of the problem, we minimize the overall pressure drop from the root of the fluid tree (P_2) to the farthest user (P_{end}), namely, $\Delta P_2 = P_2 - P_{\text{end}}$. After some algebra we obtain

$$\Delta P_2 = \frac{f}{\pi^2}\frac{\dot{m}_2^2 L_0}{\rho}\left(\frac{3/2}{r_{i2}^5} + \frac{12 + \beta^{5/2}}{16^2 r_{i1}^5}\right) \tag{7.121}$$

where $\beta = (r_{i1}/r_{i0})^2$ is shorthand for a numerical value provided by Eq. (7.115) or Eq. (7.117). Recall that at the first-construct level we used two alternative water-space constraints, (7.114) and (7.116), and the geometric optimization results were quite similar (cf. Figs. 7.35a and b.) For this reason, we continue only with the constraint of type (7.116) in which we fix the volume of all the ducts of the A_2 construct, namely, $\pi L_0\left[3r_{i2}^2 + r_{i1}^2(6 + 8/\beta)\right] =$ constant. By varying r_{i2} and r_{i1} subject to this constant, we minimize ΔP_2 and find the optimal relative size of the central duct of the A_2 construct

$$\left(\frac{r_{i2}}{r_{i1}}\right)_{\text{opt}} = \left(\frac{3 + 4/\beta}{24 + \beta^{5/2}}\cdot 2^9\right)^{1/7} = 2^{6/7} \tag{7.122}$$

Noteworthy is the inequality $(r_{i2}/r_{i1})_{\text{opt}} > (r_{i1}/r_{i0})_{\text{opt}}$, which states that the step change in duct size is more abrupt at the second-construct level than at the first-construct level.

The second part of the analysis of A_2 is concerned with the temperature (T_{end}) of the water stream received by the farthest elemental user, and the maximization of this temperature subject to the constrained total volume of thermal insulation allocated to the A_2 construct. The analysis yields an expression for the dimensionless overall temperature drop

$$\theta_2 = \frac{T_{\text{end}} - T_\infty}{T_2 - T_\infty} = \theta_1 \exp\left(-\frac{5N_0/8}{\ln R_2}\right) \tag{7.123}$$

where $\theta_1(N_0, \tilde{V}_1)$ is a function that is available numerically based on the optimization of the A_1 construct (cf. Fig. 7.35b). The total insulation volume (V_2) constraint for the A_2 construct can be written as

$$\tilde{V}_2 = \frac{V_2}{\pi r_{i0}^2 L_0} = 3\left(\frac{r_{i2}}{r_{i0}}\right)^2 (R_2^2 - 1) + 4\tilde{V}_1 \tag{7.124}$$

where r_{i2}/r_{i0} is a numerical factor known from the minimization of pressure, for example, $r_{i2}/r_{i0} = (r_{i2}/r_{i1})(r_{i1}/r_{i0}) = 2^{11/7}$ if the water-space constraint used is the total volume of the ducts.

In summary, the end-temperature function $\theta_2(N_0, \tilde{V}_1, R_2)$ of Eq. (7.123) and the constraint $\tilde{V}_2(\tilde{V}_1, R_2)$ of Eq. (7.124) define θ_2 as a function of N_0, $\tilde{V}_2$, and R_2. By maximizing θ_2 with respect to R_2 at constant N_0 and $\tilde{V}_2$ we obtain the insulation radii ratio $R_{2,opt}$ and maximized end temperature reported in Figs. 7.36a and b.

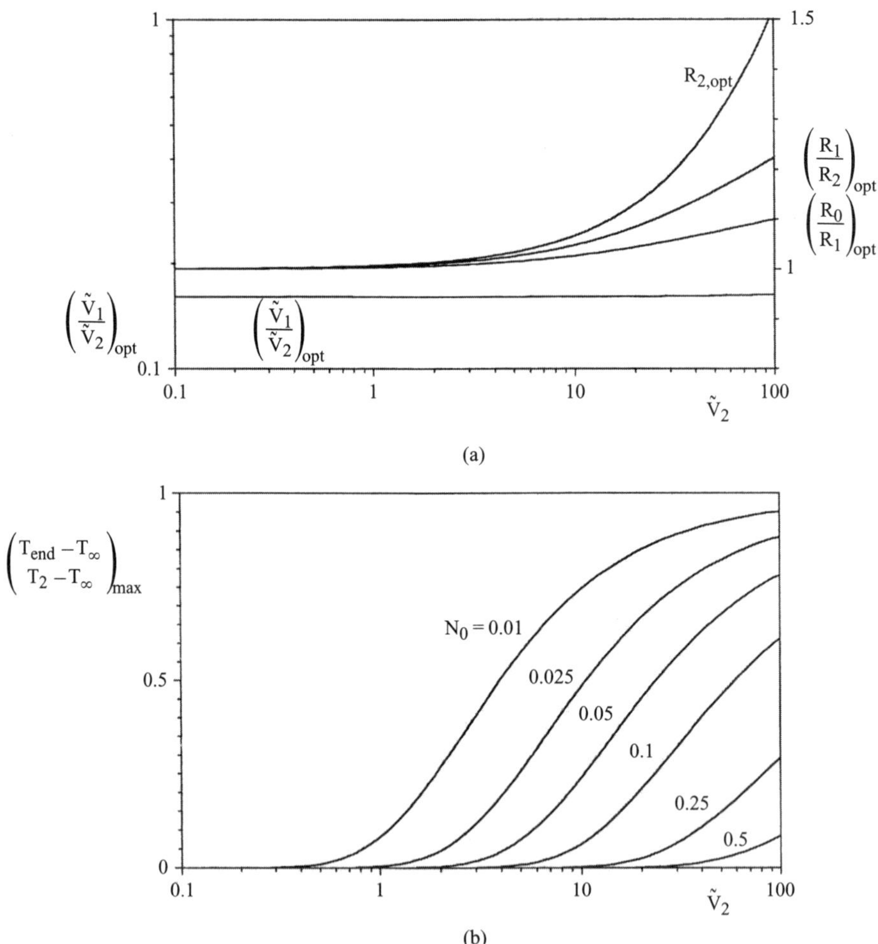

Figure 7.36 The optimal ratios of insulation radii for the second construct (A_2) shown in Fig. 7.34, and the maximized end temperature of the second construct [13].

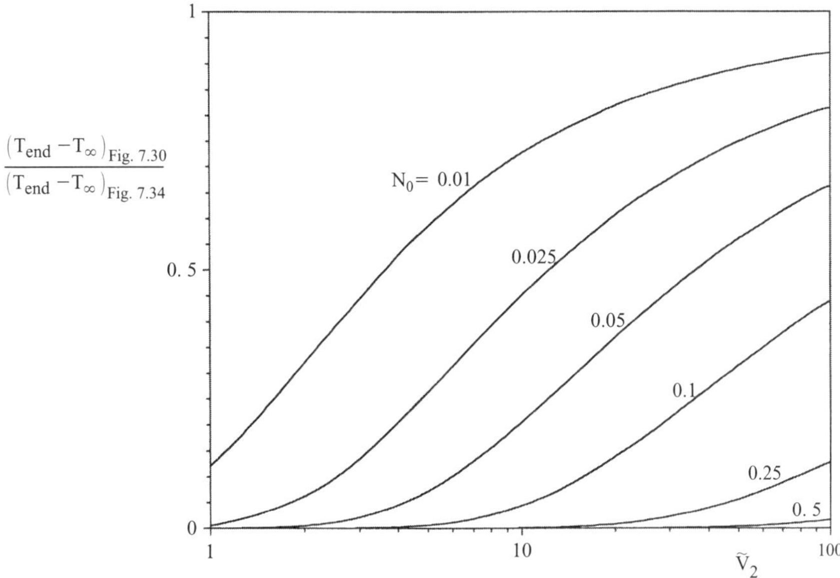

Figure 7.37 Comparison between the maximized water stream temperatures delivered to the farthest users in the string-shaped construct of Fig. 7.30 and the tree-shaped first construct (A_2) of Fig. 7.34, for the same territory covered by the construct, and the same total amount of insulation [13].

The relative superiority of the tree-shaped design (Fig. 7.34) over the string design (Fig. 7.30) is documented in Fig. 7.37. Here, we show the dimensionless end temperatures produced by the two optimized schemes on the same basis—the same covered territory (A_2) and the same amount of insulation used in the entire construct (V_2). It is clear that the tree-shaped design is superior, because the end-user water temperature in the scheme of Fig. 7.34 is consistently higher than in the coiled string arrangement of Fig. 7.30. The two schemes have nearly the same performance in the limit of plentiful insulation material ($\tilde{V}_2 > 10^2$) and high water-flow rate ($N_0 < 10^{-2}$).

7.6.5 Tree Network Generated by Repetitive Pairing

The sequence of square-shaped constructs used beginning with Fig. 7.34 is an assumption, not a result of optimization. To see whether a better rule of assembling small constructs into larger constructs can be found, consider the area doubling sequence shown in Fig. 7.38. Each area construct is obtained by putting together two constructs of the immediately smaller size. The area supplied with hot water increases in the sequence $A_0 = L_0{}^2$, $A_1 = 2L_0{}^2$, $A_2 = 2^2L_0{}^2, \ldots, A_i = 2^iL_0{}^2$, and the shape of the area alternates between square and rectangular. The elemental area that starts the sequence in Fig. 7.38 is the same as in Fig. 7.34, namely, A_0. We will see that the second construct of Fig. 7.42 covers the same area ($4L_0{}^2$) as

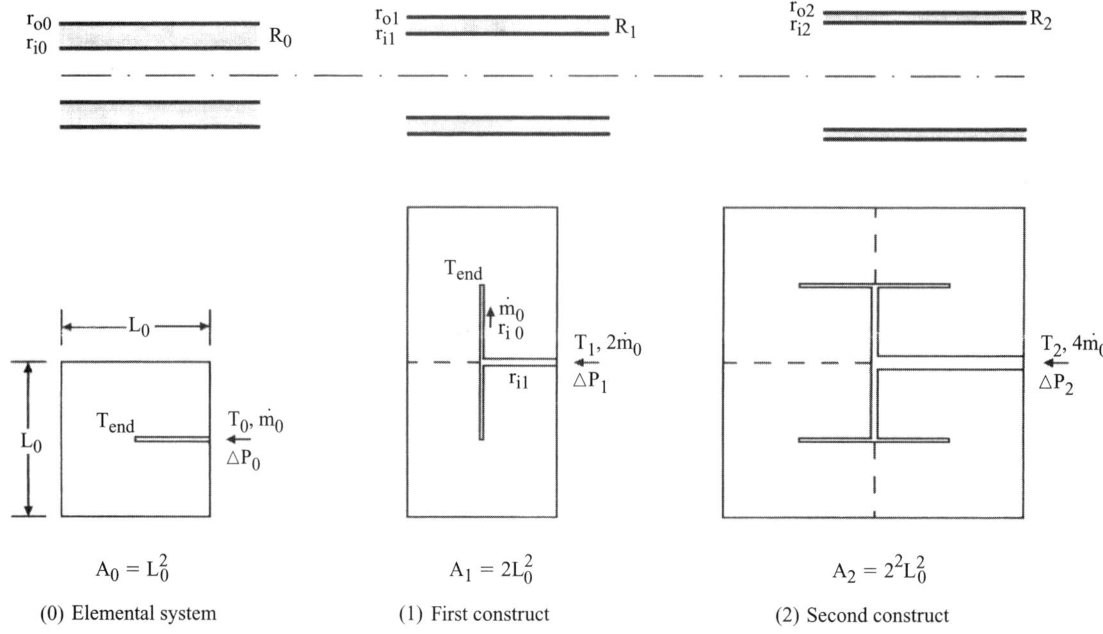

Figure 7.38 Sequence of area constructs obtained by pairing smaller constructs [13].

the first construct of Fig. 7.34. One objective of this section is to see which area construction sequence serves the farthest (end) user of the $4L_0{}^2$ territory better, Fig. 7.34 or Fig. 7.38?

For brevity, we report only the results of the minimization of water flow resistance and loss of heat to the ambient. The analysis follows step-by-step the analysis detailed in the preceding section. For the first construct of Fig. 7.38, in place of (7.115) and (7.117) we obtain

$$(r_{i1}/r_{i0})_{\text{opt}} = 2^{1/2} \tag{7.125}$$

$$(r_{i1}/r_{i0})_{\text{opt}} = 2^{3/7} \tag{7.126}$$

which correspond to invoking the duct wall material constraint and, respectively, the duct volume constraint. Equations (7.118) and (7.119) are replaced by

$$\theta_1 = \frac{T_{\text{end}} - T_\infty}{T_1 - T_\infty} = \exp\left(-\frac{N_0}{\ln R_0} - \frac{N_0}{2\ln R_1}\right) \tag{7.127}$$

$$\tilde{V}_1 = \frac{V_1}{\pi L_0 r_{i0}^2} = \frac{1}{2}\left(\frac{r_{i1}}{r_{i0}}\right)^2 (R_1^2 - 1) + (R_0^2 - 1) \tag{7.128}$$

The results for the optimal distribution of insulation and maximized end-user water temperature are reported in Figs. 7.39a and b. The choice of constraint, Eq. (7.125)

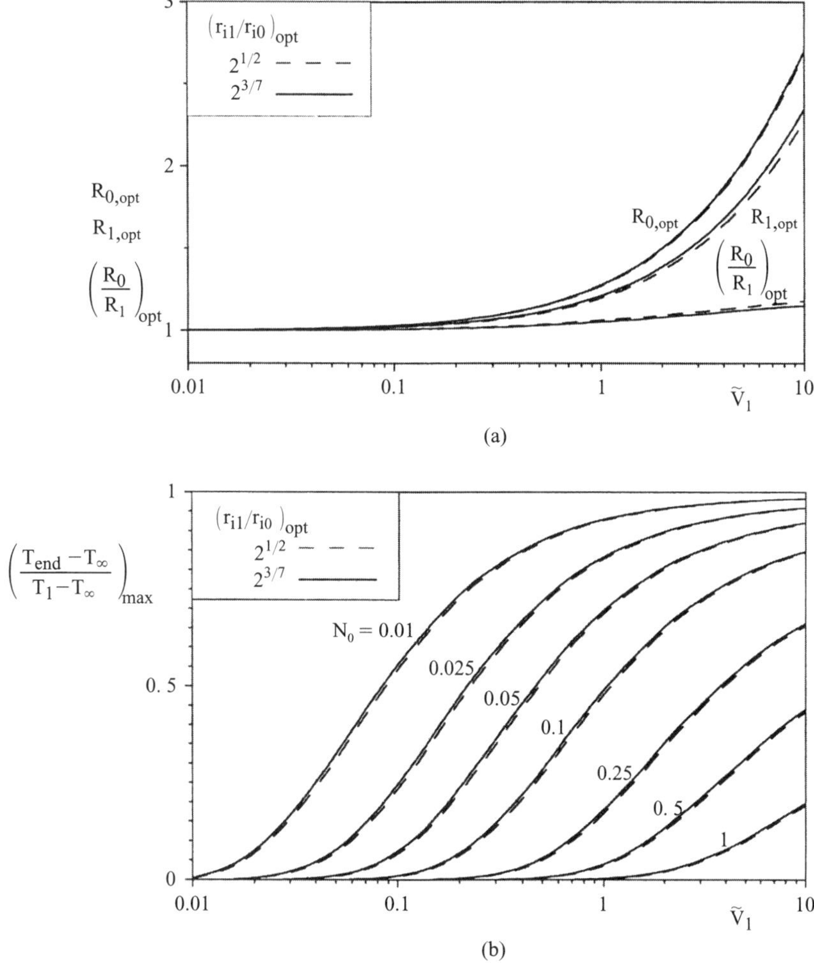

Figure 7.39 The optimal ratios of insulation radii for the first construct (A_1) shown in Fig. 7.38, and the maximized end temperature of the first construct [13].

or Eq. (7.126), has a small effect; therefore, we proceed by using only the duct volume constraint.

At the second-construct level, in place of Eq. (7.122) we obtain $(r_{i2}/r_{i1})_{opt} = 2^{3/7}$. The user temperature and total insulation volume are

$$\theta_2 = \frac{T_{end} - T_\infty}{T_2 - T_\infty} = \theta_1 \exp\left(-\frac{N_0}{2 \ln R_2}\right) \tag{7.129}$$

$$\tilde{V}_2 = \frac{V_2}{\pi L_0 r_{i0}^2} = \left(\frac{r_{i2}}{r_{i0}}\right)^2 (R_2^2 - 1) + 2\tilde{V}_1 \tag{7.130}$$

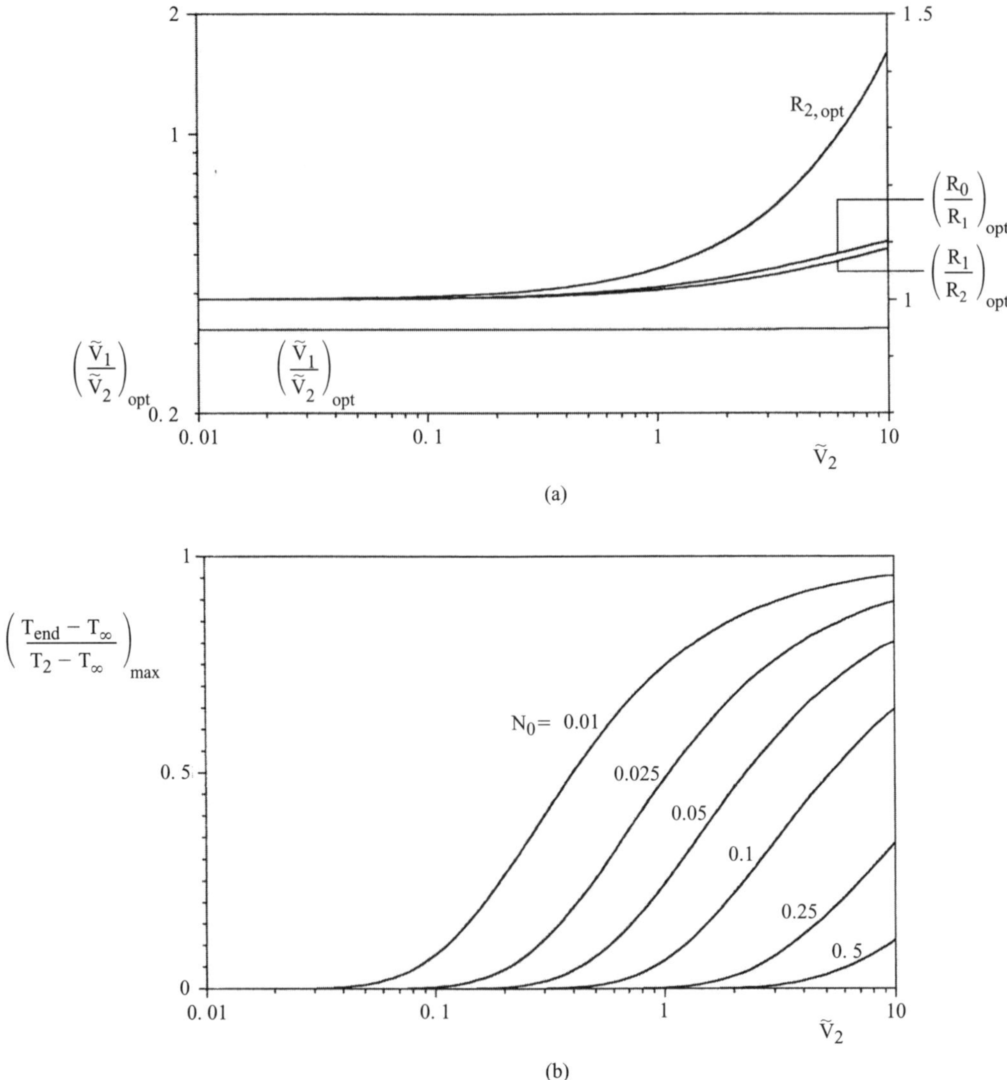

Figure 7.40 The optimal ratios of insulation radii for the second construct (A_2) shown in Fig. 7.38, and the maximized end temperature of the second construct [13].

The optimal distribution of the finite amount of insulation is reported in Fig. 7.40a, and the maximized user water temperature in Fig. 7.40b.

The numerical values and trends are similar to what we saw earlier. More to the point, we can compare the doubling sequence (Fig. 7.38) against the sequence of square areas (Fig. 7.34). In Fig. 7.41 we show the ratio of the maximized end temperatures of Figs. 7.35b and 7.40b, with the observation that both figures refer to the same construct size ($4L_0^2$), and that the insulation volume V_1 of Fig. 7.35b is

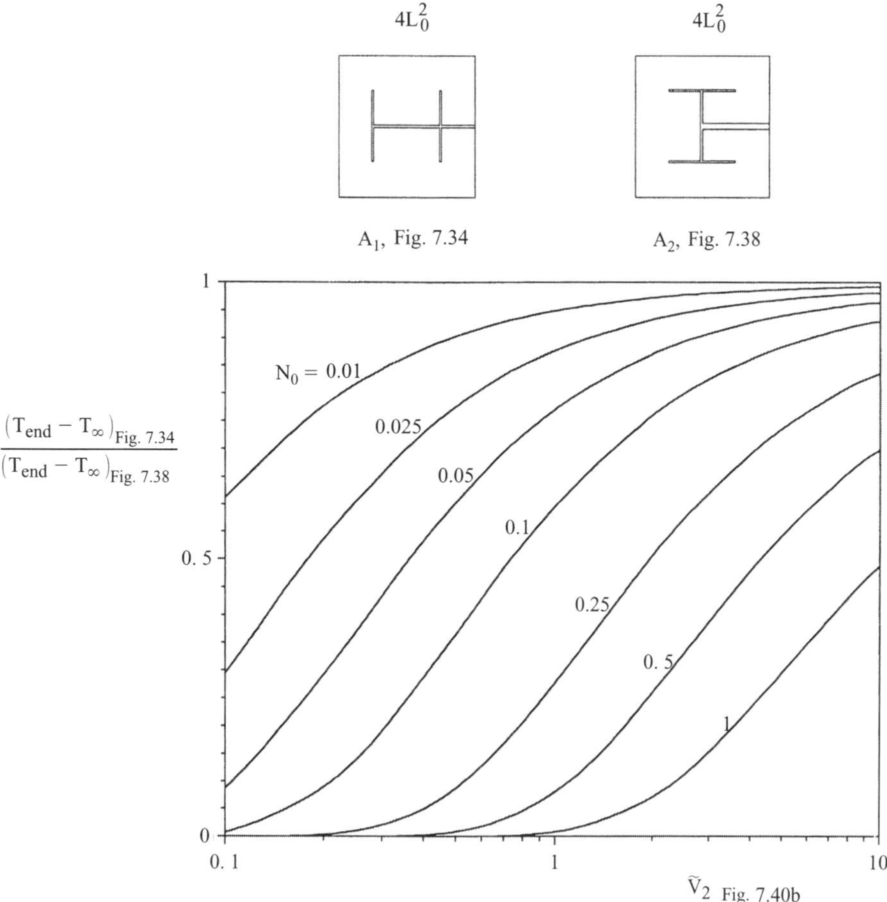

Figure 7.41 Comparison between the maximized end-user water temperature on the $4L_0^2$ construct, according to the construction sequences of Fig. 7.34 and Fig. 7.38 [13].

set equal to V_2 of Fig. 7.40b. The total duct volume is the same in the two designs sketched above the graph. Note that r_{i0} of Fig. 7.35b is not the same as the r_{i0} of Fig. 7.40b, and, consequently, the dimensionless volumes $\tilde{V}_{1,\,\mathrm{Fig.\,7.35b}}$ and $\tilde{V}_{1,\,\mathrm{Fig.\,7.40b}}$ are not the same. The comparison shown in Fig. 7.41 allows us to conclude that the tree structure generated by repeated pairing (A_2, Fig. 7.38) is superior to the square structure (A_1, Fig. 7.34). The temperature of the hot water received by the end user in Fig. 7.38 (A_2) is consistently higher. The trends and the domain in which the two schemes perform at nearly the same level are similar to what we saw in the comparison between the square tree and the coiled string designs in Fig. 7.37.

We pursued this comparison to an even higher level of assembly—the construct of size $16{L_0}^2$—on the basis of the same amount of insulation material and total

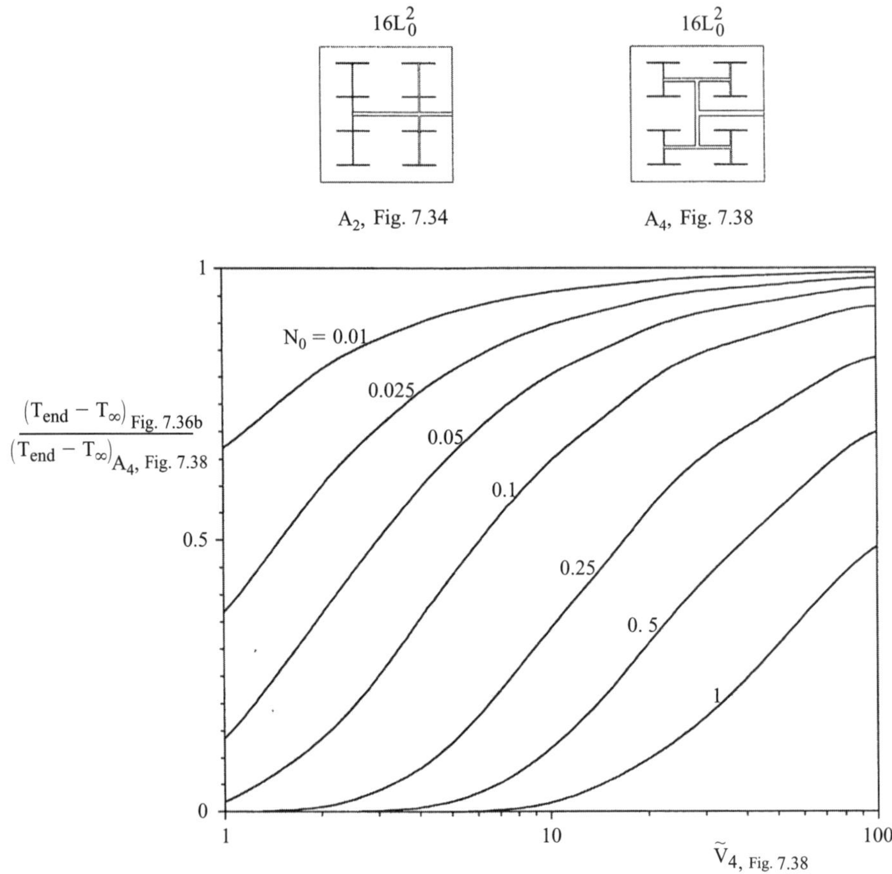

Figure 7.42 Comparison between the maximized end-user water temperature on the $16L_0^2$ construct, according to the construction sequences of Fig. 7.34 and Fig. 7.38 [13].

duct volume. The two structures are illustrated in the upper part of Fig. 7.42, with the observation that the tree generated by repeated pairing (A_4) is not shown in Fig. 7.38. Once again, the tree design based on the sequence of Fig. 7.38 outperforms the design based on Fig. 7.34.

By comparing Fig. 7.42 and Fig. 7.41 on the same basis (the same L_0, total pipe volume, total amount of insulation), it can be shown that the difference in the global performance of the two types of trees (Fig. 7.34 and Fig. 7.38) diminishes as each optimized tree structure becomes more complex. The global performance becomes progressively less sensitive to the actual layout of the tubes, provided that the distributions of tube step sizes and shells of insulation material have been optimized. When the optimized tree structure becomes more complex, it also becomes more robust with respect to changes in the tree pattern.

Another useful property of the pairing sequence (Fig. 7.39) is that each user receives hot water at the same temperature. The $\dot{m}_0$ stream received by each user passes through the same sequence of insulated tubes. This is a very useful property, because in other tree structures such as Fig. 7.34, the users located closer to the root of the tree receive warmer streams than the farther users. The tree designs of Fig. 7.38 deliver hot water to the territory uniformly—uniformly in space and in temperature.

We verified these conclusions by constructing Fig. 7.41 and Fig. 7.42 in another way. The first way was to assemble each construct (e.g., $4L_0{}^2$ in Fig. 7.41) by placing together previously optimized smaller systems (constituents). The optimized features determined for the smaller areas were preserved in the larger construct. The alternative was to begin with the needed (large) construct, and to let all the geometric parameters vary freely. Along this route, we optimized every geometric feature and arrived at exactly the same final structure as during the first approach. These alternative calculations of the optimized structures compared in Fig. 7.41 and Fig. 7.42 verify the accuracy of the results presented in these figures.

The chief conclusion of this section is that the use of geometric form (shape, structure) is an effective route to achieving high levels of global performance under constraints. The brute force approach of delivering hot water by using lots of insulation and flow rate (small N) is not economical. Much faster progress toward the goal of global performance maximization can be made by recognizing and treating the topology of the flow system as the main unknown of the problem.

7.6.6 One-by-One Tree Growth

Consider now the alternative of adding new users to an existing structure, while not having the means to reoptimize the structure after the new users have been added [37]. Furthermore, the addition of each new user is decided from the point of view of the user, not from the point of view of maximizing global performance. Each new user is connected to the tree network in the place that maximizes the user's benefit, namely, the temperature of the water stream that the new user receives.

In this new approach we must have a start, and for this we choose the optimized A_2 construct of Fig. 7.38, which is reproduced in Fig. 7.43. The A_2 construct has the area $4L_0{}^2$. It also has the property that its four users receive water at the same temperature, $T_1 = T_2 = T_3 = T_4$. The overall pressure drop has been minimized according to the pipe size ratios r_{i2}/r_{i1} and r_{i1}/r_{i0}. The total amount of insulation has been distributed optimally according to the insulation radii ratios R_0, R_1 and R_2.

Consider now the decision to add a new elemental user ($\dot{m}_0 L_0^2$) to the A_2 construct. Because of symmetry, we recognize only three possible positions (Fig. 7.43a, b, c). For the pipe that connects the new user to the existing A_2 construct, we use the pipe size and insulation design of the elemental system of A_2, namely, r_{i0} and R_0. In going from the A_2 construct of Fig. 7.38 to the constructs of Fig. 7.43, the overall mass flow rate has increased from $4\dot{m}_0$ to $5\dot{m}_0$. The addition of the new

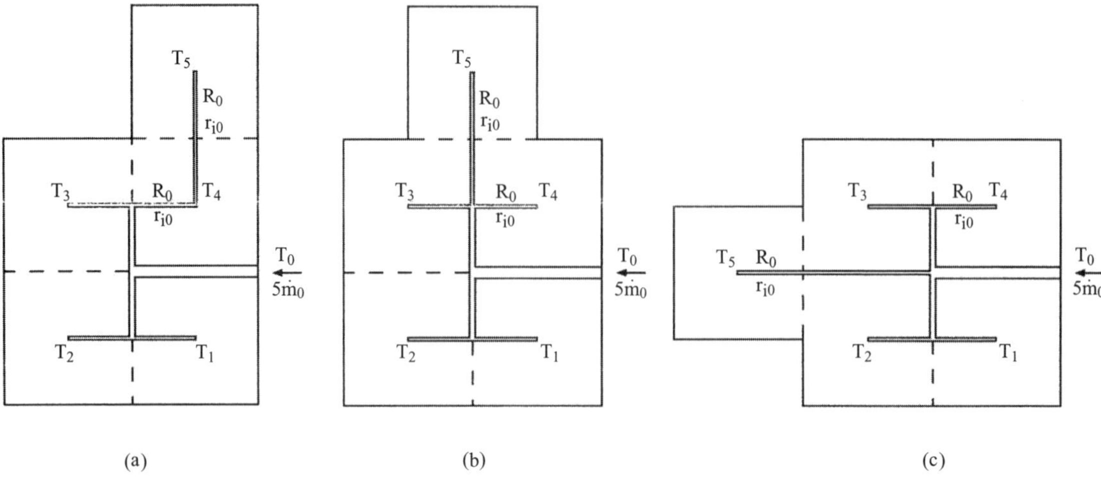

Figure 7.43 Three possible ways of attaching a new user to the optimized A_2 construct of hot water pipes of Fig. 7.38 [37].

user disturbs the optimally balanced distribution of resistances and temperatures, which was reached based on the analysis presented in the preceding section.

To deliver the hottest water from the network to the new user we must use the shortest pipe possible. By analyzing the user water temperature in the three configurations identified in Fig. 7.43, we find that $T_{5,\mathrm{b}} > T_{5,\mathrm{a}} > T_{5,\mathrm{c}}$. For example, for the configuration of Fig. 7.43a it can be shown based on an analysis of the same type as in Eqs. (7.111) and (7.112) that the temperature of the hot water received by the new user (T_5) is

$$\frac{T_5 - T_\infty}{T_0 - T_\infty} = \exp\left(-\frac{5N_0}{2\ln R_0} - \frac{N_0}{3\ln R_1} - \frac{2N_0}{5\ln R_2}\right) \tag{7.131}$$

The corresponding expressions for the hot water temperature of the new user in the configurations of Figs. 7.43b and c are

$$\frac{T_5 - T_\infty}{T_0 - T_\infty} = \exp\left(-\frac{2N_0}{\ln R_0} - \frac{N_0}{3\ln R_1} - \frac{2N_0}{5\ln R_2}\right) \tag{7.132}$$

$$\frac{T_5 - T_\infty}{T_0 - T_\infty} = \exp\left(-\frac{3N_0}{\ln R_0} - \frac{2N_0}{5\ln R_2}\right) \tag{7.133}$$

These T_5 equations and the corresponding expressions for the temperatures of the hot water streams received by the existing four users ($T_1, \ldots, T_4$) are plotted in Fig. 7.44 for the same elemental number N_0. At first sight, the temperature of the delivered water increases as the total mount of insulation ($\tilde{V}_2$) increases. The graphs, however, tell a more important story: they show not only how the choice of grafting (a,b,c) affects the temperature T_5, but also how the new (fifth) water stream affects the temperatures felt by the previous users.

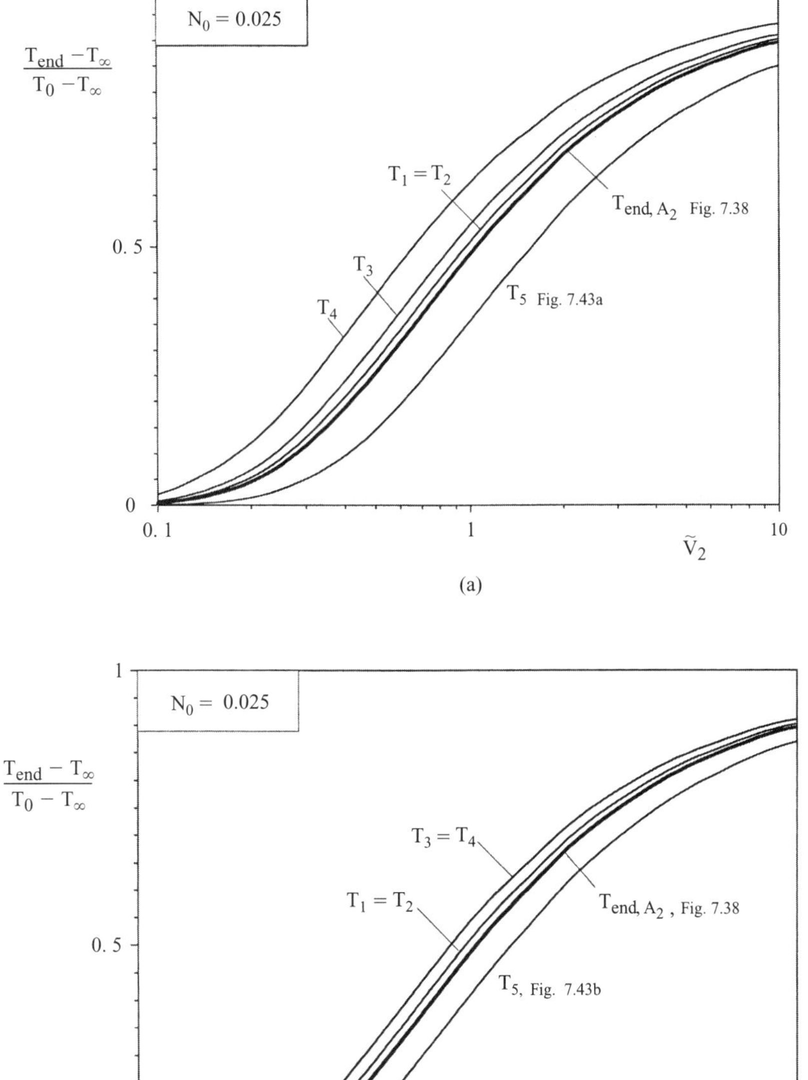

Figure 7.44 The temperatures of the hot water delivered to the five users in the configurations of Fig. 7.43a, b, and c [37].

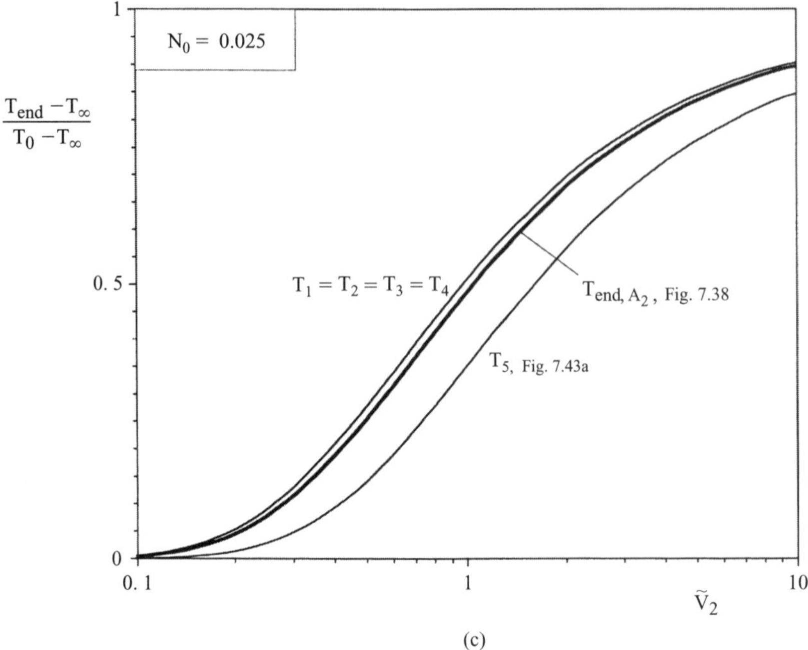

Figure 7.44 (*Continued*)

Before the fifth user was added, the constructal design of A_2 delivered water of the same temperature to the four users. Note the single curve drawn for $T_{end,A2,Fig.7.38}$ in each of the graphs of Fig. 7.44. This curve serves as reference in each of the designs of Fig. 7.43, where the temperatures of the existing users are altered by the insertion of a fifth user. In the configuration of Fig. 7.43a and Fig. 7.44a, the fifth-user temperature drops below the original temperature of hot water delivery. The temperatures of the four older users increase, the largest increases being registered by the users that are situated the closest to the newly added user. It is as if the new user insulates (or shields) its closest neighbors from cold water. It does so by drawing a larger stream of hot water in its direction and in the direction of its immediate neighbors.

These features are also visible in the performance of the configurations of Figs. 7.43b and c (cf. Fig. 7.44b and c), but the differences between the water temperatures of the five users are not as great as in the case of Fig. 7.43a (Fig. 7.44a). It is interesting that the best configuration (Fig. 7.43b and Fig. 7.44b), where the water received by the new user is the hottest, is also the configuration in which all the users receive water at nearly the same temperature. This finding is in line with the seemingly general conclusion that the optimal design of a complex flow

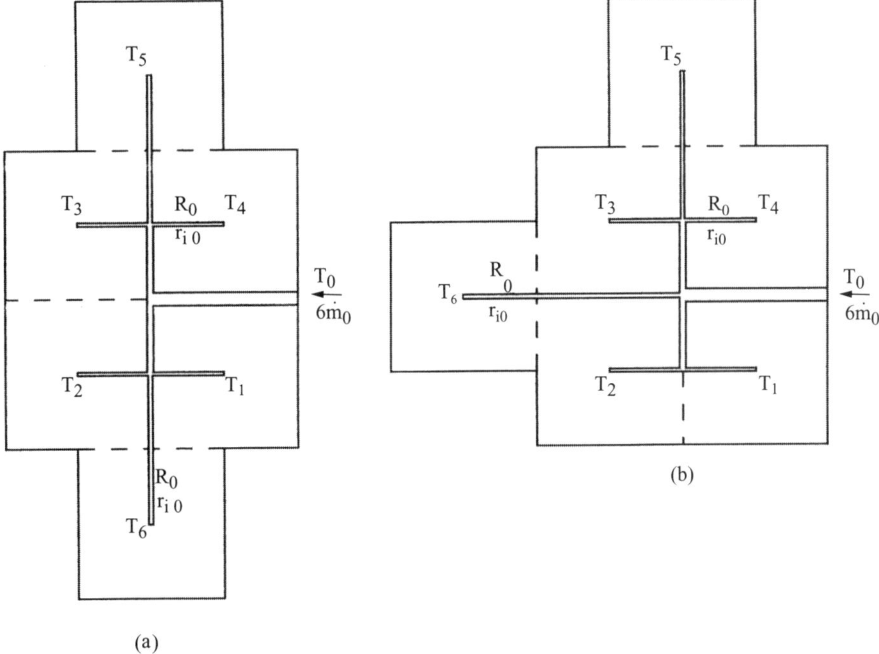

Figure 7.45 Two ways of attaching a sixth user to the construct selected in Fig. 7.43b [37].

system is the one where the imperfections are distributed as uniformly as possible [9]. In the class of problems treated here, by imperfections we mean the decrease of the hot water temperature, which is due to the loss of heat to the ambient across the insufficient (finite) amount of insulation.

This point is stressed further by Fig. 7.44c, which shows that the water temperatures of the four existing users are affected in equal measure by the insertion of the fifth user. The symmetry in this case is explained by the fact that in Fig. 7.43c the new user is attached to the hub of the previously optimized A_2 construct, at equal distances from the existing four users.

In conclusion, we retain the design of Fig. 7.43b, and proceed to the next problem, which is the placing of a new (sixth) user in the best spot on the periphery of the five-user construct. The two most promising choices are shown in Fig. 7.45. They are the most promising because in each case the new user is attached to a source with relatively high temperature. In other words, unlike in Fig. 7.43 where we considered without bias all the possible ways of attaching the fifth user, now and in subsequent steps of growing the structure we expedite the place-selection process by using conclusions and trends learned earlier. For brevity, we omit the analytical description of the type shown in Eqs. (7.131) through (7.133).

The performance curves plotted in Fig. 7.46 reinforce some of the earlier trends. Figure 7.46a shows that when the new user is placed symmetrically relative to the fifth user, Fig. 7.45a, symmetry is preserved in the temperature distribution over the entire tree. Note T_6 equals T_5. Symmetry works in favor of making the distribution of temperature (and heat loss, imperfection) more uniform.

The competing configuration (Fig. 7.45b) has the new user attached to the center of the original A_2 construct, in the same place as in Fig. 7.43c. Its performance is documented in Fig. 7.46b. The configuration of Fig. 7.45b is inferior to that of Fig. 7.45a because the temperature T_6 of Fig. 7.46b is consistently lower than the corresponding temperature of Fig. 7.46a. Furthermore, the configuration of Fig. 7.45b is inferior because it enhances the nonuniformity in the temperature of the water received by all the users.

In sum, we retain the configuration of Fig. 7.45a for the construct with six users. It is important to note that this is the first instance where the symmetrical placement of the even-numbered user emerges as the best choice. This conclusion was tested (one new user at a time) up to the 12th new user, beyond which it was adopted as a rule for expediting the optimized growth of structure. In other words, each even-numbered new user was placed symmetrically relative to the preceding odd-numbered user (the placement of which was optimized).

Figure 7.47 shows the three most promising positions that we tried for a new seventh user. Temperature distribution charts (such as Fig. 7.46) were developed for each configuration, but are not shown. On that basis we found that the seventh-user temperature T_7 decreases, in order, from Fig. 7.47b to Fig. 7.47a, and finally to Fig. 7.47c. We retained the configuration of Fig. 7.47b, and proceeded to the selection of the best place for an eight user. The best position for the eight user is the symmetric arrangement shown in Fig. 7.48a. The rest of Fig. 7.48 shows the optimal positioning of subsequent users up to the sixteenth. The rule that is reinforced by these choices is that the better position for a new user is that one that requires a short connection to that side of the existing construct where the water delivery temperatures are higher.

7.6.7 Complex Flow Structures Are Robust

We are now in a position to compare the performance of designs based on one-by-one growth (Fig. 7.48) with the constructal designs obtained by repeated doubling (Fig. 7.38). The most complex design of the one-by-one sequence (Fig. 7.48e) is compared with its constructal counterpart in Fig. 7.49a. The comparison is done on the same basis, that is, the same total amount of insulation material, and the same serviced territory ($16L_0{}^2$). We see that the excess temperature of the latest (16th) user in Fig. 7.48e is consistently less than the excess temperature felt by each of the users in A_4 construct generated by doubling (recall that in Fig. 7.38 each user receives water at the same temperature).

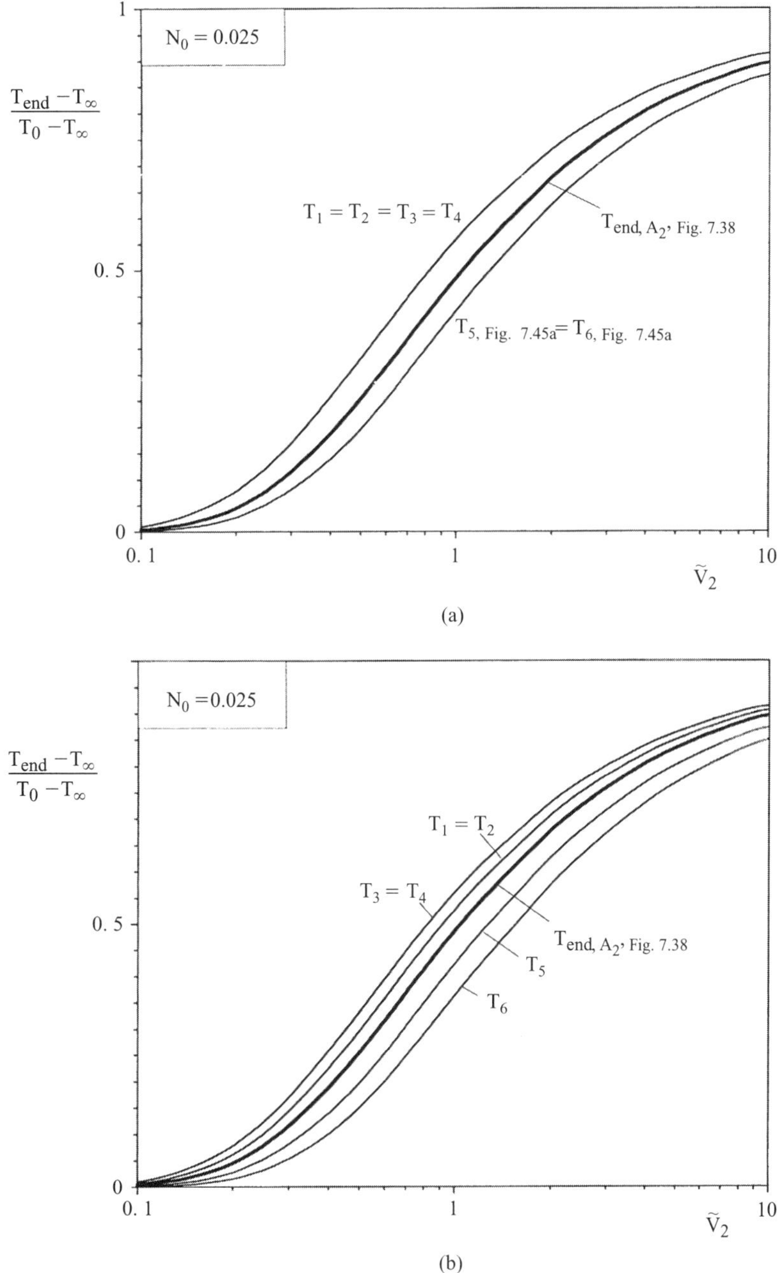

Figure 7.46 The temperatures of the hot water delivered to the fifth and sixth users in the configurations of Figs. 7.45a and b [37].

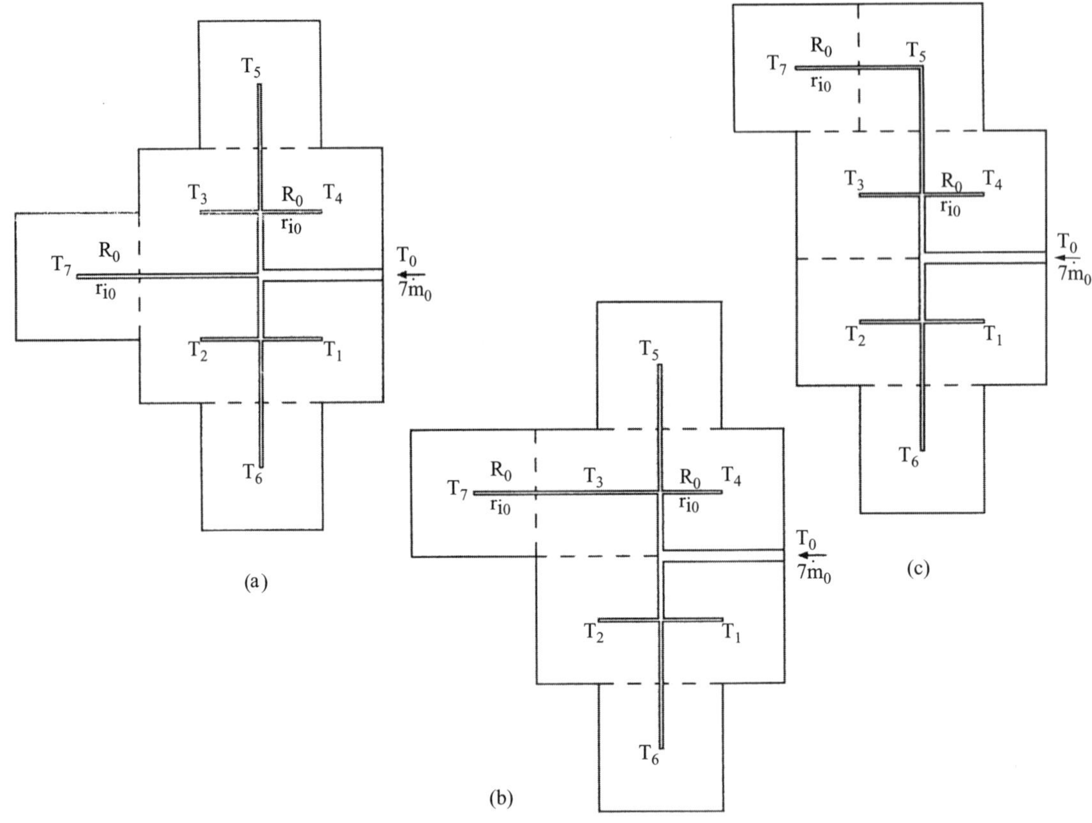

Figure 7.47 Three ways of placing a seventh user in the construct selected in Fig. 7.45(a) [37].

The discrepancy between the performances of the two designs compared in Fig. 7.49a diminishes as the total amount of insulation ($\tilde{V}_{A_{16}}$, Fig. 7.48e) increases, and as N_0 decreases. The smaller N_0 corresponds to better insulation materials (smaller k) and denser populations of users (smaller L_0). An important question is how this discrepancy depends on the complexity of the flow structure. Is it always important to strive for the best (e.g., Fig. 7.38), or will an on-the-spot optimization (e.g., Fig. 7.48) be sufficient?

We examined this question by changing the complexity of the structure, and repeating the test of Fig. 7.49a. The new structures are simpler, with only eight users. The one-by-one construction of Fig. 7.48a is compared with the A_3 design of the sequence shown in Fig. 7.38. The comparison shown in Fig. 7.49b is for the same territory ($8L_0^2$) and total amount of insulation material. The trends are the same as what we saw in Fig. 7.49a. The excess water temperature ($T_8 - T_\infty$) of the last user added to Fig. 7.48a is consistently lower than what the users of the constructal design receive.

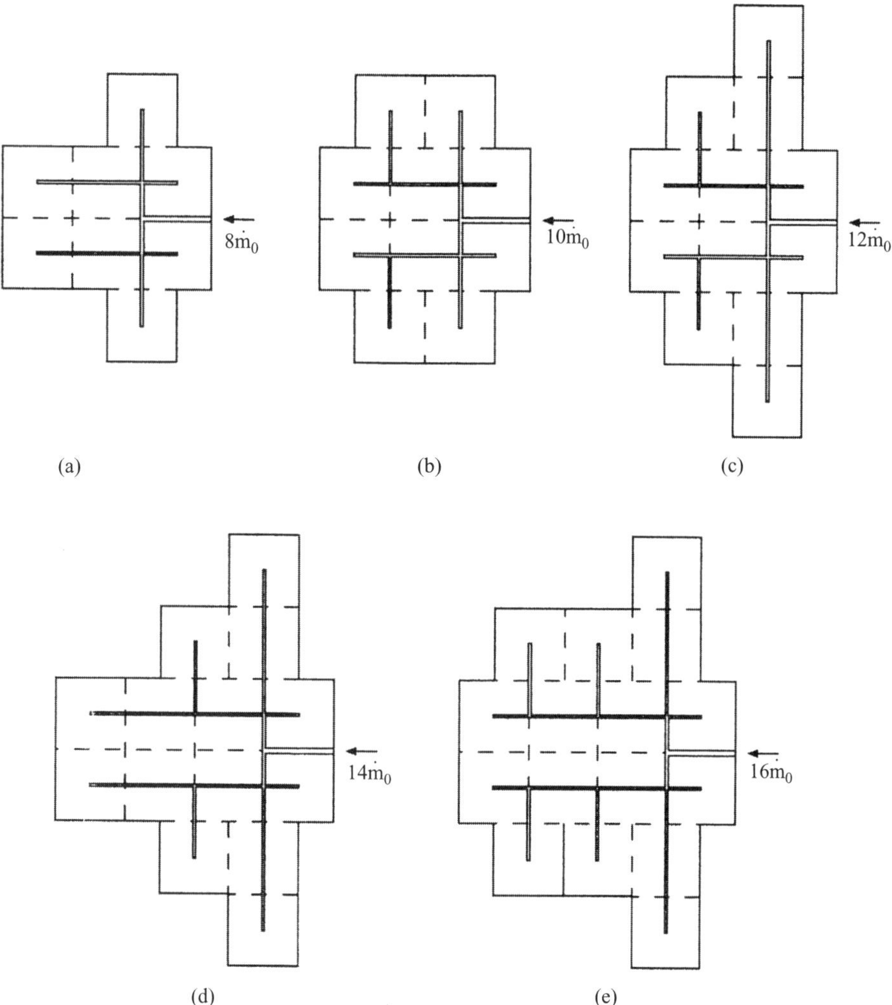

Figure 7.48 The best configurations for the constructs with users from 8 to 16 [37].

An important observation is that the ratio $(T_8 - T_\infty)/(T_{\text{end}} - T_\infty)$ of Fig. 7.49b is consistently lower than the corresponding ratio $(T_{16} - T_\infty)/(T_{\text{end}} - T_\infty)$ of Fig. 7.49a. The total amount of insulation material and the serviced territory are the same in both figures. The same territory means that $16L^2{}_{0,\text{Fig.7.49a}} = 8\text{L}^2{}_{0,\text{Fig.7.49b}}$ or that the L_0 scale of the constructs of Fig. 7.49b exceeds by a factor of $2^{1/2}$ the L_0 dimension of the constructs of Fig. 7.49a. In both designs, the individual user receives the flow rate $\dot{m}_0$.

This observation is reinforced by Fig. 7.50a, which shows the same comparison for a structure with only four users. The constructal design is A_2 of Fig. 7.38. The corresponding one-by-one design is shown on the left side. The total insulation volume ($\tilde{V}$) and serviced territory ($4L_0^2$) is the same in both designs. The one-by-one structure approaches the performance of the constructal design as N_0 decreases and $\tilde{V}$ increases. To test the effect of complexity, which decreases from Fig. 7.49a to b and, finally, to Fig. 7.50a, we must put all the designs on the same $\tilde{V}$ and A basis. This means that the territory covered by four users in the designs of Fig. 7.50a is covered by 16 users in the designs of Fig. 7.49a. Consequently, the L_0 scale in Fig. 7.50a is twice the L_0 scale of Fig. 7.49a. The same proportionality exists between the respective N_0 values.

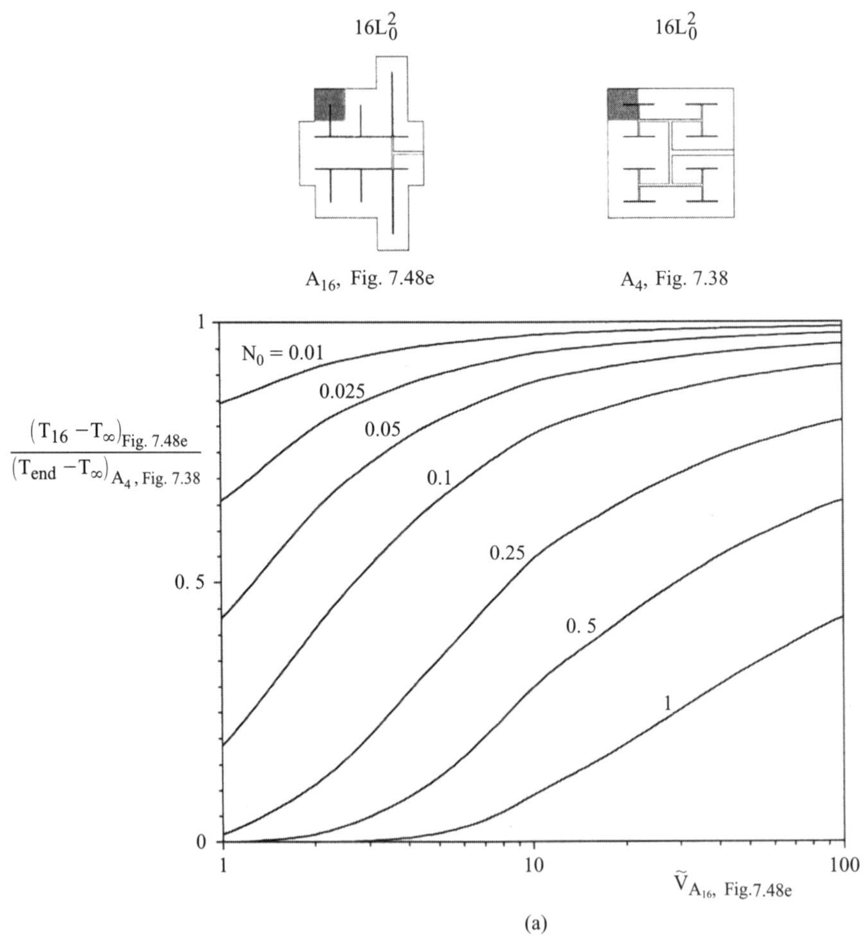

Figure 7.49 (a) Comparison between the excess temperature of the water received by the last user added to the structure of Fig. 7.48e, and the corresponding temperature in the design based on the sequence of Fig. 7.38. (b) Comparison between the excess temperature of the water received by the last user added to the structure of Fig. 7.48b, and the corresponding temperature in the design based on the sequence of Fig. 7.38 [37].

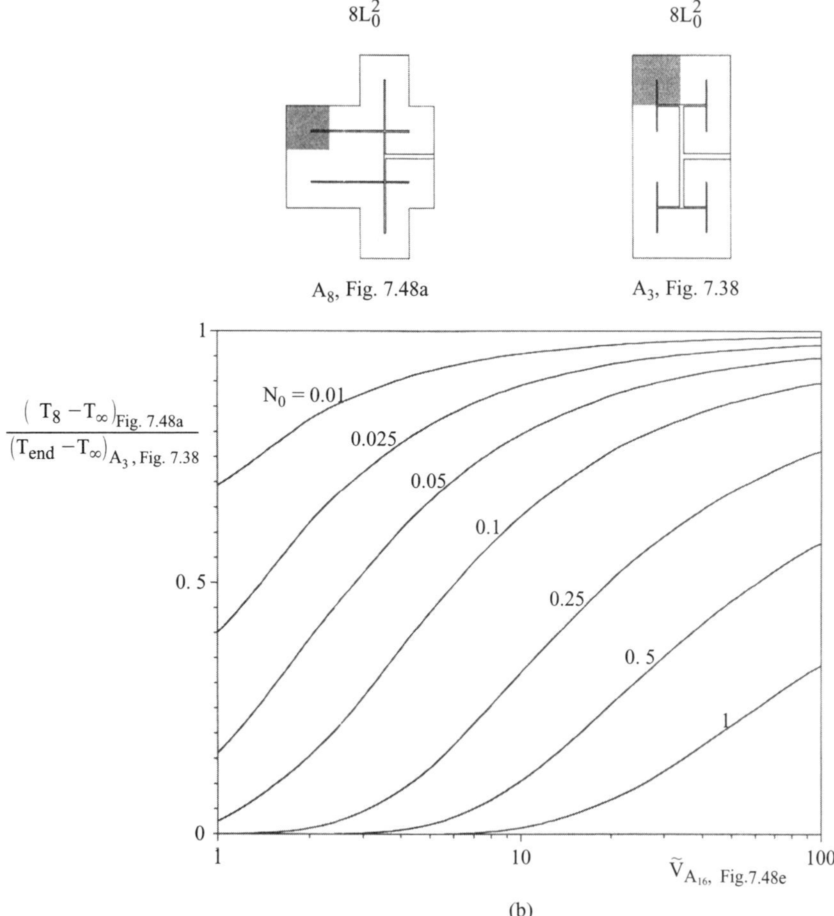

Figure 7.49 (*Continued*)

In summary, the performance level of the structure increases as its complexity increases. Fig. 7.50b brings together the one-by-one designs of Fig. 7.49 and Fig. 7.50. They cover the same area A, which means that the L_0 scale of the users in Fig. 7.48e ($n = 16$) is the shortest. The $n = 16$ design is used as reference. The three designs use the same amount of insulation material. The k and $\dot{m}_0$values are such that the N_0 value [Eq. (7.112)] based on the L_0 length of the $n = 16$ case of Fig. 7.49a is $N_0 = 0.01$. Figure 7.50b shows that the end-user water excess temperature increases as the complexity of the structure (n) increases.

A major limitation of the one-by-one design is the memory that is built into the system. Memory is reinforced at every step. What was built prior to the insertion of the last user is retained. This is why in the one-by-one sequence of Fig. 7.48, there are only three sizes of pipes and their insulations: the three sizes that were inherited in the first step, from the A_2 construct of Fig. 7.38. In a complete optimization of a multiscale flow structure, there is an optimal pipe size for every level of branching and coalescence. This means that the existence of only three pipe sizes in structures as complex as those exhibited in Fig. 7.48 is a departure from the best possible performance.

In conclusion, the design of the insulated structure is robust with respect to how the insulation is distributed over all the pipes. The performance of the structure (how well it services the end user) is relatively insensitive to how finely the distribution of pipe sizes and insulation radii is optimized.

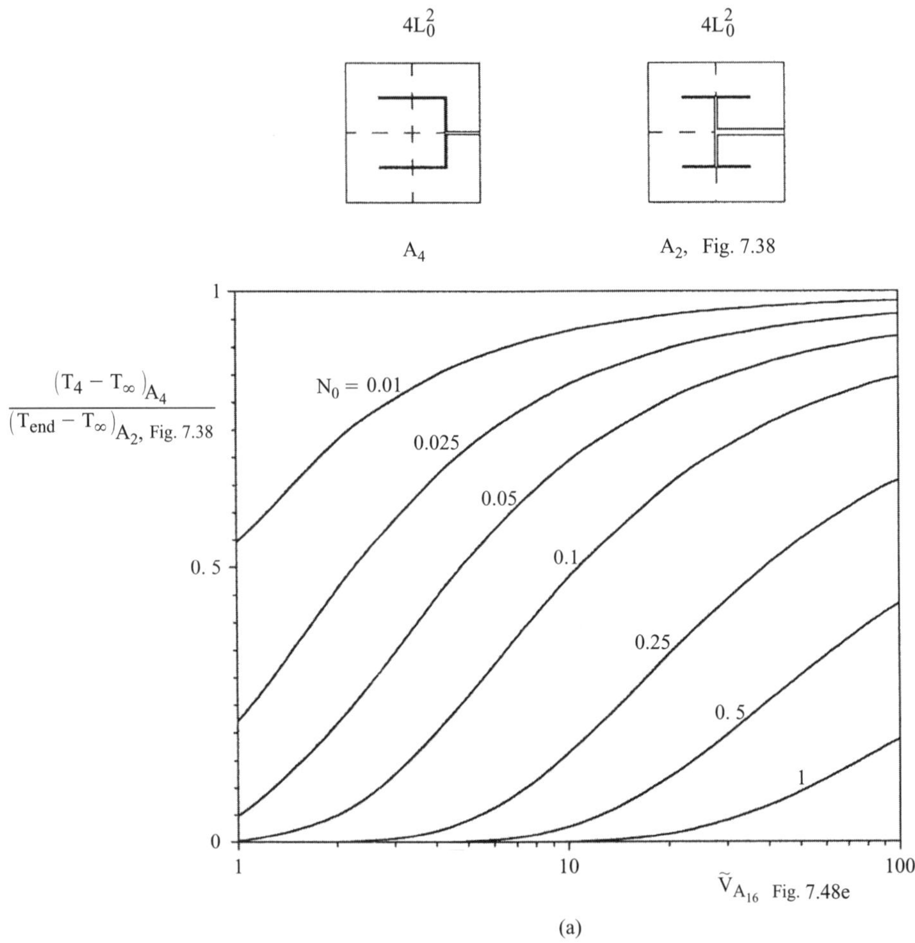

Figure 7.50 (a) Comparison between the excess temperature of the water received by the last user added to a structure with four users, and the corresponding temperature in the design (A_2) based on the sequence of Fig. 7.38. (b) Comparison between the one-by-one designs of Fig. 7.49 and Fig. 7.50, showing how the performance improves as the complexity (n) increases [37].

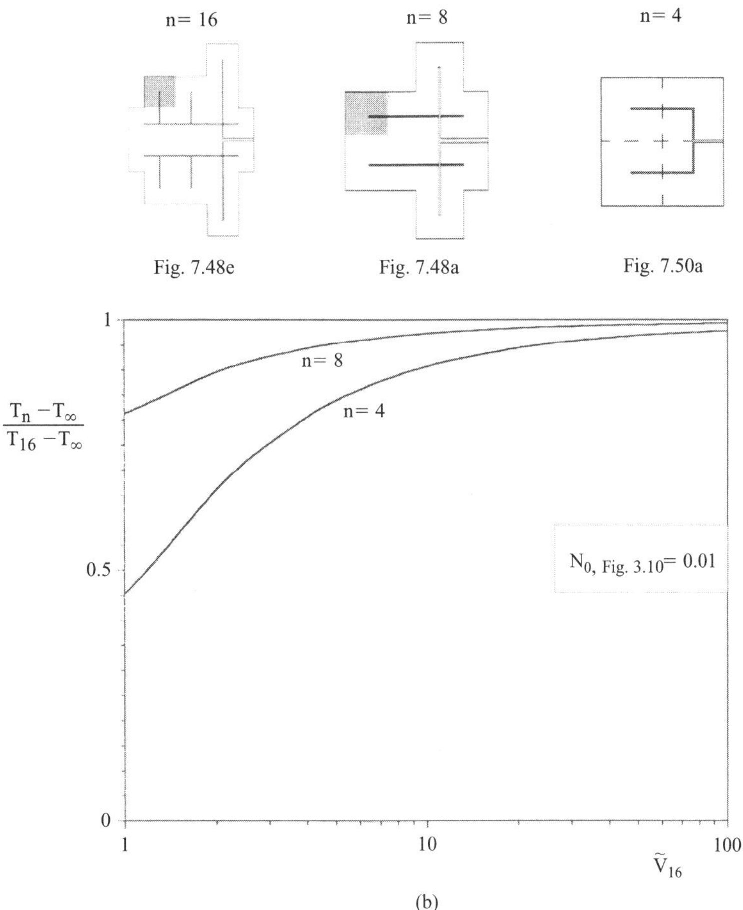

Figure 7.50 (*Continued*)

REFERENCES

1. W. Wechsatol, S. Lorente, and A. Bejan, Dendritic convection on a disc. *Int J Heat Mass Transfer*, Vol. 46, 2003, pp. 4381–4391.
2. A. Bejan and S. Lorente, The constructal law and the thermodynamics of flow systems with configuration. *Int J Heat Mass Transfer*, Vol. 47, 2004, pp. 3203–3214.
3. A. Bejan and M. R. Errera, Convective trees of fluid channels for volumetric cooling. *Int J Heat Mass Transfer*, Vol. 43, 2000, pp. 3105–3118.
4. A. Bejan, *Advanced Engineering Thermodynamics*, 2nd ed. (ch. 13). New York: Wiley, 1997.
5. A. Bejan, *Convection Heat Transfer*, 3rd ed. (ch. 3). New York: Wiley, 2004.

6. W. Wechsatol, S. Lorente, and A. Bejan, Optimal tree-shaped networks for fluid flow in a disc-shaped body. *Int J Heat Mass Transfer*, Vol. 45, 2002, pp. 4911–4924.
7. S. Lorente, W. Wechsatol, and A. Bejan, Tree-shaped flow structures designed by minimizing path lengths. *Int J Heat Mass Transfer*, Vol. 45, 2002, pp. 3299–3312.
8. W. Wechsatol, S. Lorente, and A. Bejan, Tree-shaped flow structures with local junction losses. *Int J Heat Mass Transfer*, Vol. 49, 2006, pp. 2957–2964.
9. A. Bejan, *Shape and Structure, from Engineering to Nature*. Cambridge, UK: Cambridge University Press, 2000.
10. A. Bejan, Dendritic constructal heat exchanger with small-scale crossflows and larger-scales counterflows. *Int J Heat Mass Transfer*, Vol. 45, 2002, pp. 4607–4620.
11. A. K. da Silva, S. Lorente, and A. Bejan, Constructal multi-scale tree-shaped heat exchangers. *J Appl Phys*, Vol. 96(3), 2004, pp. 1709–1718.
12. J. C. Ordonez, A. Bejan, and R. S. Cherry, Designed porous media: Optimally nonuniform flow structures connecting one point with more points. *Int J Thermal Sciences*, Vol. 42, 2003, pp. 857–870.
13. W. Wechsatol, S. Lorente, and A. Bejan, Tree-shaped insulated designs for the uniform distribution of hot water over an area. *Int J Heat Mass Transfer*, Vol. 44, 2001, pp. 3111–3123.
14. A. Bejan, A general variational principle for thermal insulation system design. *Int J Heat Mass Transfer*, Vol. 22, 1979, pp. 219–228.
15. A. Bejan, *Entropy Generation through Heat and Fluid Flow* (p. 181). New York: Wiley, 1982.
16. A. Bejan, The tree of convective heat streams: Its thermal insulation function and the predicted 3/4-power relation between body heat loss and body size. *Int J Heat Mass Transfer*, Vol. 44, 2001, pp. 699–704.
17. R. K. Shah and D. P. Sekulic, *Fundamentals of Heat Exchanger Design*. Hoboken, NJ: Wiley, 2003.
18. V. A. P. Raja, T. Basak, and S. K. Das, Heat transfer and fluid flow in a constructal heat exchanger. Proceedings of Fifth International Conference on Enhanced, Compact and Ultra-Compact Heat Exchangers: Science, Engineering and Technology, R. K. Shah, M. Ishizuka, T. M. Rudy and V. V. Wadekar, Eds., Engineering Conferences International, Hoboken, NJ, September 2005.
19. V. A. P. Raja, T. Basak, and S. K. Das, Thermal performance of a multi-block heat exchanger designed on the basis of Bejan's constructal theory. *Int J Heat Mass Transfer*, Vol. 51, 2008, pp. 3582–3594.
20. Y. Azoumah, N. Mazet, and P. Neveu, Constructal network for heat and mass transfer in a solid-gas reactive porous medium. *Int J Heat Mass Transfer*, Vol. 47, 2004, pp. 2961–2970.
21. Y. Azoumah, P. Neveu, and N. Mazet, Constructal design combined with entropy generation minimization for solid-gas reactors. *Int J Thermal Sciences*, Vol. 45, 2006, pp. 716–728.
22. Y. Azoumah, P. Neveu, and N. Mazet, Optimal design of thermochemical reactors based on constructal approach. *AIChE Journal*, Vol. 53, 2007, pp. 1257–1266.
23. S. Zhou, L. Chen, and F. Sun, Constructal entropy generation minimization for heat and mass transfer in a solid-gas reactor based on triangular element. *J Phys D: Appl Phys*, Vol. 40, 2007, pp. 3545–3550.
24. V. D. Zimparov, A. K. da Silva, and A. Bejan, Thermodynamic optimization of tree-shaped flow geometries. *Int J Heat Mass Transfer*, Vol. 49, 2006, pp. 1619–1630.

25. V. D. Zimparov, A. K. da Silva, and A. Bejan, Constructal tree-shaped parallel flow heat exchangers. *Int J Heat Mass Transfer*, Vol. 49, 2006, pp. 4558–4566.
26. L. Luo, Y. Fan, and D. Tondeur, Heat exchangers: From micro- to multi-scale design optimization. *Int J Energy Research*, Vol. 31, 2007, pp. 1266–1274.
27. L. Luo, Y. Fan, W. Zang, X. Yuan, and N. Midoux, Integration of constructal distributors to a mini crossflow heat exchanger and their assembly configuration optimization. *Chemical Engineering Science*, Vol. 62, 2007, pp. 3605–3619.
28. R. S. Matos, T. A. Laursen, J. V. C. Vargas, and A. Bejan, Three-dimensional optimization of staggered finned circular and elliptic tubes in forced convection. *Int J Thermal Sciences*, Vol. 43, 2004, pp. 477–487.
29. T. Bello-Ochende, L. Liebenberg, and J. P. Meyer, Constructal cooling channels for microchannel heatsinks. *Int J Heat Mass Transfer*, Vol. 50, 2007, pp. 4141–4150.
30. J. V. C. Vargas, A. Bejan, and D. L. Siems, Integrative thermodynamic optimization of the crossflow heat exchanger for an aircraft environmental control system. *J Heat Transfer*, Vol. 123, 2001, pp. 760–769.
31. J. V. C. Vargas and A. Bejan, Thermodynamic optimization of finned crossflow heat exchangers for aircraft environmental control systems. *Int J Heat Fluid Flow*, Vol. 22, 2001, pp. 657–665.
32. A. Bejan, Constructal theory: Tree-shaped flows and energy systems for aircraft. *J Aircraft*, Vol. 40, 2003, pp. 43–48.
33. M. Falempe, Une plate-forme d'enseignement et de recherche du procédé de cogeneration chaleur-force par voie de vapeur d'eau. *Revue Générale de Thermique*, Vol. 383, 1993, pp. 642–651.
34. M. Falempe and B. Baudoin, Comparaison des dix méthodes de résolution des réseaux de fluides à usages énergétiques. *Revue Générale de Thermique*, Vol. 384, 1993, pp. 669–684.
35. A. Barreau and J. Moret-Bailly, Présentation de deux méthodes d'optimisation de réseaux de transport d'eau chaude à grande distance. *Entropie*, Vol. 75, 1977, pp. 21–28.
36. B. Plaige, Le chauffage urbain en Pologne. *Chauffage Ventilation, Conditionnement d'Air*, Vol. 9, 1999, pp. 19–23.
37. W. Wechsatol, S. Lorente, and A. Bejan, Development of tree-shaped flows by adding new users to existing hot water pipes. *Int J Heat Mass Transfer*, Vol. 45, 2002, pp. 723–733.
38. J. Padet, *Fluides en Écoulement, Méthodes et Modèles*. Paris: Masson, 1991.
39. D'.A.W. Thompson, *On Growth and Form*. Cambridge, UK: Cambridge University Press, 1942.
40. D. L. Cohn, Optimal systems: I. The vascular system. *Bull Math Biophys*, Vol. 16, 1954, pp. 59–74.
41. E. R. Weibel, *Morphometry of the Human Lung*. New York: Academic Press, 1963.
42. N. MacDonald, *Trees and Networks in Biological Models*. Chichester, UK: Wiley, 1983.
43. R. J. Chorley, S. A. Schumm, and D. E. Sugden, *Geomorphology*. London: Methuen, 1984.
44. I. Rodriguez-Iturbe and A. Rinaldo, *Fractal River Basins*. Cambridge, UK: Cambridge University Press, 1997.
45. P. Meakin, *Fractals, Scaling and Growth Far from Equilibrium*. Cambridge, UK: Cambridge University Press, 1998.

PROBLEMS

7.1. A panel of tubes in a steam superheater consists of three parallel tubes, which are supplied with T_0 steam from a bottom plenum, Fig. P7.1. The heated streams are collected in the upper plenum and discharged at the temperature T_{out}. The total heating rate ($q = 2q_1 + q_2$) is fixed: it is provided by flames from a burner. The total mass flow rate ($\dot{m} = 2\dot{m}_1 + \dot{m}_2$) is also fixed. Because the burner is not as wide as the three-tube assembly, the inner tube receives a greater share of the heat transfer rate, $q_2/q_1 > 1$. Close inspection of the design shows that a system of bottom valves is used in order to adjust the tube flow rates ($\dot{m}_2/\dot{m}_1$) such that all the tube outlets have the same temperature ($T_1 = T_2$).

(a) Is the maintenance of $T_1 = T_2$ beneficial for the purpose of improving (i) thermodynamic performance, (ii) mechanical integrity, or (i) and (ii)?

(b) If the heat transfer rate imbalance is $q_2/q_1 = \alpha$, what is the required mass flow rate $\dot{m}_2/\dot{m}_1$ as a function of α?

(c) It is proposed to eliminate the system of valves, and to maintain the required flow imbalance $\dot{m}_2/\dot{m}_1$ by using an inner tube with a diameter larger than the diameter of the outer tubes. Note that without the valves the plenum-to-plenum pressure drop is the same for all the tubes. Assume that the flow through each tube is in the fully developed fully rough turbulent regime. Determine the required ratio D_2/D_1 as a function of α.

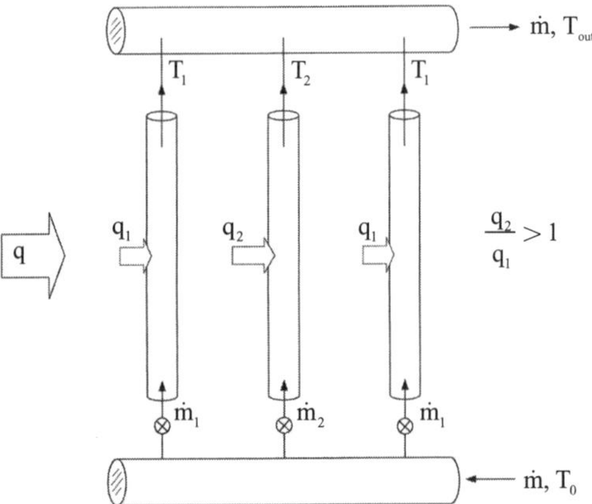

Figure P7.1

7.2. Derive the equations above Eq. 7.85. As a guide, use the analysis presented in Ref. [11].

8

VASCULARIZED MATERIALS

8.1 THE FUTURE BELONGS TO THE VASCULARIZED: NATURAL DESIGN REDISCOVERED

The current literature reviewed in Refs. [1] through [3] reveals a surge of interest in bio-looking designs of flow architectures that promise superior properties, for example, distributed and high-density heat and mass transfer. Chief among the new architectures that are being proposed are the tree-shaped (dendritic) designs. A significant stimulus for this new direction is the emergence of constructal theory as a means to explain biological and geophysical design, and as a method for developing new concepts for engineered flow architectures.

Tree-shaped flow structures have multiple scales that are distributed nonuniformly through the flow space. Tree flows are everywhere in natural flow systems, and their occurrence can be deduced based on a physics principle (the constructal law: for a flow system to persist in time [to survive] it must evolve in such a way that it provides easier and easier access to the currents that flow through it). The constructal law has become an addition to the thermodynamics of nonequilibrium systems: the thermodynamics of flow systems with configuration [4, 5].

Natural porous flow structures also exhibit multiple scales and nonuniform distribution of length scales through the available space. Can such heterogeneous flow structures be derived from the same principle of maximization of flow access?

Several classes of natural flow configurations are fingerprints of the action of the constructal law. For example, tubes with round cross section are the constructal architecture for flowing from point to point (section 3.1). The corresponding constructal architecture for flowing from one point (source or sink) to an infinity of points (line, area, and volume) is a tree (Chapter 4). The tree architecture has multiple scales, which are arranged hierarchically and distributed nonuniformly through the available flow space. The large channels are few and the small channels are many. Nonuniform distribution of scales means that on the available flow domain there is one location for the trunk and another for the canopy. The two locations are highly distinct.

The drive to discover dendritic flow architecture is fueling a *vascularization revolution* in the development of smart materials. Future smart materials promise entirely new functionality (e.g., self-healing, self-cooling [6–11]) which rests squarely on the ability to bathe volumetrically—at every point—solid bodies that perform more traditional functions (mechanical loading, sensing, actuating, morphing). Constructal design concepts serve the vascularization needs of new smart structures ideally, because trees are the architectures for maximum access between a point and a volume. If a single stream is to touch a volume at every point, that stream must invade the volume as a tree, in the same way that the bronchial tree supplies air to the entire lung volume. In turn, vascularized smart materials give constructal design new purpose: tree designs will have to be configured in ways that serve optimally more than one objective.

8.2 LINE-TO-LINE TREES

In this section we explore the properties and performance of a novel dendritic flow architecture proposed in Ref. [12]. The vision is to connect two parallel lines (or two parallel planes) with trees that alternate with upside-down trees (Fig. 8.1). The resulting dendritic pattern connects the bottom boundary of the flow domain with the top boundary (Fig. 8.2).

Figure 8.1 Trees that alternate with upside down trees form a dendritic structure that connects one plane with another parallel plane. See also Color Plate 2.

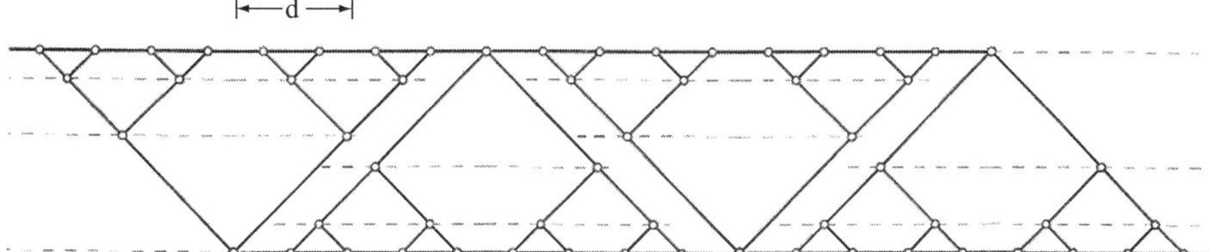

Figure 8.2 Tree architecture for connecting the points of one line with the points of another line [12].

This alternating sequence of point-to-line trees constitutes a vasculature between the two parallel boundaries of the designed porous body. The fluid flows in the same direction through all the trees, e.g., upward in Fig. 8.2. This type of vascularization (line-to-line trees) establishes a multiscale *designed porous medium* between the long parallel boundaries of the vascularized body.

The maximization of flow access between the points of one line and the points of a parallel line can be viewed as a sequence of point-to-line flow access maximization problems. The building block with which Fig. 8.2 is constructed was proposed by Lorente et al. [13], where it was based on optimally shaped rectangular areas, as shown in Fig. 4.13 in section 4.5. Because the pressure drop is proportional to the duct length, the rectangular shape d/c was chosen such that the length of the duct PQ that cuts across the fixed area A_0 is minimum. This yielded the shape $d/c = 2$, which led to 90-degree angles between tributaries, and to collinear ducts on the extremities of the V-shaped tree structure.

The 90-degree angles deduced in Fig. 4.13 are an approximation of the best "equilibrium flow structure" [4] that could be traced between one point and the many points of a line. To see this, consider the building block described in Problem 4.2 and abandon the assumption that the stem L_1 and the extreme branch L_2 are collinear. In general, α_1 is not the same as α_2. The Y construct of Fig. P4.2 is equivalent to a construct that is two-layers thick in Fig. 4.13.

Assume further that all the tubes are round and with Poiseuille flow, and that the svelteness number Sv is high such that pressure losses at the junctions can be neglected. In this case, the minimization of the pressure drop across the entire Y-shaped construct (subject to fixed total tube volume) yields the well known Hess-Murray law, according to which the ratio of successive tube diameters is $D_1/D_2 = 2^{1/3}$, regardless of the way in which the tubes are arranged on the area S.

The optimization of the tube layout is next, and is subjected to holding the area S fixed, while the shape of the area S may vary. After the optimized D_1/D_2 ratio is substituted into the global pressure drop expression, the global pressure drop is proportional to the geometric flow-resistance group

$$R = L_1 + 2^{1/3} L_2 \tag{8.1}$$

There are two degrees of freedom in the morphing of Fig. P4.2, the angles α_1 and α_2, or one angle and the aspect ratio of S. The minimization of R subject to $S =$ constant is performed numerically, and the results are

$$\alpha_1 = 40.86^\circ \qquad \alpha_2 = 53.11^\circ \tag{8.2}$$

and the minimized global resistance factor is

$$R = 1.3235\, S^{1/2} \tag{8.3}$$

How much better is the bent-Y structure (Fig. P4.2) relative to the 45-degree structure of Fig. 4.13? The answer is readily obtainable from Eq. (8.1), in which we substitute $L_2 = L_1/2$ and, after some algebra, $L_1 = (2\, S/3)^{1/2}$:

$$R = (1 + 2^{-2/3})(2/3)^{1/2} = 1.3309\, S^{1/2} \tag{8.4}$$

Comparing Eqs. (8.3) and (8.4), we see that the performance of the simpler (minimum-length) structure of Fig. 4.13 approaches within 0.5 % the performance of the more flexible structure optimized in Fig. P4.2. Such conclusions are reached often in constructal theory: nonequilibrium flow architectures come reasonably close to the equilibrium flow architectures. They come close in terms of performance, even though they may look different.

Because of the comparison made above, we retain the simpler building blocks sketched in Fig. 4.13, and with them we explore several ways in which to maximize line-to-line flow access (Fig. 8.2). Unlike in Fig. 4.13, we assume a large number of bifurcation levels ($i = 1, 2, \ldots, n$), as shown in Fig. 8.3. Because of symmetry about the bisector of the 2α angle, the tube lengths decrease by a factor of 1/2, from the largest (L_0), to $L_1 = L_0/2$, $L_2 = L_1/2$, etc. The smallest length scale is the smallest tube length,

$$L_n = 2^{-n}\, L_0 \tag{8.5}$$

or the distance between the two ends of two neighboring L_n tubes,

$$d = 2\, L_n \sin\alpha \tag{8.6}$$

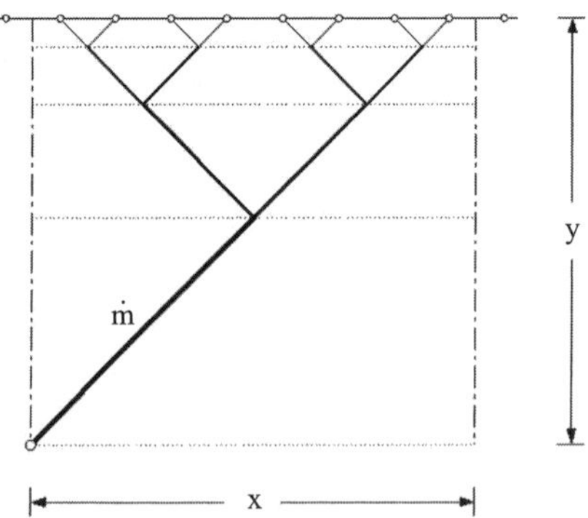

Figure 8.3 One of the point-to-line trees of Fig. 8.2 [12].

In this analysis, we carry α as a parameter, although according to the preceding discussion the value of α should be 45 degrees.

The pressure drop along one tube of length L_i and diameter D_i is

$$\Delta P_i = \dot{m}_i \frac{8}{\pi} \text{Po}\nu \frac{L_i}{D_i^4} \tag{8.7}$$

where Po is the Poiseuille constant (e.g., Po = 16 for round tubes), which appears in the formula for the friction factor,

$$f_i = \frac{\text{Po}}{\text{Re}_{D_i}} \tag{8.8}$$

and $\text{Re}_{D_i} = U_i D_i/\nu$ with $U_i = \dot{m}_i/(\rho\pi\, D_i^2/4)$. Mass conservation at every junction requires that $\dot{m}_i = 2\dot{m}_{i+1}$, where it is again assumed that the tubes are sufficiently slender so that the asymmetry of the Y junction does not affect the splitting of $\dot{m}_i$ into two equal streams $\dot{m}_{i+1}$. After using the ratios for diameters, lengths and mass flow rates indicated above, the total pressure drop from the open end of the L_0 tube to the open ends of the L_n tubes, becomes

$$\Delta P = \sum_{i=1}^{n} \Delta P_i = \dot{m}_0 \frac{8}{\pi} \nu \text{Po} \frac{L_0}{D_0^4} \sigma_1 \tag{8.9}$$

where $\sigma_1 = 1 + 2^{-2/3} + \cdots + 2^{-2n/3} = [1 - (2^{-2/3})^{n+1}]/(1 - 2^{-2/3})$. The total tube volume occupied by the tree flow is

$$V = \frac{\pi}{4}(D_0^2 L_0 + 2D_1^2 L_1 + \cdots + 2^n D_n^2 L_n) = \frac{\pi}{4} D_0^2 L_0 \sigma_1 \tag{8.10}$$

The largest length scale (L_0) is related to the vertical dimension of the tree (y) by

$$y = (L_0 + L_1 + \cdots + L_n)\cos\alpha = L_0 \sigma_2 \cos\alpha \tag{8.11}$$

where $\sigma_2 = 2\,[1 - 2^{-(n+1)}]$. The horizontal dimension (x) of the area occupied by the tree projection is

$$x = 2^n d = 2\, L_0 \sin\alpha = \frac{2}{S_2} y \tan\alpha \tag{8.12}$$

where 2^n is the number of L_n tubes that reach the upper end of the construct. Eliminating L_0 and D_0 between Eqs. (8.9) through (8.11), we obtain

$$\Delta P = \dot{m}_0 \frac{\pi}{2} \text{Po} \frac{\nu}{V^2} \left(\frac{\sigma_1 y}{\sigma_2 \cos\alpha} \right)^3 \tag{8.13}$$

We question how effective the tree structure of Fig. 8.3 is relative to a well-known reference architecture: an array of N equidistant parallel tubes, each of length y and diameter D. This classical structure carries the same total flow rate $\dot{m}_0$ in the same total tube volume ($V = N \frac{\pi}{4} D^2 y$) and over the same area $xy/2$. The structure has one degree of freedom, the tube diameter D, or the number of parallel tubes:

$$N = \frac{4V}{\pi D^2 y} \tag{8.14}$$

The pressure drop along this structure (ΔP_{ref}) is the same as the pressure drop along a single tube, cf. Eq. (8.7), through which the flow rate now is $\dot{m}_0/N$,

$$\Delta P_{\text{ref}} = \frac{\dot{m}_0}{N} \frac{8}{\pi} \text{Po}\nu \frac{y}{D^4} \tag{8.15}$$

Eliminating D by using Eq. (8.14) we obtain

$$\Delta P_{\text{ref}} = \dot{m}_0 \frac{\pi}{2} \text{Po} \frac{\nu}{V^2} N y^3 \tag{8.16}$$

The tree-shaped structure of Fig. 8.3 has a smaller flow resistance than the parallel channels when $\Delta P < \Delta P_{\text{ref}}$, or, using Eqs. (8.13) and (8.16), when

$$N > \left(\frac{\sigma_1}{\sigma_2 \cos \alpha} \right)^3 \tag{8.17}$$

The right side of this inequality is a number on the order of 1. In conclusion, as the reference structure becomes finer (i.e., as N increases), the tree-shaped design of Fig. 8.3. becomes more attractive.

This conclusion can be read as a statement of how fine the tree structure must be such that it is preferable to the reference design. For a more practical comparison, assume that the smallest dimension that can be manufactured (d) is the same in both architectures; that is, the d spacing of Fig. 8.3 is the same as the spacing between parallel tubes. This means that the number of parallel channels that occupy the area $y \times (x/2)$ is $N = 2^n/2$, and when $\alpha = 45$ degrees the inequality (8.17) becomes approximately

$$\frac{\Delta P}{\Delta P_{\text{ref}}} \cong \frac{14}{2^n} < 1 \tag{8.18}$$

We conclude that when the number of branching levels is four or larger, the tree-shaped architecture offers greater access to the flow that permeates through the porous structure of thickness y. The superiority of the tree design increases fast as n increases: when $n = 7$, the ratio $\Delta P/\Delta P_{\text{ref}}$ is as low as 1/10.

8.3 COUNTERFLOW OF LINE-TO-LINE TREES

In this section we show how line-to-line trees can be used for configuring counterflow heat exchangers with greater volumetric density of heat transfer [14]. The line-to-line trees form a periodic pattern on the elemental volume $X \times Y \times Z$, where Z is the dimension of the solid body in the direction perpendicular to the plane $X \times Y$. This elemental volume is shown in greater detail in the lower part of Fig. 8.4. We analyze the performance of a vasculature consisting of two $X \times Y$ elemental

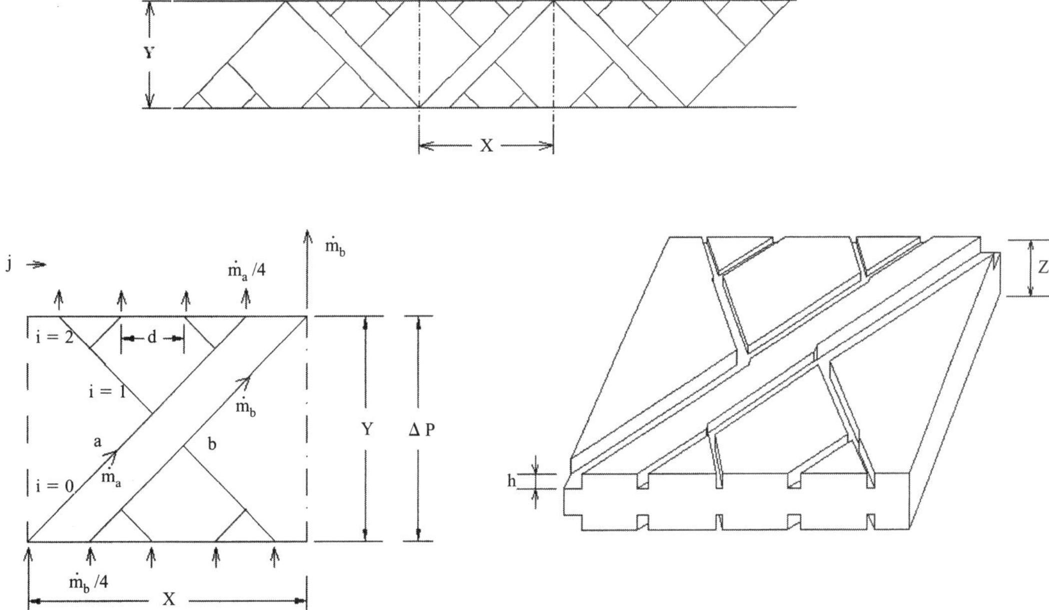

Figure 8.4 Line-to-line tree channels in counterflow. Each tree has two levels of pairing or bifurcation ($n = 2$), and, consequently, the flow rate through the trunk is approximately four times the flow rate through each of the smallest channels of the canopy [14].

flow structures, one machined into the upper surface of the *XYZ* volume, and an identical line-to-line structure machined into the underside of the *XYZ* volume. We then compare Fig. 8.4 with the corresponding design in which the line-to-line trees are replaced by parallel and equidistant channels (Fig. 8.5).

The smallest length scale of the flow architecture (d) is assumed specified (fixed). This length is the distance between two adjacent ports on the line boundary of the vascularized body. By fixing d, we fix the appearance of the vascularized body when seen from the outside.

From the point of view of manufacture, square or rectangular cross-sectional channels (e.g., Figs. 8.4 and 8.5) can be machined more conveniently into conducting materials. The problem of using square-channel cross-sections is that after geometrical optimization the depths of fluid channels at different pairing levels are different (S_i, S_{i-1}, Fig. 8.6a), which leads to additional difficulty and cost in manufacture. Rectangular cross-sections (Fig. 8.6b) that have the same depth become the preferable choice in design. In this case, the depth h is an additional degree of freedom.

The analysis focuses on the tree network on one face of the vascularized slab. For simplicity, in the first two drawings of Fig. 8.4 we indicated only the centerlines of

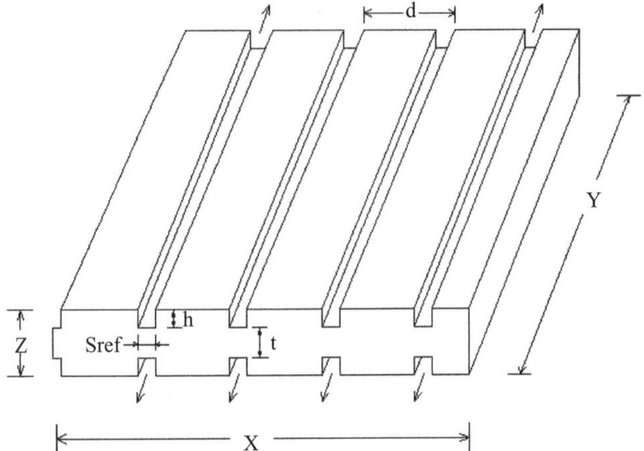

Figure 8.5 Volume element vascularized with two sets of parallel channels in the counterflow [14].

the flow channels, without displaying the cross-sectional dimensions. In a real tree-shaped design, there are special features (bifurcations, junctions) that may cause additional flow resistance as well as flow nonuniformity. These "local losses" are neglected in this analysis. As shown in section 1.2.2, the local junction and bifurcation losses can be neglected when the svelteness of the tree architecture [15]

$$\mathrm{Sv} = \frac{Y}{V_f^{1/3}} \tag{8.19}$$

is greater than 10 in an order of magnitude sense. Here V_f represents the total volume occupied by the flow channels on one $X \times Y$ face of the element, such that $V_f^{1/3}$ accounts for the internal length scale of the flow architecture. The svelteness is the ratio of the external flow length scale (Y) divided by the internal flow length scale. The effect of local losses is reduced further by smoothing the transitions between subsequent channels of different sizes.

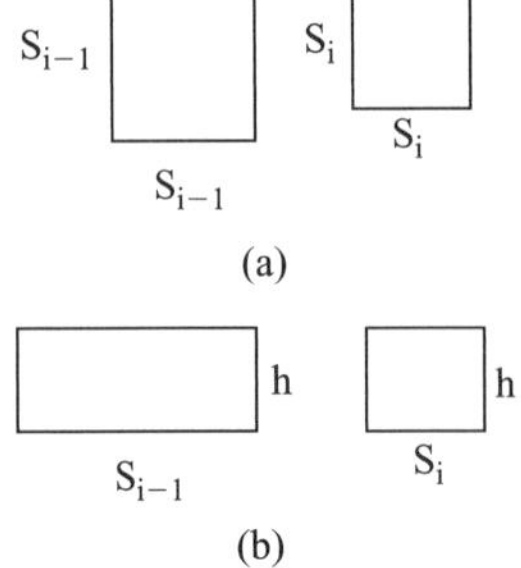

Figure 8.6 The change in channel cross-section size from mother to daughter channels: (a) channels have the same shape (square); (b) channels have the same depth (h) [14].

Four constraints are specified: the total volume $V = XYZ$, the total flow volume V_f, the elemental length scale d, and h. For a given number of pairing levels (n) and smallest length scale (d), the dimensions of the $X \times Y$ face follow from the geometric relations

$$X = (2^n + 1/2)d \quad Y = (2^n - 1/2)d \tag{8.20}$$

Furthermore, because $V = XYZ$, to fix V means to fix Z. The flow volume V_f constraint may also be expressed in dimensionless form as the fixed porosity of the composite material,

$$\phi = \frac{V_f}{V} \tag{8.21}$$

According to the Hess-Murray law, if all the channel cross-sections are round, then the ratio of successive tube diameters is $2^{1/3}$. If all the channels have square cross-sections (side S_i, Fig. 8.6a), then the ratio of successive square dimensions is

$$C_s = S_{i-1}/S_i = 2^{1/3} \tag{8.22}$$

where $i = 1, 2, \ldots, n$. This also means that successive square cross-sectional areas change by a factor of $2^{2/3}$. Because the channel cross-sections are rectangular ($S_{i-1} \times h, S_i \times h, \ldots$, Fig. 8.6b), the ratio C_S has to be optimized, and its value is close to $2^{2/3}$[14].

As in the preceding section, 90 degree angles between tributaries are adopted because they are an adequate approximation of the best architecture. From this we obtain

$$L_{i-1}/L_i = 2 \tag{8.23}$$

where $i = 1, 2, \ldots, n$, and

$$L_n/d = 2^{-1/2} \tag{8.24}$$

The smallest thickness of the solid between two channels (t, Fig. 8.5) is another constraint, which is set by mechanical strength requirements. The depth h is related to t by

$$h = \frac{Z - t}{2} \tag{8.25}$$

If $t_{\min}$ is the allowable minimum solid thickness, then Eq. (25) yields $h_{\max}$ that is the maximum range of values that the channel depth h can have.

The analysis is based on assuming fully developed laminar flow, negligible pressure drop at junctions, and fluid with constant properties. The flow in the tree-shaped channels can be oriented in two directions, either from root to canopy (tree "a") or from canopy to root (tree "b"). The assumption of negligible junction losses means that the flow direction does not matter, therefore $\dot{m}_a = \dot{m}_b$. To calculate the resulting flow rate when ΔP is specified, it is sufficient to analyze only one tree, namely tree "a" (Fig. 8.4). Mass conservation requires

$$\dot{m} = \dot{m}_a + \dot{m}_b = 2\dot{m}_a \tag{8.26}$$

where $\dot{m}$ is the total mass flow rate through the vascularized system of width X, and $\dot{m}_a$ is the flow rate through one tree. The pressure drop of fully developed laminar flow along one channel of length L_i, width S_i and depth h is

$$\Delta P_i = \frac{\nu}{2}\dot{m}_{ai}\frac{\text{Po}_i L_i (S_i + h)^2}{S_i^3 h^3} \tag{8.27}$$

where Po_i is the Poiseuille constant for fully developed laminar flow:

$$\left(f\text{Re}_{D_h}\right)_i = \text{Po}_i \tag{8.28}$$

The index i is the rank of the channel in the tree hierarchy: $i = 0$ is for the trunk, and $i = n$ for the most peripheral channels in the canopy. The Poiseuille constant depends only on the shape of the channel cross-section. In the present case (Fig. 8.6b), the value of Po_i is a function of the ratio of S_i/h, and is provided by Ref. [16]. The total pressure drop from $y = 0$ to $y = Y$ is

$$\Delta P = \sum_{i=0}^{n} \frac{\nu}{2}\dot{m}_{ai}\frac{\text{Po}_i L_i (S_i + h)^2}{S_i^3 h^3} \tag{8.29}$$

The flow rate is the same through channels of the same rank, therefore $C_S = S_{i-1}/S_i$ and Eqs. (8.23) and (8.29) yield

$$\dot{m}_a = \frac{2\,\Delta P\, S_0^3\, h^3/(\nu L_0)}{\sum_{i=0}^{n} 2^{-2i}\,\text{Po}_i\, C_S^{3i}\,(C_S^{-i}\, S_0 + h)^2} \tag{8.30}$$

The trunk width S_0 is determined from the flow volume constraint

$$V_f = 4\sum_{i=0}^{n} 2^i L_i S_i h \tag{8.31}$$

which can be further written using the C_S definition and Eq. (8.23):

$$V_f = 4L_0 S_0 h \frac{1 - C_s^{-(n+1)}}{1 - C_s^{-1}} \tag{8.32}$$

The reference flow structure consists of parallel channels with the same ΔP, V and V_f. The mass flow rate is

$$\dot{m}_{\text{ref}} = \left(2^n + \frac{1}{2}\right)\frac{2\,\Delta P\, S_{\text{ref}}^3 h^3}{\nu \text{Po}_{\text{ref}} Y (S_{\text{ref}} + h)^2} \tag{8.33}$$

where the reference width S_{ref} is determined from

$$V_f = 2\left(2^n + \frac{1}{2}\right) Y S_{\text{ref}} h \tag{8.34}$$

The imposed pressure drop ΔP is a fixed and specified parameter because in microchannel technology the driving pressures have reached levels so high that they threaten the mechanical integrity of the sandwiches of multiple plates with microchannels. A certain pressure drop is also associated with the pump characteristics (e.g., size) that are required by the larger system in which the vascularized solid is used. In sum, the maximum pressure (or ΔP) plays in the fluid mechanics of microchannels the same role that the maximum temperature (hot spot) plays in cooling with microchannels. Traditionally, microchannels were analyzed and designed by assuming that the fluid flow rate is given (specified). The importance of the fixed ΔP as a design and operational parameter has become clear relatively recently [17, 18], when it was shown that the overall pressure drop ΔP can be nondimensionalized as the group [cf. Eq. (3.35)]:

$$\text{Be} = \frac{\Delta P Y^2}{\mu \alpha} \tag{8.35}$$

The highest pressure (hence ΔP) is a limiting parameter in the direction of miniaturizing channels with forced flow. The maximum pressure is dictated by the mechanical strength of the seals in the flow network. Even at lower pressures ΔP (and not $\dot{m}$) is the likely constraint, for example, in air cooled packages where the fan is located downstream of the package (in this case ΔP cannot exceed 10^5 Pa). The relative flow resistance of a tree-shaped flow structure is represented by the ratio

$$r = \frac{\dot{m}}{\dot{m}_{\text{ref}}} \tag{8.36}$$

If $r > 1$, trees provide greater flow access than parallel channels, and are more attractive.

Equations (8.30) and (8.33) show that for constant properties and fixed pressure drop the total mass flow rate is only a function of geometry. The geometrical parameters are d, n, Z (or V), ϕ, C_S, and h. Among these, the factor C_S accounts for the step change in cross-sectional area between successive pairing levels, which will affect the pressure drop distribution among the various pairing levels. The variation of h will change the shape of the cross-sections of fluid channels. Because the square cross sections have a smaller C_f compared with the rectangular cross-sections, it is a good idea to make S_i/h approach 1 as closely as possible. There is a trade-off between C_S and h, which yields minimum global resistance for tree-shaped architecture. Note that for a fixed Z, to vary h means to vary the mechanical strength of the wall between two channels. For parallel channels with specified ϕ, the flow rate is the largest when h is selected to form square cross-sections. In

Table 8.1 Parameters used in the analysis of section 8.3 [14]

Case	d (mm)	Z (mm)	n	ϕ	C_S	h (mm)
A	10	2.585	2	0.05	1.3–1.8	0.3, 0.5, 0.7, 0.9
B	10	2.98	3	0.05	1.3–1.8	0.4, 0.7, 1.0, 1.2
C	10	3.388	4	0.05	1.3–1.8	0.5, 0.8, 1.1, 1.4
D	10	4.982	5	0.05	1.3–1.8	0.6, 1.0, 1.4, 1.8, 2.2
E	10	3	2	0.02–0.11	1.587	0.2, 0.4,0.6,0.8,1.0,1.2
F	10	3	3	0.02–0.11	1.587	0.2, 0.4,0.6,0.8,1.0,1.2
G	10	3	4	0.02–0.11	1.587	0.2, 0.4,0.6,0.8,1.0,1.2
H	10	3	5	0.02–0.11	1.587	0.2, 0.4,0.6,0.8,1.0,1.2

summary, tree-shaped structures have more degrees of freedom to morph (n, C_S) than parallel channels.

In Ref. [14] we investigated the effect of geometry (n, ϕ, C_S, h) on the relative flow access ratio r, when the pressure drop number is held constant at $\text{Be} = 10^6$. Table 8.1 summarizes the parameters used in the analysis. Figure 8.7a shows the effect of ϕ on r for $n = 2$ and several channel depths h. Here we cannot discern a trend: r increases or decreases depending on the values of ϕ and h. But the changes in r are quite small.

Figure 8.7b shows the effect of ϕ on r for $n = 3$ and several channel depths h. When h is very small, which means more slender cross-sections and higher C_{fi}, the effect of ϕ on r is very limited. In the range $h = 0.2$–1.2 mm, r decreases with the increase in ϕ, although when $h = 0.2$ mm, the variation of r is quite weak. If we calculate the corresponding svelteness number Sv, Eq. (8.19), we find that in the range $h = 0.2$–1.2 mm, the ratio r is large when the svelteness number is large and the porosity is small. Tree-shaped structures are preferable when space is at a premium, too precious to be allocated to flow channels (e.g., $\text{Sv} > 9.72$ for $h = 1$ mm, and $\text{Sv} > 8.62$ for $h = 1.2$ mm). When h is very small, the effect of Sv on r is weak. A design with a very small h will cause large flow resistance due to the large slenderness of the channel cross-sections.

Figure 8.7b also shows that a design domain with $r > 1$ emerges when h is large and ϕ is small. Here tree-shaped structures offer greater access to the flow than parallel channels. The attractiveness of tree-shaped structures depends not only on the pairing number n but also on porosity ϕ and depth h (or the cross-sectional shape).

Figures 8.7c and 8.7d show the effect of ϕ on r, for $n = 4$ and 5. The conclusions are the same as in Fig. 8.7b; that is, two regions exist, one with $r > 1$ and the other with $r < 1$. The new aspect is that when n increases, the region with $r > 1$ grows, and the achievable maximum r becomes larger. The larger the pairing number n, the smaller the resistance of the tree-shaped structure relative to parallel channels. This is in agreement with the conclusion of section 8.2. Note, however, that even

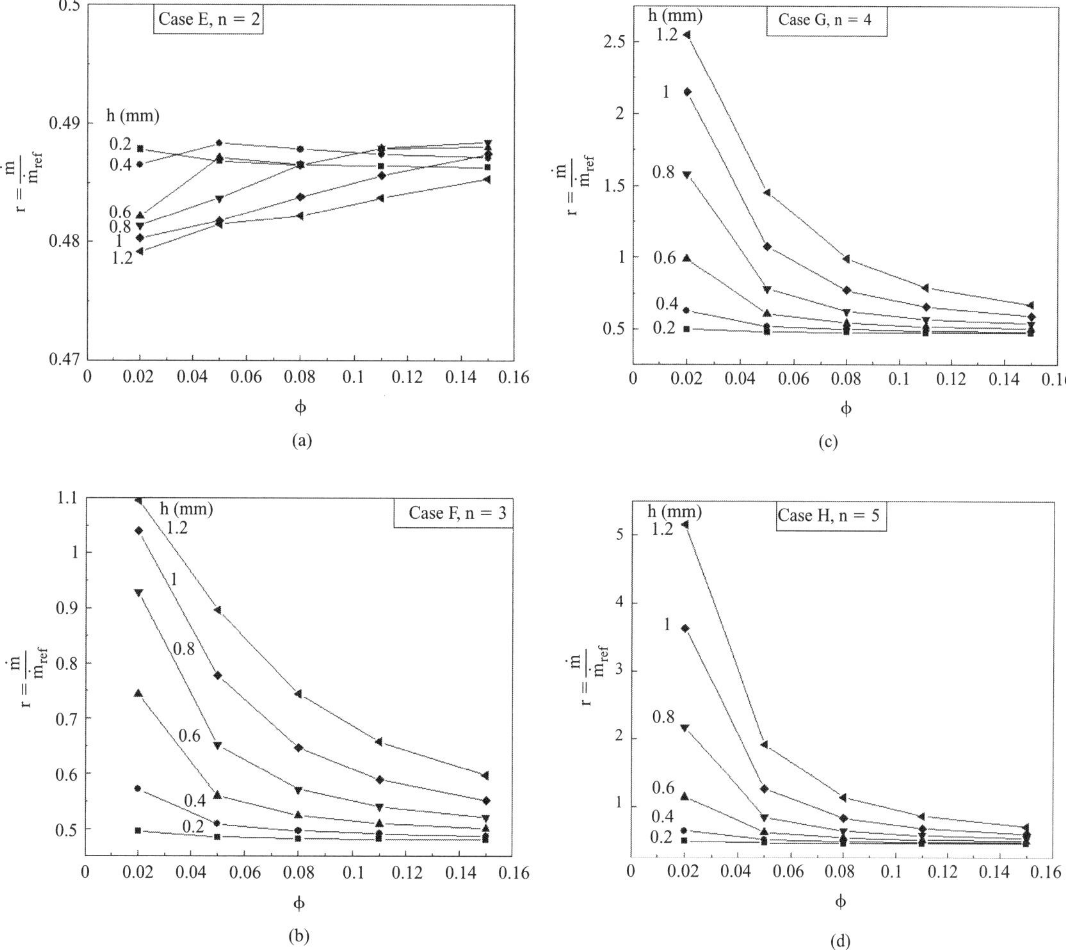

Figure 8.7 The effect of tree architecture porosity (ϕ) and channel depth (h) on the optimized ($C_s = 2^{2/3}$) global flow rate $\dot{m}$ [14].

when n is large (Fig. 8.7d) trees perform worse than parallel channels ($r < 1$) if an improper h or ϕ is used.

In the above analysis, the upper limit for h is ruled by Eq. (8.25). In an actual design, Z and $t_{\min}$ are constraints that determine $h_{\max}$, although changes in Z or t do not alter the above conclusions. If h is too large, it leads to slender cross-sections for all fluid channels, and the flow rate is small in both tree-shaped structures and parallel channels. In that limit even a larger r may not be a good design in practice. To illustrate this, Fig. 8.8 gives the mass flow rates for case G of Table 8.1. When $\phi = 0.02$ and 0.05 (for parallel channels), we find the $h_{\max}$ and $h_{\mathrm{ref\,max}}$ values that correspond to maximum flow rates: $\dot{m}_{\max}$ for tree-shaped

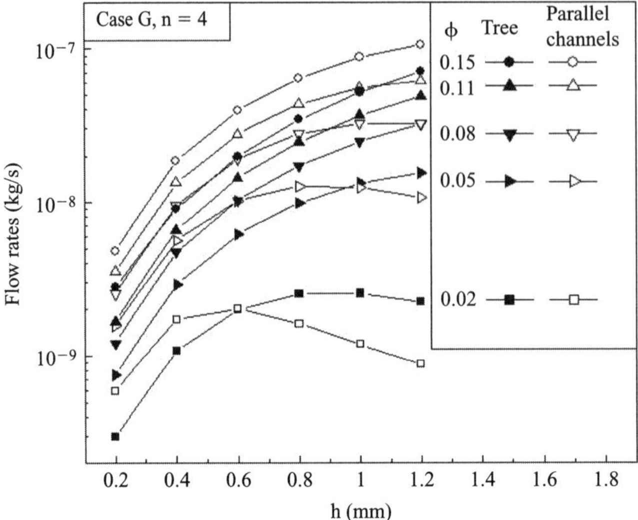

Figure 8.8 The effect of channel depth (h) and porosity (ϕ) on the flow rates through the trees and parallel channels structures [14].

structures and $\dot{m}_{\text{ref max}}$ for parallel channels. When $h > h_{\text{max}}$ and $h > h_{\text{ref max}}$, an increase in h will lead to decreases in both $\dot{m}_{\text{max}}$ and $\dot{m}_{\text{ref max}}$, although r is still increasing. Due to the constraints posed by Z and t_{min}, the values of h_{max} or $h_{\text{ref max}}$ may equal the largest h given by Eq. (8.25), as in the cases shown in Fig. 8.8 for $\phi = 0.05$ (for trees), 0.08, 0.11, or 0.15. For parallel channels, $h_{\text{ref max}}$ corresponds to a square cross-section. When h becomes much larger than $h_{\text{ref max}}$, the flow rate will decrease significantly. For tree-shaped structures, an h that is too large (when $h > h_{\text{max}}$) is detrimental to flow access. In conclusion, in design it is recommended to monitor both the dimensionless mass flow rate r and the dimensional mass flow rate.

The comparison between the configurations of Figs. 8.4 and 8.5 was continued in Ref. [14] for water flowing through the network by using fully numerical simulations. Three-dimensional laminar flow was assumed in the entire tree architecture and its many bifurcations. It was shown that tree vascularization is more attractive than parallel channels when the number of bifurcation levels increases, the global porosity of the vascularized body decreases, and the global svelteness (Sv) of the flow architecture increases. The nonuniformity (maldistribution) of flow rates through ramifications of the same rank becomes nonnegligible when the pressure drop number Be exceeds 10^9. The optimal step in the sizes of cross-sectional areas from one channel to the channel of the next rank is closely approximated by $2^{2/3}$ even when the cross section is not square or round. The agreement between analysis and numerical simulation and optimization was good.

8.4 SELF-HEALING MATERIALS

The new technological development that stimulated the work outlined next is the invention of structural composite materials with distributed (encapsulated) healing fluids such as epoxy [6,7]. When the material is overworked and tiny cracks appear in it, the tiny capsules break and the healing agent fills and fuses the cracks. It was demonstrated that after a certain healing time the composite regains its mechanical strength properties.

The filling of cracks with fluid from microcapsules is a one-time healing process. The future development of self-healing composites calls for the use of *networks* of healing agent, so that the entire volume of the structural composite is protected against volumetric cracking. The crack position is not known a priori; in fact, several cracks may form at different sites simultaneously and repeatedly. One-time healing is not the solution. Needed is a network that is configured and distributed optimally through the composite so that it provides healing agent where and when self-healing is needed. These composites are to be designed especially for high porosity so that the healing fluid can be delivered effectively.

The function of the flow architecture is to provide flow access between a volume of composite loaded with healing liquid and one or more sites where cracks may develop. For simplicity, the crack length scale d is assumed known and fixed. The crack site is modeled as a sphere of diameter d, although the crack itself may be shaped as a thin lens (or coin) of diameter d. The orientation of the lens is not known, hence the more conservative d-sphere site model. If the flow network can access the d-sphere, then it will access the crack that resides in the sphere.

The composite material is modeled as a square slab of size $L \times L$ and thickness d, where $d \ll L$. On the $L \times L$ area we must configure a network of channels that contain healing agent. These channels must be close enough to each other so that they intersect the cracks (the d-spheres) wherever they may occur. The volume fraction occupied by all the channels is fixed,

$$\phi = \frac{V_c}{V} = \frac{\text{total channel volume}}{\text{total volume}} \tag{8.37}$$

where V_c is the volume of all the channels and $V = L^2 d$ is the total volume of the composite material. The parameter ϕ is the porosity of the composite material (V) viewed as a designed porous medium.

When one crack forms, the healing liquid from the neighboring network flows into the crack. This is an area-point flow. According to constructal theory (Chapter 4), maximum flow access from the area to the point is provided by a tree-shaped structure in which the crack site is the end (root point) of the tree trunk. This structure can be configured optimally as a tree only if the position of the crack is known a priori.

8.4.1 Grids of Channels

Because the number and locations of the cracks are not known, the need to feed every crack leads to the generation of numerous area-point tree flows, all oriented differently and superimposed on the $L \times L$ plane. This superposition is approximated by a grid with channels of several thicknesses. The generation of the grid architecture is analogous to the emergence of urban grids of streets for city traffic [19,20], which were the first patterns generated based on the constructal law.

The fluid flow is modeled in three ways [8,9]. The more realistic model, which is mentioned briefly at the end of this section, is based on the assumption that the liquid is slightly compressible, pressurized, and stored in the grid channels. The fluid is single phase, isothermal, and with the density ρ at the pressure P. When the grid is charged with this liquid, the pressure is raised to $P + \Delta P$, and the liquid density is $\rho + \Delta\rho$. We assume that $\Delta\rho \ll \rho$, such that $\Delta\rho = c\Delta P$, where c is a constant factor that is proportional to the isothermal compressibility κ of the liquid, $c = \rho\kappa$, where $\kappa = \rho^{-1}\,(\partial\rho/\partial P)_T$. When a crack forms, the pressure at the crack site drops instantly from $P + \Delta P$ to P. The liquid expands, and the installed pressure difference ΔP drives liquid from the surrounding channels into the crack. The compressibility of the fluid is a key parameter that will have to be known before design and implementation begin.

The simpler models are based on the assumption that the fluid is incompressible [8,9]. The flow from the grid to the crack site is approximated as the flow from many (distributed) point sources and one sink. In one model, the point sources are distributed uniformly over the slab area. In the other, the point sources are distributed equidistantly along the perimeter of the slab area. In both models, the grid discharges its fluid into the crack site.

Consider the $L \times L$ domain of Fig. 8.9, which is covered by a grid with square loops $d \times d$. A single crack is located in the center of the large square, and serves as sink. The flow originates from liquid sources placed in the nodes of the grid. If $N = L/d$ is the number of $d \times d$ squares counted in one direction, then the number of sources is $(N + 1)^2 - 1$. The liquid mass flow rate delivered by one source to the grid is a constant, $\dot{m}_s$.

The flow along every channel is in the laminar fully developed (Poiseuille) regime. With reference to the notation defined in Fig. 8.9, if "k" is the channel that links the source (i, j) with the neighboring source $(i + 1, j)$, then the mass flow rate along the k channel is

$$\dot{m}_k = \frac{\pi D_k^4 (P_{i,j} - P_{i+1,j})}{128\,\nu\, d} \tag{8.38}$$

This expression holds for a channel with round cross-section and diameter D_k, but its use is more general. For channels with other cross-sectional shapes, the factor $\pi/128$ is replaced by factors of the same order of magnitude, and D_k is replaced by the hydraulic diameter of the cross-section. The optimized grid architectures

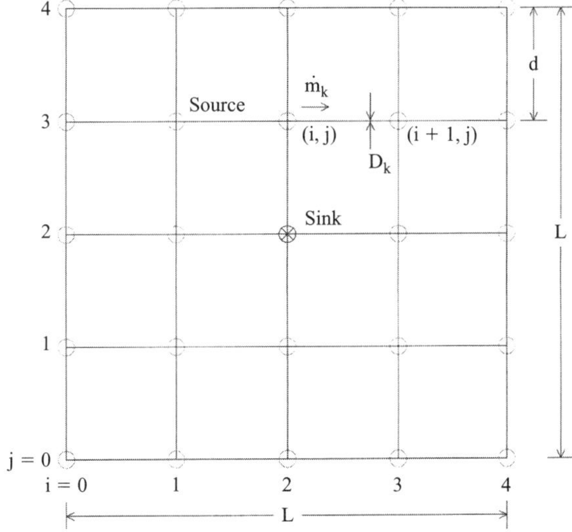

Figure 8.9 Square grid of channels with mass flow sources at every node except the center, which is the sink [9].

developed in Ref. [9] are valid for any cross-sectional shape provided that the cross-sectional shape is the same for all the channels regardless of size.

Another key assumption is that the pressure losses at the junctions are negligible. This is the case when the channels are sufficiently slender so that the entire flow architecture is svelte, Sv $\geq$ 10. In the case of Fig. 8.9, the svelteness Sv is defined as

$$\text{Sv} = \frac{\text{external length scale}}{\text{internal length scale}} = \frac{L}{V_c^{1/3}} = \left(\frac{N}{\phi}\right)^{1/3} \left(\frac{L}{D}\right)^{2/3} \tag{8.39}$$

Grid svelteness is assured when there are many nodes ($L \gg d$) and ϕ is finite.

A steady-state pressure field ($P_{i,j}$) is established when the steady flow produced by all the sources $[(N+1)^2 - 1]\,\dot{m}_s$ flows into the sink. Let $P_{\max}$ be the highest of all the node pressures $P_{i,\,j}$. The sink pressure is $P_{\min}$. The global flow resistance overcome by the grid-to-point flow is $(P_{\max} - P_{\min})/\left[(N+1)^2 - 1\right]\dot{m}_s$. The minimization of this global quantity leads to the discovery of grid architecture.

The work of changing the geometry and calculating the global flow resistance was performed numerically and in dimensionless form. The scale of the mass flow rate through a channel is set by the source strength $\dot{m}_s$, hence the dimensionless mass flow rate through a channel:

$$\tilde{\dot{m}}_k = \frac{\dot{m}_k}{\dot{m}_s} \tag{8.40}$$

The scale of the channel diameter is related to the total channel volume V_c. For example, in Figs. 8.10(1) and (2) we considered the crack positions "1" and "2".

There are two channel sizes, D_1 and D_2, such that

$$\frac{D_2}{D_1} = \gamma < 1 \tag{8.41}$$

The spacing between two channels of the same size is $2d$. When L/d is even, N is even and the number of horizontal and vertical D_1 channels inside the large square is N. Because in Figs. 8.10(1) and (2) the center of the large square is the intersection of two D_2 channels, the total number of D_2 channels is $N + 2$. The length of one channel is $L = Nd$. The total duct volume is

$$V_c = N^2 d \frac{\pi}{4} D_1^2 + (N+2)Nd \frac{\pi}{4} D_2^2 \tag{8.42}$$

If the center of the large square is the intersection of two D_1 channels, as in Figs. 8.10(3) and (4), then D_1 and D_2 switch places in Eq. (8.42). In conclusion, Eq. (8.42) indicates that the scale of D_k is $D_{\text{scale}} = (4V_c/\pi N^2 d)^{1/2}$, such that the dimensionless channel sizes are

$$(\tilde{D}_1, \tilde{D}_2) = \frac{(D_1, D_2)}{(4V_c/\pi N^2 d)^{1/2}} \tag{8.43}$$

After using the dimensionless notation (8.43) and the γ ratio of Eq. (8.41), we can write Eq. (8.42) as

$$\tilde{D}_1 = \left[1 + \gamma^2 \left(1 + \frac{2}{N}\right)\right]^{-1/2} \tag{8.44}$$

This shows how the degree of freedom of the grid (γ) influences the channel size $\tilde{D}_1$ and, from this, $\tilde{D}_2 = \gamma \tilde{D}_1$. Finally, the pressure difference scale follows from Eq. (8.38),

$$\Delta P_{\text{scale}} = \frac{128 \nu \dot{m}_s d}{\pi D_{\text{scale}}^4} \tag{8.45}$$

so that the dimensionless excess pressure over the $L \times L$ field is defined as

$$\tilde{P} = \frac{P - P_{\min}}{\Delta P_{\text{scale}}} \tag{8.46}$$

By writing Eq. (8.38) for all the node-to-node channels of the grid, and adding to this system the equations that express mass conservation at every node, we obtain a system of equations for the pressures at nodes that serve as sources. We solve this system for many grid configurations, and identify the peak excess pressure $\tilde{P}_{\max}$, which is the dimensionless global flow resistance for the entire flow architecture. The starting condition is $\tilde{P}_{i,j} = 1$ at all the nodes, except at the crack site where it is maintained at $\tilde{P} = 0$ starting with $\tilde{t} = 0$. From this starting assumption we use the system of equations to develop a new estimate for the pressure field. The

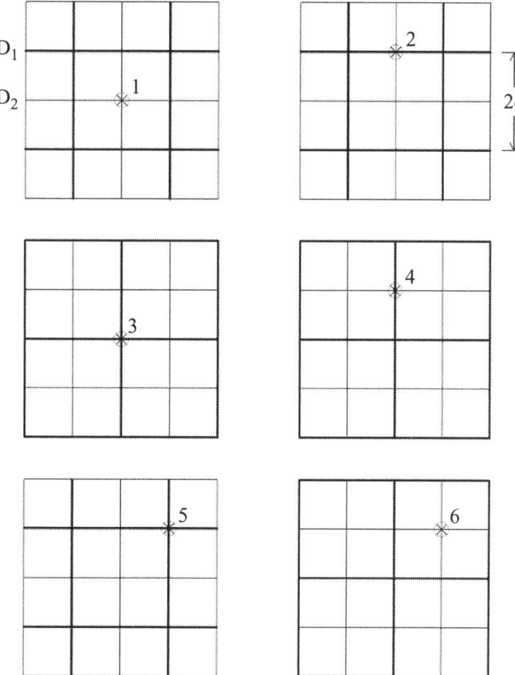

Figure 8.10 Grids of channels with two sizes (D_1, D_2), and six sink positions [9].

iterations continue until the criterion $\left|\tilde{P}_{new} - \tilde{P}_{old}\right| < 10^{-10}$ is met at every node (i, j).

The grids may differ in how they touch the crack site (the sink) (Fig. 8.10). When the sink is in or near the center of the large square, there are six ways of catching the crack in the net formed by the grid channels. The assumed grid configurations can also vary with respect to the ratio D_2/D_1 and the size of the domain ($N \times N$).

Figure 8.11 shows how the global flow resistance varies with D_2/D_1, $N \times N$ and the position of the central sink. The flow resistance $\tilde{P}_{max}$ is referenced to the equivalent calculation in the limit where there is only one channel size ($D_1 = D_2$). Several trends become visible. First, there is no optimal D_2/D_1 when the sink is at the intersection of two thin channels, Fig. 8.10(1) and Fig. 8.10(6). In this case the best design is the one with one channel size, $D_1 = D_2$. For the remaining four cases shown in Fig. 8.10, we found that $\tilde{P}_{max}$ can be minimized: the best designs have D_2/D_1 ratios that decrease as N increases. The minimized value of $\tilde{P}_{max}$ decreases monotonically as N increases.

The results of the optimizations performed in Fig. 8.11 are condensed in Fig. 8.12. The optimized grid with two channel sizes is robust: its performance does not change much as the position of the sink changes from Fig. 8.10(2) to 8.10(4) and 8.10(5). The benefits of optimizing the D_2/D_1 ratio are greater when N is larger.

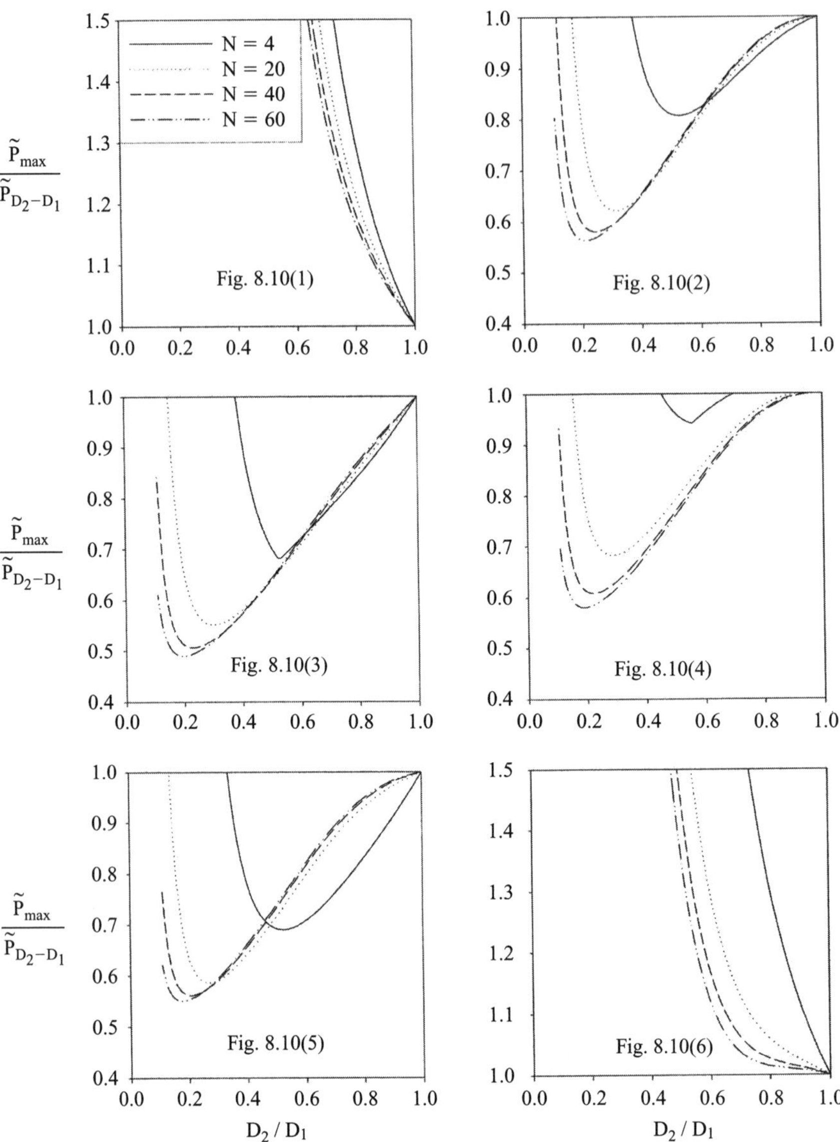

Figure 8.11 The optimization of the ratio D_2/D_1 for the grids shown in Figs. 8.9 and 8.10.

The robustness of this conclusion was tested by optimizing in the same manner the arrangement shown in Fig. 8.13. Instead of placing fluid sources at every node (Fig. 8.9), now the sources are placed only in the peripheral nodes. The sink may have several positions near the center of the $N \times N$ square, e.g., positions 7, ..., 14. The center of the large square is at the intersection of two thin channels D_2. The design conclusions are the same as in Fig. 8.11. When the sink is at the intersection

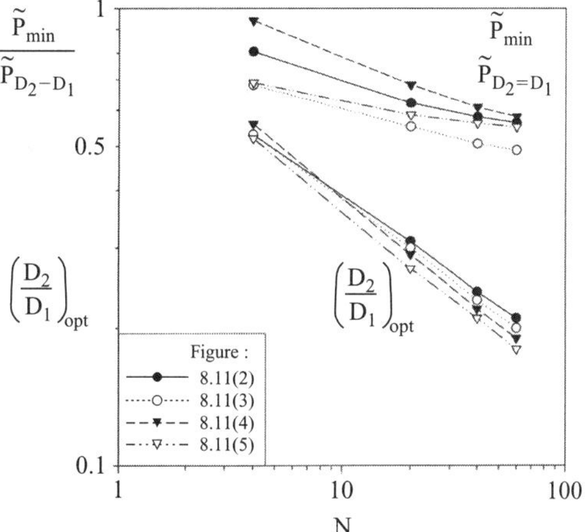

Figure 8.12 The optimized ratio of channel sizes and the minimized global flow resistance for the configurations shown in Figs. 8.9 and 8.10.

of two thin channels (positions 13, 14), the best design is the one with $D_1 = D_2$. All the other cases can be optimized by varying the ratio D_2/D_1.

The physical implementation of the grid scheme for filling cracks relies on fluids pressurized and stored in grids, not on fluid sources. The fluid that invades the cracks comes from the pressurized grid, and the flow is due to the time-dependent relaxation of the pressure field. The measure of performance is not the minimized volume-point flow resistance, but the minimized crack-filling time, that is, the time of relaxation of the grid pressure field.

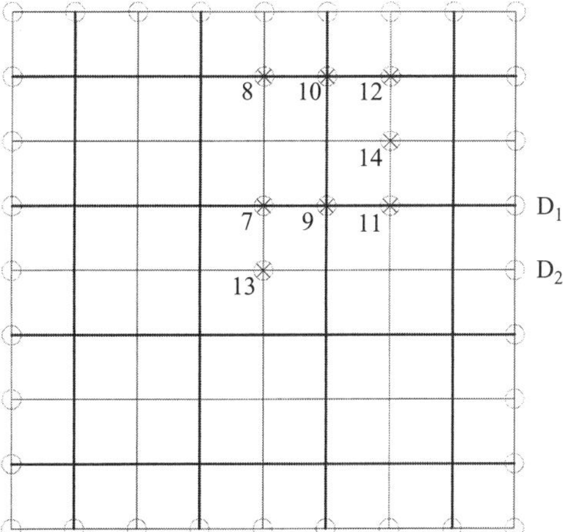

Figure 8.13 Square grid of channels with mass flow sources along the perimeter, and with two thin channels intersecting in the center [9].

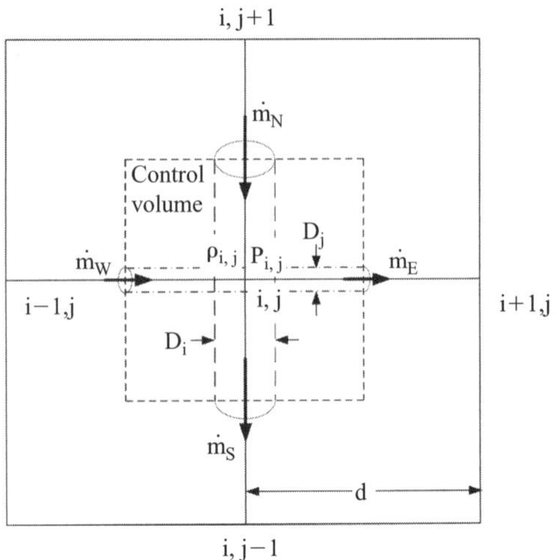

Figure 8.14 Control volume containing the intersection of two channels with slightly compressible fluid [9].

The numerical formulation [9] is based on the control volume drawn around node (i, j) in Fig. 8.14. This node is at the intersection of channels D_i and D_j, where in general $D_i \neq D_j$. Because the fluid density increases slightly with the pressure, each control volume $(P_{i,j}, \rho_{i,j})$ acts as a mass reservoir, the inventory of which is time dependent. Mass conservation in the control volume requires

$$\frac{d}{dt}\left(\rho_{i,j} V_{i,j}\right) = \dot{m}_N + \dot{m}_W - \dot{m}_S - \dot{m}_E \tag{8.47}$$

where $d\rho_{i,j}/dt = c\ dP_{i,j}/dt$, and $V_{i,j}$ is the volume of the channels present in the control volume. The channel sizes are nondimensionalized by using the D_{scale} expression shown above Eq. (8.43). The scale of $V_{i,j}$ is $D^2_{\text{scale}} d$, therefore the dimensionless duct volume is $\tilde{V}_{i,j} = V_{i,j}/(D^2_{\text{scale}} d)$. The channels pressure difference scale is $\Delta P_{\text{scale}} = P_0 - P_\infty$, where P_0 is the initial (uniform) pressure of the grid, and P_∞ is the pressure imposed at $t = 0^+$ at the sink (the crack site). The instantaneous dimensionless excess pressure is

$$\tilde{P} = \frac{P - P_\infty}{P_0 - P_\infty} \tag{8.48}$$

The flow resistance between two adjacent nodes continues to be modeled as Poiseuille flow with constant properties, in accordance with Eq. (8.38). By substituting the scales of D, $V_{i,j}$ and P in Eqs. (8.38) and (8.48) we find that the time scale of the area-point transient flow is

$$t_{\text{scale}} = \frac{128}{\pi} \frac{c \nu d^2}{D^2_{\text{scale}}} \tag{8.49}$$

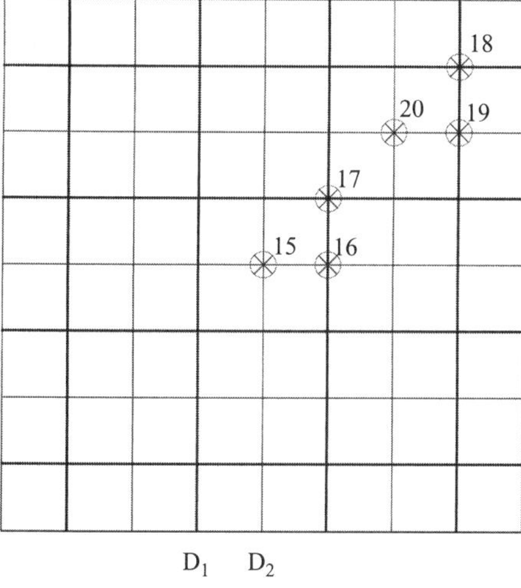

Figure 8.15 Positions of crack sites on a grid with slightly compressible fluid, and with two thin channels intersecting in the center of the domain [9].

The dimensionless time is $\tilde{t} = t/t_{\text{scale}}$. The dimensionless Eqs. (8.38) and (8.48) are written for every node, and they permit the calculation of the pressure field history, $\tilde{P}_{i,j}(\tilde{t})$. The initial pressure is $\tilde{P} = 1$; the final pressure (at $\tilde{t} \to \infty$) tends to $\tilde{P} = 0$ throughout the domain. We define the characteristic depressurization time of the grid as the time when the highest pressures decrease by 90% from their initial values. At any time $\tilde{t}$, the highest pressures are located in the four corners of the square domain, because the corners are the farthest from the sink, which is in the center of the square. Therefore, the characteristic time $\tilde{t}_c$ is defined by setting: $\tilde{P}_{\text{corners}}\left(\tilde{t}_c\right) = 0.1$.

Six configurations (numbered 15 through 20 in Fig. 8.15) were simulated and optimized. The center of the $N \times N$ domain is at the intersection of two thin channels. The sink positions 15 through 17 are set relative to the center of the $N \times N$ domain. This means that the assumed sink positions do not change relative to the center when N increases. The sink positions 18-20 are set relative to the corner of the $N \times N$ square, and do not change as N increases. The results show that when the sink is touched by two thin ducts (cases 15 and 20) the best decision is to use one channel size ($D_1 = D_2$). This, however, is the exception. The rest of the results show that when at least one thick channel touches the sink there is an optimal ratio D_2/D_1, and with it comes a 1/2 reduction in the characteristic time of grid discharge. In future design and implementation, the crack filling time will have to be compared with the fluid (epoxy) hardening time, all in a more realistic model.

We repeated the work of Fig. 8.15 by making D_1 and D_2 switch places, such that the center of which is marked by the intersection of two thick ducts. When the sink is touched by two thin ducts, the best grid has $D_1 = D_2$. When the sink is touched by one or two thick ducts, the best grid is the one with no thin ducts

($D_2 = 0$): this choice is impractical because it constitutes a grid with loops of size $2d \times 2d$, which are too large to capture the crack sites.

8.4.2 Multiple Scales, Loop Shapes, and Body Shapes

We investigated systematically the advantages of endowing the grid flow architecture with more freedom to morph [21]. The four ways to increase design freedom are: multiscale grids (one, two, three, and four diameter sizes), multishape loops (square, triangular, hexagonal, rhombic), multishape bodies (hexagonal, square, rhombic), and vascularization with grids vs. trees. We showed that significant gains in global flow access are achieved as the number of optimized channel diameters increases.

The global flow resistance decreases to 1/3 of its original value when the number of optimized channel sizes is 3 or 4. We considered three loop shapes (Fig. 8.16). Each loop is sized so that it is circumscribed to a disc of diameter d, which represents the crack dimension. The slab that housed these grids had three shapes: square, rhombic and hexagonal. With these loops and body shapes we formed five combinations, which are sketched in Fig. 8.17:

SS–SL: square slab with square loops
RS–TL: rhombic slab with triangular loops
RS–RL: rhombic slab with rhombic loops
HS–TL: hexagonal slab with triangular loops
HS–RL: hexagonal slab with rhombic loops

These five architectures have several common features: the size (area A) of the slab, the total volume occupied by all the channels (V_c), the disc inscribed in each

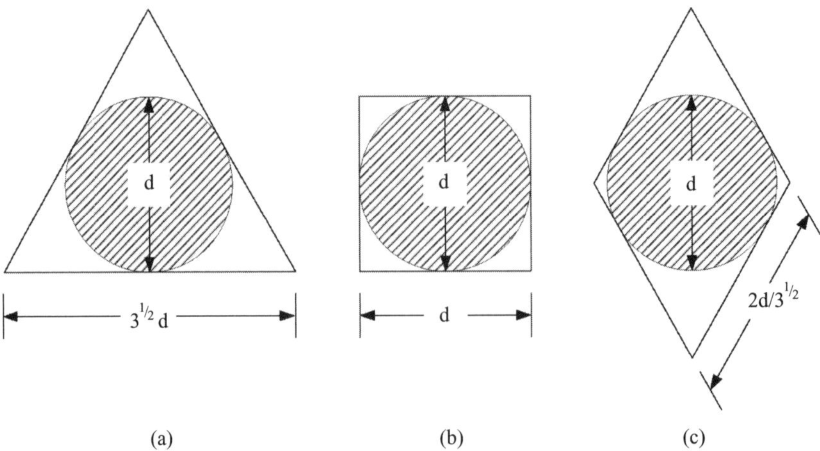

Figure 8.16 Three loop shapes that capture the same crack site [21].

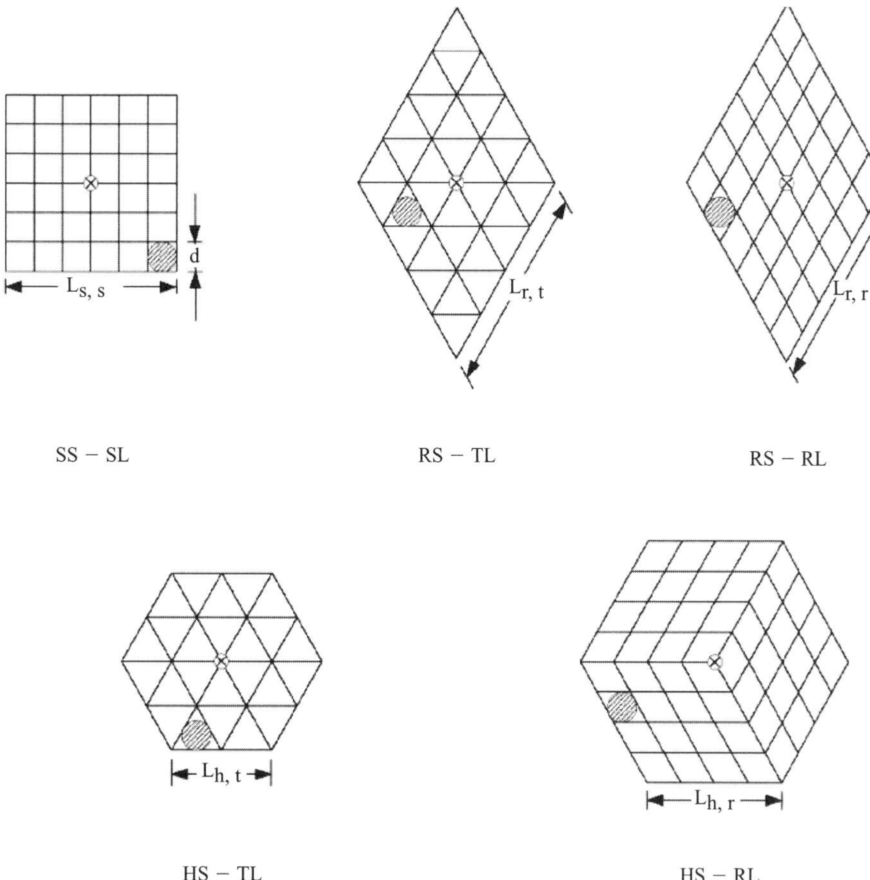

Figure 8.17 Five configurations of grids with three loop shapes on three slab shapes [21].

loop (d), and the fact that each grid has channels of a single size (D). Because of the slab size and channel volume constraints, the outer dimensions of the slabs are not exactly the same (note $L_{s,s} \neq L_{r,t}$, etc.), and the channel thickness varies from one configuration to the next.

In all the flow simulations there was only one crack site serving as sink, and it was located in the center of the slab. Steady flow was simulated by placing equal mass flow sources at every node in the grid, including the nodes on the perimeter (cf. section 8.5). All the fluid generated by the mass sources at the nodes finds its way to the center of the slab (the sink). The flow is steady and is driven by the pressure field established in the grid. The minimum pressure ($P_{\min}$) is at the sink, and the maximum ($P_{\max}$) occurs at several points on the perimeter (e.g., in the corner of the square SS–SL, Fig. 8.17). The overall (maximum) pressure difference is $\Delta P = P_{\max} - P_{\min}$. The mass flow rate generated by all the mass sources on a given A is $\dot{m}$. The flow rate $\dot{m}$ is the total mass flow rate that reaches

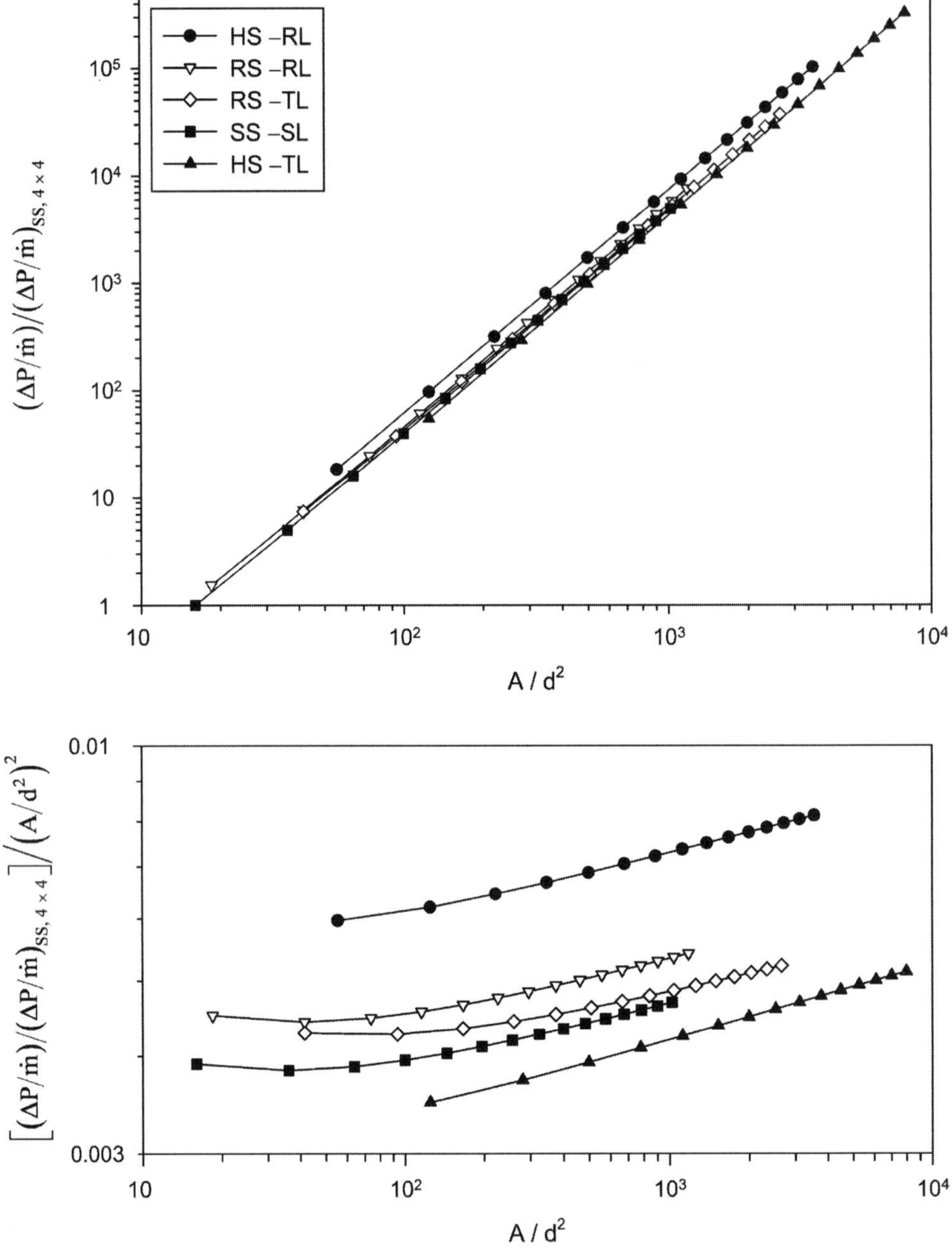

Figure 8.18 The relative global resistance of the five designs shown in Fig. 8.17, when the size of the slab is specified [21].

the sink. To compare the global flow resistances ($\Delta P/\dot{m}$) of the five designs is to compare their peak ΔP values. In Fig. 8.18 we report these values in dimensionless form, where $\Delta P_{SS,4\times4}$ is the overall pressure difference for the first configuration (SS–SL) when A contains 4×4 square loops of size d^2. The abscissa A/d^2 is a dimensionless measure of the size of the slab. The figure shows that in all five designs the overall pressure difference increases approximately in proportion with $(A/d^2)^2$. This is highlighted at the bottom of Fig. 8.18, where the ordinate values have been divided by $(A/d^2)^2$.

Differences exist between the global flow resistances offered by the five configurations. The lowest resistance belongs to the hexagonal slab with triangular loops (HS–TL). Next are the square slab with square loops (SS–SL) and the rhombic slab with triangular loops (RS–TL). The worst is the hexagonal slab with rhombic loops (HS–RL).

Several important conclusions follow from Fig. 8.18. The global flow performance is better when the loops are triangular. This is in agreement with the conclusion reached in Ref. [8] based on an analysis in which the grid was modeled as a homogeneous porous medium with Darcy flow. When the slab size A/d^2 is fixed, the $\Delta P/\dot{m}$ value of the square slab with square loops is larger by a factor of 1.3 than the $\Delta P/\dot{m}$ of the hexagonal slab with triangular loops. This difference in overall flow resistance is consistent with the conclusion based on the Darcy-flow model [8], where the factor was 2.

8.4.3 Trees Matched Canopy to Canopy

Mating trees canopy to canopy was proposed as a novel volume-bathing architecture in Refs. [20] and [22]. The objective of the mating trees is to provide healing fluid to one or more sites in steady state. The self-healing composite is a slab with rectangular face ($H \times L$), and the flow layout is two dimensional (Fig. 8.19). The stream enters the $H \times L$ domain through a trunk of hydraulic diameter D_2, and splits into several thinner channels of diameter D_1. These thinner channels form the canopy of the first tree. Further downstream, the flow from the D_1 channels is reconstituted into a single stream that exits the domain through a channel of diameter D_2, which is the trunk of the second tree. The canopy (D_2) shared by the two trees bathes every area element of the $H \times L$ domain.

The crack length scale d is known and fixed. Each crack site is modeled as a sphere of diameter d. The composite material is a slab of thickness d. The tree networks must be fine enough so that they touch all the cracks wherever they may occur. This is why the layout of the two trees is such that at least one channel passes through the center of a square element $d \times d$. The flow rate through the entire network ($\dot{m}$) is steady and much larger than the flow rate needed to fill the cracks that may form. Consequently, the flow through the network is modeled as steady and incompressible.

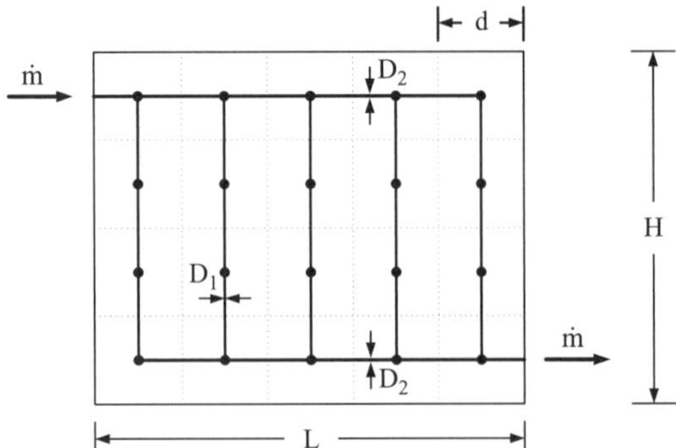

Figure 8.19 Rectangular domain with square elements bathed by two trees mated canopy to canopy [10].

Another constant during the search for optimal flow configurations is the volume fraction occupied by all the channels, cf. Eq. (8.37). Channel diameters are expected to be in the 10–100-μm range, with Reynolds numbers of order 1. The pressure drop along a channel with length L_i, diameter D_i and mass flow rate $\dot{m}_i$ is

$$\Delta P_i = C \frac{\dot{m}_i L_i}{D_i^4} \tag{8.50}$$

where C is a constant factor, for example, $C = 128\nu/\pi$ if the duct cross-section is round. Equation (8.50) continues to be valid for other cross-sectional shapes (with slightly different values for C) provided that D_i is the hydraulic diameter of channel "i," and that the shape of the channel cross-section does not change from one channel to the next.

Figure 8.20 shows the smallest domain for which we can illustrate the search for the best tree architecture. The domain to be bathed has six $d \times d$ elements. The centers of the elements are indicated with black circles. There are two ways in which to bathe the six elements with D_1 and D_2 channels: 2×3 in configuration a, and 3×2 in configuration b. For each configuration, we minimized the global

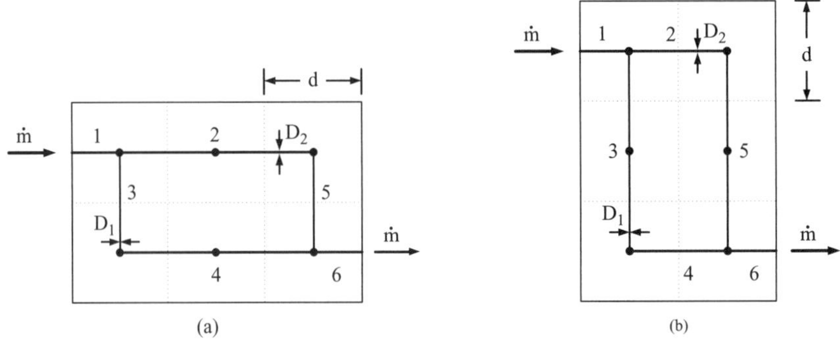

Figure 8.20 The two configurations for bathing six elements [10].

Table 8.2 The optimal design of two trees matched canopy to canopy on a rectangular domain composed of 6 elements [10].

Configuration	$D_1^*\left(\frac{d}{V_c}\right)^{1/2}$	$D_2^*\left(\frac{d}{V_c}\right)^{1/2}$	$\frac{D_1^*}{D_2^*}$	$\frac{\Delta P^*}{C\,d\,\dot{m}}\left(\frac{V_c}{d}\right)^2$	$\left[\frac{\Delta P^*}{C\,d\,\dot{m}}\left(\frac{V_c}{d}\right)^2\right]_{D_1=D_2}$
2 × 3, Fig. 8.20a	0.403	0.436	0.925	74.54	75.56
3 × 2, Fig. 8.20b	0.405	0.454	0.891	72.61	75.56

flow resistance $\Delta P/\dot{m}$ with respect to the diameter ratio D_1/D_2. The results are summarized in Table 8.2, where the asterisk means "optimized," and V_c is the total volume of the channels. The table shows that the configuration of Fig. 8.20b is better than the configuration of Fig. 8.20a. The table also shows the global flow resistance for the reference design in which there is only one channel size, $D_1 = D_2$. Relative to the reference design, the optimal configuration of Fig. 8.20b offers a 4% reduction in global flow resistance.

The search for optimal tree-tree configurations continued with larger $H \times L$ domains. The next class contained 12 elements. The total size of the domain is fixed, $HL = 12d^2$; however, the shape H/L can have four values: 3/4, 4/3, 2/6, and 6/2. There are two channel sizes, D_1 and D_2, the ratio of which was optimized subject to fixed total channel volume (V_c). The analysis was conducted as in the preceding example, and is omitted. The best turns out to be the configuration $H/L = 4/3$.

A pattern is beginning to emerge. The best shape of a domain with 12 elements is similar to the best shape when there are only 6 elements ($H/L = 3/2$, Table 8.2). The optimal ratios D_1^*/D_2^* are also similar, 0.802 for 12 elements vs. 0.891 for 6 elements.

This pattern can be used to abbreviate the search for optimal tree layouts on larger domains. When there are 24 elements to be bathed by the two mating trees, there are 6 ways to configure the elements: 4 × 6, 6 × 4, 3 × 8, 8 × 3, 2 × 12, and 12 × 2. We know to expect an optimal H/L ratio of order 1, but slightly greater than 1. This makes the case 6 × 4 the expected best; however, numerical minimization of $\Delta P/\dot{m}$ performed for all the cases confirmed this expectation. The optimal D_1/D_2 ratio is 0.738.

In Ref. [10], we continued with the optimization of an even larger structure, $N = 48$. The optimal cases are put on display in Table 8.3, which shows an additional similarity. If we write N for the total number of $d \times d$ elements ($HL = Nd^2$), then $V_c = \phi V$, $V = HLd = Nd^3$, and the global flow resistance can be nondimensionalized as

$$\psi = \frac{\Delta P^* \phi^2 d^3}{C\dot{m}} \tag{8.51}$$

We found that this new dimensionless version of the minimized global flow resistance is almost insensitive to changes in N. Table 8.3 also shows the reduction in

Table 8.3 Optimal layouts of two trees mated canopy to canopy on a rectangular domain [10].

Configuration	$\frac{D_1^*}{D_2^*}$	$\frac{H}{L}$	$\frac{\Delta P^*}{C d \dot{m}}\left(\frac{V_c}{d}\right)^2$	N	$\frac{\Delta P^*}{C \dot{m}} \phi^2 d^3$	Reduction
3 × 2, Fig. 8.20b	0.891	3/2	72.61	6	2.017	3.9%
4 × 3	0.802	4/3	324.46	12	2.253	13.2%
6 × 4	0.738	3/2	1372.5	24	2.383	22.9%
8 × 6	0.651	4/3	5385.3	48	2.337	39.1%

global flow resistance obtained in going from the $D_1 = D_2$ structure to the (D_1^*/D_2^*) structure. The relative reduction increases as N increases. In conclusion, the use of vascularization becomes more attractive as the size of the bathed volume increases.

Table 8.3 shows that the optimized external shape (H/L) turns out to be a number close to 1 when $6 \leq N \leq 48$. This observation suggests a short cut in the search for optimal flow architect ures: choose from the beginning square assemblies $(H/L = 1)$, which means using numbers such as $N = 3^2, 4^2, 5^2 \ldots$ For each such case it is still necessary to optimize the D_1/D_2 ratio. The results are reported in Table 8.4. They are consistent with the optimization results summarized in Table 8.3, in fact, all these results trace a ragged line of "best performance" in the $\psi - N$ domain of Fig. 8.21. Above this line reside the designs that are not optimized. The objective is to discover new configurations that may push this line lower.

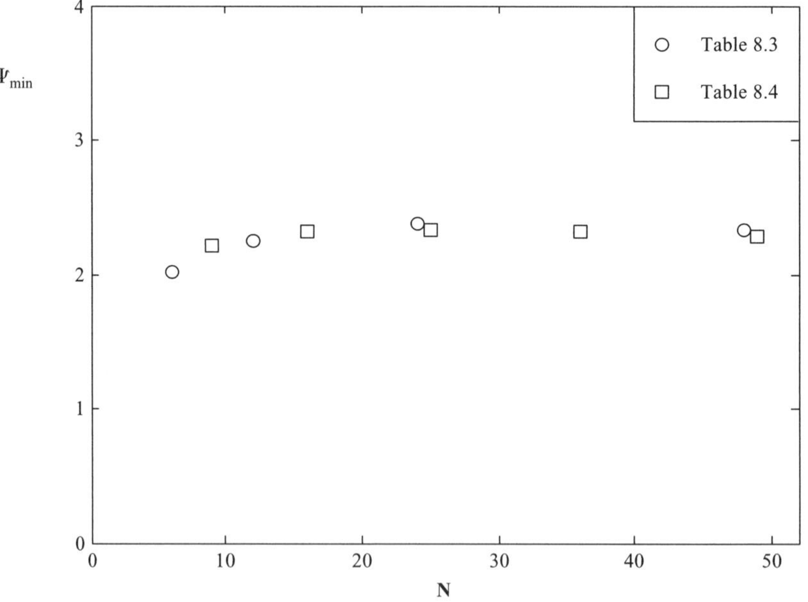

Figure 8.21 The minimized global flow resistance of the designs summarized in Tables 8.3 and 8.4 [10].

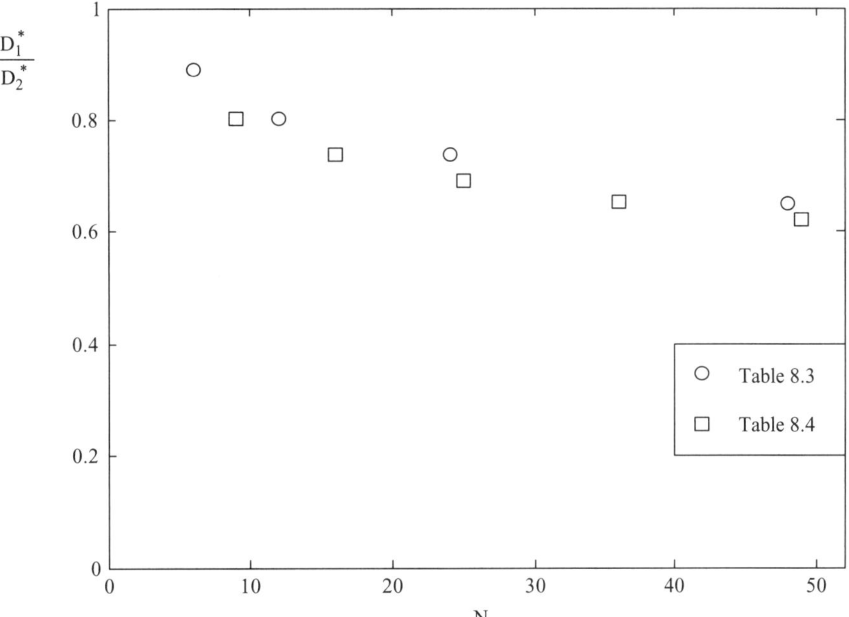

Figure 8.22 The optimal ratio of channel diameters of the designs summarized in Tables 8.3 and 8.4 [10].

Figure 8.22 shows another ragged line, which corresponds the optimized D_1/D_2 ratios of all the designs reported in Tables 8.3 and 8.4. There is a clear trend: D_1^*/D_2^* decreases slowly as N increases. Important is the observation that the D_1^*/D_2^* values of Tables 8.2 and 8.3 are correlated: they depend in the same way on N. In sum, by using a shortcut such as $H = L$, one can achieve a level of performance that is essentially the same as the best performance.

It is possible to extend to even larger N values this line of work by performing an analysis based on the assumption that the D_1 channels are so numerous that the variation of the mass flow rate along the D_2 channels varies continuously, not

Table 8.4 Optimal designs of two trees matched canopy to canopy when $H/L = 1$ [10].

Configuration	$D_1^* \left(\frac{d}{V_c}\right)^{1/2}$	$D_2^* \left(\frac{d}{V_c}\right)^{1/2}$	$\frac{D_1^*}{D_2^*}$	$\frac{\Delta P^*}{C\,d\,\dot{m}}\left(\frac{V_c}{d}\right)^2$	$\left[\frac{\Delta P^*}{C\,d\,\dot{m}}\left(\frac{V_c}{d}\right)^2\right]_{D_1=D_2}$	$\frac{\Delta P^*}{C\dot{m}}\phi^2 d^3$
3 × 3	0.304	0.379	0.803	179.57	205.87	2.217
4 × 4	0.227	0.307	0.739	594.00	765.45	2.321
5 × 5	0.181	0.262	0.691	1460.2	2125.3	2.336
6 × 6	0.151	0.231	0.653	3004.2	4914.5	2.318
7 × 7	0.130	0.209	0.621	5488.6	10022.0	2.286

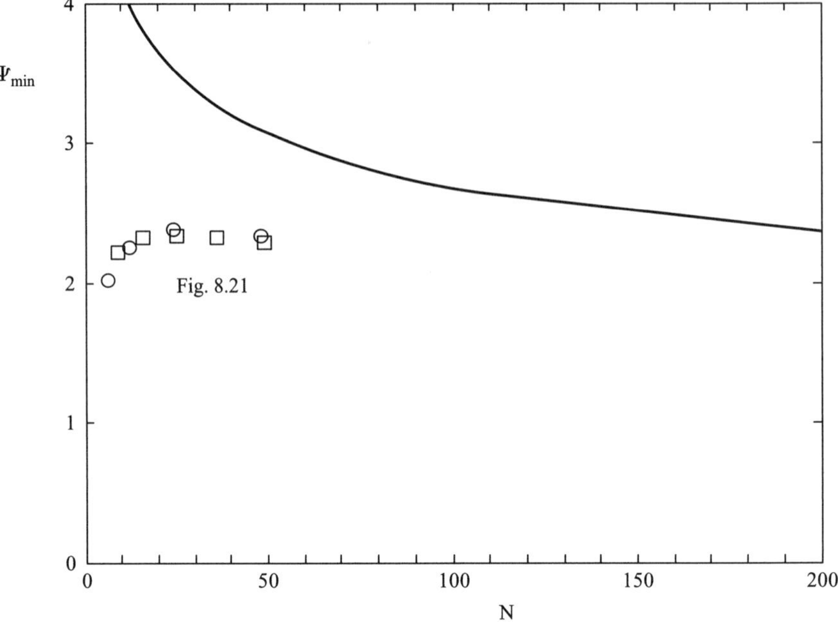

Figure 8.23 The minimized global flow resistance when N is large [10].

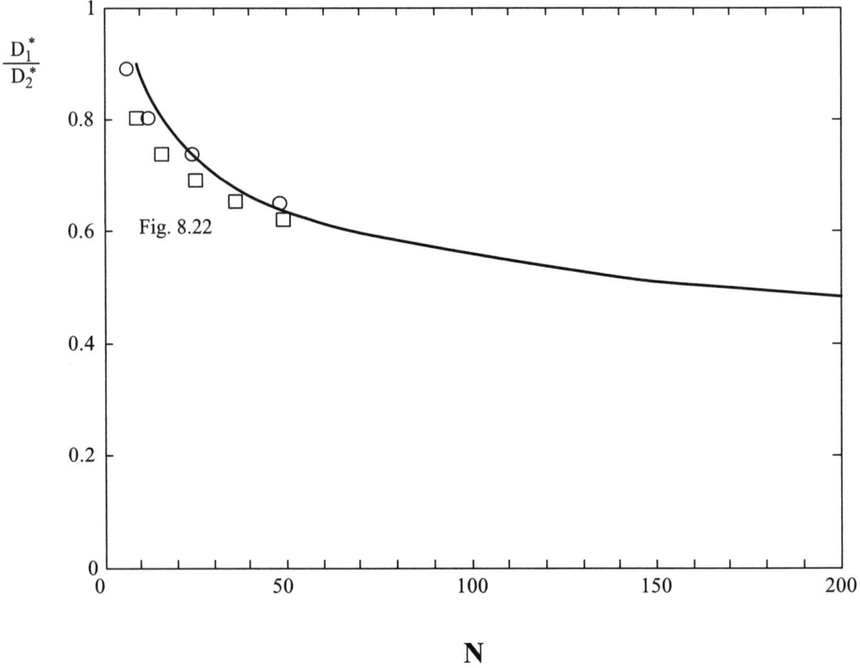

Figure 8.24 The optimal ratio of channel sizes when N is large [10].

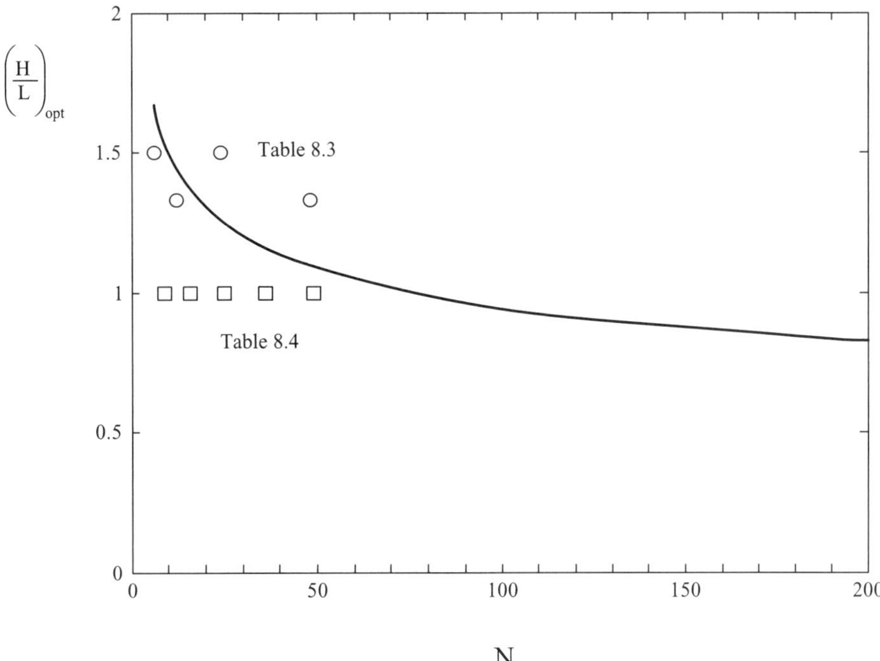

Figure 8.25 The optimal shape of the rectangular domain when N is large [10].

stepwise. This analytical approach is available in Ref. [10] and is not detailed here. There are two degrees of freedom when the overall size (N) is fixed: H/L and D_1/D_2. The global performance indicator ψ can be minimized with respect to H/L and D_1/D_2, independently. The results of this analytical formulation are reported in Figs. 8.23 through 8.25.

The ψ function minimized with respect to both H/L and D_1/D_2 is shown as a continuous curve in Fig. 8.23, next to the case-by-case results discussed earlier in this section. The agreement between the two sets of results is good. As N increases, ψ continues on its plateau, and starts to decrease slowly.

The optimized ratio of channel sizes is shown as a continuous curve in Fig. 8.24. The agreement between the analytical solution and the case-by-case designs is excellent. The same conclusion holds for the external aspect ratio of the assembly, H/L, which is reported in Fig. 8.25. The dimensionless flow resistance ψ is proportional to the actual resistance ($\Delta P/\dot{m}$) divided by the size of the assembly (N^2), namely $\psi = [\Delta P/(C\, d\, \dot{m})]\,(V_c/d)^2\,/N^2$. Note that ψ is the total resistance averaged over the $H \times L$ domain and expressed for (allocated to) a single $d \times d$ flow element. See also the earlier ψ definition [Eq. (8.51)]. The fact that ψ decreases slightly as N increases means that it is advantageous to bathe more elements by using the same flow structure.

8.4.4 Diagonal and Orthogonal Channels

We extended the search for trees matched canopy to canopy, and increased the freedom to morph in several directions: channel orientations (diagonal vs. orthogonal), channel sizes, and system sizes [23]. We found that tree-tree configurations provide greater access when diagonal channels are combined with orthogonal channels and when multiple and optimized channel sizes are used.

For example, in Fig. 8.26 we compare on the basis of the same volume constraints two design concepts: (a) diagonal channels combined with peripheral orthogonal channels and (b) orthogonal channels, as in section 8.7. The channels are of one thickness, D. The two concepts perform as shown in Fig. 8.27, for which the global flow resistance ψ is defined in Eq. (8.51). The size of the square flow system varies from 3×3 to 10×10. Orthogonal channels offer less resistance when the system is smaller than 7×7. Larger systems offer less resistance when configured with diagonal channels.

Diagonal channels were pursued further by allowing the channel volume to be distributed optimally between channels with two sizes (D_1, D_2), and three sizes (D_1, D_2, D_3). Examples of canopy-canopy flows with diagonal channels of three sizes are illustrated in Fig. 8.28. The performance of design (a) of Fig. 8.28 is reported in the upper graph of Fig. 8.29. The use of three channel diameters leads to a ψ reduction of roughly 10 percent in the range $3 \leq N \leq 10$, where $N = L/d$ (note that in the preceding section N is defined as HL/d^2). The reduction is relative to the corresponding design (Fig. 8.28a) in which there are just two channel sizes, i.e. ($D_2 = D_3$). Once again, we find that freedom is good for design. Flow architectures consisting of trees matched canopy to canopy [22,23] are described further in Refs. [24] through [26].

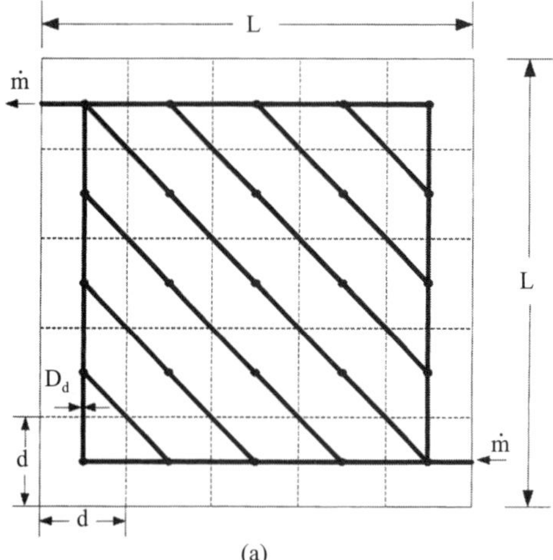

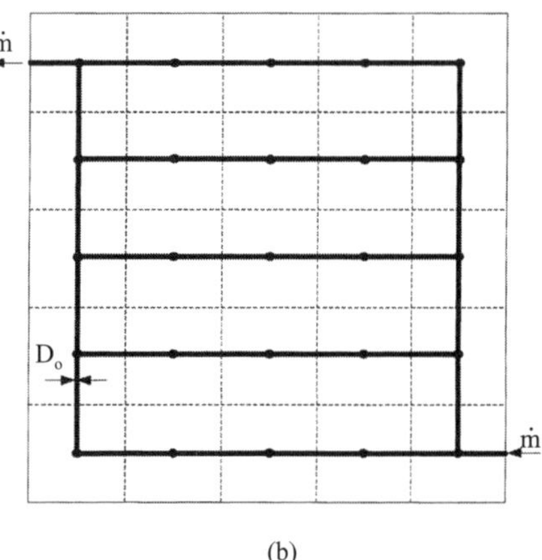

Figure 8.26 Trees matched canopy to canopy on square flow domains with square elements ($d \times d$), and channels with one diameter (D): (a) diagonal channels, (b) orthogonal channels [23].

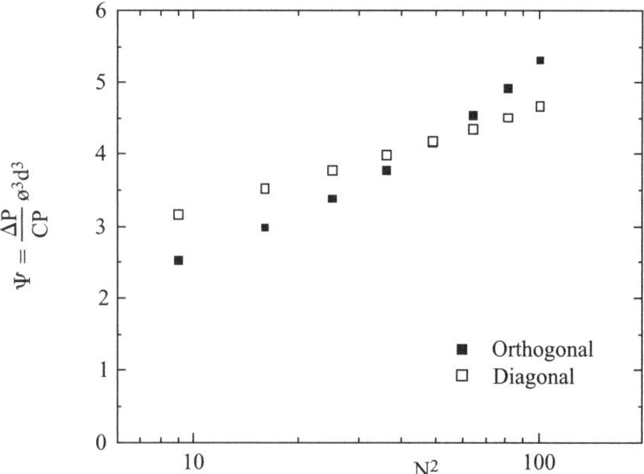

Figure 8.27 The global flow resistances of the flow architectures shown in Fig. 8.26a and b [23].

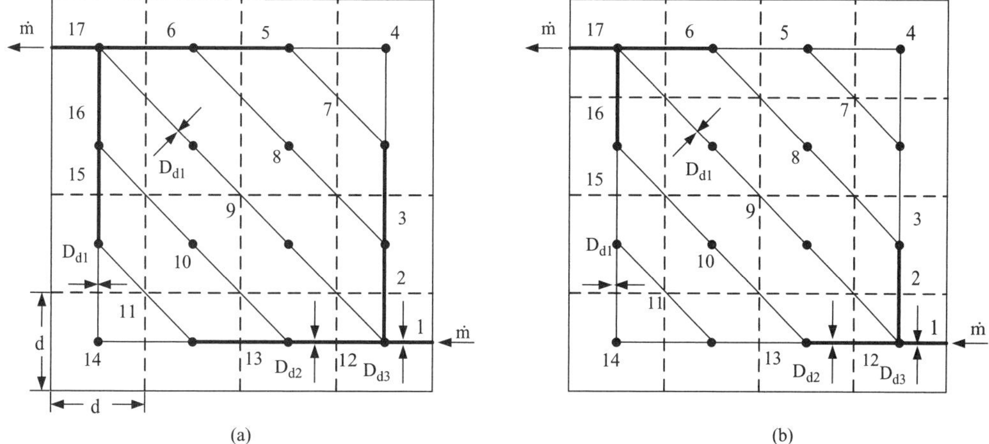

Figure 8.28 Tree-tree configurations with diagonal channels with three sizes (D_1, D_2, D_3) [23].

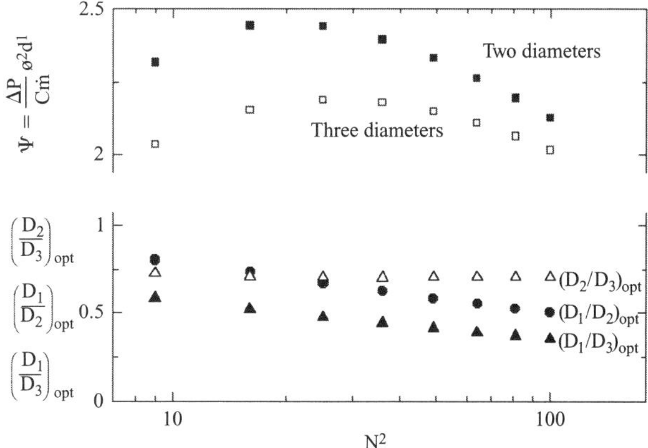

Figure 8.29 The minimized flow resistance and optimized ratios of channel sizes for the configurations shown in Fig. 8.28 [23].

8.5 VASCULARIZATION FIGHTING AGAINST HEATING

In this section we consider the challenge to vascularize a solid body with the objective of building into the body structure the function of self-cooling. The body may experience sudden or steady heating from one side. The need is to send a coolant to the heated spots fast, efficiently and reliably. For this, the body design needs vascularization, which must carry the coolant from one entry point to the finite-size area that needs cooling.

Here, we begin with the most fundamental formulation of this configuration [11], and model the whole problem in one direction (Fig. 8.30): the solid body is in the middle, the heat attack comes from the left, and the coolant is forced from the right. The solid slab of thickness L and thermal conductivity k_s is heated from one side with uniform heat flux q''. The other side is insulated. The slab is cooled by a single-phase fluid that flows along L, against the direction of q''. The flow is driven by a fixed pressure difference ΔP. The fluid flows through parallel slots of width D, one slot for each interval H along the wall. The porosity of the wall ($\phi = D/H$) is fixed.

The highest temperature ($T_{\max}$) occurs in hot spots located on the wall surface that receives q''. The lowest temperature ($T_{\min}$) is the fluid inlet temperature. The objective is to determine the internal spacing D such that the maximum excess temperature ($\Delta T = T_{\max} - T_{\min}$) is minimum. In other words, we are interested in identifying the internal flow architecture that guarantees the least nonuniform distribution of temperature in the solid. The flow structure is destined to remain imperfect, with cold spots on the surface entered by the fluid, and hot spots on the surface bombarded with q''. The objective is to identify the least imperfect structure possible—the best tapestry of highs and lows.

The analysis is an application of the intersection of asymptotes method (sections 3.3 through 3.5), which consists of developing analytically the relationship between the global objective (ΔT) and the varying geometry (D) in the two extremes, $D \to 0$ and $D \to \infty$, and using these asymptotes to argue that ΔT can be minimized

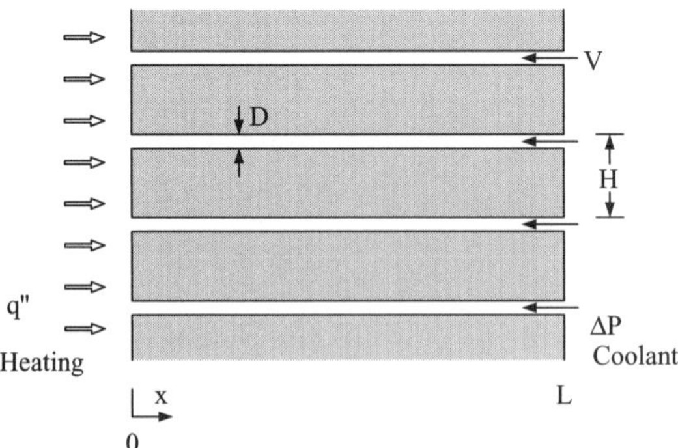

Figure 8.30 Two-dimensional parallel channels across a slab with heat flux from one side and forced flow of coolant from the other side [11].

by selecting an intermediate D. The optimal D value is located approximately by intersecting the asymptotes.

The small-D limit also means $H \to 0$ because the porosity is fixed. The L layer can be treated as a fluid saturated porous medium with Darcy flow. The uniform volume averaged velocity V points in the negative x direction,

$$V = \frac{K}{\mu} \frac{\Delta P}{L} \tag{8.52}$$

The permeability K associated with Poiseuille flow through D-thin fissures is [27]

$$K = \frac{D^2}{12} \tag{8.53}$$

The temperature distribution across the L is obtained by solving the energy equation

$$-V \frac{dT}{dx} = \alpha \frac{d^2T}{dx^2} \tag{8.54}$$

subject to the boundary conditions

$$q'' = -k \frac{dT}{dx} \quad \text{at} \quad x = 0 \tag{8.55}$$

$$T \to T_{\min} \quad \text{as} \quad x \to \infty \tag{8.56}$$

The thermal diffusivity of the saturated porous medium is defined as $\alpha = k/(\rho c_P)_f$, where $(\rho c_P)_f$ is the heat capacity of the fluid, and k is the effective thermal conductivity of the porous medium with the fluid filling its pores. Because the fluid-filled spaces are parallel to the solid parts, and because both are parallel to the direction of heat flow, the effective thermal conductivity is [27]

$$k = \phi k_f + (1 - \phi) k_s \tag{8.57}$$

The solution to Eqs. (8.54) through (8.56) is

$$T - T_{\min} = \frac{q''\alpha}{kV} e^{-Vx/\alpha} \tag{8.58}$$

and shows that the effect of q'' propagates into the porous structure to a depth x of order α/V. This means that the boundary condition (8.56) holds when the penetration depth is smaller than the slab thickness,

$$\frac{\alpha}{V} < L \tag{8.59}$$

Finally, from Eq. (8.58) we find the maximum excess temperature $\Delta T = T_{\max} - T_{\min}$, which occurs on the $x = 0$ surface:

$$\Delta T = \frac{q''\alpha}{kV} \tag{8.60}$$

After using Eqs. (8.52) through (8.53), this result becomes

$$\Delta T = 12 \frac{q'' \mu L}{D^2 \Delta P (\rho c_P)_f} \tag{8.61}$$

The small-D conclusion is that the maximum excess temperature ΔT increases as $D \rightarrow 0$. We are interested in smaller ΔTs, and because of this we turn our attention to how ΔT depends on D in the opposite limit.

The large-D limit means that D (or $H = D/\phi$) is sufficiently large that the coolant flows through the channels without experiencing a significant rise in temperature. The dominant thermal resistance between the surface with imposed heat flux q'' and the channel surface (the heat sink) is due to heat conduction in square chunks of solid of dimension $H/2$. One such element is shown in Fig. 8.31. The left side of the square is heated with the flux q'', the top side is isothermal at $T_{\min}$, and the bottom and right sides are adiabatic. The square shape of the domain is not assumed: this shape follows from the energy equation for steady conduction in a rectangular domain,

$$\frac{\partial^2 T}{\partial x^2} + \frac{\partial^2 T}{\partial y^2} = 0 \tag{8.62}$$

which requires that $\Delta T / L_x^2 \sim \Delta T / (H/2)^2$, or that the path of heat conduction in the x direction (L_x, Fig. 8.31) must have the same length scale as the path in the y direction,

$$L_x \sim \frac{H}{2} \tag{8.63}$$

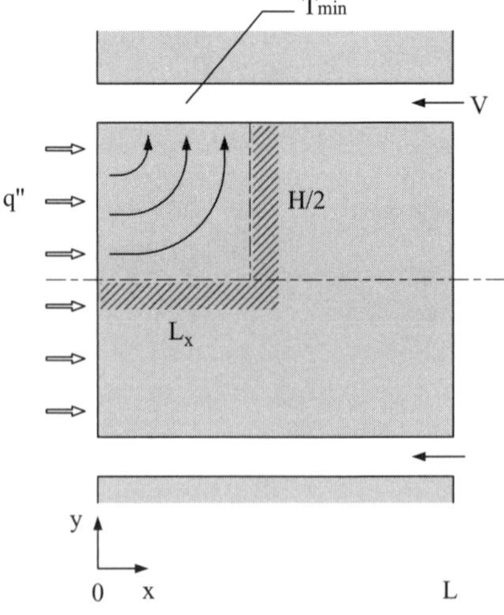

Figure 8.31 The "square" pattern of heat conduction in the large-D limit [11].

The conservation of heat current through the $L_x \times (H/2)$ element requires

$$q''\frac{H}{2} \sim k_s \frac{H}{2}\frac{\Delta T_x}{L_x} \sim k_s L_x \frac{\Delta T_y}{H/2} \tag{8.64}$$

These relations can be used to estimate the overall temperature difference scale

$$\Delta T \sim \Delta T_x + \Delta T_y \sim \frac{q''}{k_s}\left(L_x + \frac{H^2}{4L_x}\right) \tag{8.65}$$

which in view of Eq. (8.63) becomes

$$\Delta T \sim \frac{q''H}{k_s} \sim \frac{q''D}{k_s\phi} \tag{8.66}$$

This result is valid when the square element $L_x \times (H/2)$ is present, that is, when the spacing $H/2$ is smaller than the slab thickness,

$$\frac{H}{2} < L \tag{8.67}$$

The conclusion drawn from Eq. (8.66) is that in the large D limit the ΔT scale increases as D increases.

The intersection of asymptotes means that we intersect Eqs. (8.61) and (8.66). Because ΔT increases in both extremes, it must be minimum at an intermediate D (or ϕH) value. This can be found approximately by intersecting the asymptotes,

$$\frac{H_{\text{opt}}}{L} \cong 2\left(\frac{3}{2}\right)^{1/3} \phi^{-2/3}\,\text{Be}^{-1/3}\left(\frac{k_s}{k_f}\right)^{1/3} \tag{8.68}$$

where Be is the dimensionless pressure drop [17, 18] [cf. Eq. (3.35)],

$$\text{Be} = \frac{\Delta P \cdot L^2}{\alpha_f \mu} \tag{8.69}$$

and $\alpha_f = k_f/(\rho c_P)_f$. The corresponding minimum ΔT is obtained by substituting H_{opt} into Eq. (8.66):

$$\Delta T_{\text{min}} \cong 2\left(\frac{3}{2}\right)^{1/3} \frac{q}{k_s} L\phi^{-2/3}\,\text{Be}^{-1/3}\left(\frac{k_s}{k_f}\right)^{1/3} \tag{8.70}$$

In the course of developing these results we made two assumptions, Eqs. (8.59) and (8.67). By using the optimal spacing (8.68), we find that Eqs. (8.59) and (8.67) require, respectively,

$$(12)^{-1/3}\phi^{2/3}\text{Be}^{1/3}\left(\frac{k_s}{k_f}\right)^{-1/3} \frac{1}{1-\phi\left(1-k_f/k_s\right)} > 1 \tag{8.71}$$

$$\left(\frac{2}{3}\right)^{1/3} \phi^{2/3}\,\text{Be}^{1/3}\left(\frac{k_s}{k_f}\right)^{-1/3} > 1 \tag{8.72}$$

In other words, conditions (8.59) and (8.67) are essentially the same, and are valid when the group $\text{Be}\phi^2 k_f/k_s$ is greater than 1. This condition is met by the cases studied numerically and discussed next. The group $\text{Be}\phi^2 k_f/k_s$ has values in the range 300 – 10^5 when $10^6 \leq \text{Be} \leq 10^8$, $10 \leq k_s/k_f \leq 100$, and $0.1 \leq \phi \leq 0.2$.

In summary, the best parallel-channels structure for suppressing the hot spots that occur on the q'' surface of the slab has the spacings shown in Eq. (8.68). The minimized hot-spot temperature $\Delta T_{\min}$ decreases further by increasing the applied pressure difference (Be) and by making the channel-to-channel spacings smaller.

An alternate configuration for fighting the heat flux by forcing coolant from the opposite side is shown in Fig. 8.32. The coolant flows through channels with round cross-section, diameter D and length L. Assume that the channels are arranged in a square pattern with the distance H between centers. Because of periodicity in the H direction, it is sufficient to optimize the configuration of a single volume element with the volume $H \times H \times L$, which has a single channel centered on its long axis of symmetry.

The analysis of this volume element is completely analogous to what we presented for the two-dimensional configuration of Fig. 8.30. In this section we highlight the differences. In place of Eqs. (8.53), (8.61), and (8.66), now we have

$$K = \frac{\phi D^2}{32}, \phi = \frac{\pi D^2}{4H^2} \tag{8.73}$$

$$\Delta T \cong \frac{32 q'' \mu L}{\phi D^2 \Delta P \, (\rho c_P)_f} \tag{8.74}$$

$$\Delta T \cong \frac{q''}{k_s} \cdot \frac{3H}{4} \tag{8.75}$$

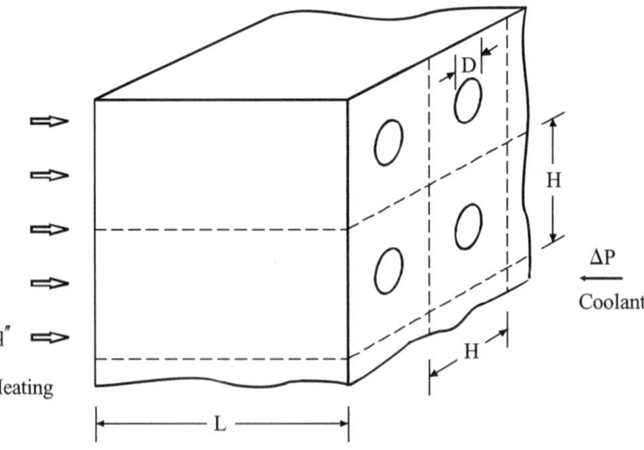

Figure 8.32 Round channels across a slab with heat flux from one side and forced flow from the other side [11].

Table 8.5 Comparison between the exponents (a, b) optimized based on numerical simulations and the analytical value [11].

	H/L, 2-D	ΔT_{min}, 2-D	H/L, 3-D	ΔT_{min}, 3-D	Analytical Results
a	1.27	0.70	0.85	0.60	2/3
b	0.21	0.56	0.17	0.65	1/3

The intersection of the asymptotes (8.74) and (8.75) yields

$$\frac{H_{\text{opt}}}{L} \cong 2\left(\frac{4}{3}\pi\right)^{1/3} \phi^{-2/3}\text{Be}^{-1/3}\left(\frac{k_s}{k_f}\right)^{1/3} \tag{8.76}$$

Important in a fundamental sense is the fact that these optimization results agree within a factor of order 1 with the corresponding results for two-dimensional channels. This means that the scaling laws derived in this section are robust. They depend on global parameters of the structure (ϕ, Be, L, q'', k_s, k_f), not on the cross-sectional shapes of the channels through which the coolant is forced to flow.

$$\Delta T_{\text{min}} \cong \frac{3}{2}\left(\frac{4}{3}\pi\right)^{1/3} \frac{q''}{k_s} L\phi^{-2/3}\text{Be}^{-1/3}\left(\frac{k_s}{k_f}\right)^{1/3} \tag{8.77}$$

Robustness is a precious quality in rules that are used for sizing complex flow architectures. We tested numerically the validity of the results in the parametric range indicated under Eq. (8.72). The numerical results are correlated very tightly by using two dimensionless groups, which should be constants on the order of 2:

$$\frac{H_{\text{opt}}}{L}\phi^a\left(\frac{k_f}{k_s}\right)^b \text{Be}^{1/3}, \qquad \frac{\Delta T_{\text{min}} k_s}{q'' L}\phi^a\left(\frac{k_f}{k_s}\right)^b \text{Be}^{1/3} \tag{8.78}$$

The theoretical exponents appearing in Eqs. (8.68), (8.70), (8.76) and (8.77) are $a = 2/3$ and $b = 1/3$. These exponents are not the best, but are close, as indicated by the exponents optimized numerically [11] and reported in Table 8.5.

8.6 VASCULARIZATION WILL CONTINUE TO SPREAD

A first-time review of a fresh direction of research is a good predictor of what lies ahead on the same route. The vascularized designs reviewed in this chapter have this quality. For example, we showed in several ways that dendritic architectures are superior to parallel channels when the smallest manufacturable scales are fine enough so that the tree-shaped structure has four or more levels of bifurcation. This lesson means that the way to fight against intense side heating with finer flow architectures is to switch from the parallel channels of Fig. 8.30 to the dendritic alternative proposed here in Fig. 8.33. An application of this new constructal design concept is shown on the right side of the figure. Contemporary gas turbines employ gas cooling to control the temperature of the blade surfaces exposed to the products

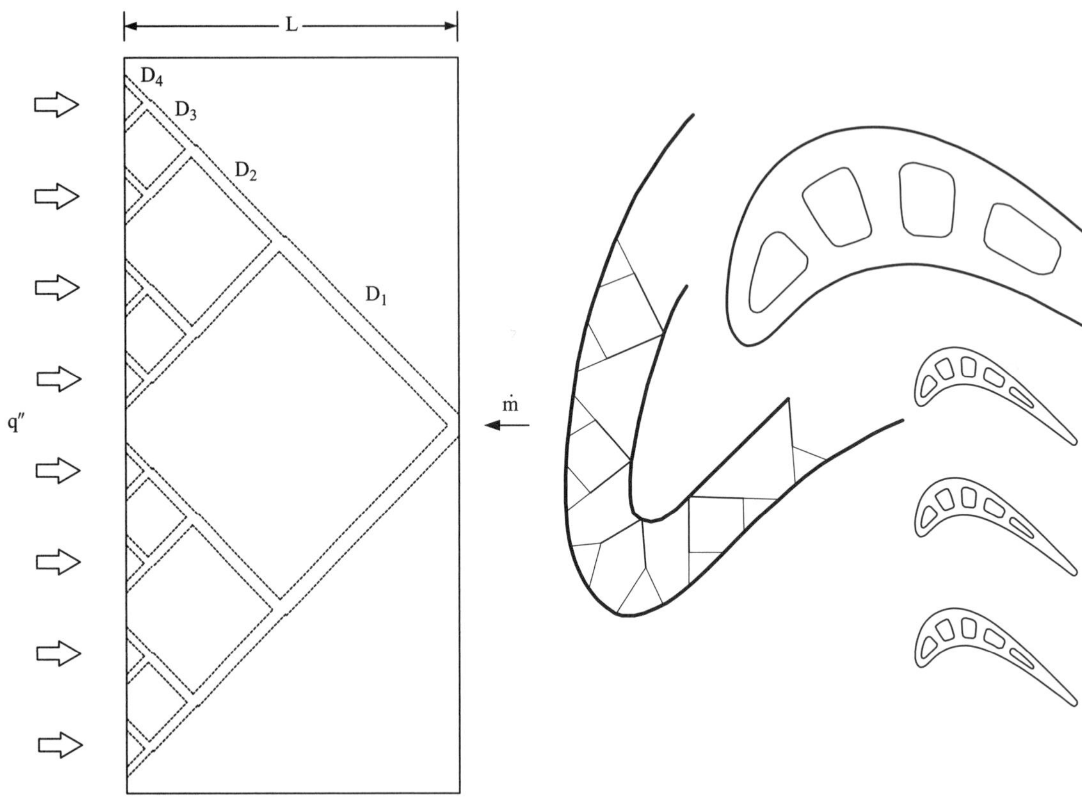

Figure 8.33 Tree-shaped channels for fighting with convection against intense side-heating [28]. Compare the dendritic channels with the parallel channels of Fig. 8.30.

of combustion (i.e., the high-*T* and high-*P* "working fluid" that comes from the combustion chamber and enters the turbine) [2]. The cooling gas comes from the core of the blade, and permeates through the vasculature that guides it to the surface that is under heat flux attack. We know to expect that this new kind of blade architecture can be optimized because we showed in the preceding section that a simpler configuration (Fig. 8.30) can be optimized.

This is the fundamental value of the pursuit of design "as science." Because design (the configuration generation phenomenon) has scientific principles that are now becoming known, it is possible to know even from an introductory course such as this where to expect *opportunities* for optimizing and discovering new configurations. How to pursue the discovery with less effort and time (i.e., with strategy) is the chief merit of learning design generation as a scientific subject. The newest work is illustrating this, from microchannels for heat sinks [29], heat exchangers [30], and fuel cells [31,32], to dendrites for cooling electronics [33–36], the modeling of living tissues vascularized with arterial and venous dendrites [37–40], self-healing materials [41] and photovoltaic cells [42].

At the end of the day, this engineering design paradigm makes a contribution to physics, to predicting nature. The structural properties uncovered in this chapter are qualitatively the same as those of *natural porous materials*, for example, the soil of the hill slope in a river drainage basin, where two scales dominate: fine porous soil with seepage and larger pores ("pipes") embedded in the fine structure. Very promising are the similarities that emerge between natural porous structures and the ones derived from principle. They shed light on the natural process that *generates* multiple scales and heterogeneity in natural flow systems such as hill slope drainage and living tissues. The fact that natural flow structures (the champions of flow perfection) have features similar to those discovered in constructal design lends confidence in pursuit of better engineering design with constructal theory.

REFERENCES

1. A. Bejan and S. Lorente, Constructal theory of generation of configuration in nature and engineering. *J Appl Phys*, Vol. 100, 2006, 041301.
2. A. Bejan, *Advanced Engineering Thermodynamics*, 3rd ed. (ch. 13). Hoboken, NJ: Wiley, 2006.
3. A. H. Reis, Constructal theory: From engineering to physics, and how flow systems develop shape and structure. *Appl Mech Reviews*, Vol. 59, 2006, pp. 269–282.
4. A. Bejan and S. Lorente, The constructal law and the thermodynamics of flow systems with configuration. *Int J Heat Mass Transfer*, Vol. 47, 2004, pp. 3203–3214.
5. A. Bejan and S. Lorente, *La Loi Constructale*. Paris: L'Harmattan, 2005.
6. S. R. White, N. R. Sottos, P. H. Geubelle, *et al.* Autonomic healing of polymer composites. *Nature*, Vol. 409, 2001, pp. 794–797.
7. D. Therriault, S. R. White, and J. A. Lewis, Chaotic mixing in three-dimensional microvascular networks fabricated by direct-write assembly. *Nature Materials*, Vol. 2, 2003, pp. 265–271.
8. A. Bejan, S. Lorente, and K.-M. Wang, Networks of channels for self-healing composite materials. *J Appl Phys*, Vol. 100, 2006, 033528.
9. K.-M. Wang, S. Lorente, and A. Bejan, Vascularized networks with two optimized channels sizes. *J Phys D: Appl Phys*, Vol. 39, 2006, pp. 3086–3096.
10. S. Kim, S. Lorente, and A. Bejan, Vascularized materials: Tree-shaped flow architectures matched canopy to canopy. *J Appl Phys*, Vol. 100, 2006, 063525.
11. S. Kim, S. Lorente, and A. Bejan, Vascularized materials with heating from one side and coolant forced from the other side. *Int J Heat Mass Transfer*, Vol. 50, 2007, pp. 3498–3506.
12. S. Lorente and A. Bejan, Heterogeneous porous media as multiscale structures for maximum flow access. *J Appl Phys*, Vol. 100, 2006, 114909.
13. S. Lorente, W. Wechsatol, and A. Bejan, Tree-shaped flow structures designed by minimizing path lengths. *Int J Heat Mass Transfer*, Vol. 45, 2002, pp. 3299–3312.
14. H. Zhang, S. Lorente, and A. Bejan, Vascularization with trees that alternate with upside-down trees. *J Appl Phys*, Vol. 101, 2007, 094904.

15. S. Lorente and A. Bejan, Svelteness, freedom to morph, and constructal multi-scale flow structures. *Int J Thermal Sciences*, Vol. 44, 2005, pp. 1123-1130.
16. R. K. Shah and A. L. London, *Advances in Heat Transfer*, Vol. 14. New York: Academic Press, 1978.
17. S. Bhattacharjee and W. L. Grosshandler, The formation of a wall jet near a high temperature wall under microgravity environment. *ASME HTD*, Vol. 96, 1988, pp. 711–716.
18. S. Petrescu, Comments on the optimal spacing of parallel plates cooled by forced convection. *Int J Heat Mass Transfer*, Vol. 37, 1994, p. 1283.
19. A. Bejan, *Advanced Engineering Thermodynamics*, 2nd ed. New York: Wiley, 1997.
20. A. Bejan, *Shape and Structure, from Engineering to Nature*, Cambridge, UK: Cambridge University Press, 2000.
21. K.-M. Wang, S. Lorente, and A. Bejan, Vascularization with grids of channels: Multiple scales, loop shapes and body shapes. *J Phys D: Appl Phys*, Vol. 40, 2007, pp. 4740–4749.
22. A. Bejan and M. R. Errera, Convective trees of fluid channels for volumetric cooling. *Int J Heat Mass Transfer*, Vol. 43, 2000, pp. 3105–3118.
23. J. Lee, S. Kim, S. Lorente, and A. Bejan, Vascularization with trees matched canopy to canopy: Diagonal channels with multiple sizes. *Int J Heat Mass Transfer*, Vol. 51, 2008, pp. 2029–2040.
24. F. Lundell, B. Thonon, and J. A. Gruss, Constructal networks for efficient cooling/heating. Second Conference on Microchannels and Minichannels, Rochester, NY, 2004.
25. M. Lallemand, F. Ayela, M. Favre-Marinet, *et al.*, Thermal transfer in microchannels: Applications to micro-exchangers. French Congress on Thermics, SFT 2005, Reims, May 30–June 2, 2005.
26. N. Kockmann, T. Kiefer, M. Engler, and P. Woias, Channel networks for optimal heat transfer and high throughput mixers. ECI International Conference on Heat Transfer and Fluid Flow in Microscale, Castelvecchio Pascoli, Italy, September 25–30, 2005.
27. A. Bejan, *Convection Heat Transfer*, 3rd ed. (p. 632). Hoboken, NJ: Wiley, 2004.
28. S. Kim, S. Lorente, and A. Bejan, Dendritic vascularization for countering intense heating from the side. *Int J Heat Mass Transfer*, Vol. 51, 2008, in press.
29. T. Bello-Ochende, L. Liebenberg, and J. P. Meyer, Constructal cooling channels for micro-channel heat sinks. *Int J Heat Mass Transfer*, Vol. 50, 2007, pp. 4141–4150.
30. V. A. P. Raja, T. Basak, and S. K. Das, Thermal performance of a multi-block heat exchanger designed on the basis of Bejan's constructal theory. *Int J Heat Mass Transfer*, Vol. 51, 2008, pp. 3582–3594.
31. J. C. Ordonez, S. Chen, J. V. C. Vargas, *et al.*, Constructal flow structure for a single SOFC. *Int J Energy Res*, Vol. 31, 2007, pp. 1337–1357.
32. C. E. Damian Ascencio, A. Hernandez Guerrero, J. A. Escobar Vargas, and C. Rubio-Arana, Three-dimensional numerical prediction of the current density in an innovative flow field pattern. Paper IMECE2007-42449, International Mechanical Engineering Congress and Exposition, Seattle, WA, November 11–15, 2007.
33. X.-Q. Wang, A. S. Mujumdar, and C. Yap, Numerical analysis of blockage and optimization of heat transfer performance of fractal-like microchannel nets. *J Electronic Packaging*, Vol. 128, 2006, pp. 38–45.
34. X.-Q. Wang, C. Yap, and A. S. Mujumdar, Laminar heat transfer in constructal microchannel networks with loops. *J Electronic Packaging*, Vol. 128, 2006, pp. 273–280.

35. X.-Q. Wang, A. S. Mujumdar, and C. Yap, Effect of bifurcation angle in tree-shaped microchannel networks. *J Appl Phys*, Vol. 102, 2007, 073530.
36. F. J. Hong, P. Cheng, H. Ge and G. T. Joo, Conjugate heat transfer in fractal-shaped microchannel network heat sink for integrated microelectronic cooling application. *Int J Heat Mass Transfer*, Vol. 50, 2007, pp. 4986–4998.
37. W. Dai, A. Bejan, X. Tang, L. Zhang, and R. Nassar, Optimal temperature distribution in a 3D triple layered skin structure with embedded vasculature. *J Appl Phys*, Vol. 99, 2006, 104702.
38. X. Tang, W. Dai, R. Nassar, and A. Bejan, Optimal temperature distribution in a three-dimensional triple-layered skin structure embedded with artery and vein vasculature. *Numerical Heat Transfer, Part A*, Vol. 50, 2006, pp. 809–834.
39. H. Wang, W. Dai and A. Bejan, Optimal temperature distribution in a 3D triple-layered skin structure embedded with artery and vein vasculature and induced by electromagnetic radiation. *Int J Heat Mass Transfer*, Vol. 50, 2007, pp. 1843–1854.
40. W. Dai, H. Wang, P. M. Jordan, R. E. Mickens, and A. Bejan, A mathematical model for skin turn injury induced by radiation heating. *Int J Heat Mass Tranfer*, Vol. 51, 2008, doi: 10.1016/j.ijheatmasstransfer.2008.01.006
41. M. R. Kessler, Self-healing: a new paradigm in materials design. *Proc IMechE. Part G: J Aerospace Engineering*, Vol. 221, Special issue paper, pp. 479–494.
42. A. M. Morega, J. C. Ordonez, P. A, Negoias, M. Morega, and R. Hovsapian, Optimal electrical disign of spherical photovoltaic cells, COMSOL Conference, Czech Technical University Prague, 26 October 2006.

PROBLEMS

8.1. Consider the fundamental two-objectives problem of providing large mechanical strength and low fluid-flow resistance in a wall vascularized with parallel channels. As shown in Fig. P8.1, the wall thickness is H and the spacing between channels is B. The channel cross-section is the rectangle $h \times b$. The volume fraction occupied by the channels is fixed, ϕ. The mass flow rate through the wall is fixed. This means that if $\dot{m}_1$ is the mass flow rate through one channel, then the ratio $\dot{m}_1/B$ is fixed. The flow through every channel is in the Poiseuille regime. The friction factor can be approximated as (Ref. [27], p. 110)

$$f \cong \frac{h^2 + b^2}{(h+b)^2} \frac{24}{\mathrm{Re}_{D_h}}$$

where Re_{D_h} is the channel Reynolds number based on hydraulic diameter. Express the end-to-end pressure difference ΔP as a function of b and h. Next express the resistance modulus (the stiffness) of the wall cross-section as a function of geometry, $W = \left(BH^3 - bh^3\right)/6H$. For simplicity, assume that $B = H$, and show that ΔP is minimum when $b = h$. Show how the minimum ΔP varies with the stiffness of the wall structure.

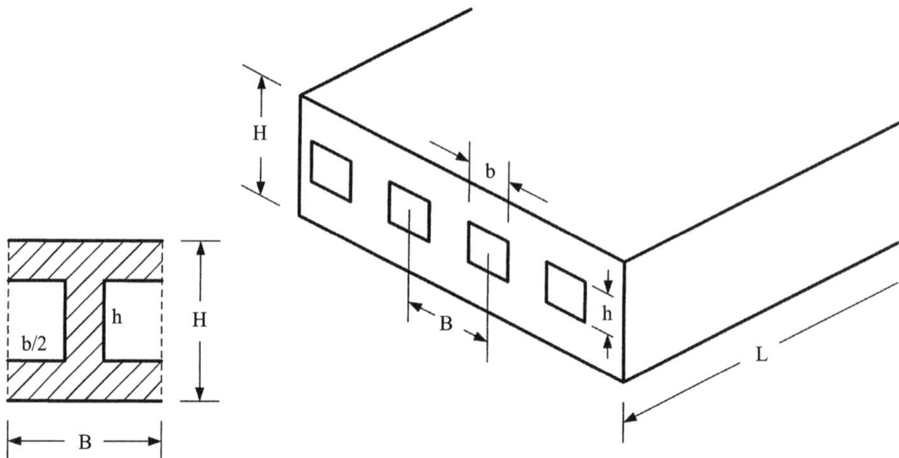

Figure P8.1

8.2. Here we determine the optimal tapering of a slender tube with Poiseuille flow and round cross-section. The tube length is L, and the axial coordinate is x. The tube diameter varies along the tube, $D(x)$. The tube volume is fixed. The stream with mass flow rate $\dot{m}_L$ enters the tube through its $x = L$ end. The mass flow rate $\dot{m}(x)$ varies along the tube because fluid leaks laterally (perpendicular to x) at a uniform rate $d\dot{m}/dx$ that is proportional to x^p, where p is a known constant. The $x = 0$ end of the tube is closed. The pressure difference maintained between $x = L$ and $x = 0$ is ΔP. Determine the tube shape $D(x)$ for which the overall flow resistance $\Delta P/\dot{m}_L$ is minimal. Hint: assume power-law solutions, $\dot{m} = ax^{1+p}$ and $D = bx^n$, and determine the constants a, b, and n.

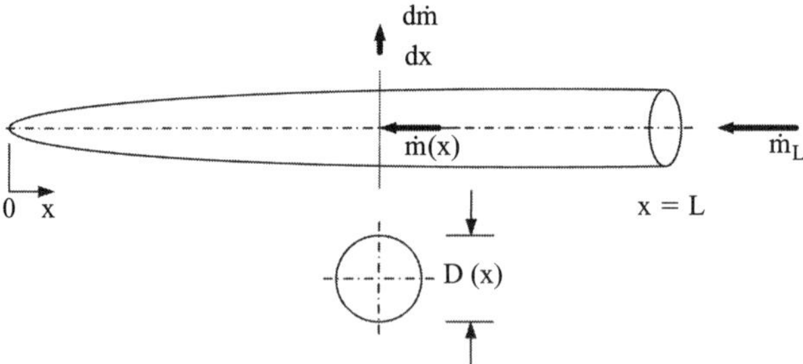

Figure P8.2

8.3. The longitudinal flow through a slender porous blade obeys the Darcy-flow model $u = (K/\mu)dP/dx$, where u is the longitudinal volume averaged fluid velocity, K is the permeability, μ is the viscosity, and dP/dx is the longitudinal pressure gradient at x. The blade has a two-dimensional geometry. The flow length is L, and the blade thickness $D(x)$ is not specified. Fixed is the area of the blade profile, $A = \int_0^L D\,dx$, or the total blade volume ($V = AW$), where W is the blade width perpendicular to A. The stream with mass flow rate $\dot{m}_L$ enters the blade through the $x = L$ end, that is, by flowing through the cross-section $D(L) \times W$. The $x = 0$ end is closed. The fluid escapes laterally at a uniform rate per unit area, $\dot{m}''$. The flow is driven by the pressure difference ΔP maintained between the $x = L$ and $x = 0$ ends of the blade. Determine the shape of the blade profile [the function $D(x)$] such that the overall flow resistance $\Delta P/\dot{m}_L$ is minimal. Hint: assume power-law solutions, $\dot{m} = ax^m$ and $D = bx^n$, and determine a, b, m, and n.

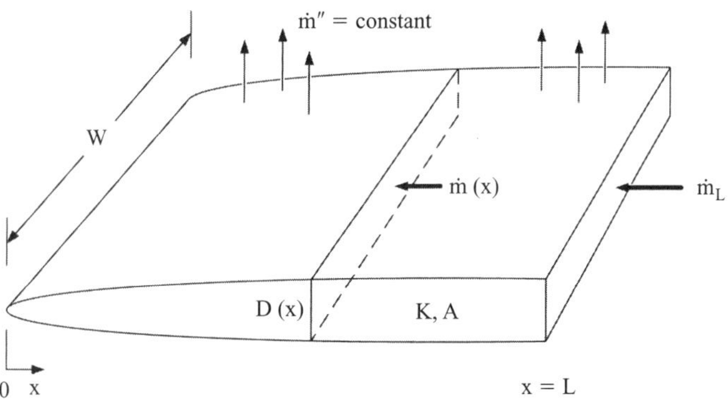

Figure P8.3

8.4. The architecture of the tree leaf can be discovered by matching the solutions of Figs. P8.2 and P8.3, as shown in Fig. P8.4. The tapered tube $D(x)$ is the nerve of volume V_{nerve} and length L. The leaf material is divided into two symmetric parts. Each part has the width $H(x)$ and the cross-sectional profile $A(x)$. Note that H corresponds to L in Fig. P8.3. The surface of leaf material seen from above in Fig. P8.4 is S. The volume of leaf material is V_{leaf}. The total volume of the structure ($V = V_{\text{nerve}} + V_{\text{leaf}}$) is fixed.

Match the solutions of the preceding two problems so that the continuity of mass flow and pressure is respected along the cuts shown in Fig. P8.4. The overall pressure difference between the nerve root and the rim of the leaf material is ΔP. Mass flows along the nerve and vertically along the leaf material. There is no flow perpendicular to cross-sections of type A. The mass flux from the leaf surface to the atmosphere is $\dot{m}''$(constant), which means that $\dot{m}'' = \dot{m}_L/(2S)$.

Show that the nerve and leaf solutions match when D is proportional to $x^{5/14}$ and H to $x^{8/7}$. Express the overall flow resistance $\Delta P/\dot{m}_L$ in terms of only V_{nerve}, V_{leaf}, S, L and K, and seek to minimize it subject to constant-V and other constraints that you consider reasonable and necessary.

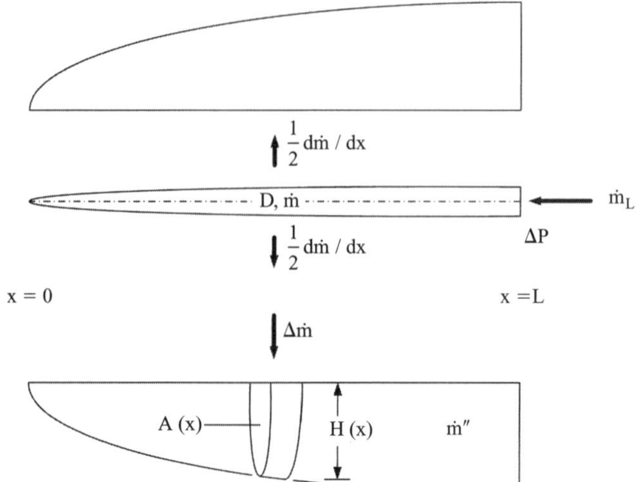

Figure P8.4

8.5. The one-nerve leaf designed in Problem 8.4 can be used as a building block (one lobe) in the construction of a larger leaf with many lobes. The construction begins with solution to Problem 8.4, which is sketched on the left side of Fig. P8.5. The length $2H(L)$ is the base of the lobe of volume V. There will be two lobes installed symmetrically on a tapered central tube $D(x)$. The side volume installed per unit length of tube is $V' = 2V/2H(L)$. The mass flow rate per unit of tube length, which is extracted laterally from the tube is $\dot{m}' = 2\dot{m}_L/2H(L)$. With V' and $\dot{m}'$ as unknown functions of x, redo the analysis of Problems 8.3 and 8.4 with the objective of minimizing the global flow resistance $\Delta P/\dot{m}$. The sum of the central tube volume and the total side volume is fixed.

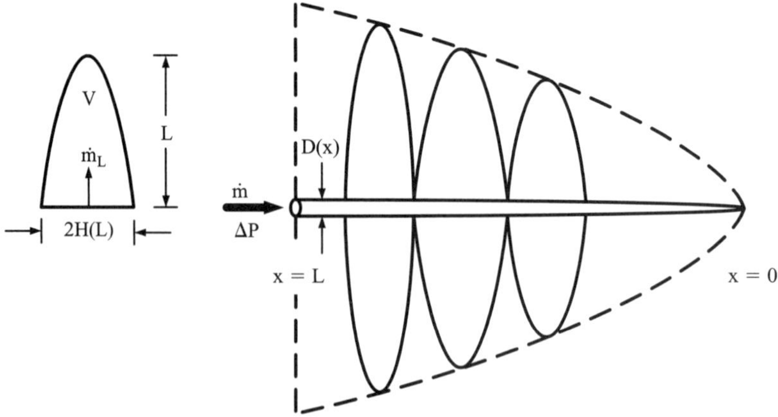

Figure P8.5

8.6. Laptops are getting thinner and more powerful than before. For the heat generated by the electronics, this is a problem that defines an important frontier for the cooling of electronics in general, at all scales. The heat generation rate is distributed nonuniformly—usually concentrated in a major component inside "the box." The available heat sink is all around: the atmosphere. How can the heat current flow with maximum access from its localized origin to the heat sink that touches the outer surface of the volume that holds all the heat-generating components?

The tree leaf constructed in Problems 8.2 through 8.5 shows the way. Not the natural leaf per se, but the principle-based sequence in which we constructed the leaf. In constructal design, we do not copy nature—we construct flow architectures for greater access, and in this way we discover flow architectures that also occur in nature.

The simplest "leaf" viewing of a laptop that generates heat at one point (in a special component) and rejects this heat to the surrounding air is a rectangular plate of length L, width $2B$ and small thickness t, in this order: $L > 2B > t$, Fig. P8.6 The heat current (q) originates from the point (T_{max}) situated in the middle of the left side. It flows by conduction along the plate of temperature $T(x,y)$, and it is rejected to the ambient of temperature T_{min}. The convective heat transfer coefficient between the plate and ambient is constant (h).

To help the flow of heat along the leaf, we install a longitudinal "nerve" of constant cross-sectional area A and length L. The shape of A is not important. The thermal conductivity of the nerve material is the same as that of the plate (k). The total volume of conducting material is fixed, $V = (A + 2Bt)\,L$. The question is how to allocate the material between the plate ($2BtL$) and the nerve (AL), so that the overall flow conductance $q/(T_{max} - T_{min})$ is maximized.

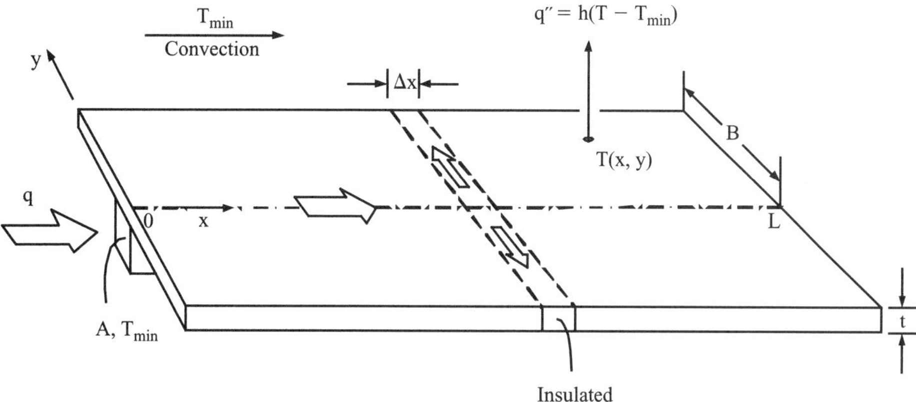

Figure P8.6

To set up the analysis, measure x along the nerve, from $x = 0$ at the origin of q, to $x = L$ at the adiabatic edge of the rectangular plate $L \times 2B$. Conduction along the nerve is continued by conduction perpendicular to the nerve, along the plate, in the y direction. The plate with nerve is a plate fin with rib. This structure can be analyzed as an assembly of lateral fins of width Δx, length B and thickness t, installed on a stem (A). All these geometrical features are x-independent, but they vary according to how we allocate V. It is known that the profile of the lateral fin (t/B) with fixed volume and maximum thermal conductance is $t/B \cong (ht/k)^{1/2}$ (cf. A. Bejan, *Heat Transfer*. New York: Wiley, 1993, pp. 65–67). The corresponding heat current through the lateral elemental fin $\Delta x \times t \times B$, from the nerve $[T(x)]$ to the surrounding air flow ($T_{\min}$) is $\Delta q \cong 1.26\, k\, \Delta x (T - T_{\min})(ht/k)^{1/2}$. At every Δx along L, the nerve loses two Δq's, one to each side of the nerve. Write the fin conduction equation for the nerve, integrate, estimate $q/(T_{\max} - T_{\min})$, and maximize this global quantity subject to the total material volume constraint.

8.7. Red blood cells flow through capillary vessels in organized fashion: equidistantly. They are separated by equal lengths of fluid, all flowing with the same longitudinal velocity. The constructal law calls for the flow configuration that flows most easily. Show that the theoretical (self-lubrication) configuration is the one sketched in Fig. P8.7a. Base your argument on a comparison between (a) and (b). Use scale analysis to estimate the wall shear stresses along the capillary of length L. Model the fluid as Newtonian with constant properties.

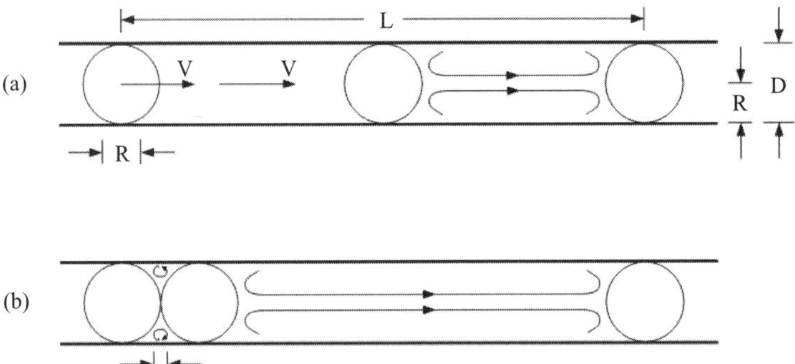

Figure P8.7

8.8. It is observed that the density of red blood cells in the bloodstream decreases when the speed (or flow rate) of the stream increases. This behavior is illustrated in Fig. P8.8 with two streams: if $V_1 > V_2$, then $\lambda_1 > \lambda_2$. Construct an argument from which to expect the proportionality $\lambda_2/\lambda_1 \sim V_2/V_1$. One way is to recognize the procession of red blood cells as a stream of oxygen, which flow longitudinally and is deposited transversally by mass diffusion across the duct wall of thickness δ. The diffusivity of O_2 through the wall material is constant (D). Match the time of mass diffusion across the wall to the time of arrival of one O_2 carrier (one cell) at the wall site where transversal diffusion occurs.

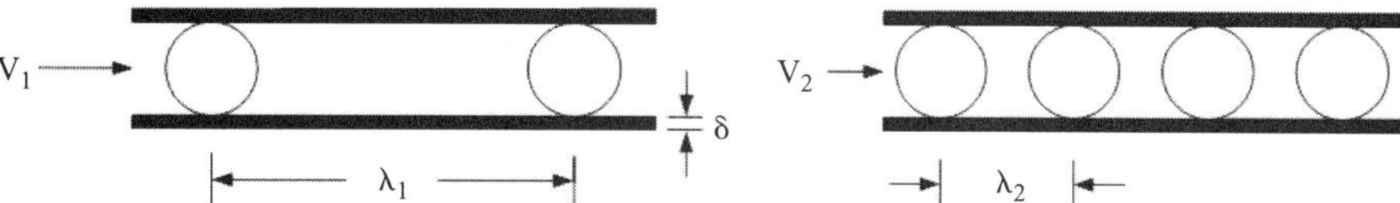

Figure P8.8

9

CONFIGURATIONS FOR ELECTROKINETIC MASS TRANSFER

The transfer of species through porous media is a broad topic that has applications in many fields and at many different scales. Remarkable is that this topic concerns as much natural systems as engineered structures [1]. Vascularized tissues do not have much in common with the filters used against aerosol particles: the first is a living structure, while the second is man-made. Yet, both are porous media through which species diffuse. Because diffusion is generally a very slow process, often it is necessary to accelerate it by means of an external electrical current, or by imposing a gradient of electrical potential, when the species are electrically charged species.

Widely varying applications of what has become known as electrokinetic processes may be found in the literature [2–7]. Scale is not a dividing issue: see, for example, the delivery of drugs by iontophoresis through the human body [8], and the dechlorination of concrete structures such as bridges contaminated in sea water [9,10]. In addition to enhancing transport, the electrical current allows us to control it by simply choosing the polarity of the electrodes.

9.1 SCALE ANALYSIS OF TRANSFER OF SPECIES THROUGH A POROUS SYSTEM

We start with the simple case of molecular diffusion. The concentration profile $c(x,t)$ of a molecular species diffusing through a nonreactive porous medium with porosity* p, is obtained by solving the continuity equation, under the assumption

*Note the use of p instead of ϕ for porosity in this Chapter.

that the representative elementary volume model (REV) is valid:

$$p\frac{\partial c}{\partial t} = -\nabla \cdot J \tag{9.1}$$

$$J = -D\nabla c \tag{9.2}$$

Here, D is the effective diffusion coefficient of the species, and J is the flux of species. In this case, Eq. (9.2) is Fick's first law of diffusion [11] while the continuity equation is also known as the Fick's second law of diffusion. Its one-dimensional form is

$$\frac{\partial c}{\partial t} = \frac{D}{p}\frac{\partial^2 c}{\partial x^2} \tag{9.3}$$

Let L be the porous medium thickness, and Δc the maximum concentration difference taking place in the material, so that $c = \Delta c$ at $x = 0$, and $c = 0$ at $x = L$. The initial conditions are $c = 0$ in $0 < x < L$, at $t = 0$. The scale analysis of Eq. (9.3) provides an estimate of the time needed by the concentration front of the diffusing species to cross the porous medium, that is, the time necessary for the side $x = L$ to have a concentration different from 0 (Fig. 9.1). In this scale analysis [12] we follow the notation proposed by in Ref. [13]. See Appendix A for more details. The order of magnitude of the left-hand side term in Eq. (9.3) is

$$\frac{\partial c}{\partial t} \sim \frac{\Delta c}{t} \tag{9.4}$$

On the right-hand side, we can write

$$\frac{D}{p}\frac{\partial^2 c}{\partial x^2} = \frac{D}{p}\frac{\partial}{\partial x}\left(\frac{\partial c}{\partial x}\right) \approx \frac{D}{p}\frac{\Delta c}{L^2} \tag{9.5}$$

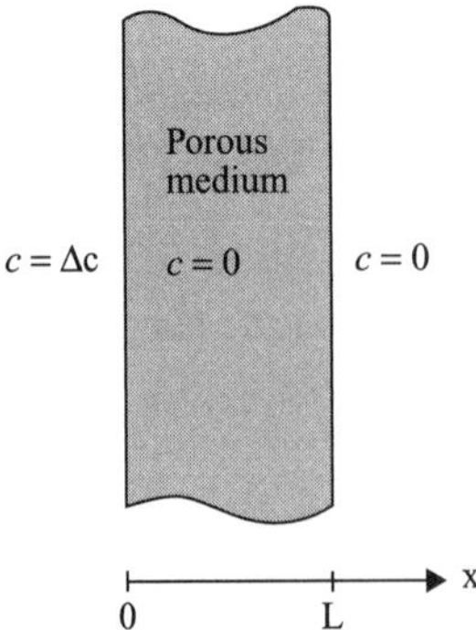

Figure 9.1 Pure diffusion: geometry, boundary, and initial conditions [12].

Equation (9.3) states that the two orders of magnitude given by Eqs. (9.4) and (9.5) are equal; therefore,

$$t \sim p\frac{L^2}{D} \tag{9.6}$$

The diffusion of ionic species is a distinct class of diffusion problems [14]. Because electrical interactions exist between the ionic species in solution, an additional term has to be included in the flux equation. This term is due to the membrane potential, that is, the potential created between the different ionic species in the pore solution [15]. Fick's first law of diffusion is replaced by the Nernst-Planck equation which has to be written for every diffusing species,

$$J_i = -D_i\left(\nabla c_i + z_i\frac{F}{RT}c_i\nabla\varphi\right) \tag{9.7}$$

where subscript i stands for the ionic species i, z is the charge number, F is the Faraday constant, R is the ideal gas constant, T is the absolute temperature, and φ is the membrane potential. Equation (9.3) becomes

$$\frac{\partial c_i}{\partial t} = \frac{D_i}{p}\frac{\partial}{\partial x}\left(\frac{\partial c_i}{\partial x} + z_i\frac{F}{RT}c_i\frac{\partial\varphi}{\partial x}\right) \tag{9.8}$$

An additional equation is required in order to close the problem. That is the current conservation equation,

$$0 = F\sum_i z_i J_i \tag{9.9}$$

where J_i represents the ionic flux through the porous medium (Eq. 9.7).

The electrical field can be calculated from Eqs. (9.7) and (9.9):

$$\frac{\partial\varphi}{\partial x} = -\frac{RT}{F}\frac{\sum_i z_i D_i(\partial c_i/\partial x)}{\sum_i z_i^2 D_i c_i} \tag{9.10}$$

The scale analysis starts with Eq. (9.10), from which we obtain

$$\frac{\Delta\varphi}{L} \sim \frac{RT}{F}\frac{D_0\Delta c/L}{D_0\Delta c} \tag{9.11}$$

where D_0 is an effective diffusion coefficient of reference. Finally, Eq. (9.11) yields

$$\Delta\varphi \sim \frac{RT}{F} \tag{9.12}$$

Equation (9.12) reveals the order of magnitude of the electrical potential created by the ionic interactions. If we assume an absolute temperature of 300 K, the electrical potential is around 26 mV, which is comparable with experimental results obtained when measuring the membrane potential corresponding to sea water attack through concrete, for example, Ref. [16]. The order of magnitude of the electrical potential given by Eq. (9.12) is used to estimate the scales of the terms in the Nernst-Planck equation, Eq. (9.8),

$$\frac{\Delta c}{t} \sim \frac{D_0}{pL}\left(\frac{\Delta c}{L}\right), \frac{D_0}{pL}\left(\frac{F}{RT}\Delta c\frac{RT}{FL}\right) \tag{9.13}$$

which, again, leads to Eq. (9.6) if the second scale on the right hand side is negligible relative to the first:

$$t \sim p\frac{L^2}{D_0} \tag{9.6}$$

There are applications in which an external electrical curent is applied. The technique is chosen in order to accelerate the penetration of a certain ionic species, or to control the ionic transport during the decontamination of the porous medium. Assuming that the porous medium is subjected to a constant current density j applied from outside, the current conservation equation [Eq. (9.9)] and the electrical field [Eq. (9.10)] become

$$j = F\sum_i z_i J_i \tag{9.14}$$

$$\frac{\partial\varphi}{\partial x} = -\frac{RT}{F}\frac{(j/F)+\sum_i z_i D_i\ (\partial c_i/\partial x)}{\sum_i z_i^2 D_i c_i} \tag{9.15}$$

We apply the scale analysis method to Eq. (9.15), and obtain

$$\frac{\Delta\varphi}{L} \sim \frac{RT}{F^2}\frac{j}{D_0\Delta c}, \frac{RT}{F}\frac{D_0\,\Delta c/L}{D_0\Delta c} \tag{9.16}$$

$$\frac{\Delta\varphi}{L} \sim \frac{RT}{F^2}\frac{j}{D_0\Delta c}, \frac{RT}{F}\frac{1}{L} \tag{9.17}$$

Consider now the example of the attack of sea water through concrete. The chloride effective diffusion coefficient through cement-based materials is 10^{-12} m^2/s in an order of magnitude sense, while Δc is 10^3 mol/m^3. The dechlorination of concrete is usually performed in order to avoid the corrosion of the steel bars of reinforced concrete [17]. The closest bars are located a few centimeters from the material surface, therefore L is of order 10^{-2} m. The order of magnitude of j is 1 A/m^2. Therefore, $j/(FD_0\Delta c) \sim 10^4\ m^{-1}$, and $1/L \sim 10^2\ m^{-1}$. From the scale analysis rules set in Ref. [13], if in the sum of two terms the order of magnitude of one term is greater than the order of magnitude of the other term, then the order of magnitude of the sum is dictated by the dominant term. This is the case in Eq. (9.16), where the external electrical field is much greater than the ionic potential.

Consequently, neglecting the right-hand side term in Eq. (9.17), we find

$$\Delta\varphi \sim \frac{RT}{F^2}\frac{Lj}{D_0\Delta c} \tag{9.18}$$

This new order of magnitude of the electrical field is introduced into the scaling of Eq. (9.8),

$$\frac{\Delta c}{t} \sim \frac{D_0}{pL}\left(\frac{\Delta c}{L}\right), \frac{D_0}{pL}\left(\frac{F}{RT}\Delta c\frac{RT}{F^2}\frac{j}{D_0\Delta c}\right) \tag{9.19}$$

Assuming that electrical effect [the second term on the right-hand side of Eq. (9.19)] is greater than the molecular diffusion effect [the first term in the right-hand side of Eq. (9.19)], we obtain

$$\frac{\Delta c}{t} \sim \frac{j}{pLF} \tag{9.20}$$

The time scale of the ionic transport is therefore

$$t \sim p\frac{LF\Delta c}{\Delta j} \tag{9.21}$$

Let t_{diff} be the time scale in the natural diffusion of ionic species [from Eq. (9.6)] and $t_{\text{elec.diff}}$ the time scale corresponding to Eq. (9.21). The ratio $t_{\text{diff}}/t_{\text{elec.diff}}$ shows the influence of the externally applied electrical current on the time necessary to an ion to cross the porous medium:

$$\frac{t_{\text{diff}}}{t_{\text{elec.diff}}} \sim \frac{Lj}{D_0F\Delta c} \tag{9.22}$$

Using the same values as the ones chosen to illustrate Eq. (9.17), we obtain $t_{\text{diff}} \sim 10^2 t_{\text{elec.diff}}$. Increasing the external electrical current will increase the ratio $t_{\text{diff}}/t_{\text{elec.diff}}$, and therefore it will decrease the time needed by the ionic species to cross the porous medium.

9.2 MODEL

The concentration profiles of the ionic species in the pore solution are obtained by solving Eq. (9.8) for each species. The link between the continuity equations for the individual species is the membrane potential given by Eq. (9.15). In this work it is assumed that the local electroneutrality condition can be applied [18]; therefore,

$$\sum_i z_i c_i = 0 \tag{9.23}$$

and the Poisson equation becomes

$$\nabla^2\varphi = 0 \tag{9.24}$$

Equation (9.8) can be rewritten as

$$\frac{\partial c_i}{\partial t} = \frac{D_i}{p}\left(\frac{\partial^2 c_i}{\partial x^2} + z_i \frac{F}{RT}\frac{\partial \varphi}{\partial x}\frac{\partial c_i}{\partial x}\right) \tag{9.25}$$

The system constituted by Eqs. (9.25) and (9.15) can be written in dimensionless form by using the dimensionless variables: $\tilde{c}_i = c_i/\Delta c$, $\tilde{D}_i = D_i/D_0$, and $\tilde{x} = x/L$. The scale analysis conducted in the preceding section gives us the orders of magnitude of the time scale, and the potential difference created in the porous medium. These two results are used to write the t, and φ variables in dimensionless form as $\tilde{t} = t/\Delta t$, and $\tilde{\varphi} = \varphi/\Delta\varphi$, where $\Delta t = pL^2B/D_0$, and $\Delta\varphi = RT/FB$. The new dimensionless group B is defined as,

$$B = \frac{F\Delta c D_0}{Lj} \tag{9.26}$$

Finally, Eqs. (9.25) and (9.15) become

$$\frac{\partial \tilde{c}_i}{\partial \tilde{t}} = \tilde{D}_i\left(B\frac{\partial^2 \tilde{c}_i}{\partial \tilde{x}^2} + z_i \frac{\partial \tilde{\varphi}}{\partial \tilde{x}}\frac{\partial \tilde{c}_i}{\partial \tilde{x}}\right) \tag{9.27}$$

$$\frac{\partial \tilde{\varphi}}{\partial \tilde{x}} = -\frac{1 + B\sum\limits_i z_i \tilde{D}_i\,(\partial \tilde{c}_i/\partial \tilde{x})}{\sum\limits_i z_i^2 \tilde{D}_i \tilde{c}_i} \tag{9.28}$$

According to Eq. (9.26), the higher the external electrical current, the lower the B value. A low B value means that the effect of the electrical interactions between the ionic species in the pore solution decreases when the current is increased, Eq. (9.28). In addition, the transport of species is due more to electrical effects than to the diffusion effects: this can be seen in Eq. (9.27), where the first term on the right-hand side becomes smaller when the external current is increased. Note that in order to be consistent with the preceding section, B is equal to 1 in the pure diffusion limit.

The system formed by Eqs. (9.27) and (9.28) was solved with a second-order finite differences scheme. The diffusion term in Eq. (9.27) was written with a Crank-Nicholson scheme, while the convection term was treated with a fully implicit upwind Lax-Wendroff scheme. Grid independence tests were performed. The dependence of the results on the time step was also checked. The grid spacing and the time step were made small enough to ensure a solution that is independent of the grid size and time step. Specifically, if the time step is Δt, then from one test to the next the Δt value was divided by 2 until the following criterion was satisfied: $|(c_{i,\delta t} - c_{i,\delta t/2})/c_{i,\delta t/2}| \le 0.1\%$.

9.3 MIGRATION THROUGH A FINITE POROUS MEDIUM

In this section we consider the penetration of an ionic species through a porous medium in the presence of an electrical current. A porous medium sample is placed between two electrolytic solutions. The height and width of the sample are sufficiently large compared with its thickness, so that the problem may be modeled as one-dimensional (Fig. 9.2). An electrical current is imposed by means of two electrodes placed between the two faces of the porous medium.

Assume that the anode is located on the right side of Fig. 9.2. The ionic species Ion-1 is an anion. Initially the porous medium and the anodic compartment are free of this ionic species. Ion-1 is combined with three other ions. The charge number of each species is given in Table 9.1, along with their effective diffusion coefficients, the boundary conditions (cathodic compartment and anodic compartment), and the initial conditions (pore solution). Note that the boundary conditions are maintained constant in time. The concentrations are specified in a nondimensional way.

At $\tilde{t} = 0$, the two compartments are filled with their respective solutions, and the electrical current is applied for a given time. Several values of the electrical current are tested. They are listed in Fig. 9.3 with their respective B values. Figure 9.3 shows the concentration profiles of Ion-1 obtained. Also plotted in Fig. 9.3 is the concentration profile corresponding to the pure diffusion problem. The flux of species is due to the contribution of several terms. Combining Eqs. (9.7) and (9.15), we obtain:

$$J_i = -D_i \left(\frac{\partial c_i}{\partial x} - z_i c_i \frac{\sum\limits_i z_i D_i \left(\partial c_i / \partial x \right)}{\sum\limits_i z_i^2 D_i c_i} - z_i c_i \frac{j/F}{\sum\limits_i z_i^2 D_i c_i} \right) \tag{9.29}$$

The first term inside the parentheses corresponds to the diffusive part. The second term is due to the ionic interactions (i.e. the membrane potential). The third term exists because an external current is applied.

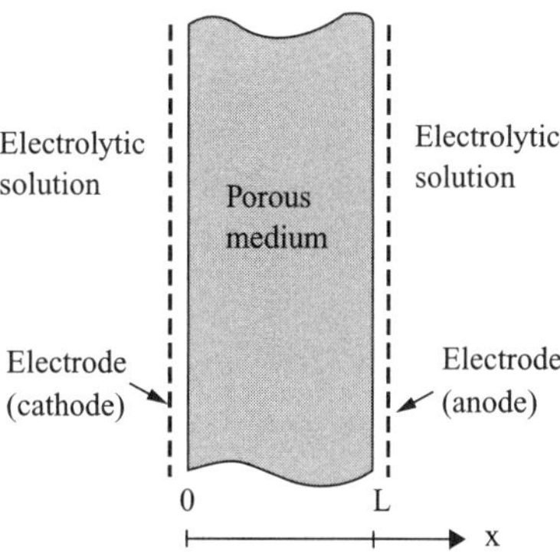

Figure 9.2 Sketch of a porous layer exposed to an external electrical current [12].

Table 9.1 Boundary and initial conditions

					Anodic Compartment	Anodic Compartment
	Charge Number	Effective Diffusion Coefficient (m^2/s)	Cathodic Compartment	Pore Solution	Migration through a finite porous medium, Fig. 9.2	Symmetric contamination, Fig. 9.7
Ion-1	−1	10^{-12}	1	0	0	1
Ion-2	−1	$2.6\ 10^{-12}$	0.01	0	0	0.01
Ion-3	+1	$0.7\ 10^{-12}$	1	0	0	1
Ion-4	+1	10^{-12}	0.01	0	0	0.01

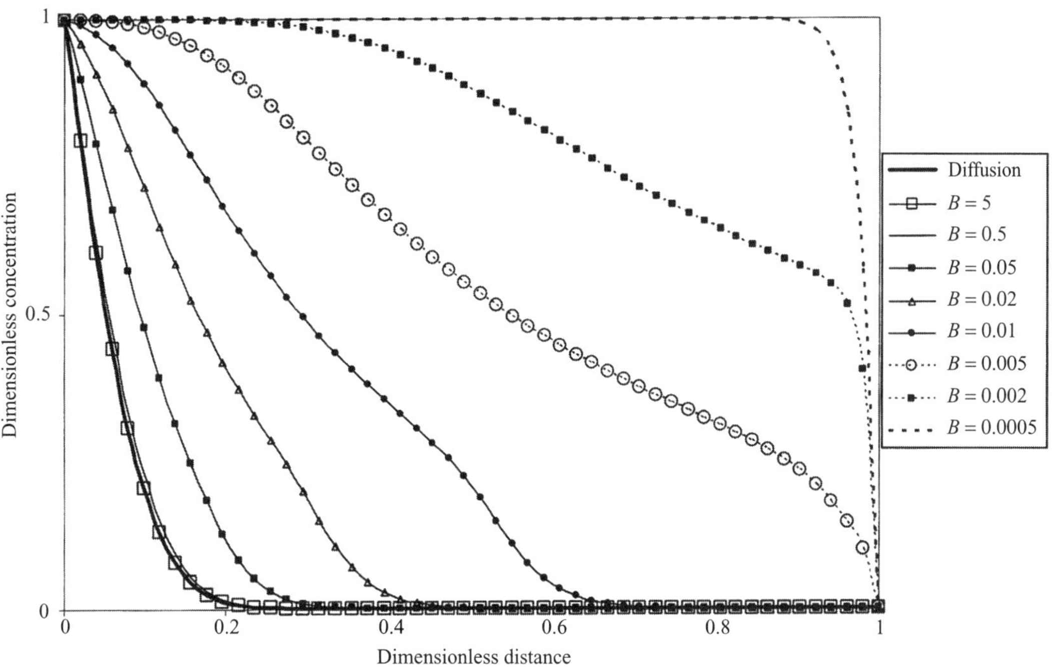

Figure 9.3 Concentration profiles during electrokinetic transfer and diffusion [12].

The purpose of the electrical current in this example is to enhance the transport of Ion-1. This can be seen in Fig. 9.3, where the higher the current density (the lower the B value), the higher the penetration on Ion-1 through the porous medium. Note that the shape of the concentration profiles changes significantly with the intensity of the current. Above $B = 0.05$, the concentration profiles are very similar to the one corresponding to pure diffusion (i.e., the limit $j = 0$). When the current density increases, the convective part of Eq. (9.27) has a greater impact on the results, as shown in the case $B = 0.01$. For this value of the current density and higher, Ion-1 has reached the opposite face of the sample.

An approximate solution to the system formed by Eqs. (9.25) and (9.15) can be obtained by assuming that the electrical current is the only driving force,

$$\frac{\partial c_i}{\partial t} = \frac{D_i}{p}\left(z_i \frac{F}{RT}\frac{\partial \varphi}{\partial x}\frac{\partial c_i}{\partial x}\right) \tag{9.30}$$

$$\frac{\partial \varphi}{\partial x} = -\frac{RT}{F}\frac{j/F}{\sum_i z_i^2 D_i c_i} \tag{9.31}$$

This is equivalent to setting the B parameter equal to zero in Eqs. (9.27), and (9.28). The system becomes

$$\frac{\partial \tilde{c}_i}{\partial \tilde{t}} = \tilde{D}_i\, z_i \frac{\partial \tilde{\varphi}}{\partial \tilde{x}} \frac{\partial \tilde{c}_i}{\partial \tilde{x}} \tag{9.32}$$

$$\frac{\partial \tilde{\varphi}}{\partial \tilde{x}} = -\frac{1}{\sum_i z_i^2 \tilde{D}_i \tilde{c}_i} \tag{9.33}$$

Figures 9.4a–c show the results obtained when the B parameter is respectively 0.1, 0.05, and 0.01. The bold curves correspond to the exact solution, Eqs. (9.27), and (9.28), while the dashed curves are the concentration profiles obtained by solving Eqs. (9.32), and (9.33). The assumption that the external electrical current is the only driving force leads to a significant discrepancy in the concentration profiles when the B parameter is higher than 0.01. Plotted in Fig. 9.4a is also the concentration profile obtained during a diffusion case when the membrane potential is due only to the interactions between the ions of the pore solution. The combination of the different effects (namely, concentration gradient, electrical potential due to both the ionic interactions, and the external electrical current) leads to the concentration profile shown with bold black line. For B values of 0.01 and

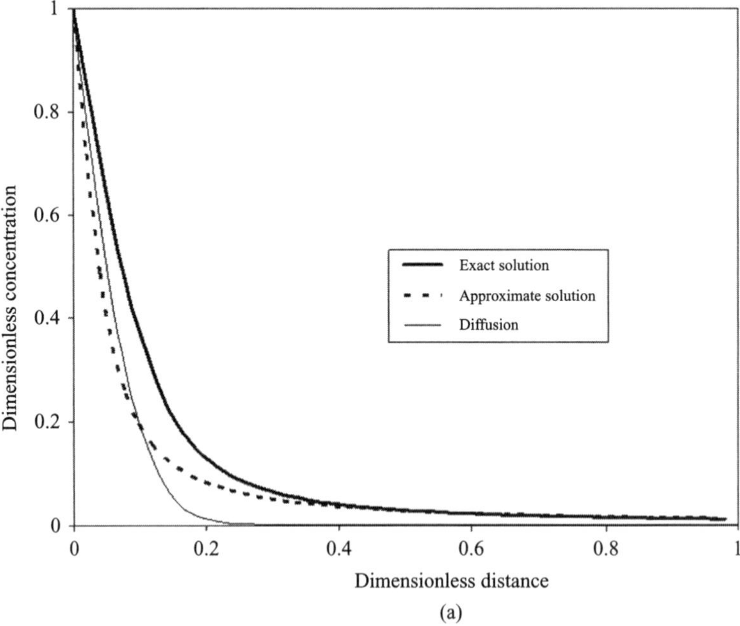

Figure 9.4 Concentration profiles obtained by solving the complete problem [Eqs. (9.27) and (9.28)] and by assuming that the external current is the only driving force [Eqs. (9.32) and (9.33)]. (a) $B = 0.1$, (b) $B = 0.05$, and (c) $B = 0.01$ [12].

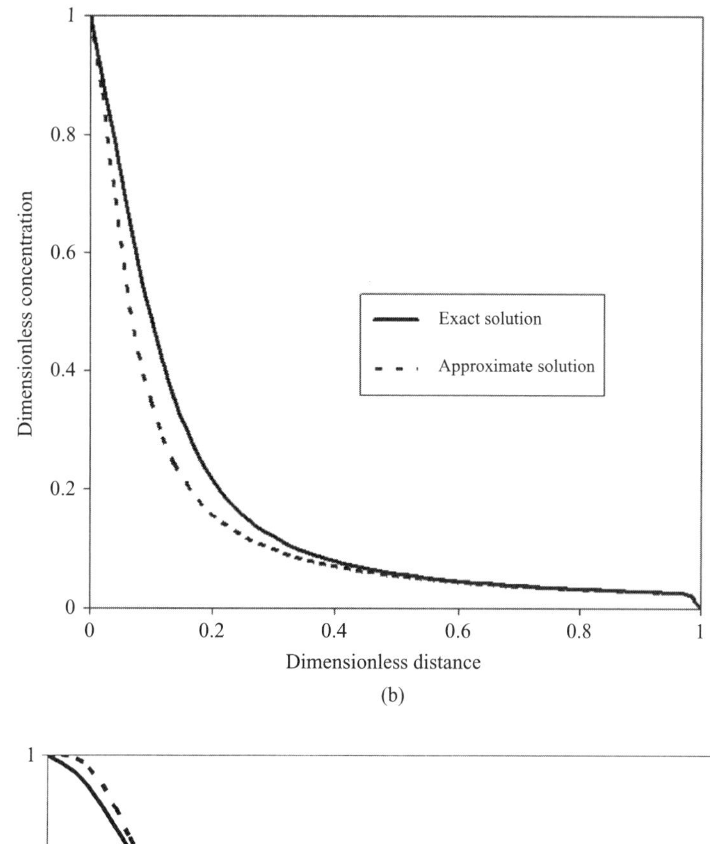

(b)

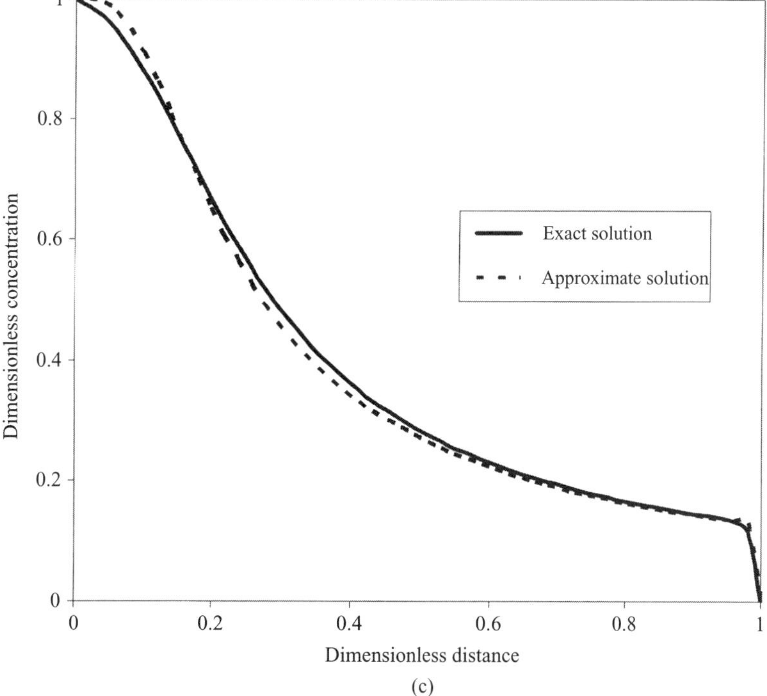

(c)

Figure 9.4 (*Continued*)

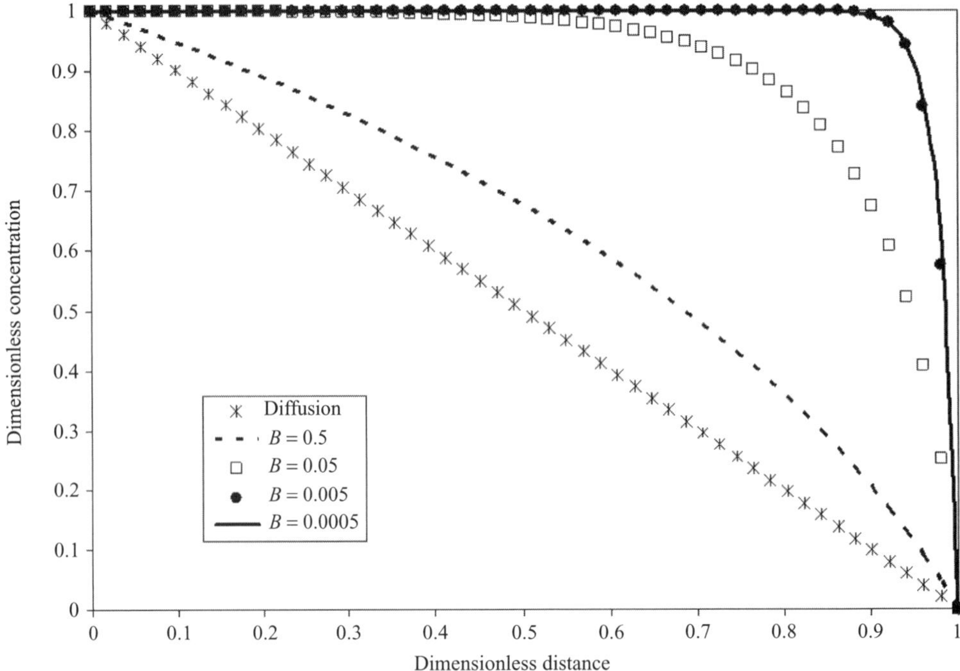

Figure 9.5 Steady-state concentration profiles during electrokinetic transfer and diffusion [12].

lower, the approximate solution becomes accurate. Accounting for the effect of concentration gradients effects is no longer necessary.

To complete the study of the impact of an external electrical current on the ionic transport, an analysis of the time necessary to reach steady state was performed. Typically, the calculations were conducted until the relative difference in concentrations between two consecutive simulations was $|(c_{k.\delta t} - c_{(k+1).\delta t})/c_{k.\delta t}| \leq 0.1$, where k represents the number of iterations. The concentration profiles corresponding to steady state are presented in Fig. 9.5 for the diffusion case and for several current densities represented by the B parameter. The profiles differ depending on the nature of the driving force. Purely diffusive transport is characterized by a linear concentration profile. The profile becomes more convex as the current density increases. When B decreases below 0.005, the shape of the profiles does not change anymore.

Figure 9.6 shows the ratio $t_{\mathrm{ss.diff}}/t_{\mathrm{ss.elec.diff}}$ for the different values of the B parameter, where $t_{\mathrm{ss.diff}}$ is the time to steady state in pure diffusion, and $t_{\mathrm{ss.elec.diff}}$ is the time needed to reach steady state in electrokinetic transfer. Lower B values correspond to the higher current densities. Therefore, decreasing B leads to higher ratios $t_{\mathrm{ss.diff}}/t_{\mathrm{ss.elec.diff}}$. When the current density approaches zero, the B parameter becomes infinite, and the curve $t_{\mathrm{ss.diff}}/t_{\mathrm{ss.elec.diff}} = f(B)$ reaches an asymptote corresponding to $t_{\mathrm{ss.diff}}/t_{\mathrm{ss.elec.diff}} = 1$.

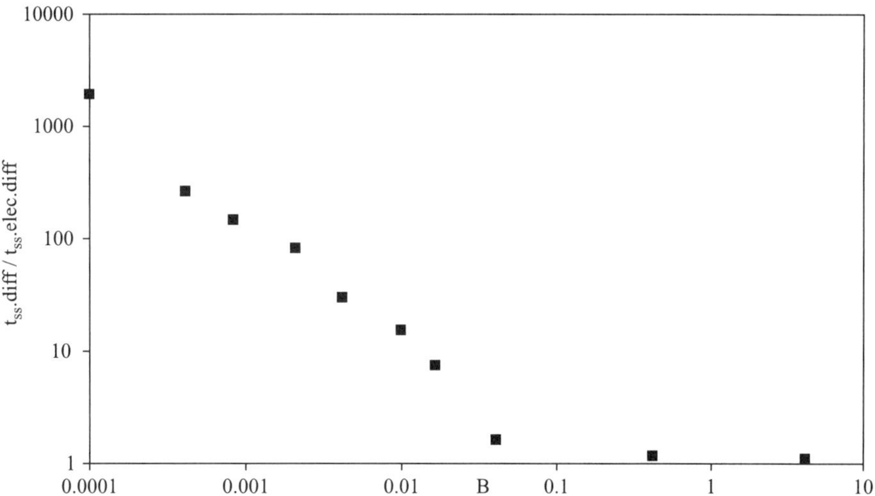

Figure 9.6 Steady-state comparison between electrokinetic transfer and diffusion [12].

9.4 IONIC EXTRACTION

Consider now the case of a porous medium saturated with a pore solution containing a contaminating anionic species. The objective is to remove this species from the porous medium. Assume that the contaminating ionic species (Ion-1) is distributed uniformly through the porous medium. The initial concentration is $\tilde{c} = 1$ throughout the medium. The environmental conditions are such that the concentration of Ion-1 is equal to zero outside the material. Because it is an ionic species, Ion-1 is associated with Ion-2, which is a counter-ion having the same charge number. And, because a concentration gradient exists between the material and the outside, at $\tilde{x} = 0$ both species tend to diffuse to the outside. This represents a leaching process. It is assumed that the time is short enough and the medium sufficiently large for its core not to be affected by the diffusion of species.

The evolution of the concentration profiles of Ion-1 with time is shown in Fig. 9.7. Figure 9.7a exhibits the results concerning the transport of Ion-1 from the porous medium towards the outer anode, in natural diffusion. The profiles have a steep slope in the beginning of the diffusion to the outside. This slope tends to become smoother as the diffusion time increases. In order to enhance the process, an electrical field can be applied.

Figure 9.7b shows the equivalent of the results of Fig. 9.7a for a given electrical current. Again, because all the data in this work are nondimensional, the electrical current is represented by the B parameter defined in Eq. (9.26). In the example of Fig. 9.7b, $B = 0.002$. At first sight, the ionic concentration varies in the same fashion as in the diffusive case: the slope of the concentration profile loses its sharpness as time increases. Yet, a closer look at Fig. 9.7b reveals an important difference. Unlike in Fig. 9.7a, where the concentration is always

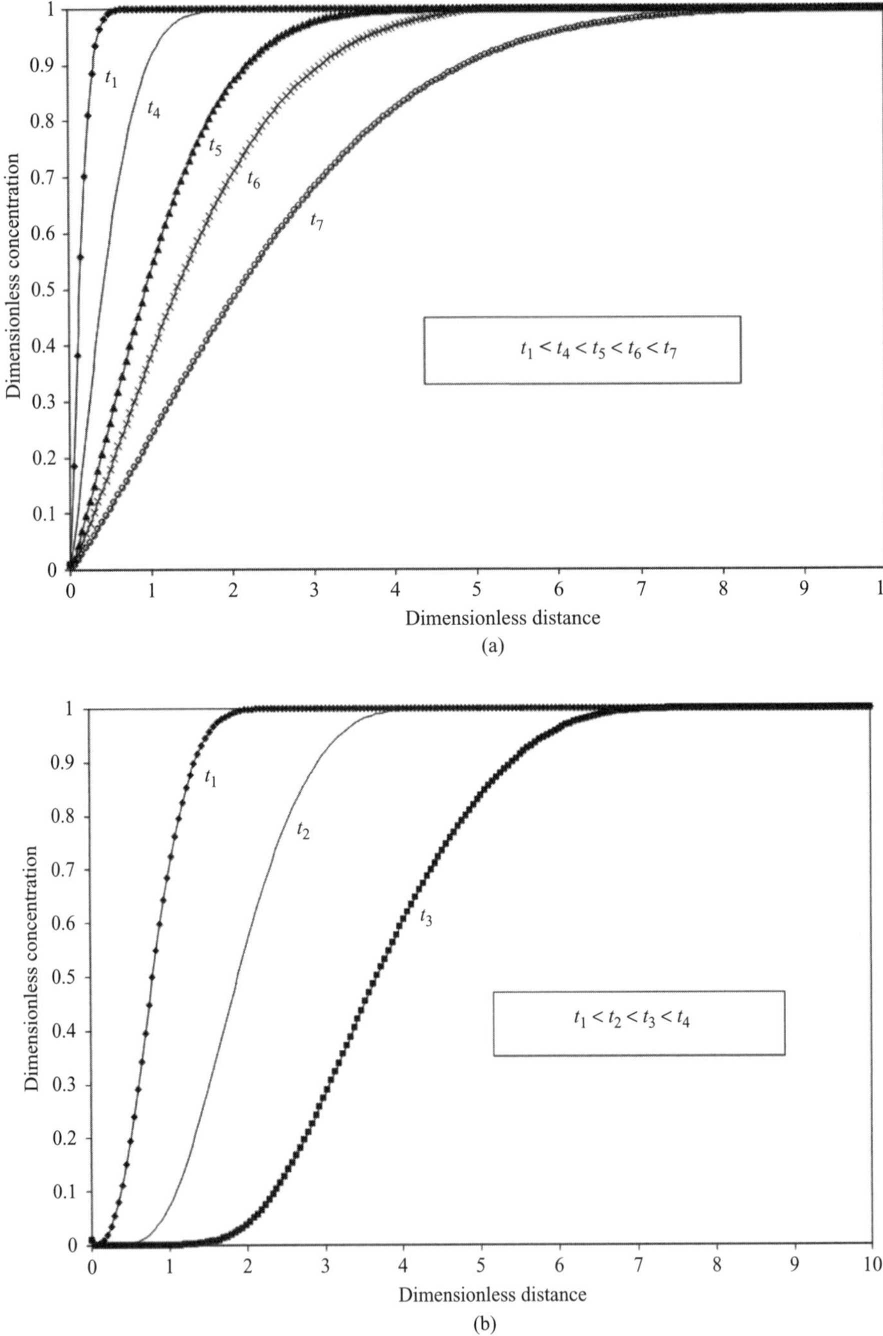

Figure 9.7 (a) Diffusion toward the outside (leaching process), and (b) electrokinetic transfer toward the outside ($B = 0.002$). $t_1 < t_2 < t_3 < t_4 < t_5 < t_6 < t_7$ [12].

higher than zero, in Fig. 9.7b the concentration reaches zero. This indicates that a complete decontamination of the material can be reached when applying an electrical field. The decontamination proceeds from the surface toward the core of the porous medium. In the electrokinetic transfer example of Fig. 9.7b, the concentration profiles evolve with time in two ways. One is the same as in Fig. 9.7a: diffusion to the outside. This is combined with a translation along the horizontal axis, which is a propagation front. This combination is due to the coupling of the two effects occurring during electrokinetic transfer: molecular diffusion and migration.

Scale analysis is also suitable for predicting the onset of steady state: Eq. (9.22) predicts the ratio of the time scales in natural diffusion and, in electrokinetic transfer. Combining Eqs. (9.22) and (9.26), we obtain

$$\frac{t_{\text{diff}}}{t_{\text{elec.diff}}} \sim \frac{1}{B} \tag{9.34}$$

Figure 9.8 is the equivalent of Fig. 9.6 where $1/B$ is on the abscissa, and the ratio $t_{\text{ss.elec.diff}}/t_{\text{ss.diff}}$ on the ordinate. The black squares are the results obtained with the numerical simulations. New is the curve of slope –1, which was obtained from Eq. (9.34). The numerical results and the scale analysis are in very good agreement. They indicate that the time to steady state decreases dramatically as the current density increases.

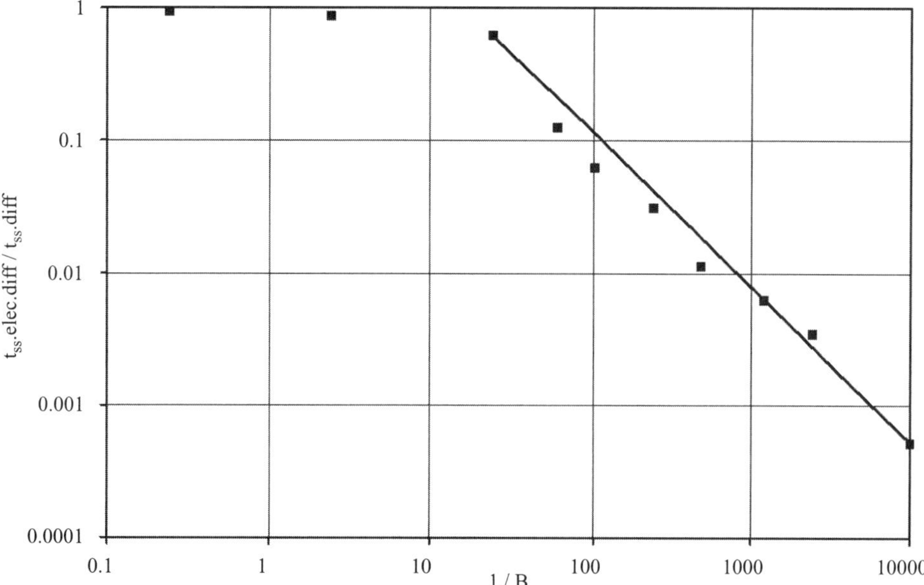

Figure 9.8 Comparison in the steady state: numerical results and scale analysis [12].

9.5 CONSTRUCTAL VIEW OF ELECTROKINETIC TRANSFER

In the beginning of an electrokinetic process aiming to force the contamination of a porous system with an ionic species, the ionic transfer is initially controlled by pure diffusion. It is only after a characteristic transition time that electrical effects become the main driving force: these effects are due to an external electrical potential difference [19]. In this section we exploit this idea in a more general way and place it in perspective with constructal theory.

Figure 9.2 shows a slab of porous material, the thickness of which is L. The dimensions of the material in the two other directions are considerably larger such that the configuration can be considered as one-dimensional. Assume that the objective is to contaminate uniformly the material with an ionic species at a concentration level Δc. The left side of the slab is in contact with a solution containing the ionic species of interest.

In the first part of this study we assume that no external electrical source is used. Therefore, it is only because of the concentration gradient created between the material and the solution that the contaminant species will slowly enter the porous medium. In the case of a saturated porous system with no pressure gradient and no interactions of the ionic species with the solid phase, the mass conservation at the scale of the material is given by Eq. (9.8). The electrical field $\partial\varphi/\partial x$ is calculated from the current law, Eq. (9.10).

The order of magnitude of the ionic concentration is Δc, while D_0 represents the diffusion coefficient of the ionic species, in an order of magnitude sense. As shown in Fig. 9.2, the thickness of the slab (L) is the length scale of the flow system. Yet, in the beginning of the process, *L is not the proper length scale of the flow field*. The variable x is kept in order to account for the strong variation (the time dependence) of the concentration gradients, which is the basis for a time-dependent scale analysis [19]. The scale analysis of the electrical field due to the membrane potential in natural diffusion can be performed, and it shows that

$$\frac{\partial\varphi}{\partial x} \sim \frac{RT}{F}\frac{1}{x} \tag{9.34}$$

Combining Eq. (9.34) with the scale analysis of Eq. (9.8), we obtain

$$\frac{\Delta c}{t} \sim \frac{D_0}{p}\frac{\Delta c}{x^2}, \frac{D_0}{p}\frac{F}{RT}\frac{\Delta c}{x}\left(\frac{RT}{F}\frac{1}{x}\right) \tag{9.35}$$

and

$$t \sim \frac{px^2}{D_0} \tag{9.36}$$

Therefore, in pure diffusion (in the absence of electrical effects), the time scale of the effects due to the concentration gradient is the same as the time scale due to the membrane potential. This is the reason why in the natural diffusion of ionic species, the membrane potential effects should not be neglected [20].

The phenomenon is quite different during an electrokinetic process, because the objective of applying an external electrical source is to accelerate the ionic transport by enhancing the second term between parenthesis in Eq. (9.8). For the sake of clarity, we start with a review of the analysis conducted in the case of electrokinetic transport under an electrical potential difference.

Let ΔU be the potential difference applied across the porous material between two electrodes sandwiching the material transversally to the x axis of Fig. 9.2. The electrical field created in the porous material is due to the contribution of the membrane potential [Eq. (9.10)] and the external electrical source, which is assumed constant, $\Delta U/L$. In an order of magnitude sense, we have

$$\frac{\partial \varphi}{\partial x} \sim \frac{\Delta U}{L} \tag{9.37}$$

such that the scale analysis of Eq. (9.8) becomes

$$\frac{\Delta c}{t} \sim \frac{D_0}{p}\frac{\Delta c}{x^2}, \frac{D_0}{p}\frac{F}{RT}\frac{\Delta c}{x}\frac{\Delta U}{L} \tag{9.38}$$

From Eq. (9.38), the time scale of the diffusive part of the ionic transport is

$$t_{\text{diffusion}} \sim \frac{px^2}{D_0} \tag{9.39}$$

while the time scale of the electrical effects is

$$t_{\text{elec\_}\Delta\text{U}} \sim \frac{RTL}{F\Delta U}\frac{p}{D_0}x \tag{9.40}$$

Note that the time scale of diffusion during an electrokinetic process Eq. (9.39) is the same as the time scale in natural diffusion Eq. (9.36) [21]. The comparison between Eqs. (9.39) and (9.40) highlights the fact that for small values of the distance x from the surface of the material the time scale of diffusion is shorter than the time scale associated with electrical effects.

Next, assume that I is a constant electrical current applied between two electrodes placed perpendicularly to the material thickness. In such a geometry the associated current density j is also a constant. From the current law [Eq. (9.14)], the equation for the electrical field becomes Eq. (9.15). Recall that unlike in diffusion, the electrical field is due to contributions from two terms: the membrane potential and the external electrical source. The scale analysis of Eq. (9.15) leads to

$$\frac{\partial \varphi}{\partial x} \sim \frac{RT}{F}\left(\frac{j}{FD_0\Delta c}, \frac{1}{x}\right) \tag{9.41}$$

In practical applications such as ionic transport through concrete structures for example, the orders of magnitude of the diffusion coefficients, concentration levels, and current densities are such that the term $j/FD_0\Delta c$ is much greater than $1/x$

when considering the material thickness. We obtain

$$\frac{\partial \varphi}{\partial x} \sim \frac{RT}{F^2} \frac{j}{D_0 \Delta c} \tag{9.42}$$

This result is then introduced into the scale analysis of the continuity equation [Eq. (9.38)], which reads

$$\frac{\Delta c}{t} \sim \frac{D_0}{p} \frac{\Delta c}{x^2}, \frac{j}{Fx} \tag{9.43}$$

Equation (9.43) provides the two time scales during electrokinetic transfer with constant current. The time scale of pure diffusion remains the same as in Eq. (9.36), while the time scale due to electrical effects is

$$t_{\mathrm{elec\_I}} \sim p \frac{F}{j} x \tag{9.44}$$

Again, for small distances x, the diffusion time scale is shorter.

The preceding results are more useful if we express them in terms of penetration depths. In natural diffusion of ionic species, even though two different mechanisms of transport are involved, one time scale rules the transport and leads to a penetration depth varying with the square root of time:

$$x \sim \left(\frac{D_0}{p}\right)^{1/2} t^{1/2} \tag{9.45}$$

In electrokinetics an additional mechanism exists (additional to diffusion), and for a constant potential difference we have

$$x \sim \frac{F \Delta U}{RTL} \frac{D_0}{p} t \tag{9.46}$$

or, for a constant current,

$$x \sim \frac{j}{Fp} t \tag{9.47}$$

Note that in both cases the greater the external electrical source, the greater the penetration depth.

Plotted in Figs. 9.9a and b are the variations of the penetration depths in time for the two mechanisms: diffusion (curve 1) and electrical effects (curve 2). For conciseness, the values are presented in nondimensional form. Equations (9.46) and (9.47) indicate that the penetration depth varies linearly with time. Therefore, on a nondimensional diagram, Eqs. (9.46) and (9.47) can be combined into a single curve. Figure 9.9b is a zoom of Fig. 9.9a focusing on the initial moments of the electrokinetic process. The highest curve represents the penetration depth due to the combination of the two mechanisms. The effect of the diffusive transport is noticeable only in the beginning of electrokinetics (see Fig. 9.9b).

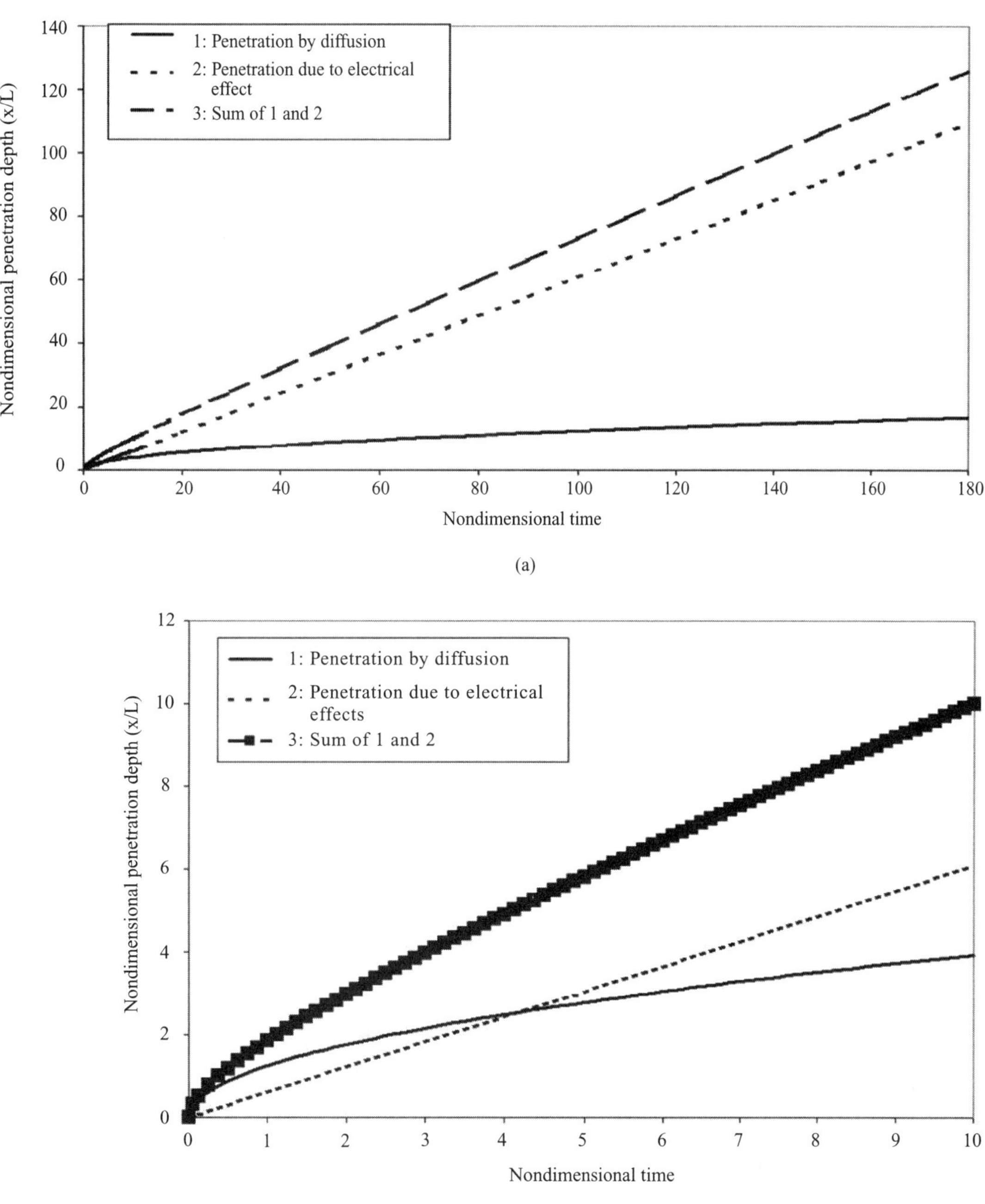

Figure 9.9 Variation of the ionic penetration in time: (b) is an expanded view of (a) [21].

Before a transition time that corresponds to the intersection of curves 1 and 2, the penetration due to diffusion is greater than the penetration due to electrical effects. It is of fundamental importance that the intersection between the two curves always exists. Depending on the ionic diffusion coefficient, the geometry of the problem, the applied potential difference or the applied electrical current, the transition time will be shorter or longer, but it is characteristic, and it is predictable. This transition time is $[RTL/(F\Delta U)]^2(p/D_0)$ in the case when an electrical potential difference is used. The transition time is $(F/j)^2 D_0 p$ when an electrical current is imposed.

9.5.1 Reactive Porous Media

Consider now the case where the porous medium interacts with the ionic species. The total concentration of ionic species is $C_T = pc + C_B$, where C_B represents the concentration of the species (mol/m^3 of porous media) that interacts with the solid matrix. This time the continuity equation is

$$p\frac{\partial c_i}{\partial t} + \frac{\partial C_{B,i}}{\partial t} = D_i\left(\frac{\partial^2 c_i}{\partial x^2} + z_i\frac{F}{RT}\frac{\partial \varphi}{\partial x}\frac{\partial c_i}{\partial x}\right) \tag{9.48}$$

The interactions of the ionic species with the porous medium can be written $C_B = C_B(c)$, where

$$\frac{\partial C_{B,i}}{\partial t} = \frac{\partial C_{B,i}}{\partial c_i}\frac{\partial c_i}{\partial t} \tag{9.49}$$

Therefore, Eq. (9.48) is replaced by

$$\frac{\partial c_i}{\partial t} = \frac{D_i}{p + \partial C_{B,i}/\partial c_i}\left(\frac{\partial^2 c_i}{\partial x^2} + z_i\frac{F}{RT}\frac{\partial \varphi}{\partial x}\frac{\partial c_i}{\partial x}\right) \tag{9.50}$$

For simplicity, assume that the interactions are linear. Hence $\partial C_{B,i}/\partial c_i$ is a constant and

$$p + \frac{\partial C_{B,i}}{\partial c_i} = K \tag{9.51}$$

We focus on the case of electrokinetics with external electrical field. The same analysis as the one previously conducted leads to

$$t_{\text{diff}} \sim K\frac{x^2}{D_0} \tag{9.52}$$

$$t_{\text{electrodiff}} \sim K\frac{RTLx}{F\Delta U\, D_0} \tag{9.53}$$

Again, the propagation is initially driven by diffusion, $x \sim (D_0/K)^{1/2}t^{1/2}$, then convection takes over and $x \sim [F\Delta U\, D_0/(RTL\,K)]t$. The depth at which the switch between the two propagation modes occurs is given by $x_{\text{transition}} = RTL/(F\Delta U)$. Note that the penetration depth of transition is the same as before, which means that

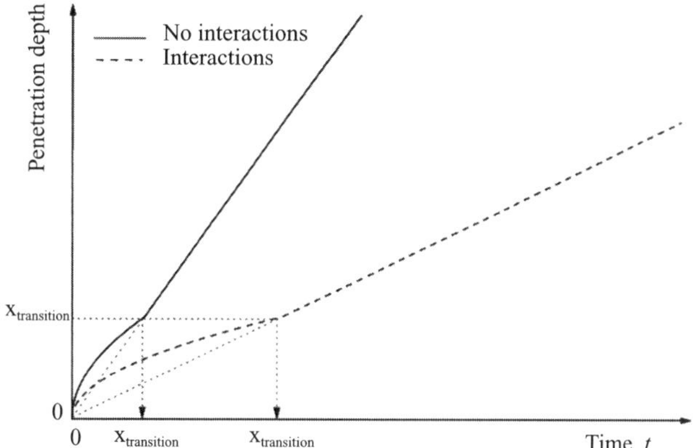

Figure 9.10 Penetration depths with inert and reactive porous media [19].

it is not affected by the existence of interactions. Whether the material is reactive or inert has no effect on the transition depth.

The time corresponding to the penetration depth of transition is

$$t_{\text{transition}} = \left(\frac{RTL}{F \Delta U} \right)^2 \frac{K}{D_0} \tag{9.54}$$

Equation (9.54) can be compared with $[RTL/(F\Delta U)]^2(p/D_0)$ which is the transition time corresponding to an inert porous medium. For an ionic species penetrating a porous medium initially free of that particular species, the term $\partial C_{B,i}/\partial c_i$ is positive. Therefore, $K > p$, and the transition time obtained for a reactive material is greater than the transition time for an inert porous medium. In other words, because of the existence of interactions the switch between diffusion and convection, (i.e., diffusion vs. electrokinetics), occurs later (see Fig. 9.10).

When the time increases and becomes greater than $t_{\text{transition}}$, the propagation rate is also affected by the nature of the material: a reactive porous medium tends to slow down the propagation rate by binding parts of the species from the pore solution. This can be seen in Fig. 9.10, where the slope of the electrokinetic part of the curve (linear) is smaller than its counterpart for the inert porous medium.

9.5.2 Optimization in Time

We recognize here an example of application of the constructal law: the increase of flow access through the selection of flow mechanism. In this problem, "flow" means the transport (current) of ionic species. With the objective of maximizing this current, the most efficient mechanism becomes the main force that drives the flow. The combination of competing mechanisms for the maximization of current access was also used as an explanation for the occurrence of dendritic agglomeration of dust particles in the atmosphere [22] and dendritic solidification [23]. The solidification process during the growth of snowflakes is due to both thermal diffusion (temperature change in the nucleation site) and growth of ice needles at constant velocity. In this example, diffusion is one of the mechanisms while the second one is convection. Note that the electrical effects from the present work in the mass conservation equation can be interpreted as the convective part of the transport, Eq. (9.8). More recently, Miguel [24, 25] also used the competition between two mechanisms to construct a predictive analogy between coral growth, bacterial colony growth and plant roots. Their development can be explained by a unifying principle: the competition between diffusive and convective mechanisms as a consequence to the maximization of access to nutrients.

From the constructal theory of maximizing current access comes the proposed strategy to use each mechanism at the moment when it is the most effective: the optimization of electrokinetics in time. Electrokinetic transfer is not a natural process like snowflakes or coral growth. It is an engineered mechanism. Therefore, it is proposed to control the electrical driving force so that the diffusive transport can enhance at best the penetration of ionic species.

Electrokinetic transfer at low electrical driving force is more sensitive to the diffusive mechanism in the early stages than a high electrical driving force electrokinetic process. An example of this kind was described by Bégué and Lorente [19], for a case where the electrical potential gradient varies from $\nabla\varphi_1$ to $\nabla\varphi_2$, with $\nabla\varphi_1 < \nabla\varphi_2$. It was shown that the combination of $\nabla\varphi_1$ in the beginning followed by $\nabla\varphi_2$ led to a higher ionic penetration through the porous material compared with the case where the higher electrical field is applied immediately (Fig. 9.11).

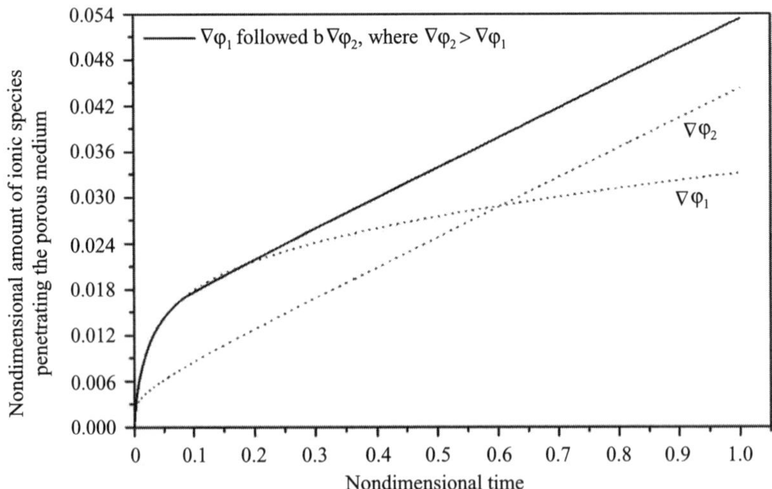

Figure 9.11 Combination of two different electrical fields leading to the optimal ionic transfer in time [19].

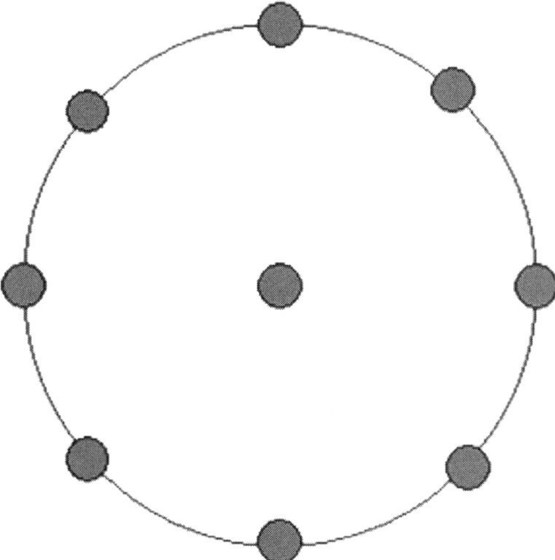

Figure 9.12 Plane view of electrodes placement in a contaminated porous material [21].

9.5.3 Optimization in Space

The optimization of an electrokinetic decontamination process can also be envisaged in space [21]. In the nuclear field, it is sometimes necessary to decontaminate "hot spots" that is, the places in the concrete structure where high radioactivity is concentrated because of the diffusion of radioelements through the material porous network. Such configurations may call for a cylindrical geometry of the decontamination. Figure 9.12 shows the view from above of a configuration of this type. An electrode is positioned in the center, while electrodes with opposite polarization are located equidistantly on a circle. The radius of the circle is the distance between the central electrode and one of the electrodes on the perimeter. The electrodes on the perimeter are connected electrically to each other, and this circuitry is external to the material.

Assume that the contaminated portion of the material is known. Therefore, the volume of material to be decontaminated is known and fixed. For simplicity, it is assumed that the decontamination depth is constant and that this depth is comparable with the length of the electrodes. Consequently, the configuration is modeled as two-dimensional (Fig. 9.12). The surface that must be decontaminated is known. We search for the optimal number of electrodes to implement for a given contamination level and for a given electrical current. Because of the symmetry of the configuration, the optimization of the number of electrodes to implement can be reduced to the optimization of α, which is the angle of the sector of radius L presented in Fig. 9.13.

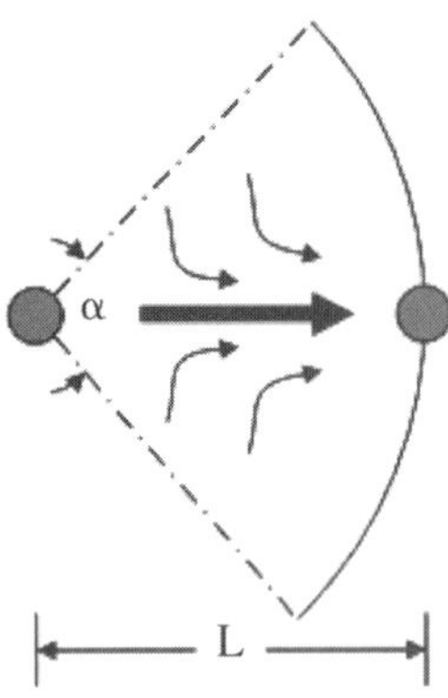

Figure 9.13 Elemental sector described by the distance between anode and cathode and the angle α [21].

Assume that the sector of angle α is contaminated uniformly. When an electrical current I is imposed between the two electrodes of Fig. 9.13, the ionic species from the entire sector is entrained and transported. In the case of cation decontamination, the transport occurs from anode (the central electrode) to cathode. A main stream forms along the radius L separating the two electrodes. From the scale analysis of the Nernst-Planck equation [Eq. (9.7)], and the assumption that diffusion is negligible compared with the electrical effects in the radial direction L (from anode to cathode), we find that the longitudinal flow rate of species $\dot{m}_{\text{longitudinal}}$ is given by

$$\dot{m}_{\text{longitudinal}} \sim D_0 \frac{F}{RT} \Delta c \nabla \varphi L \alpha \tag{9.55}$$

where $L\alpha$ is the transversal area crossed by the flux from electrode to electrode. The order of magnitude of the current density is

$$j \sim \frac{I}{L\alpha} \tag{9.56}$$

Combining Eqs. (9.55) and (9.56) with Eq. (9.42), we obtain

$$\dot{m}_{\text{longitudinal}} \sim \frac{I}{F} \tag{9.57}$$

The associated time scale can be obtained from the scale analysis of the law of conservation of mass, Eq. (9.8):

$$t_{\text{longitudinal}} \sim \frac{\Delta c F L^2 \alpha}{I} \tag{9.58}$$

In the direction perpendicular to this main direction of transport, the flux is mainly diffusive (Fig. 9.13). Yet the strength of the convective flux is such that close to the main stream of ionic flux the diffusive transport exhibits a longitudinal component. Such a configuration possesses analogies in nature. See, for example, the flow of ground water close to a river bed (the *infero-flux* phenomenon).

The order of magnitude of the transversal flow rate is also provided by the scale analysis of the Nernst-Planck equation,

$$\dot{m}_{\text{transversal}} \sim \frac{D_0 \Delta c}{\alpha} \tag{9.59}$$

The corresponding time scale is obtained from the mass conservation

$$t_{\text{transversal}} \sim \frac{L^2 \alpha^2}{D_0} \tag{9.60}$$

Next, we write that the optimal geometry is the one that corresponds to the equipartition of times of transport [23]. In other words, the angle of the sector is optimal when the time scale of longitudinal transport is of the order of magnitude of the time scale of transversal transport. Therefore,

$$\alpha_{\text{optimal}} \sim \frac{\Delta c F D_0}{I} \tag{9.61}$$

The same optimal angle can be calculated by writing that the longitudinal and transversal flow rates are equal. The result obtained in Eq. (9.61) may seem counterintuitive because it indicates that the optimal angle decreases if the current increases. If the current increases, the amount of ionic species extracted per unit time longitudinally increases. The only possibility to extract the same amount transversally is to increase the concentration gradient. Because the contamination level is fixed (Δc), increasing the concentration gradient means to decrease the angle α.

The above analysis can be repeated for the case where the electrical potential difference ΔU is fixed instead of the current. The corresponding electrical field is, in an order of magnitude sense, $\Delta U/L$, leading to

$$\dot{m}_{\text{longitudinal}} \sim D_0 \frac{F}{RT} \Delta c \Delta U \alpha \tag{9.62}$$

The transversal ionic flow rate is given by Eq. (9.59). Setting the two ionic flow rates equal gives the optimal angle α

$$\alpha_{\text{optimal}} \sim \frac{RT}{F \Delta U} \tag{9.63}$$

Again, increasing the electrical effects (ΔU in this case) means to decrease the value of α.

REFERENCES

1. A. Bejan, L. Dincer, S. Lorente, A. F. Miguel, and A. H. Reis, *Porous and Complex Structures in Modern Technologies*. New York: Springer, 2004.
2. R. F. Probstein and R. E. Hicks, Removal of contaminants from soils by electric fields. *Science*, Vol. 260, 1993, pp. 498–503.

3. K. S. Dickenson, M. R. Ally, C. H. Brown, M. I. Morris, and M. J. Wilson-Nichols, Demonstration recommendations for accelerated testing of concrete decontamination methods. Department of Energy, 1995.
4. D. W. DePaoli, M. T. Harris, I. I. Morgan, and M. R. Ally, Investigation of electrokinetic decontamination of concrete. *Symposium on Separation Science and Technology for Energy Applications*, Vol. 32, 1997, pp. 387–404.
5. D. B. Sogorka, H. Gabert, and B. Sogorka, Emerging technologies for soils contaminated with metals—electrokinetic remediation. *Hazardous and Industrial Waste*, Vol. 30, 1998, pp. 673–685.
6. F. Frizon, P. Thouvenot, J. P. Ollivier, and S. Lorente, Description of the radiological decontamination by electromigration in saturated concrete: A multi species approach. Waste Management Conference, Tucson, AZ, 2002.
7. F. Frizon, S. Lorente, J. P. Ollivier, and P. Thouvenot, Modeling the decontamination by electromigration of a porous medium. *J Porous Media*, Vol. 7, No. 3, 2004, pp. 213–227.
8. S. Hsieh, *Drug Permeation Enhancement, Theory and Applications*. New York: Marcel Dekker, Inc., 1994.
9. J. Benson and H. A. Eccles, A novel technique for the decontamination of concrete and steel. *I Mech E Conf Transactions*, Vol. 7, 1995, pp. 363–369.
10. J. C. Oreillan, G. Escadeillas and G. Arliguie, Electrochemical chloride extraction: Efficiency and side effects. *Cement and Concrete Research*, Vol. 34, 2004, pp. 227–234.
11. A. Fick, On liquid diffusion. *The London, Edinburgh, and Dublin Philosophical Magazine and Journal of Science*, 1855, pp. 30–39.
12. S. Lorente and J. P. Ollivier, Scale analysis of electrodiffusion through porous media. *J Porous Media*, Vol. 9, No. 4, 2006, pp. 307–320.
13. A. Bejan, *Convection Heat Transfer*, 3rd ed. New York: Wiley, 2004.
14. J. H. Masliyah, *Electrokinetic Transport Phenomena*. AOSTRA Technical Publication Series #12, Alberta, Canada, 1999.
15. A. Révil, Ionic diffusivity, electrical conductivity, membrane and thermoelectric potentials in colloids and granular porous media: A unified model. *J Colloid Interface Sc.*, Vol. 212, 1999, pp. 503–522.
16. J. Arsenault, Etude des mécanismes de transport des ions chlore dans le béton en vue de la mise au point d'un essai de migration. PhD thesis, National Institute of Applied Sciences, Toulouse, France, Laval University, Québec, Canada, 1999.
17. G. Fajardo, G. Escadeillas and G. Arliguie, Electrochemical chloride extraction (ECE) from reinforced concrete specimens containing chloride penetrated from "artificial" sea water. *Corrosion Science*, Vol. 48, No. 1, 2006, pp. 110–125.
18. I. Rubinstein, *Electro-diffusion of Ions*. Philadelphia: SIAM: 1990.
19. P. Bégué and S. Lorente, Migration versus diffusion through porous media: Time dependent scale-analysis. *J Porous Media*, Vol. 9, No. 7, 2006, pp. 637–650.
20. S. Lorente, D. Voinitchi, P. Bégué-Escaffit, and X. Bourbon, The single-valued diffusion coefficient for ionic diffusion through porous media. *J Applied Physics*, Vol. 101, No. 2, 2007, 024907.
21. S. Lorente, Constructal view of electrokinetic transfer through porous media. *J Phys D: Appl Phys*, Vol. 40, 2007, pp. 2941–2947.

22. A. H. Reis, A. F. Miguel, and A. Bejan, Constructal theory of particle agglomeration and design of air-cleaning devices. *J Phys D: Appl Phys*, Vol. 39, No. 10, 2006, pp. 2311–2318.
23. A. Bejan, *Shape and Structure, from Engineering to Nature*. Cambridge, UK: Cambridge University Press, 2000.
24. A. F. Miguel, Constructal pattern formation in stony corals, bacterial colonies and plant roots under different hydrodynamics conditions. *J. Theoretical Biology*, Vol. 242, No. 4, 2006, pp. 954–961.
25. A. F. Miguel, Shape and complexity in living systems. In *Along with Constructal Theory*, A. Bejan, S. Lorente, A.F. Miguel and A.H. Reis, Presses de l'Université de Lausanne, 2007.

10

MECHANICAL AND FLOW STRUCTURES COMBINED

10.1 OPTIMAL FLOW OF STRESSES

Flow systems with configuration do not exist alone. They are confined and supported by solid structures that provide strength and rigidity. The supporting structure morphs and mates with the flow structure. In this chapter we teach how to combine thermo-fluid objectives with mechanical objectives in the pursuit of configuration. Examples are cavernous brick walls for maximum thermal insulation stiffness, support beams that must be strong and resistant to sudden heating (e.g., fire, thermal attack), and the natural design of trees bending in the wind.

The key proposal is to view mechanical configurations in the same way that we treat flow configurations: mechanical structures are networks (connections with minimum volume) through which stresses "flow" from components to their neighbors. We showed that flow strangulation is not good for performance: it is the *uniform* distribution of flow strangulation that is good for performance (Problems 1.1 and 1.2). The solid-structure equivalent of this principle is that concentrations of maximum stresses are not good for performance. The best use of mechanical support material is achieved when *the maximum allowable stresses are distributed uniformly through the available material.* This design principle is taken for granted, as something that is intuitively obvious. Our objective is to show that this principle follows from the constructal law, and that the application of the principle is an important sector in constructal design.

The simplest illustration of this principle is provided by Fig. 10.1. An elastic bar of length L transmits the tensile load F from one of its ends to the other. The cross-sectional area of the bar is not uniform: the smaller cross-section (A_1) occurs over a partial length (L_1). In this narrower region the stresses reach their maximal

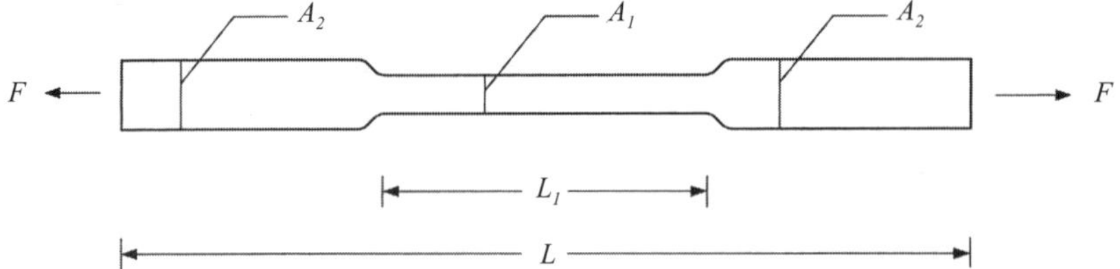

Figure 10.1 Elastic bar in tension and with nonuniform cross-sectional area.

allowable level, F/A_1. The stresses are lower in the thicker portions, F/A_2. The total volume of bar material is $A_1L_1 + A_2(L - L_1)$. This volume is minimal when the bar is shaped such that $A_1 = A_2$: in this configuration the "flow of stresses" is not strangulated. Every material element works as hard as the hardest-working element. Throughout the bar, the tensile stresses are at the highest permissible level (Problem 10.1).

The same principle governs the generation of configuration in more complex structures. In Fig. 10.2 the compressive force F_1 is transmitted vertically by a Y-shaped construct of three elastic bars connected by four frictionless hinges. The top hinge and the bottom hinges are fixed (i.e., z and $2x$ are known constants). The vertical position of the bifurcation (y) may vary. This means that there is an infinity of possible Y shapes. The particular shape for which the total amount of material is minimal is the case $L_1 = 0$, where the Y becomes shaped as a V, or a Λ. In this special case the stresses reach their maximal allowable level throughout the structure (Problem 10.2). The global performance of the structure is the best because the imperfections (the stress strangulations) are distributed uniformly through the solid construct. This finding is key: from the infinity of possible drawings (i.e., out of nothing) the principle identifies of the drawing.

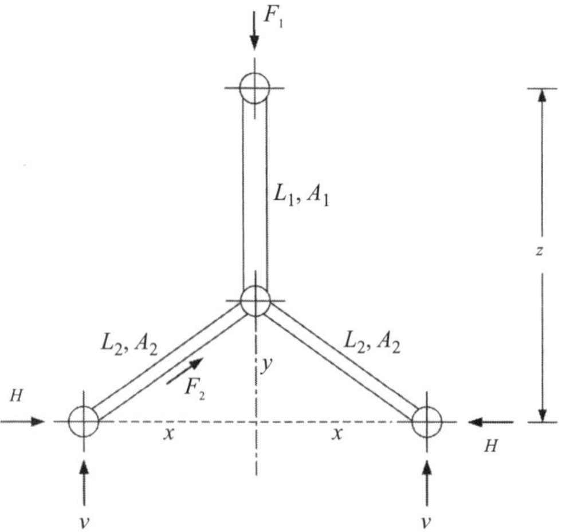

Figure 10.2 Y-shaped construct in vertical compression.

We wish to thank our students and colleagues who have contributed to the art that we have collected in this set: S. Kim, K.-M. Wang, J. Lee, T. Bello-Ochende, L. A. O. Rocha, W. Wechsatol, S. Périn, and A. K. da Silva.

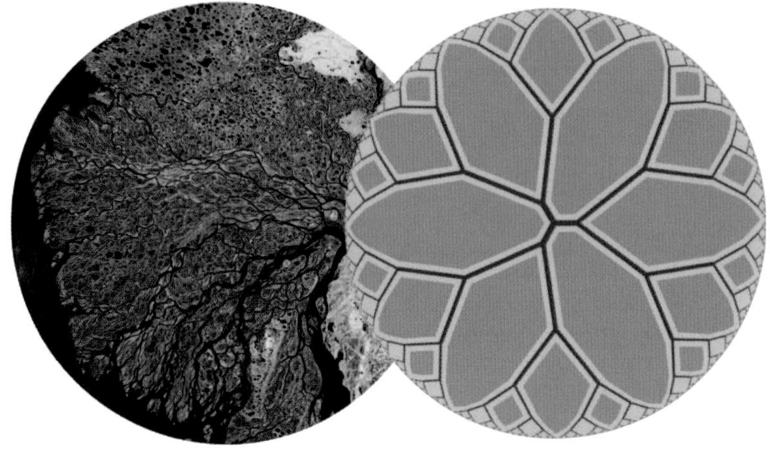

PLATE 1 Pattern in nature is flow. The two sides of the constructal theory coin: with constructal theory we predict and explain the occurrence of design in nature, and we design with strategy (faster, and for greater flow performance) devices for human use. The left side shows the delta of the river Lena in northern Siberia. The right side shows the constructal tree architecture that connects the center with ports on the disc perimeter.

PLATE 2 The invention of line-to-line trees illustrates the two sides of constructal theory (Plate 1). One side is the expectation that the tree is the natural design for achieving greatest flow access between one point and a volume. The other side is the strategy for achieving greater flow access between two parallel lines (or two parallel planes): the flow space must be filled with trees that alternate with upside-down trees. Natural porous media exhibit multiscale flow structures consistent with the multiple scales and performance of the line-to-line architecture.

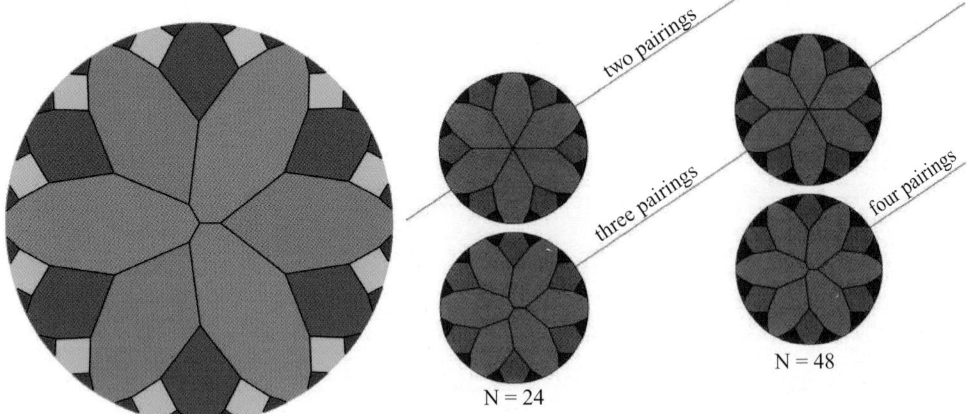

PLATE 3 Complexity is simple. Pattern is "pattern" precisely because it is simple enough for us to comprehend, discuss, remember, and reproduce. In the design of laminar flow between the center and 48 ports on the circular rim, least global resistance is offered by trees with three ducts in the center and only four levels of bifurcation, or pairing. We often hear that the road to perfection in natural design points toward greater and greater complexity. Not true. We can find drawings much more complicated than these trees, but their performance would be inferior.

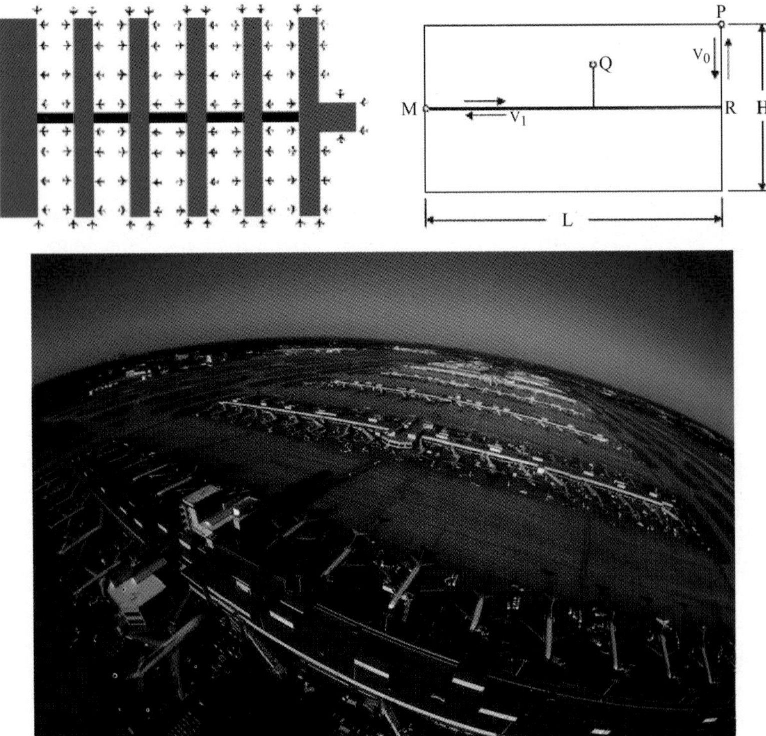

PLATE 4 Two ways to flow are better than one. A large airport without a train, or without walking, cannot compete on the same area with the design that combines walking with riding on a vehicle. The Atlanta airport is a modern illustration of the seed from which all forms of urban and natural flow networks have grown. On a fixed area ($A = HL$) with variable shape (H/L) and two speeds (walking V_0, and train V_1), the time of travel from P to M (or from all points Q to M, averaged over A) is minimal when the shape is $H/L = 2V_0/V_1$. The walking time (PR) is equal to the riding time (RM). The long and fast travel is balanced with the short and slow travel. (Bottom photo: Ron Sherman/ Photographer's Choice/ Getty Images).

PLATE 5 The movement of people, vehicles and goods in a city is a superposition of point-area and area-point tree flows. (Photo courtesy of the City of Toulouse.)

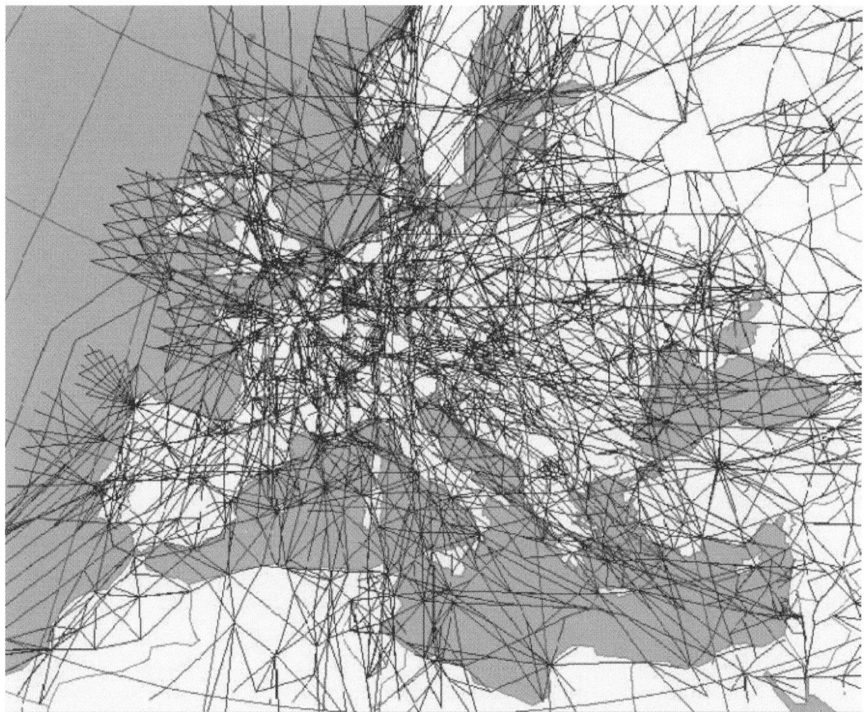

PLATE 6 Tapestry of air mass-transit over Europe. The burning of jet fuel is for moving people and goods: this flow is hierarchical and nonuniformly distributed. Large centers and thick channels are allocated to numerous smaller channels. The fine channels are allocated to area elements (yellow) that are covered by ground movement—people, and all the animate and the inanimate flows of the environment. The time to travel long and fast (along the red) is comparable with the time to travel short and slow (across the yellow).

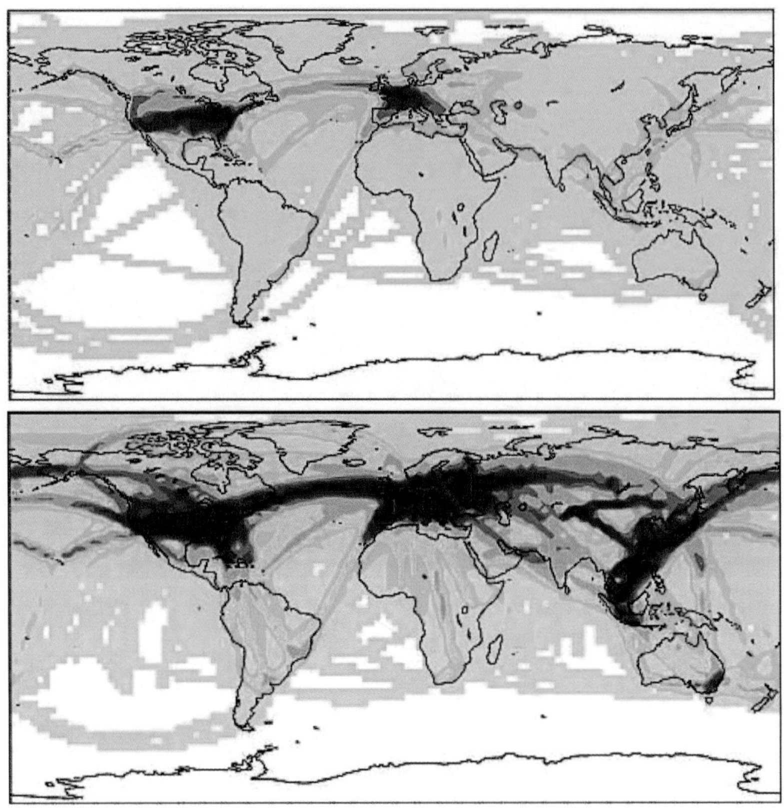

PLATE 7 Where aircraft flew in 1992 (top) and where aircraft will fly in 2050 (bottom): the persistent contrail coverage for the aviation fleet. (From Ref. [15] of ch. 11, with permission from Springer.)

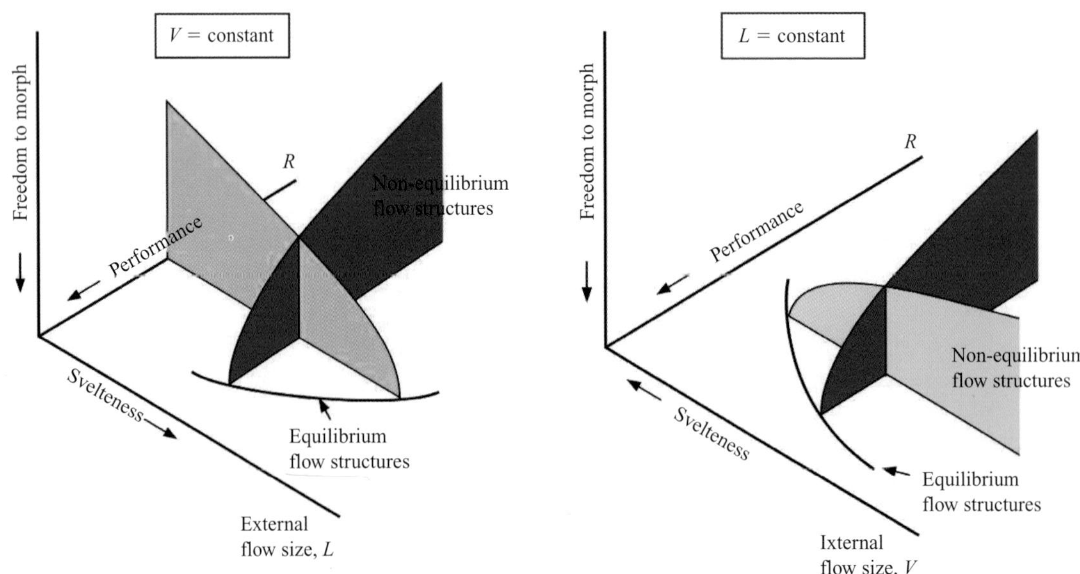

PLATE 8 Survival by increasing performance (red), increasing svelteness (green), and increasing flow territory (blue). The world of flow configurations of a morphing flow system when the internal flow volume is fixed (left side) and when the external flow size is fixed (right side). The freedom to change the flow configuration increases downward. The bottom plane houses the equilibrium flow structures.

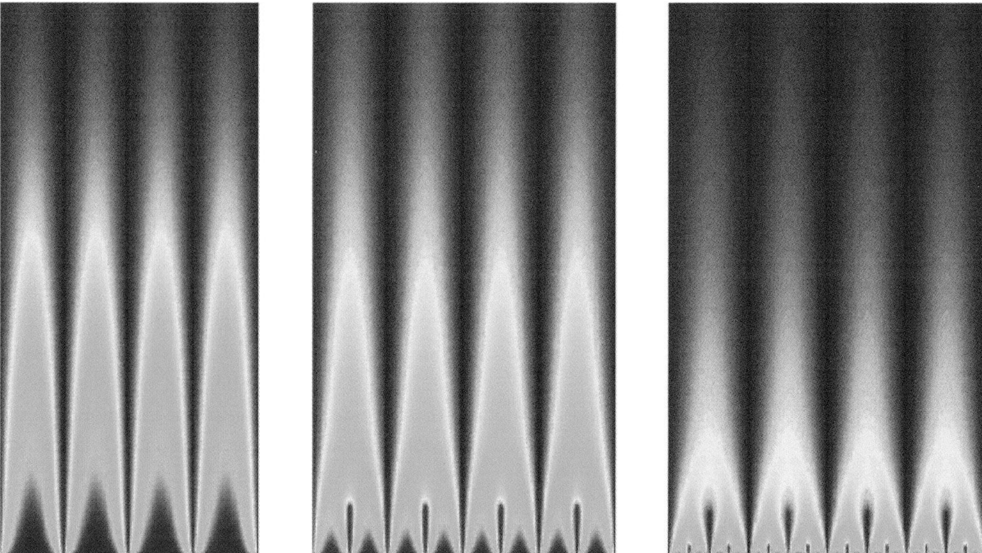

PLATE 9 Temperature distributions in multiscale packages of vertical parallel plates cooled by natural convection. The coolant (blue) enters from the bottom. The red areas indicate the boundary layers along vertical plates with uniform maximum-allowable temperature. The plates are covered on both sides by electronics that generate heat, which is swept upward by the flow. The objective is to change and select the configuration so that the "packing" (the heat generation rate per unit volume) is maximum. Three designs are shown: each image is cut from a larger stack containing many plates. The first design contains plates that stretch from bottom to top. The spacing between plates has been optimized for maximum packing. The heat-transfer density is increased by inserting smaller heat-generating plates in the blue entrance regions, and optimizing all the spacings. In the middle design, plate inserts with one (shorter, optimized) length scale are used. In the third design, plate inserts with two length scales are used. The global heat transfer density increases from left to right.

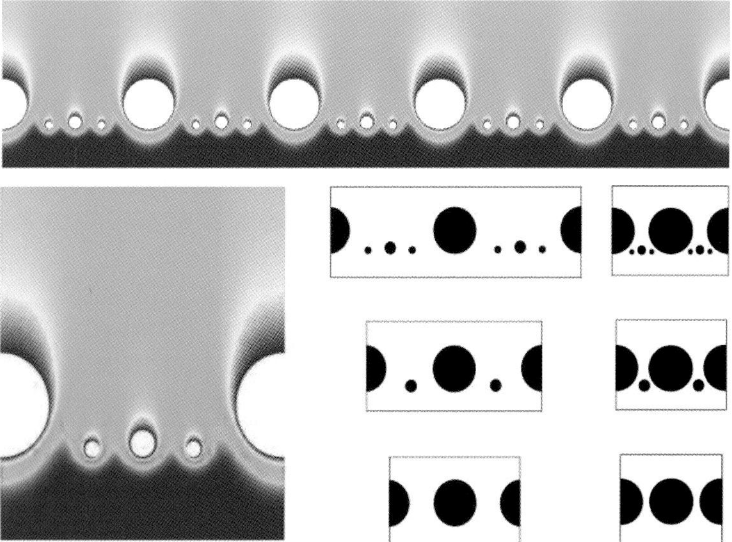

PLATE 10 Rows of cylinders with one, two and three diameters. The cooling is by natural convection, with cold fluid entering from below. The heat-transfer density increases with the number of optimized length scales.

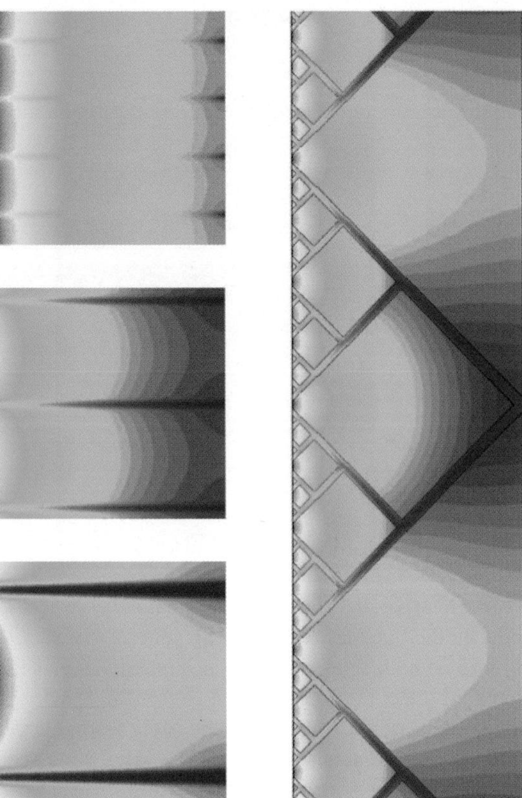

PLATE 11 Vascularized two-dimensional slabs heated with uniform heat flux on the left face and cooled by fluid that perfuses from the right. The long dimensions of the slab are two: vertical and perpendicular to the figure. The left column shows three designs with the coolant flowing through parallel equidistant channels. The spacing between channels increases from the first design to the third. The hot spots are the least pronounced in the second design, which reveals the constructal spacing that should be used for such vasculatures. The right column shows a slab with embedded tree-shaped vasculature.

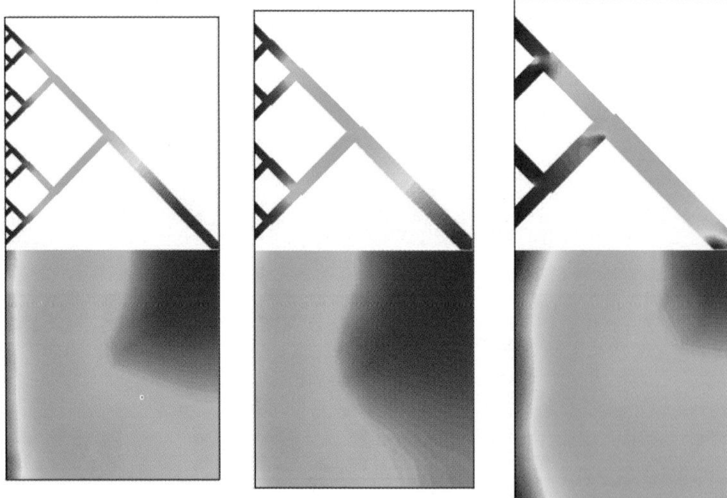

PLATE 12 In a slab with tree-shaped vasculatures the coolant pushes the hot spots under the surface attacked by intense heating. There is a special number of pairing levels in the tree architecture so that the canopy has the best "fineness" for suppressing the hot spots. The best design is in the middle, and has three pairing levels. The dendrites show the distribution of pressure in the midplane of the channels of the dendrites. The color panes show the temperature distribution in the midplane between two successive dendrites stacked in the direction perpendicular to the figure.

PLATE 13 Vascularized square slabs cooled by tree-shaped flow channels embedded under its top surface. Heat is generated uniformly on the bottom surface of the square slab. The coolant enters through the center and exits through ports on the square perimeter. The hot spots occur on the bottom surface. From left to right, the dendritic designs have one, two, and three levels of pairing.

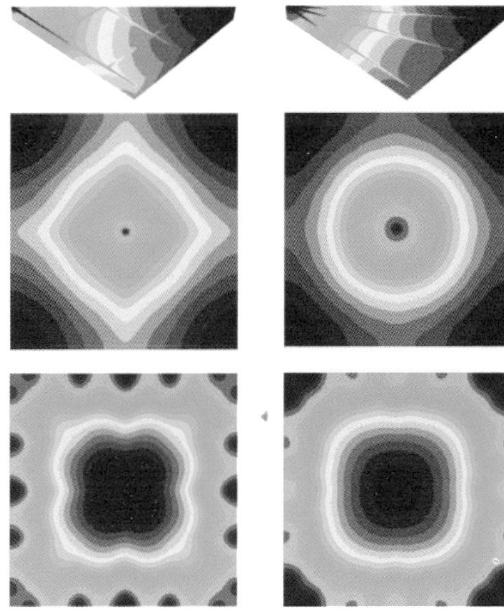

PLATE 14 Vascularized square slabs cooled volumetrically with embedded square-grid and radial channels. The upper figures show 1/8 slices of the slab. The left corner of the slice is the center of the square slab. Heat flux impinges uniformly on the bottom surface of the slab. The coolant (blue) enters either through the center of the slab (as in the upper images) or through the peripheral ports. The four bottom squares show the temperature distribution on the bottom surface of the slab. The images on the left are for slabs cooled with square-grid architectures. The images on the right are for cooling with radial channels. The upper two squares are for coolant entering through the center. The bottom images show the temperature distribution when the coolant enters from the periphery.

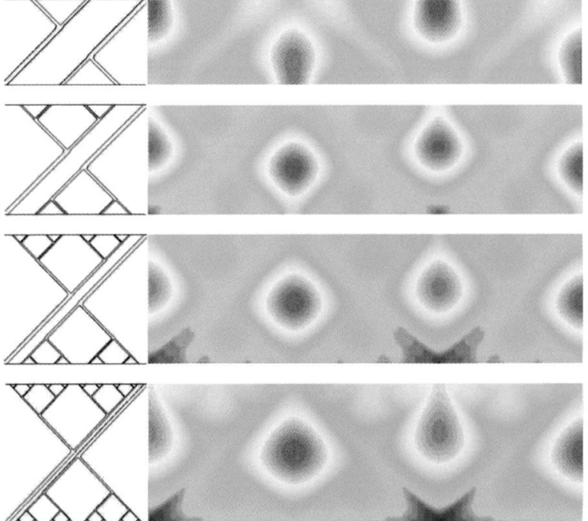

PLATE 15 Vascularized long plates with uniform heating landing perpendicularly on the figure and coolant flowing from the bottom edge to the top edge. The vasculature design is based on the alternating line-to-line trees concept proposed in Plate 2. The dendrites have one, two, three and four levels of pairing.

PLATE 16 Round patches of vegetation in muddy terrain at low tide. The pattern is reminiscent of Fig. 11.7. Why the similarities? What is the global flow access that the configuring of vegetation is facilitating?

PLATE 17 Had there been two design objectives, mechanical strength and pumping water from ground to atmosphere, the Eiffel Tower might have been the Eiffel Tree.

PLATE 18 The many and the few, together.

10.2 CANTILEVER BEAMS

The constructal design of solid structures began with the *evolution* of a cantilever beam toward an architecture in which the maximum allowable stresses are distributed uniformly [1]. If the beam has a uniform solid cross-section, then the stresses cannot be distributed uniformly. The route to configurations that use less material consists of geometric changes that force more and more of the beam material to be stressed at the highest allowable level.

Why the time evolution toward less structural material (svelteness) is the same as the time arrow of the constructal law is the subject of sections 4.2 and 11.7. The axiom of uniform (maximal) stresses is a consequence of the constructal law.

The evolution of the beam architecture is summarized in Figs. 10.3 and 10.4. The analytical details can be found in Ref. [1], where this beam evolution scenario was first formulated. Start with the top configuration, Fig. 10.3a, in which the beam is a horizontal rod with round cross-section of diameter D_1 and length L_1. The beam is loaded at the right end with the vertical force P. Assume that the beam is sufficiently slender such that the vertical end deflection (δ, under P) is due to pure bending. In the upper half of the rod cross-section the stresses (s) are tensile, and

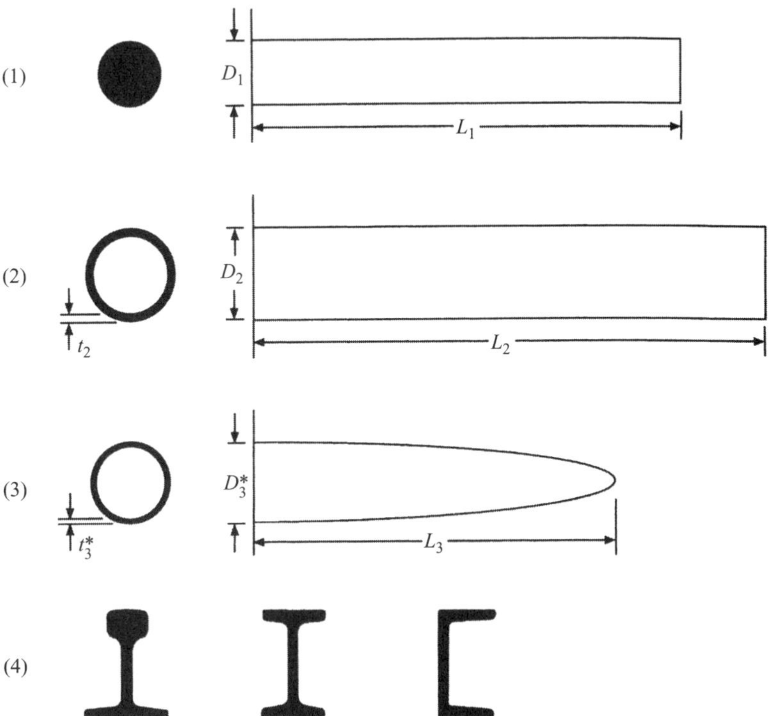

Figure 10.3 (1) Solid cantilever beam with round cross-section, (2, 3) reductions in material requirement via structure (hollowing) and combined shape and structure, and (4) practical beam cross-sections that represent improved internal structure [1].

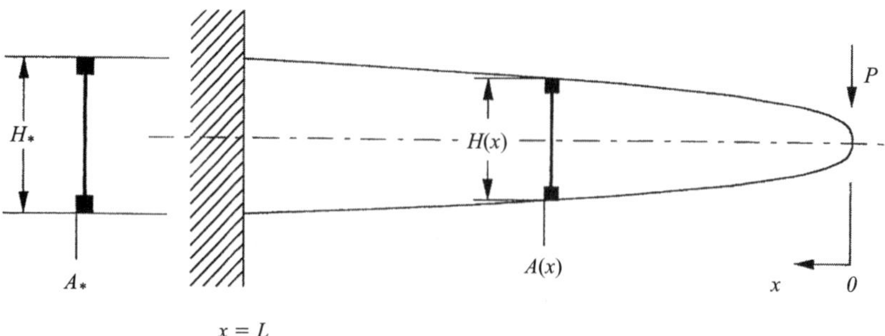

Figure 10.4 Cantilever beam consisting of dorsal and ventral fibers stressed uniformly and to the maximum.

in the lower half they are compressive:

$$s(x, y) = \frac{M(x)}{I_1} y \tag{10.1}$$

Here y is measured vertically away from the midplane of the rod, and $M = Px$ is the bending moment in the vertical cut made at the distance x from the free end. The area moment of inertia of the solid round cross-section A is

$$I_1 = \iint_A y^2 dA \frac{\pi D_1^4}{64} \tag{10.2}$$

At any given x, the highest stresses occur in the top and bottom fibers (the dorsal and ventral fibers),

$$s_{max}(x) = \frac{Px}{I_1} y_{max} \tag{10.3}$$

where $y_{max} = D_1/2$. The largest of all the s_{max} values is the maximum allowable stress, $s_{ma} = \max(s_{max})$. In the present configuration, s_{ma} occurs in the plane where x is the largest: $x = L_1$ (i.e., in the plane of implantation). There, the top and bottom fibers are stressed to the maximal allowable level, which is specified

$$s_{ma} = \frac{PL_1}{I_1} \frac{D_1}{2} = \frac{32PL_1}{\pi D_1^3} \tag{10.4}$$

The specified vertical deflection of the free end is given by the general formula [1]

$$\delta = \int_0^L \frac{M}{EI(x)} \frac{\partial M}{\partial P} dx \tag{10.5}$$

where L is the beam length, $x = L$ is the plane of implantation, and E is the modulus of elasticity. In the case of Fig. 10.3(1), we set $L = L_1$, $I = I_1$, and $M = Px$, and Eq. (10.5) yields

$$\delta = \frac{PL_1^3}{3EI_1} \tag{10.6}$$

The figure of merit of this beam design is the volume of beam material used:

$$V_1 = \frac{\pi}{4} D_1^2 L_1 \tag{10.7}$$

After using Eqs. (10.4) and (10.6) to eliminate D_1 and L_1 we obtain

$$V_1 = 12 \frac{EP\delta}{s_{\text{ma}}^2} \tag{10.8}$$

The improved designs shown in frames (2) and (3) of Fig. 10.3 follow from the idea that in the solid rod (1) most of the material is stressed at levels lower than s_{ma}. The least stressed material is located near the free end (at small x) and in the "core" near the horizontal midplane (at small y).

The core material has been removed in the second design, Fig. 10.3(2). The beam cross-section is independent of axial position. It has the diameter D_2, wall thickness t_2, and length L_2. It can be shown that in this new design, the counterparts of Eqs. (10.4), (10.5), and (10.8) are ([1], Problem 10.6)

$$s_{\text{ma}} = \frac{PL_2}{\pi R_2^2 t_2} \tag{10.9}$$

$$\delta = \frac{PL_2^3}{3EI_2} \tag{10.10}$$

$$V_2 = 2\pi R_2 t_2 L_2 = 6 \frac{EP\delta}{s_{\text{ma}}^2} \tag{10.11}$$

with $I_2 = \pi R_2^3 t_2$. Comparing Eqs. (10.8) and (10.11) we see that the tubular design is dramatically better than the solid rod design because $V_2/V_1 = 1/2$. The relative dimensions of the two designs are $D_2/D_1 = (4\varepsilon)^{-2/5}$ and $L_2/L_1 = (4\varepsilon)^{-1/5}$, where $\varepsilon = t_2/(D_2/2)$ represents the relative thickness of the tube wall. The cross sections shown in Figs. 10.3(1) and (2) have been drawn to scale based on the assumption that $\varepsilon = 0.1$. The volume of beam material V_2 does not depend on the assumed ε, as long as $\varepsilon \ll 1$.

The design achieves its objective even better when it experiences simultaneously the development of internal structure (hollow tube) and external shape (tapered body). In the design of Fig. 10.3(3) the beam diameter D_3 varies with the longitudinal position in such a way that the stress is s_{ma} at every point of the top and bottom fibers of the beam. If $\varepsilon = t_3/(D_3/2) \ll 1$, then the same analysis yields ([1], Problem 10.7)

$$s_{\text{ma}} = \frac{Px}{\pi R_3^2 t_3} \tag{10.12}$$

$$\delta = \frac{3(\pi\varepsilon)^{1/3} s_{\text{ma}}^{4/3} L_3^{5/3}}{5EP^{1/3}} \tag{10.13}$$

$$V_3 = \int_0^{L_3} 2\pi R_3 t_3 dx = \frac{6(\pi\varepsilon)^{1/3} P^{2/3} L_3^{5/3}}{5 s_{\text{ma}}^{2/3}} \tag{10.14}$$

Equation (10.12) shows that in this design R_3 varies proportionally with $x^{1/3}$, because $t_3 = \varepsilon R_3$. After eliminating L_3 between Eqs. (10.13) and (10.14), we find that

$$V_3 = 2\frac{EP\delta}{s_{ma}^2} \tag{10.15}$$

Comparing Eqs. (10.15) and (10.11), we conclude that the external shaping of the hollow beam reduces the material requirement to one-third: $V_3/V_2 = 1/3$. The simultaneous shaping and structuring (hollowing) of the beam reduces the material requirement to one sixth: $V_3/V_1 = 1/6$.

Figure 10.3(3) was drawn to scale based on the assumption that $\varepsilon = 0.1$. It has been shown that $L_3/L_1 = (5/9)^{3/5}(4\varepsilon)^{-1/5}$ and $D_3^*/D_1 = (5/9)^{1/5}(4\varepsilon)^{-2/5}$, where D_3^* represents the value of D_3 in the plane of implantation.

The design can be improved further, by noting that in Fig. 10.3(3), the material situated in the vicinity of the equator of the cross-section is not stressed. By minimizing the amount of such material, one arrives at cross-sectional shapes where most of the stressed material is concentrated at distances far from the center, for example, the I-beam and rail profile shown in Fig. 10.3(4). The ultimate design along this evolutionary course is the combination of the I profile with the optimally tapered design of Fig. 10.3(3). The result is the beam shown in Fig. 10.4, where one dorsal fiber of cross-section $A(x)$ is attached by a rigid web of vanishing thickness and height $H(x)$ to a ventral fiber of cross-section $A(x)$. The stress level is uniform (s_{ma}) in both fibers, tension in the top fiber, and compression in the bottom fiber. In place of Eqs. (10.3), (10.5), and (10.7), we write the rotation equilibrium condition, end deflection, and total beam volume

$$s_m AH = Px \tag{10.16}$$

$$\delta = \int_0^L \frac{2Px^2dx}{EH^2A} \tag{10.17}$$

$$V = \int_0^L 2Adx \tag{10.18}$$

The beam geometry is not unique, because $A(x)$ and $H(x)$ can be specified independently. For example, if we assume the family of designs represented by the power-law shapes,

$$A(x) = A_* \left(\frac{x}{L}\right)^m \tag{10.19}$$

$$H(x) = H_* \left(\frac{x}{L}\right)^n \tag{10.20}$$

where A_* and H_* are the dimensions in the plane of implantation ($x = L$), then Eqs. (10.16) through (10.18) become

$$s_{\mathrm{ma}} A_* H_* = PL, \qquad m + n = 1 \tag{10.21a,b}$$

$$\delta = \frac{2PL^3}{(3 - 2n - m)EH_*^2 A_*} \tag{10.22}$$

$$V = \frac{2LA_*}{m + 1} \tag{10.23}$$

By using Eqs. (10.21a,b), we substitute H_* and n into Eq. (10.22), and obtain

$$\delta = \frac{2s_{\mathrm{ma}}^2 LA_*}{(1 + m)EP} \tag{10.24}$$

Finally, by eliminating LA_* between Eqs. (10.23) and (10.24), we find that the figure of merit of the design is independent of the assumed shape (m):

$$V = \frac{EP\delta}{s_{\mathrm{ma}}^2} \tag{10.25}$$

The conclusion is that the material required by the designs of the class shown in Fig. 10.4 is half of the material needed in Fig. 10.3(3). Three examples of designs that perform at this ultimate level are drawn to scale in Fig. 10.5. There is an infinity

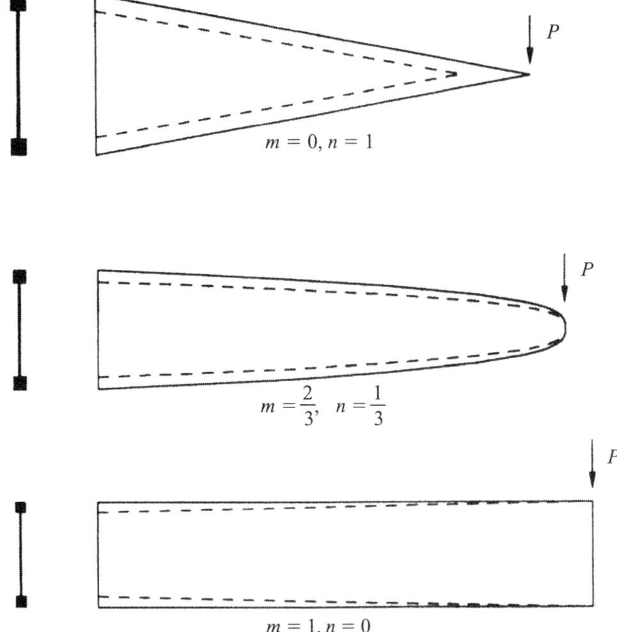

Figure 10.5 Three members of the class of designs represented by Fig. 10.4.

of such designs, depending on the value chosen for one parameter, m. They all look different, but they perform at the same level. This is a characteristic noted very early in constructal theory [2], where most of the design evolution phenomenon was illustrated for tree-shaped flow networks. The conclusion was that diversity of architectures coexists with uniqueness of (high) performance, as marginally better designs replace older ones. With reference to tree-shaped flow arhitectures, this conclusion was summarized as follows ([1], p. 64):

> These relatively unimportant improvements tell a very important story: The tree design is *robust* with respect to various modifications in its internal structure. This means that the global performance of the system is relatively insensitive to changes in some of the internal geometric details. Trees that are not identical have nearly identical performance. The robustness of the tree flows are never identical geometrically. They do not have to be, if the maximization of global performance is their guiding principle. The ways in which the details of natural trees may differ from case to case are without number because, unlike in the constructions presented in this chapter, the number of degrees of freedom of the emerging form is not constrained. Local details differ from case to case because of unknown and unpredictable local features such as the heterogeneity of the natural flow medium and the history and lack of uniformity of the volumetric flow rate that is distributed over the system. The point is that the global performance is predictable, and the principle that takes the system to this level of performance is deterministic.

10.3 INSULATING WALL WITH AIR CAVITIES AND PRESCRIBED STRENGTH

In this section we get to the heart of the "animal", and focus on the close relationship between the internal structure of a flow system and its global thermal and mechanical performance. The fundamental question is how to select the size of air cavities in walls or wall elements (e.g., bricks) for the purpose of maximizing the global resistance to heat transfer through the wall, while preserving the mechanical strength of the wall [3].

There is a large volume of descriptive research on heat transfer through specified (frozen) enclosure configurations that are heated from one side and cooled from the other. This body of fundamental heat transfer was summarized most recently in Ref. [4]. The problem we consider is fundamentally different: it is about the optimization of geometry for a specific purpose, in this case, the maximization of the global performance of the side-heated wall with internal air cavities.

Lorente [5] and Lartigue et al. [6] considered two objectives at the same time: thermal and mechanical performance. They documented the behavior of the global resistance when the side walls of the cavity are deformed (bowed inward) so that the cavity profile resembles a concave lens. Deformations of this kind are commonly found in the air cavities between two glass panes in cold and windy climates. These studies showed that the global resistance to heat transfer increases significantly, in spite of the fact that the narrowing of the cavity mid-section suppresses the flow and its natural convection effect.

Even though Refs. [5] and [6] have dealt with the minimization (not the maximization) of the global thermal resistance, they are seminal because they documented the strong relationship that exists between global performance and cavity geometry. This is the relationship that we explore in this section. Our objective this time is the evolution of the wall with internal cavities as an *insulation* system that must also perform adequately as a strong mechanical structure.

Why should we expect to find an optimal cavity size when we design a cavernous wall as an insulation system? Consider the two-dimensional wall configuration shown in Fig. 10.6. Its overall dimensions are fixed—the thickness L, the height H, and the width W perpendicular to the plane of Fig. 10.6. There are n vertical air-filled cavities of thickness t_a, which are distributed equidistantly over the wall thickness L. This means that there are $(n + 1)$ slabs of solid wall material (e.g., brick) of individual thickness t_b, which are also distributed equidistantly. We characterize the air and brick (terra cotta) composite by using the air volume fraction ϕ, which

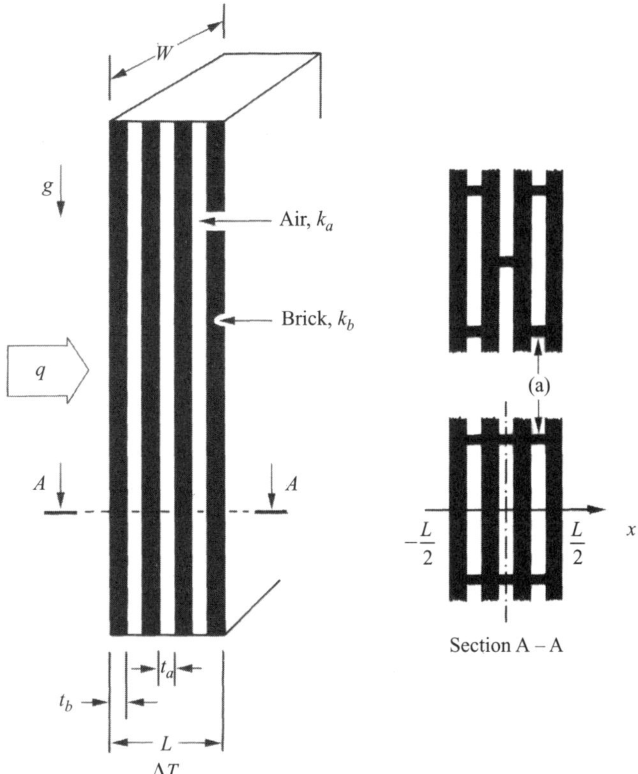

Figure 10.6 Vertical insulating wall with alternating layers of solid (brick) material and air [3].

along with the wall volume HLW is a global design parameter:

$$\phi = nt_a/L \tag{10.26}$$

$$1 - \phi = (n+1)t_b/L \tag{10.27}$$

The overall thermal resistance of this composite is the sum of the resistances of air and brick layers. If the heat transfer across each air space is by pure conduction, then the thermal resistance posed by each air space is $t_a/(k_a HW)$, where k_a is the thermal conductivity of air. Similarly, the resistance of each layer of brick material is $t_b/(k_b HW)$. The overall resistance is

$$R = \frac{nt_a}{k_a HW} + \frac{(n+1)t_b}{k_b HW} \tag{10.28}$$

or, after using Eqs. (10.26) and (10.27),

$$R = \frac{\phi L}{k_a HW} + \frac{(1-\phi)L}{k_b HW} \tag{10.29}$$

This R expression shows that the thermal performance of the composite does not depend on the varying geometry, that is, on how many air spaces and slabs of brick we use. This is correct, but only when the air space is ruled by pure conduction, that is, when the thickness t_a is smaller than the thickness of the laminar natural convection boundary layers that would line the vertical walls of each cavity [4]. The order of magnitude criterion for small t_a is

$$t_a < H\mathrm{Ra}_{H,\theta}^{-1/4} \tag{10.30}$$

where $\mathrm{Ra}_{H,\theta}$ is the Rayleigh number based on the height (H) and temperature difference (θ) across one air cavity,

$$\mathrm{Ra}_{H,\theta} = g\beta H^3\theta/(\alpha\nu) \tag{10.31}$$

The difference θ is smaller than the overall difference ΔT that is maintained across the entire system (Fig. 10.6). In the case of air and brick material, the two thermal conductivities are markedly different ($k_b/k_a \sim 20 \gg 1$), and this means that the overall ΔT is essentially the sum of the temperature differences across all the air cavities:

$$\Delta T \cong n\theta \tag{10.32}$$

Putting Eqs. (10.30) through (10.32) together, we see that the insensitivity of R to varying the internal structure (n), Eq. (10.29), can be expected only when the number of air spaces is large enough:

$$n^{5/4} > \phi\frac{L}{H}\mathrm{Ra}_{H,\Delta T}^{1/4} \tag{10.33}$$

In this inequality, $Ra_H, \Delta T$ is based on the overall temperature difference, $Ra_H, \Delta T = g\beta H3\Delta T/(\alpha\nu)$, and is a known constant because H and ΔT are specified global parameters.

If the number of air spaces is smaller than in Eq. (10.33), the natural convection effect decreases the resistance posed by each air space, and the overall R value is greater than in Eq. (10.28). This is why a large enough n, or a small enough t_a, is desirable from a thermal insulation standpoint. However, a large n is detrimental to the mechanical stiffness (rigidity) of the wall assembly. When ϕ is prescribed, the stiffest wall is the one where all the solid material is placed in the outermost planes, that is, the wall where two t_b-thin slabs sandwich a single air space. The stiffest wall is the worst thermal insulator, because it contains the thickest air space, which is traveled by the largest natural convection heat and fluid-flow currents. The optimal internal structure of the wall (n) results from the competition between thermal performance and mechanical performance.

The mechanical strength of the wall, or the resistance to bending and buckling in the plane of Fig. 10.6 is controlled by the area moment of inertia of the horizontal wall cross-section [cf. Eq. (10.2)]:

$$I_n = \int_{-L/2}^{L/2} x^2 W \, dx \tag{10.34}$$

The cross-section over which this integral is performed is sketched in the details shown on the right side of Fig. 10.6. The area element $W dx$ counts only the solid parts of the cross-section, namely, the t_b-thick slabs of brick material. For the sake of simplicity, in this calculation we neglect the transversal ribs [labeled (a) in Fig. 10.6] that connect the t_b slabs so that the wall cross-section rotates as a plane during pure bending. We assume that the transversal ribs use considerably less material than the t_b slabs. Their role is the same as the role of the central part (the web) of the I profile of an I-beam (Figs. 10.3d) and 10.4. In fact, the cross-section of the cavernous wall structure is an assembly of I-beam profiles that have been fused solidly over the top and bottom surfaces of the I shape. In practice, the ribs (a) are more commonly arranged in a staggered pattern, as shown in the upper-right part of Fig. 10.6.

In the case of a wall with vanishing cavities ($\phi = 0$) the area moment of inertia is maximum and equal to $L^3W/12$. We use this value as reference in the nondimensionalization of I_n,

$$\tilde{I}_n = \frac{I_n}{L^3 W/12} \tag{10.35}$$

where the subscript n indicates the number of air gaps. We evaluate the integral (10.34) case by case, assuming in each case that the cross-section is symmetric

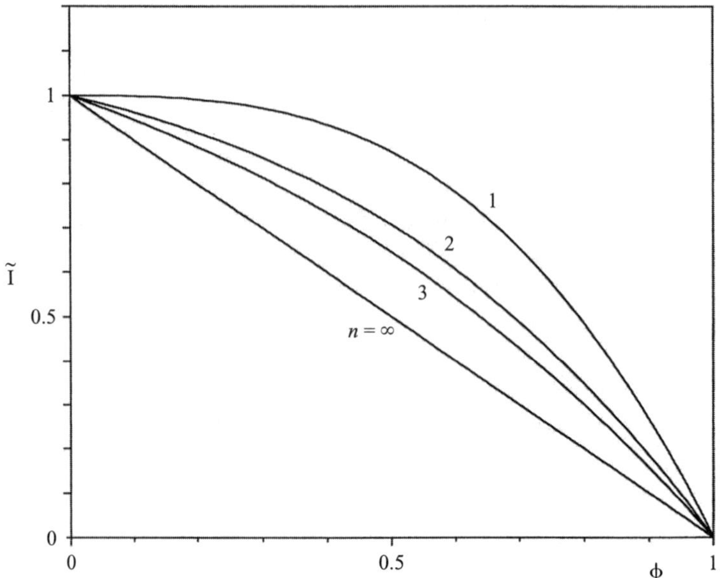

Figure 10.7 The area moment of inertia of the wall cross-section, as a function of the number of air gaps and air volume fraction [3].

about $x = 0$,

$$\tilde{I}_1 = 1 - \phi^3 \tag{10.36}$$

$$\tilde{I}_2 = 1 + \left(\frac{1-\phi}{3}\right)^3 - \left(\frac{1+2\phi}{3}\right)^3 \tag{10.37}$$

$$\tilde{I}_3 = 1 + \left(\frac{3-\phi}{6}\right)^3 - \left(\frac{\phi}{3}\right)^3 - \left(\frac{\phi+1}{2}\right)^3 \tag{10.38}$$

$$\tilde{I}_\infty = 1 - \phi \tag{10.39}$$

These results are displayed in Fig. 10.7. The stiffness is larger when n and ϕ are smaller.

Another way of looking at this relation is presented in Fig. 10.8. When the stiffness is constrained, $\tilde{I}_n$ = constant, for each geometry (n) that the designer might contemplate there is one value of ϕ that the wall composite must have. In such cases the ϕ value is larger when the number of air gaps is smaller. Less structural (solid) material is needed when there are fewer air gaps.

When the effect of natural convection cannot be neglected, the overall thermal resistance formula (10.28) has the form

$$R = \frac{nt_a}{k_a HW \mathrm{Nu}} + \frac{(n+1)t_b}{k_b HW} \tag{10.40}$$

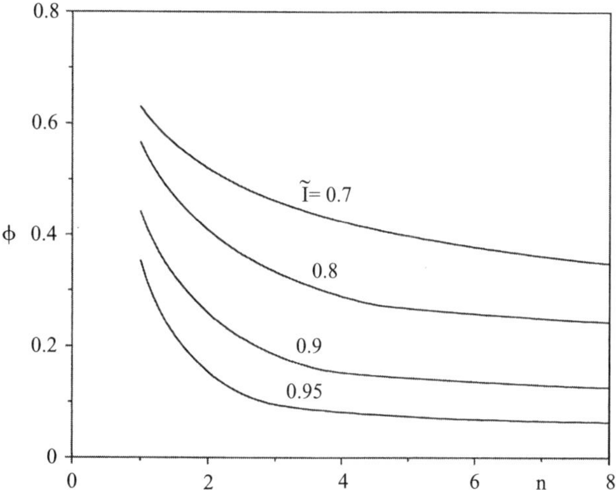

Figure 10.8 The relation between air volume fraction and number of air gaps when the area moment of inertia of the wall cross-section is constrained [3].

In the denominator of the first term (the contribution of all the air gaps), Nu is the overall Nusselt number that expresses the relative heat-transfer augmentation effect due to natural convection in a single air space:

$$\mathrm{Nu} = \frac{q_{\text{actual}}}{q_{\text{conduction}}} \tag{10.41}$$

Several correlations of experimental and numerical Nu values have been reported [4]; however, they cannot be used in the reported forms because they refer only to the high-Rayleigh number, or the convection dominated regime (Nu $\geq$ 2). More appropriate in the present search for geometry is a Nu function that covers smoothly the entire range of possibilities, from conduction (small t_a) to convection (large t_a). The solution we chose is based on the summary presented in Fig. 5.8 of Ref. [4], which shows that the most frequently used high-Ra correlations are represented well by an analytical expression [7] derived based on boundary layer theory (Ref. [4], p. 256):

$$\mathrm{Nu} = 0.364 \frac{t_a}{H} \mathrm{Ra}_{H,\theta}^{1/4} \quad (\text{when Nu} \geq 2) \tag{10.42}$$

It is worth pointing out that Eq. (10.42) is consistent with the pure conduction criterion (10.30). In other words, Eq. (10.42) holds when Eq. (10.30) fails. Next, we joined the high-Ra asymptote with the pure conduction asymptote (Nu = 1) by using the technique of Churchill and Usagi [8]:

$$\mathrm{Nu} = \left[1 + \left(0.364 \frac{t_a}{H} \mathrm{Ra}_{H,\theta}^{1/4}\right)^m\right]^{1/m} \tag{10.43}$$

For the curve-smoothing exponent, we chose $m = 3$. In summary, the overall resistance formula (10.40) can be nondimensionalized by using as reference scale the resistance across a completely solid wall $[L/(k_b\, HW)]$, and converting Ra_H,θ into $\mathrm{Ra}_H,\Delta T$ via Eqs. (10.31) and (10.32):

$$\tilde{R} = \frac{R}{L/(k_b HW)} = \frac{k_b}{k_a}\phi\left[1 + \left(0.364 n^{-5/4}\phi\frac{L}{H}\mathrm{Ra}_{H,\Delta T}^{1/4}\right)^m\right]^{-1/m} + 1 - \phi \tag{10.44}$$

The overall resistance $\tilde{R}$ emerges as a function of the variable geometric parameters n and ϕ, the fixed parameters k_b/k_a and the global parameter

$$b = \frac{L}{H}\mathrm{Ra}_{H,\Delta T}^{1/4} \tag{10.45}$$

The geometric parameters n and ϕ are related through the global stiffness constraint ($\tilde{I}$ = constant) displayed in Fig. 10.8.

In conclusion, when the stiffness constraint is invoked, the global resistance $\tilde{R}$ depends on only one geometric parameter, ϕ or n. This dependence is illustrated in Fig. 10.9, which shows that $\tilde{R}$ can be maximized with respect to the number of air cavities. The $\tilde{R}$ maximum shifts toward larger n values (more numerous and narrower air gaps) as b increases. The $\tilde{R}$ maximization illustrated in Fig. 10.9, was repeated for other $\tilde{I}$ values in the range 0.7–0.95.

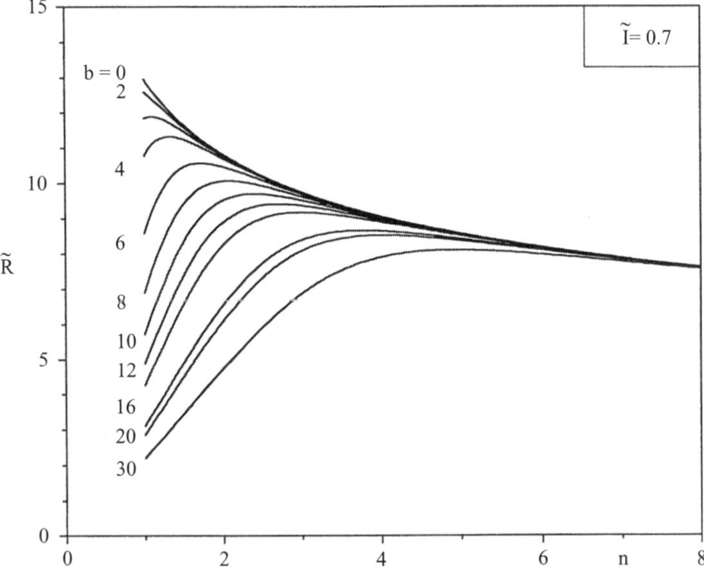

Figure 10.9 The overall thermal resistance as a function of the number of air gaps when the external parameters b and $\tilde{I}$ are fixed [3].

Let $\tilde{R}_{max}$ and n_{opt} denote the coordinates of the peak of one of the b = constant curves plotted in Fig. 10.9. The maximum thermal resistance $\tilde{R}_{max}$ $(b, \tilde{I})$ deduced from Fig. 10.9 and from similar calculations for other $\tilde{I}$ values, is reported in Fig. 10.10. Larger b values represent stronger natural convection, and this is reflected in smaller $\tilde{R}_{max}$ values. Larger $\tilde{I}$ values belong to stiffer walls that use more solid material (Fig. 10.8), and, consequently, $\tilde{R}_{max}$ decreases as $\tilde{I}$ increases.

The optimal number of air gaps (n_{opt}) that corresponds to the $\tilde{R}_{max}$ results of Fig. 10.10 is reported in Fig. 10.11. Fewer air gaps are better when the natural convection effect is weak (small b), and when the required stiffness approaches that of the solid wall ($\tilde{I} = 1$).

In conclusion, the internal structure of a cavernous wall can be derived from the competition between the thermal insulation and mechanical strength functions of the wall. This combination of two functions, thermal and mechanical, is new in an optimization at such a simple and fundamental level. Previous studies of walls with air enclosures have dealt only with the thermal insulation characteristics of various wall structures [4].

The combined *flow and strength* search for geometry, which was proposed in Ref. [3] and illustrated in this section, can be applied in other fields where mechanical structures must carry loads while posing least resistance to internal and external flows. This combination may be carried further into the design of structural elements for vehicles, which, like the wall of Fig. 10.6, can be conceptualized and morphed into geometric forms with more than one function.

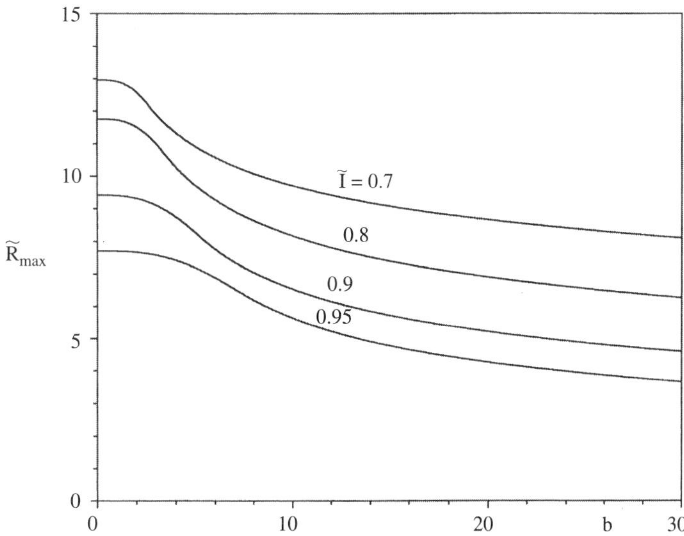

Figure 10.10 The maximized wall thermal resistance as a function of the global natural convection parameter b and the global stiffness parameter $\tilde{I}$ [3].

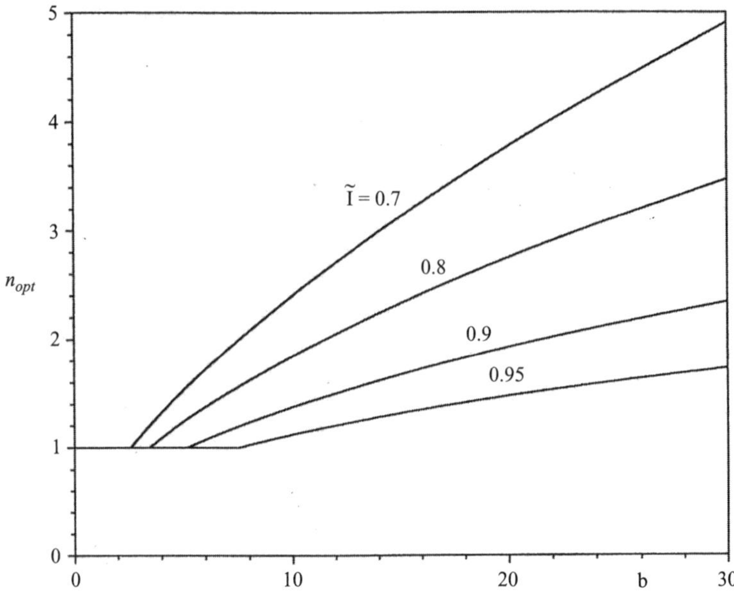

Figure 10.11 The optimal number of air gaps as a function of the global natural convection parameter b and the global stiffness parameter $\tilde{I}$ [3].

10.4 MECHANICAL STRUCTURES RESISTANT TO THERMAL ATTACK

In this section we take the "flow and strength" constructal method in a new direction: systems that must be mechanically strong and, at the same time, must retain their strength and integrity during thermal attack [9]. Mechanical structures become weaker and may collapse if they are exposed to intense heating. If the time in which they become weak is short, the chances that the fire rescue team can save the occupants are reduced. The collapse of the World Trade Center is a reminder of how dangerous the effect of sudden intense heating can be. Large buildings, highway overpasses and industrial installations are vulnerable.

The classical approach to providing a structure with thermal resistance against intense heating is by coating the structure with a protective layer *after* the structure has been designed [10]. Gosselin et al. [9] proposed to change the conceptual approach to optimal structures, away from the single-objective lessons of the past, and in line with the two-objective morphing of structures proposed in Chapter 7. We illustrate this approach by using two classes of structures exposed to sudden heating: beams in pure bending, and beams of concrete reinforced with steel. In both cases the solid structure is penetrated by time-dependent conduction heating. We show that the mechanical and thermal objectives compete, and that this competition gives birth to the geometry of the system.

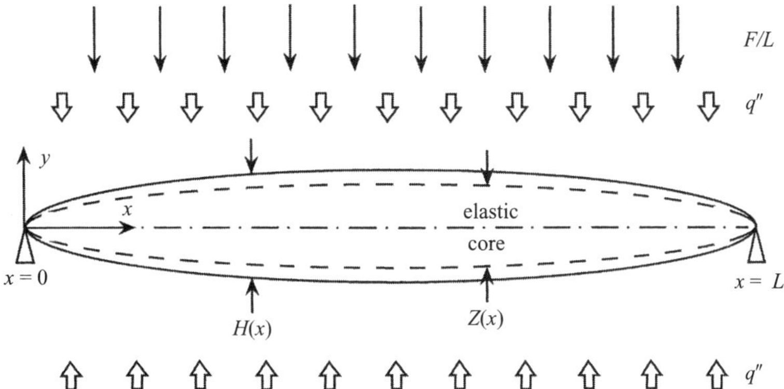

Figure 10.12 Beam in bending with uniform loading and sudden heating from above and below [9].

10.4.1 Beam in Bending

Consider a beam simply supported at each end (Fig. 10.12). The beam geometry is two-dimensional, with the length L and symmetric profile $H(x)$. The total load F' [Nm^{-1}] is distributed uniformly over the beam length L. The force F' is expressed per unit length in the direction perpendicular to the plane of Fig. 10.12. The weight of the beam is assumed negligible in comparison with the load. The beam profile is sufficiently slender so that its deformation in the y-direction is due mainly to pure bending.

The beam is initially isothermal at the ambient temperature T_∞, where it behaves elastically throughout its volume. The modulus of elasticity is E, which for simplicity is assumed constant. Thermal attack means that at the time $t = 0$ the beam is exposed on both surfaces to the uniform heat flux q''. Temperatures rise throughout, but they rise faster in the subskin regions (Fig. 10.13). These are the first regions where the material behavior changes from elastic to plastic. The last to undergo this change is the core region of thickness $Z(x)$, in which the material behaves elastically.

The total bending moment in a constant-x cross-section is

$$\frac{M}{W} = \frac{F'L}{2}\frac{x}{L}\left(1 - \frac{x}{L}\right) \tag{10.46}$$

where W is the beam length in the z-direction, which is perpendicular to the plane of Fig. 10.12. This moment is balanced by the moment due to the tensile and compressive stresses (σ) that are present in the cross-section. When σ is less than the yield stress σ_y, the material behaves elastically. The yield stress decreases as the local temperature increases. For simplicity, we assume a linear model for the

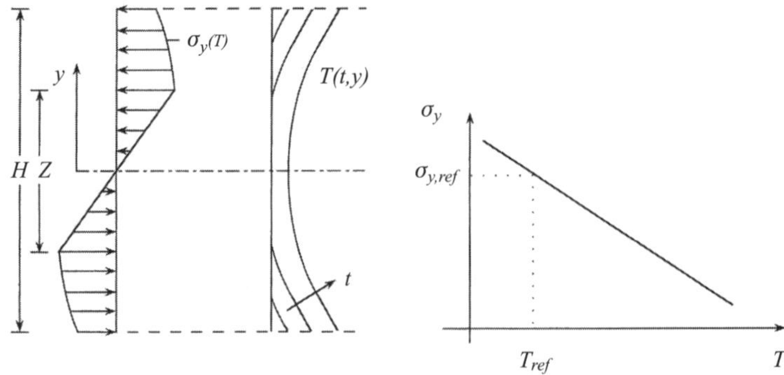

Figure 10.13 The stresses in the elastic and thermoplastic regions of the beam cross-section [9].

temperature effect on σ_y,

$$\frac{\sigma_y}{\sigma_{y,\text{ref}}} = 1 - \beta(T - T_{\text{ref}}) \tag{10.47}$$

where the β coefficient is a property of the material, and T_{ref} is a reference temperature, such that $\sigma_{y,\text{ref}} = \sigma_y(T_{\text{ref}})$. The reference temperature was set equal to the ambient temperature, $T_{\text{ref}} = T_\infty$ and $\sigma_{y,\text{ref}} = \sigma_{y,\infty}$. In the elastic core the stresses vary linearly,

$$\frac{\sigma}{\sigma_y} = \frac{y}{Z/2} \tag{10.48}$$

where σ_y is the yield stress at $y = \pm Z/2$, which is associated with the instantaneous temperature at that location, T, cf. Eq. (10.47). We assume that in the peripheral regions outside $y = \pm Z/2$ the material is perfectly plastic, so that σ is equal to $\sigma_y(T)$, where T is the local temperature.

The first drawing in Fig. 10.13 summarizes qualitatively the distribution of stresses in the cross-section, at a time when plastic regions are present, $Z < H$. In this model we accounted for the fact that in the beginning there is a time interval when the entire beam is elastic, and the maximum stress ($\sigma_{\max}$, at $y = \pm H/2$) is still below the yield stress. During this initial time interval the beam deflection is constant in time. The moment formed by the stresses in the beam cross-section,

$$\frac{M(x,t)}{W} = \int_{-H/2}^{H/2} \sigma(t,x,y)y\,dy \tag{10.49}$$

leads to a two-term expression that accounts for the elastic and plastic regions,

$$\frac{M}{W} = \frac{1}{6}\sigma_y[T(t,x,y=Z/2)]Z^2 + 2\sigma_{y,\text{ref}} \int_{Z/2}^{H/2} (1 - \beta\Delta T)y\,dy \tag{10.50}$$

where

$$\sigma_y[T(t,x,y=Z/2)] = \sigma_{y,\mathrm{ref}}[1 - \beta \Delta T(t,x,y=Z/2)] \tag{10.51}$$

$$\Delta T(t,x,y) = T(t,x,y) - T_\infty \tag{10.52}$$

Equations (10.46) and (10.50) can be combined to pinpoint the location of the elastic-plastic interface, $Z(t,x)$, for a specified beam profile $H(x)$, and temperature distribution $T(t,x,y)$.

Consider next the beam deflection in the y-direction. The local radius of curvature ρ of the deformed beam is

$$\rho(t,x) = \frac{EZ(t,x)}{2\sigma_{\max}} \tag{10.53}$$

As a first approximation, for small deflections the position of the neutral line [$y = \delta(x)$] can be written as

$$\frac{d^2\delta}{dx^2} = \frac{1}{\rho} \tag{10.54}$$

In the absence of a plastic zone, the stress in the outer fibers ($y = \pm H/2$) is

$$\sigma_{\max} = \frac{3F'}{H^2} x \left(1 - \frac{x}{L}\right) \tag{10.55}$$

which must be used in Eq. (10.53). However, when a plastic zone is present, the maximum stress is reached at the plastic-elastic interface, $\sigma_{\max} = \sigma_y[T(Z/2)]$. The position of the neutral line is obtained by integrating Eq. (10.54) twice. The boundary conditions are $\delta = 0$ at $x = 0$, and $\delta = 0$ at $x = L$. The maximal deflection occurs in the midplane, $\delta_m = -\delta(x = L/2)$.

10.4.2 Maximization of Resistance to Sudden Heating

The amount of beam material is fixed, and, in view of the two-dimensional geometry of Fig. 10.12, the area of the profile is also fixed,

$$A = \int_0^L H(x)dx \tag{10.56}$$

Gosselin et al. [9] considered many profile shapes, for example, Eq. (10.62) in the next section. For every assumed shape, they calculated numerically the time evolution of the maximal deflection, $\delta_m(t)$. The objective is to identify the shape for which δ_m is the smallest at a given t. This shape is the most resistant to thermal attack.

The numerical work was conducted in dimensionless terms by using the dimensionless variables:

$$\tilde{x} = \frac{x}{L}, \quad \tilde{y} = \frac{y}{H/2}, \quad \tilde{t} = \frac{\alpha t}{(H/2)^2}, \quad \tilde{Z} = \frac{Z}{H}, \quad \tilde{H} = \frac{H}{L}, \quad \tilde{A} = \frac{A}{L^2}$$

$$\tilde{\beta} = \beta \frac{q''L}{2k}, \quad \tilde{\sigma} = \frac{\sigma}{F'/L}, \quad \tilde{\delta} = \frac{\delta E}{2F'}, \quad \Delta\tilde{T} = \Delta T \frac{2k}{qL} \tag{10.57}$$

The local beam temperature is known from Fourier analysis [11], under the assumption that the beam profile is slender so that conduction in the x-direction is negligible:

$$\Delta\tilde{T} = \tilde{H}\left[\tilde{t} + \frac{\tilde{y}^2}{6} - \frac{1}{6} - 2\sum_{n=1}^{\infty} \frac{(-1)^n}{n^2\pi^2} e^{-n^2\pi^2\tilde{t}} \cos(n\pi\tilde{y})\right] \tag{10.58}$$

The infinite sum in the square brackets is important only in the beginning, and vanishes rapidly for $\tilde{t} \geq 1$. Equations (10.50), (10.54), and (10.56) become

$$\tilde{x}(1-\tilde{x}) = \tilde{\sigma}_{y,\text{ref}}\tilde{H}^2 \left\{\frac{\tilde{Z}^2}{3}[1 - \tilde{\beta}\Delta\tilde{T}(\tilde{Z}, \tilde{x}, \tilde{t})] + \int_{\tilde{Z}}^{1} [1 - \tilde{\beta}\Delta\tilde{T}(\tilde{x}, \tilde{y}, \tilde{t})]\tilde{y}\,d\tilde{y}\right\} \tag{10.59}$$

$$\frac{d^2\delta}{d\tilde{x}^2} = \frac{\tilde{\sigma}_y[T(Z/2)]}{\tilde{Z}(x,t)} \tag{10.60}$$

$$\tilde{A} = \int_0^1 \tilde{H}(\tilde{x})\,d\tilde{x} \tag{10.61}$$

To start with, we considered a family of beam shapes that are smooth and thicker in the middle (e.g., Fig. 10.12):

$$\tilde{H} = C[\tilde{x}(1-\tilde{x})]^m \tag{10.62}$$

The shape parameters C and m are related through the size constraint (10.61):

$$\frac{A/L^2}{C} = \int_0^1 [\tilde{x}(1-\tilde{x})]^m\,d\tilde{x} \tag{10.63}$$

The geometry is characterized by one shape parameter (m), which plays the role of degree of freedom, and by three construction parameters: $\tilde{A}$, β and $\tilde{\sigma}_{y,\text{ref}}$. The latter is defined as

$$\tilde{\sigma}_{y,\text{ref}} = \frac{\sigma_y(T_\infty)}{F'/L} \tag{10.64}$$

The calculation of $\tilde{\delta}_m(\tilde{t})$ is performed from $\tilde{t} = 0$ until the elastic core disappears at a location $\tilde{x}$. The model constructed in the preceding section is not valid when the elastic core is absent.

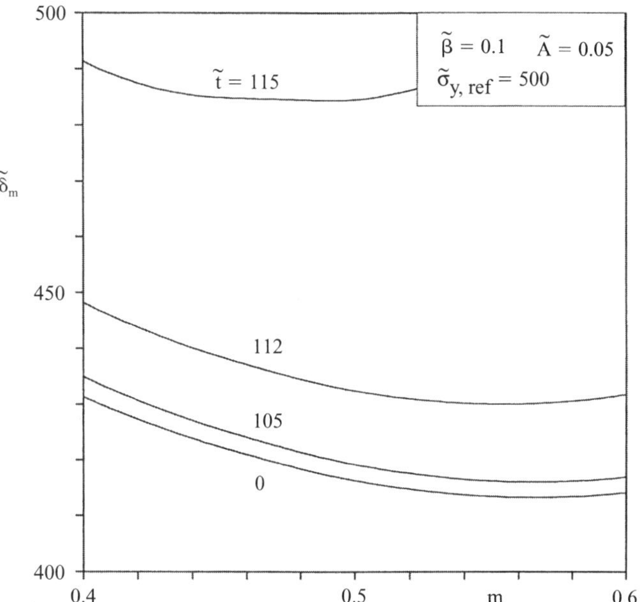

Figure 10.14 The minimization of the mid-length deflection by selecting the beam shape parameter m [9].

The numerical example given in Fig. 10.14 shows that the deflection increases in accelerated fashion in time, and that $\tilde{\delta}_m$ can be minimized by selecting the shape parameter m. This is the discovery: the beam geometry can be selected in such a way that the beam as a whole is most resistant to thermal attack. This is a result for how the whole beam performs—a global result—because $\tilde{\delta}_m$ is a global feature. All the strained fibers contribute to $\tilde{\delta}_m$.

The influence of geometry on performance is described further in Fig. 10.15, where $\tilde{\delta}_m(\tilde{t})$ has been plotted for three m values. Because the objective is to achieve the smallest $\tilde{\delta}_m$, we conclude that the best shape (m) changes as the time increases. The intersecting $\tilde{\delta}_m(\tilde{t})$ curves indicate that m_{opt} decreases as $\tilde{t}$ increases. This decrease accelerates in time, as shown in Fig. 10.16. The same figure shows that the minimal mid-plane deflection $\tilde{\delta}_{m,\min}$, which corresponds to the optimally changing shape m_{opt} $(\tilde{t})$, also accelerates in time. If $\tilde{t}$ denotes the prescribed lifetime of the beam—the time in which it must withstand the thermal attack—then for every $\tilde{t}$ there exists an optimal beam shape.

Important in Fig. 10.15 and Fig. 10.16 are the short times, where deflections are small and comparable with deflections based on the assumption that thermal attack is absent. In this limit there is a definite beam shape that is optimal. This is also the limit in which the model constructed in section 10.4.1 is valid.

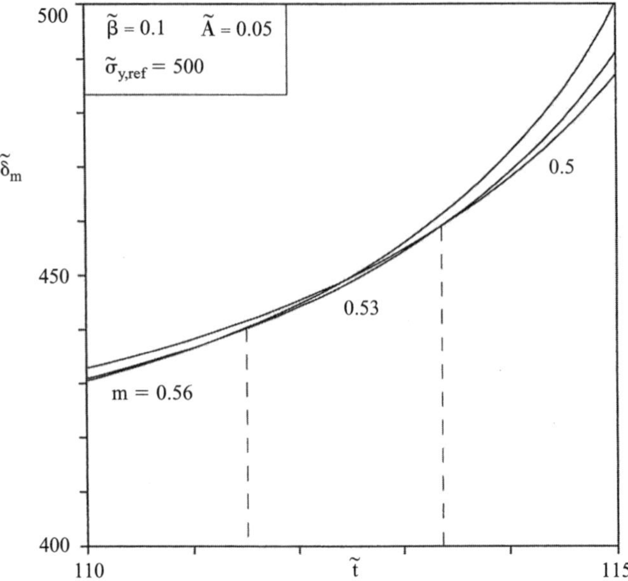

Figure 10.15 The effect of the beam lifetime ($\tilde{t}$) on the minimization of the mid-length deflection [9].

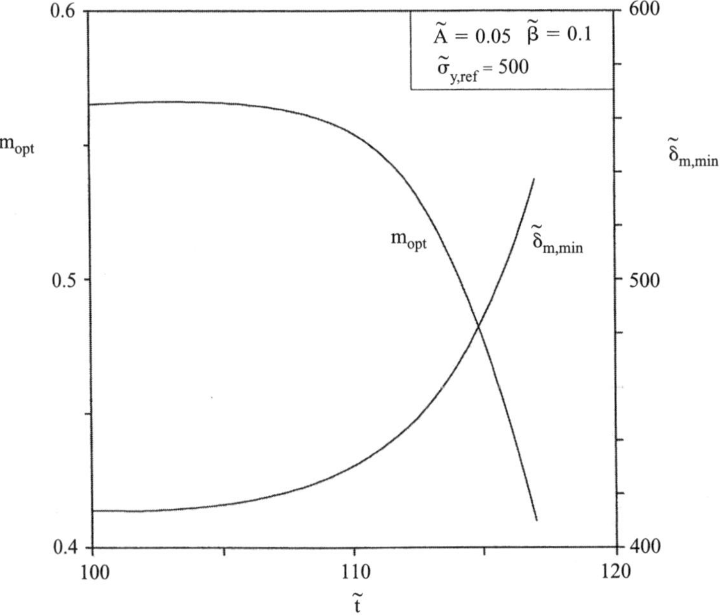

Figure 10.16 The optimized beam shape parameter and the minimized mid-length deflection [9].

10.4.3 Steel-Reinforced Concrete

Here we rely on the same idea in the search for the internal structure of a beam of concrete reinforced with steel bars. The objective is maximal survivability to thermal attack. The beam is in pure bending, and its cross-section is shown in Fig. 10.17. The steel bars run in the direction perpendicular to the figure and are modeled as a slab with cross-section $h_s \times b$. The beam is loaded such as the steel slab is in tension, while the concrete situated above the neutral line is in compression.

Thermal attack is modeled as a uniform heat flux (q''), which is imposed suddenly on the periphery of the beam cross-section. The most critical part that is vulnerable under thermal attack is the steel, therefore in the simplest model we focus on the q''-heating that is applied on the bottom of the cross-section, which is the closest to the steel. A layer of concrete of thickness λ protects the steel against the thermal wave driven by q''.

The thickness λ plays an important role. In order for the beam to support a large load, λ must be small: the steel must be positioned as far as possible from the top of the beam cross section. However, a high resistance to thermal attack requires a large λ. The competition between these two requirements generates the design, the drawing.

A competition exists because the beam design must meet two objectives, mechanical strength and thermal resistance. There are two constraints, the area of the

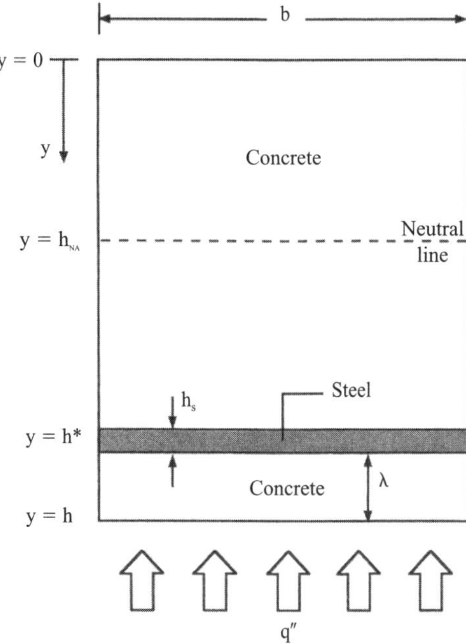

Figure 10.17 The cross-section of a beam of concrete reinforced with steel and heated suddenly from below [9].

beam cross-section

$$A = hb \tag{10.65}$$

and the cross-sectional area of the steel, $A_s = h_s b$. Alternatively, the steel constraint can be expressed as the area fraction occupied by all the steel in the cross-section,

$$\phi = \frac{A_s}{A} = \frac{h_s}{h} \tag{10.66}$$

Because steel is expensive, it is reasonable to assume that $h_s \ll h$, or $\phi \ll 1$. The distance from the top of the beam to the midline of the steel cross-section is

$$h^* = h - \lambda - \frac{h_s}{2} \tag{10.67}$$

In accordance with the classical model of a reinforced beam [12–14], we assume that the beam is loaded entirely in the elastic regime, with concrete in compression ($0 < y < h_n$), and steel in tension. The moduli of elasticity of concrete and steel are E_c and E_s, respectively. The stiffness of the cross-section is characterized by

$$\overline{EI} = E_s A_s (h^* - h_n)^2 + E_c b \frac{h_n^3}{3} \tag{10.68}$$

where h_n denotes the position of the neutral line, which is given by

$$\frac{E_s}{E_c} A_s (h^* - h_n) = b \frac{h_n^2}{2} \tag{10.69}$$

10.4.3a Heating from Below.

Heating the beam from the bottom is a process of unidirectional time-dependent conduction in a heterogeneous medium containing two materials, concrete of thermal conductivity k, and steel. The assumption that the amount of steel is small ($\phi \ll 1$) justifies the use of a conduction model in which the steel is represented by a line drawn at $y = h^*$. In a more realistic model in which the finiteness of h_s and the high thermal conductivity of steel are taken into account, the advancing thermal wave would be characterized by a narrow and relatively isothermal region of thickness h_s. This "flat spot" of the instantaneous temperature profile is neglected in the present model, for which the conduction equation and boundary and initial conditions are

$$\frac{\partial T}{\partial t} = \alpha \frac{\partial^2 T}{\partial y^2} \tag{10.70}$$

$$\frac{\partial T}{\partial y} = -\frac{q''}{k} \quad \text{at} \quad y = h \tag{10.71}$$

$$\frac{\partial T}{\partial y} = 0 \quad \text{at} \quad y = 0 \tag{10.72}$$

$$T = T_\infty \quad \text{at} \quad t = 0 \tag{10.73}$$

The analytical solution for the temperature field is [11]

$$\frac{T(y,t)-T_\infty}{q''h/k} = \frac{\alpha t}{h^2} + \frac{1}{2}\left(\frac{y}{h}\right)^2 - \frac{1}{6} - 2\sum_{n=1}^{\infty} \times \frac{(-1)^n}{(n\pi)^2} e^{-\frac{\alpha t}{h^2}(n\pi)^2} \cos\left(n\pi \frac{y}{h}\right) \tag{10.74}$$

Because the elastic modulus of steel decreases monotonically as the temperature increases, the heating process has the effect of decreasing the beam stiffness, in accordance with Eq. (10.68). We account for the coupling between the changing temperature field and the global stiffness of the beam by using the relative (dimensionless) elastic modulus [13]

$$\tilde{E}_s = \frac{E_s(T)}{E_s(20\,^\circ\mathrm{C})} \tag{10.75}$$

The reference elastic modulus was set at $E_s\,(20\,^\circ\mathrm{C}) = 200$ GPa, which is representative of low-carbon steel and high-carbon steel. The relative modulus for steel is [13]

$$\tilde{E}_s = \left\{\begin{array}{ll} 1 & T \le 100\,^\circ\mathrm{C} \\ 1.10 - 0.001T & 100\,^\circ\mathrm{C} \le T \le 500\,^\circ\mathrm{C} \\ 2.05 - 0.0029T & 500\,^\circ\mathrm{C} \le T \le 600\,^\circ\mathrm{C} \\ 1.39 - 0.0018T & 600\,^\circ\mathrm{C} \le T \le 700\,^\circ\mathrm{C} \\ 0.41 - 0.0004T & 700\,^\circ\mathrm{C} \le T \le 800\,^\circ\mathrm{C} \\ 0.27 - 0.000225T & 800\,^\circ\mathrm{C} \le T \le 1200\,^\circ\mathrm{C} \\ 0 & 1200\,^\circ\mathrm{C} \le T \end{array}\right\} \tag{10.76}$$

where T is expressed in °C. The elastic modulus of concrete was assumed insensitive to temperature changes, and was set at $E_c = 20$ GPa. The thermal properties of concrete are $k = 1.44\ \mathrm{Wm^{-1}K^{-1}}$ and $\alpha = 6.92 \times 10^{-7}\ \mathrm{m^2s^{-1}}$. Numerical simulations were performed for a beam with these material properties and $A = 0.3\ \mathrm{m^2}$, $\phi = 0.03$ and $q'' = 2 \times 10^4\ \mathrm{Wm^{-2}}$.

The cross-section geometry has two degrees of freedom, the aspect ratio h/b and the protective thickness λ. In the first phase of numerical simulations we fixed h/b and varied λ. Figure 10.18 shows the evolution of the global stiffness in time, as the heat wave expands into the beam. Most resistant to this softening effect are beams with thicker protective layers. Such beams are also the weakest when not under thermal attack.

Assume that the lifetime (t) for the survival of the beam under the effect of q'' is specified. Reading Fig. 10.18 at constant t we see that there is an optimal $\lambda_{opt}(t)$ where the beam stiffness is maximal $\overline{EI}_{max}(t)$. In other words, for a beam to support its load with maximal stiffness at the end of its life under thermal attack, it must be designed with an optimal thickness for its protective layer: λ_{opt} is the trade-off between the mechanical and thermal objectives recognized at the start of section 10.4.3. The optimal thickness of the protective layer is shown in Fig. 10.19.

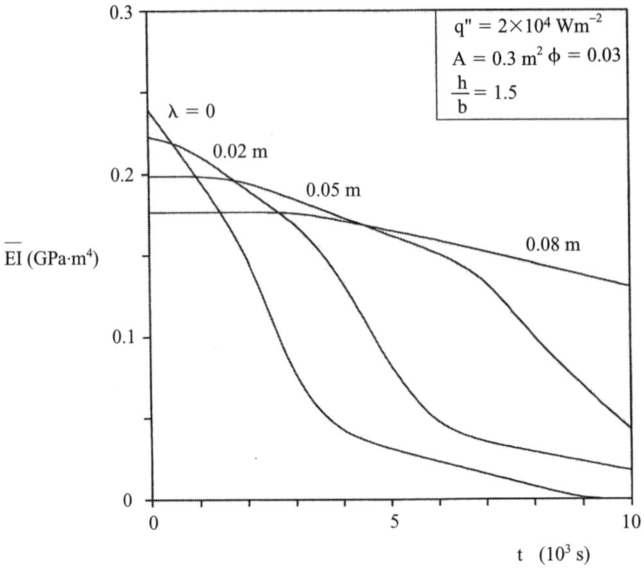

Figure 10.18 The evolution of the beam global stiffness during heating from below, when the beam aspect ratio is constrained [9].

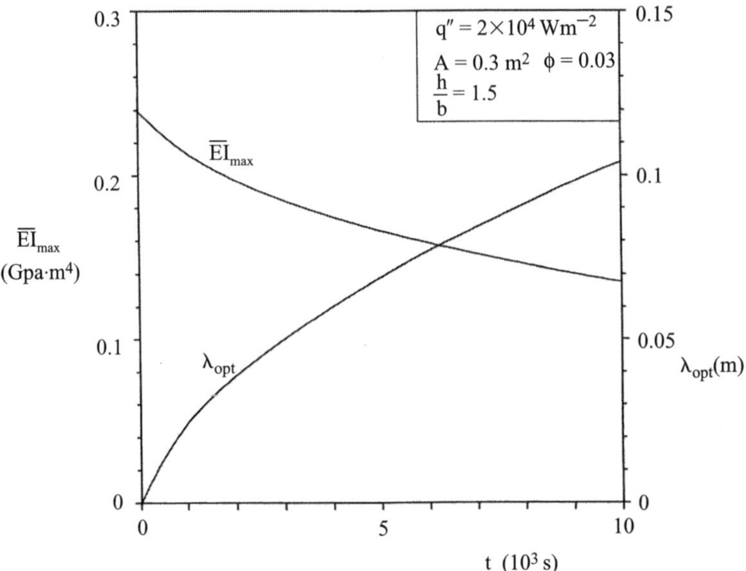

Figure 10.19 The optimal protective layer thickness and maximal stiffness for a specified lifetime ($\tilde{t}$) under thermal attack [9].

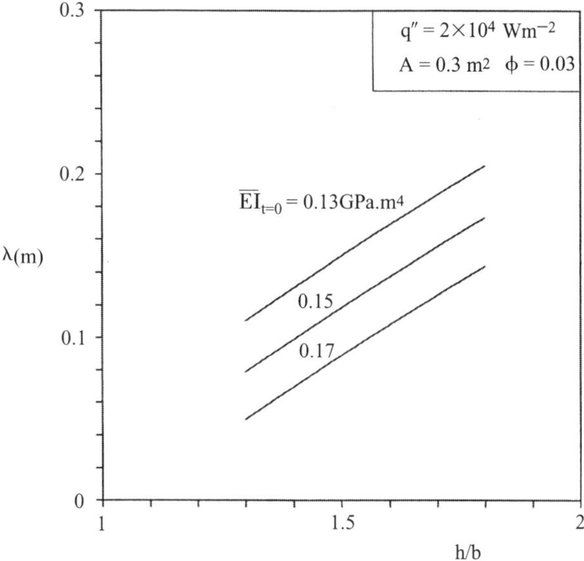

Figure 10.20 The relationship between protective layer thickness and cross-sectional aspect ratio when the global stiffness in the absence of heating is specified [9].

An alternative way to exploit the mechanical and thermal trade-off is by relaxing the assumption that the aspect ratio of the cross-section is fixed. In this case λ and h/b may vary. If the global stiffness of the beam (before heating) is specified by design, $\overline{EI}_{t=0}$, then λ is a function of h/b. This function is illustrated in Fig. 10.20, where λ is almost proportional to h/b, and almost inversely proportional to $\overline{EI}_{t=0}$.

In summary, when $\overline{EI}_{t=0}$ is constrained the geometry of the beam cross-section has only one degree of freedom, λ or h/b (e.g., Fig. 10.21). The global stiffness decreases monotonically as the q''-heating process progresses. The decrease in $\overline{EI}$ is slower when h/b is larger. This makes sense, because taller cross-sections have thicker protective layers. At a specified lifetime t, $\overline{EI}(t)$ increases monotonically as h/b increases. This behavior is unlike in Fig. 10.18, where an optimal λ was found.

There are at least three reasons that limit the push toward larger h/b and λ values. First, during a real thermal attack scenario q'' acts all around the beam cross-section. When h/b is large, the h-tall side surfaces play a significant role in the heating of the steel bars, and because of this the unidirectional conduction model no longer applies. The second reason is that tall beam cross-sections (large h/b) require large head room. The rooms and buildings in which such beams are used must be tall. Finally, if the beam width b is too small, then, contrary to the model of Fig. 10.17, it may be impossible to place the steel bars with enough spacing between them in one single row—impossible to embed them securely so that they would cling to the surrounding concrete.

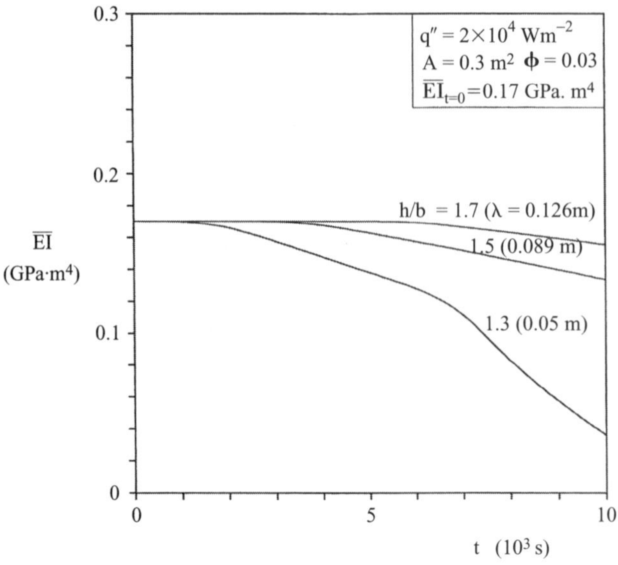

Figure 10.21 The evolution of the beam global stiffness during heating from below, when the initial stiffness is constrained [9].

10.4.3b Heating from All Sides.

The unidirectional model of Fig. 10.17 has the merit that it is the simplest to shed light on the opportunity to discover the geometry of the reinforced beam. A more realistic model of the beam cross-section is presented in Fig. 10.22. The beam height and width are h and b. The size of the cross-sectional area A is constrained, Eq. (10.65). There are n round steel bars of diameter D. The spacing between two adjacent bars is d. The area fraction occupied by steel in the cross-section is

$$\phi = \frac{n\pi D^2}{4A} \tag{10.77}$$

The slab-shaped region occupied by the n bars is surrounded by a protective layer of concrete. It is assumed that the distance from this region to the nearest heated surface is the same (λ) in each of the three directions from which heating is threatening the structure: from below and from the sides.

An important construction requirement is that each steel bar must continue to cling to concrete when it is in tension. This requirement calls for positioning the bars sufficiently far from each other. Design rules have been developed for meeting this requirement [12]. For this reason, in the model of Fig. 10.22 we require that the bar-to-bar distance d must be greater than or equal to D. If the required bars (required by the specified ϕ) do not fit in a single row, then they must be placed equidistantly in two rows.

The description of the temperature history in the beam cross-section is available analytically in the limit $\phi << 1$, when the only conductive material that matters is

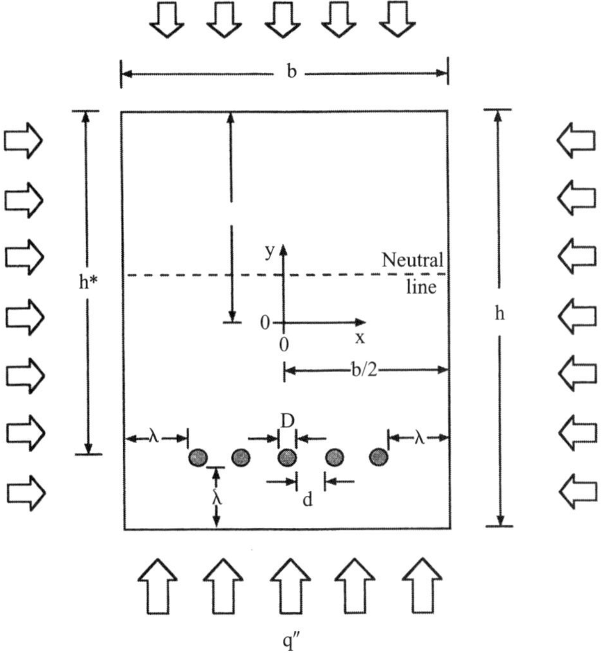

Figure 10.22 The cross-section of a beam of concrete reinforced with round steel bars and heated suddenly on all sides [9].

concrete. The temperature history due to q''-heating imposed at $t = 0$ on all four boundaries is [11]

$$\begin{aligned}\theta(t, x, y) &= T(t, x, y) - T_\infty \\ &= [T(t, x) - T_\infty]_x + [T(t, x) - T_\infty]_y\end{aligned} \tag{10.78}$$

where

$$\begin{aligned}[T(x, t) - T_\infty]_x &= \left(\frac{q''b}{2k}\right) \times \\ &\times \left[\frac{\alpha t}{(b/2)^2} + \frac{1}{2}\left(\frac{x}{b/2}\right)^2 - \frac{1}{6} - 2\sum_{n=1}^{\infty} \frac{(-1)^n}{(n\pi)^2} e^{-\frac{\alpha t}{(b/2)^2}(n\pi)^2} \cos\left(n\pi \frac{x}{b/2}\right)\right]\end{aligned} \tag{10.79}$$

$$\begin{aligned}[T(y, t) - T_\infty]_y &= \left(\frac{q''h}{2k}\right) \times \\ &\times \left[\frac{\alpha t}{(h/2)^2} + \frac{1}{2}\left(\frac{y}{h/2}\right)^2 - \frac{1}{6} - 2\sum_{n=1}^{\infty} \frac{(-1)^n}{(n\pi)^2} e^{-\frac{\alpha t}{(h/2)^2}(n\pi)^2} \cos\left(n\pi \frac{y}{h/2}\right)\right]\end{aligned} \tag{10.80}$$

The origin of the x-y coordinates is placed in the center of the beam cross-section (see Fig. 10.22). The temperature of each bar cross-section is assumed uniform and equal to the concrete temperature at the x-y location of the bar center.

It is possible to combine this bidirectional time-dependent conduction model with the elasticity-temperature model of Eqs. (10.75) and (10.76), and to calculate the position of the neutral line and the time-evolution of the global stiffness of the beam. This approach would limit the results to beams and lifetimes (under attack) when every element of the beam is still stressed in the elastic domain. Gosselin et al. [9] chose a more realistic and general strength model, to take advantage of the refinements contributed by the heat-transfer model. To calculate the maximal deflection of a beam in a general way, that is, without assuming that the elastic regime prevails throughout, it is necessary to calculate the stresses at every point in the three-dimensional reinforced beam. Then the deflection differential equation has to be solved. A simpler version of this approach is available. To characterize the strength of the beam in a global sense, one can use (instead of deflection, or $\overline{EI}$) the nominal moment [12]

$$M_n = A_s \bar{f}_s \bar{h}\left(1 - \frac{0.59\phi \bar{f}_s}{f_c'}\right) \tag{10.81}$$

In this equation, $\bar{f}_s$,$f_c{}'$ and $\bar{h}$ represent the strength of steel, the strength of concrete, and the y-position of the bars measured from the top of the beam:

$$\bar{f}_s = \frac{1}{n}\sum_{i=1}^{n} f_s(T_i) \tag{10.82}$$

$$\bar{h} = \frac{\sum_{i=1}^{n} h_i^* f_s(T_i)}{\sum_{i=1}^{n} f_s(T_i)} \tag{10.83}$$

The overbar indicates an average made over all the steel bars that are present. The strength of the concrete $f_c{}'$ has been evaluated at the average temperature of the beam cross-section:

$$T_{avg} = \frac{1}{A}\iint_A T\,dA \tag{10.84}$$

The effect of the local temperature on the strength of steel and concrete is taken into account by using the dimensionless factors [13]

$$\tilde{f}_s = \frac{f_s(T)}{f_s(20\,^\circ\text{C})}, \qquad \tilde{f}_c = \frac{f_c'(T)}{f_c'(20\,^\circ\text{C})} \tag{10.85}$$

where the reference strengths for steel and concrete are f_s (20 °C) = 500 MPa and $f_c{}'$ (20 °C) = 35 MPa. The influence of temperature on $\tilde{f}_s$ and $\tilde{f}_c$ is described by the functions

$$\tilde{f}_s = \begin{Bmatrix} 1 & T \leq 350\,°\text{C} \\ 1.899 - 0.00257T & 350\,°\text{C} \leq T \leq 700\,°\text{C} \\ 0.24 - 0.0002T & 700\,°\text{C} \leq T \leq 1200\,°\text{C} \\ 0 & 1200\,°\text{C} \leq T \end{Bmatrix} \tag{10.86}$$

$$\tilde{f}_s = \begin{Bmatrix} 1 & T \leq 350\,°\text{C} \\ 1.067 - 0.00067T & 100\,°\text{C} \leq T \leq 400\,°\text{C} \\ 1.44 - 0.0016T & 400\,°\text{C} \leq T \leq 900\,°\text{C} \\ 0 & 900\,°\text{C} \leq T \end{Bmatrix} \tag{10.87}$$

where T is expressed in °C.

According to the model of Fig. 10.22, the beam cross-section has three degrees of freedom: h/b, λ and n. For each assumed geometry (h/b, λ, n), we monitor the evolution of the nominal moment strength in the presence of heating from all sides, $M_n(t)$. The constrained parameters are A, ϕ and q''. We also fixed the initial nominal moment, leaving only two degrees of freedom, h/b and n.

Figure 10.23 shows the relation between the aspect ratio h/b and the protective layer thickness λ for a specified initial strength and number of bars. The curves for $n > 1$ exhibit jumps as the aspect ratio increases. These jumps are due to discrete changes in the internal structure—the way the bars are positioned in the beam cross-section. For instance, when the aspect ratio is high, in order to satisfy the

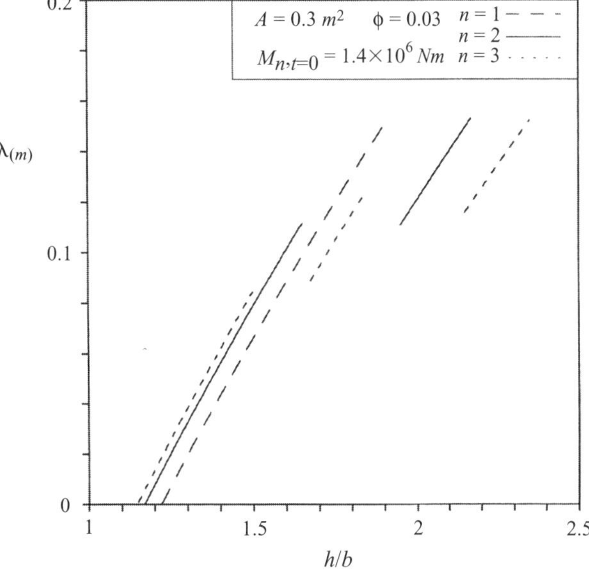

Figure 10.23 The relation between the thickness of the protective layer and the aspect ratio of the beam cross-section, for a fixed initial strength and various numbers of steel bars [9].

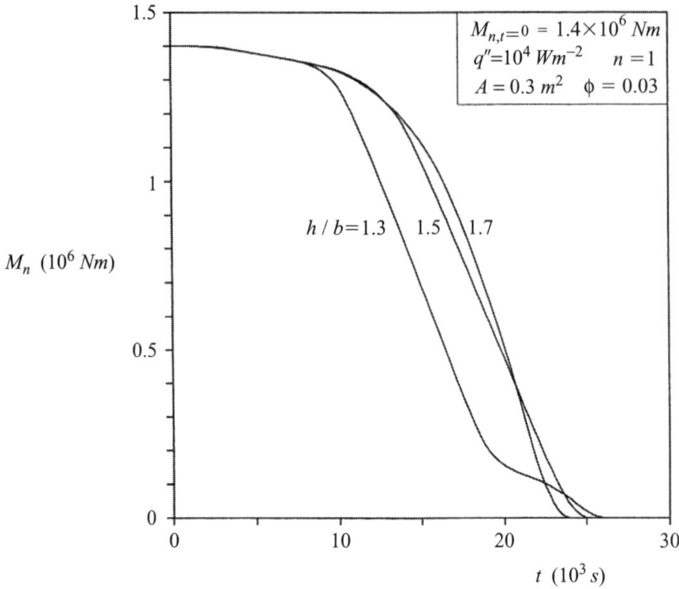

Figure 10.24 The evolution of the nominal strength of a beam reinforced with one steel bar [9].

condition $d = D$ (see section 10.4.5) the designer is forced to place many bars in a single row.

In the absence of thermal attack, all the beam designs described by the curves of Fig. 10.23 are equivalent. They provide the same nominal moment at $t = 0$. However, they behave differently in case of thermal attack. For example, Fig. 10.24 shows the time evolution of M_n for a one-bar beam. At a given time, it appears that there is an optimal aspect ratio h/b for maximal strength. Similar calculations have been performed for $n = 2$ and 3. The results are shown in Figs. 10.25 and 10.26. The existences of an optimal design for a given time is again confirmed.

Finally, all the optimal shapes and structures are reported in Fig. 10.27. Even though the optimal beam shape (h/b) varies depending on how many bars are embedded in the beam, the beam performance under thermal attack does not appear to be affected significantly by the number of bars.

In conclusion, multiobjective systems are numerous and diverse, and to address simultaneously their objectives calls for truly interdisciplinary research. In section 10.4, we illustrated the interdisciplinary approach by showing that shapes and structures of beams can be optimized to face thermal attack. Examples of optimized shapes were the beam profile and cross-sectional aspect ratio. Optimized internal structure was the position of the steel bars in concrete. The optimal architecture of the multiobjective system is a consequence of the competition between objectives.

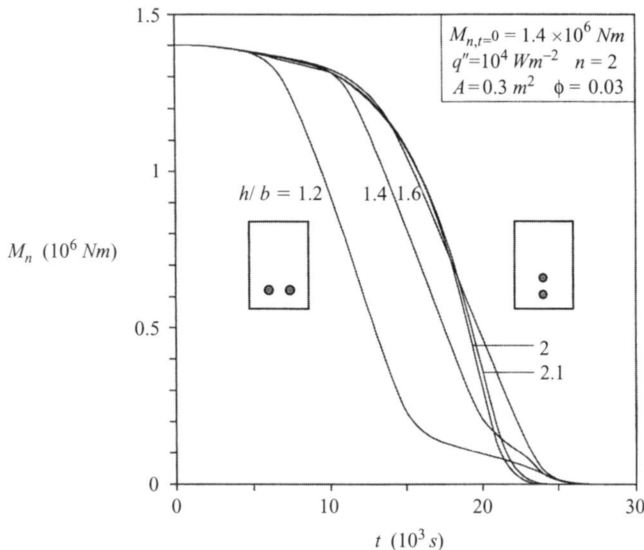

Figure 10.25 The evolution of the nominal strength of a beam reinforced with two steel bars [9].

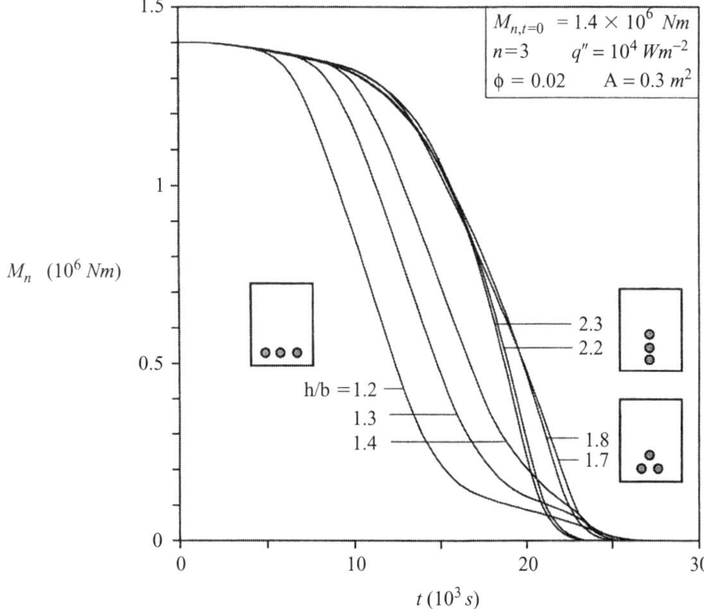

Figure 10.26 The evolution of the nominal strength of a beam reinforced with three steel bars [9].

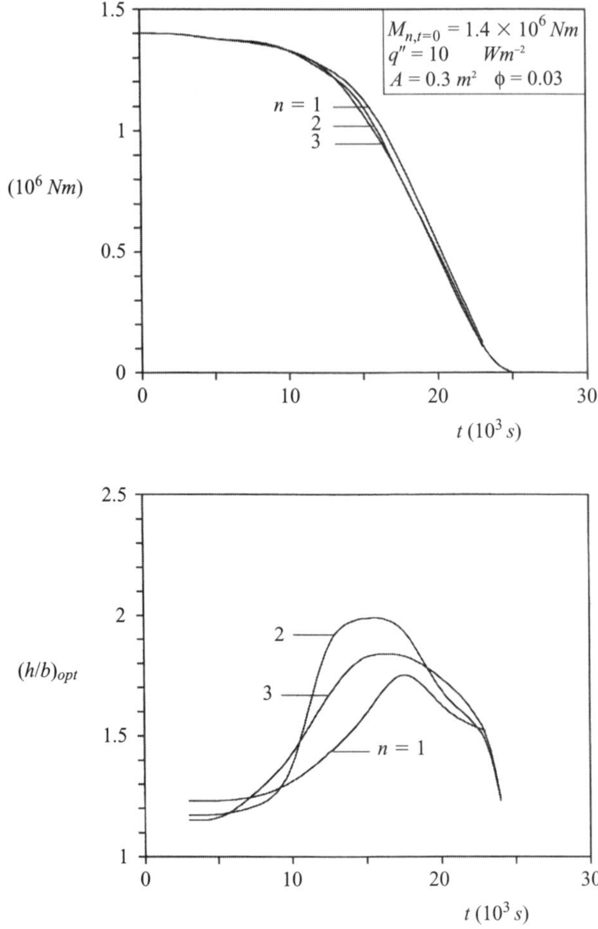

Figure 10.27 The evolution of the maximum nominal strength and the corresponding optimal aspect ratio of the cross-section [9].

10.5 VEGETATION

According to constructal theory, "plants (vegetation) are completely analogous to snowflakes. They occur and survive because they facilitate rapid ground-air mass transfer" [15] (i.e., rapid evaporation). Earlier, constructal theory showed that dendritic crystals such as snowflakes are the most effective heat-flow configurations for achieving rapid solidification [16, 17]. The same mental viewing was used by Miguel [18] to explain the variations in the morphology of stony corals, bacterial colonies, and plant roots. In this section we rely on the lessons of constructal design in order to *construct* based on a single principle the main features of a plant, from root to canopy.

10.5.1 Root Shape

The plant root is a conduit shaped in such a way that it provides maximum access for the ground water to escape above ground, into the trunk of the plant. The ground water enters the root through all the points of its surface. In the simplest possible description, the root is a porous solid structure shaped as a body of revolution (Fig. 10.28). The shape of the body $[L, D(z)]$ is not known, but the volume is fixed:

$$V = \int_0^L \frac{\pi}{4} D^2 dz \tag{10.88}$$

The flow of water through the root body is in the Darcy regime. The permeability of the porous structure in the longitudinal direction (K_z) is greater than the permeability in the transversal direction (K_r). Anisotropy is due to the fact that the woody vascular tissue (the xylem) is characterized by vessels and fibers that are oriented longitudinally.

We assume that the (L, D) body is sufficiently slender so that the pressure inside the body depends mainly on longitudinal position, $P(r, z) \cong P(z)$. This slenderness assumption is analogous to the slender boundary layer assumption in boundary layer theory [4]. In Darcy flow, the z volume averaged longitudinal

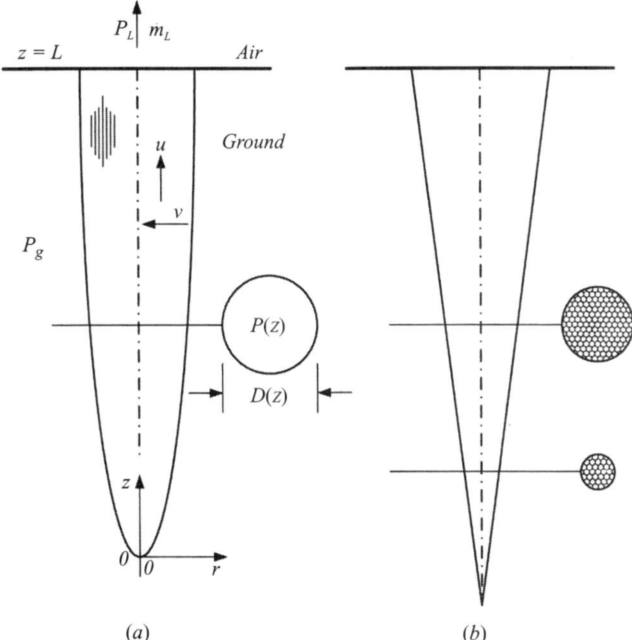

Figure 10.28 (a) Root shape with power-law diameter; (b) constructal root design: conical shape and longitudinal tubes with constant (z-independent) diameters and density, and z-independent u and v.

velocity is given by

$$u = -\frac{K_z}{\mu}\frac{dP}{dz} \tag{10.89}$$

where μ is the fluid viscosity. Because of the $P(r, z) \cong P(z)$ assumption, for the transversal volume averaged velocity v (oriented toward negative r) we write approximately

$$v \cong \frac{K_r}{\mu}\frac{P_g - P(z)}{D/2} \tag{10.90}$$

The ground-water pressure (P_g) outside the body is assumed constant. This means that in this simplest model the hydrostatic pressure variation with depth P_g (z) is assumed to be negligible, and that the root sketched in Fig. 10.28 can have any orientation relative to gravity. Ground level is indicated by $z = L$: here the pressure is P_L, and is lower than P_g. Throughout the body, $P(z)$ is lower than P_g, and the radial velocity v is oriented toward the body centerline.

The conservation of water flow in the body requires

$$d\dot{m} = \rho\pi D v dz \tag{10.91}$$

where $\dot{m}$ is the longitudinal mass flow rate at level z:

$$\dot{m} = \rho\frac{\pi}{4}D^2 u \tag{10.92}$$

Equations (10.91) and (10.92) yield

$$\frac{d}{dz}(D^2 u) = 4vD \tag{10.93}$$

Summing up, the three equations (10.89), (10.90), and (10.93) should be sufficient for determining $u(z)$, $v(z)$ and $D(z)$ when the length L is specified. Here, the challenge is of a different sort (much greater). We must determine the shape [L, $D(z)$] that allows the global pressure difference ($P_g - P_L$) to pump the largest flow rate of water to the ground level,

$$\dot{m}_L = \rho\frac{\pi}{4}D^2(L)\,u(L) \tag{10.94}$$

subject to the volume constraint (10.88). Instead of trying a numerical approach or one based on variational calculus, here we use a much simpler method. We assume that the unknown function $D(z)$ belongs to the family of power-law functions

$$D = bz^m \tag{10.95}$$

where b and m are two constants. We also make the assumption that the function $P(z)$ belongs to the family represented by

$$\frac{P_g - P(z)}{\mu/K_z} = az^n \tag{10.96}$$

where a and n are two additional constants. When we substitute assumptions (10.95) and (10.96) into Eqs. (10.89) and (10.90), and then substitute the resulting u and v expressions into Eq. (10.93), we obtain two compatibility conditions for the assumptions made in Eqs. (10.95) and (10.96):

$$m = 1 \tag{10.97}$$

$$b^2 n(n+1) = 8\frac{K_r}{K_z} \tag{10.98}$$

The volume constraint (10.88) yields a third condition,

$$b^2 L^3 = \frac{12}{\pi} V \tag{10.99}$$

A fourth condition follows from the statement that the overall pressure difference is fixed, which in view of Eq. (10.96) means that

$$\frac{P_g - P_L}{\mu / K_z} = aL^n, \text{ constant} \tag{10.100}$$

Finally, the mass flow rate through the $z = L$ end of the body is [cf. Eq. (10.94)]

$$\begin{aligned} \dot{m}_L &= \rho \frac{\pi}{4} (bL)^2 \frac{K_z}{\mu} \left(\frac{d(P_g - P)}{dx} \right)_{z=L} \\ &= \rho \frac{\pi}{4} b^2 an\, L^{n+1} \end{aligned} \tag{10.101}$$

for which $b(n)$ and $L(n)$ are furnished by Eqs. (10.98) and (10.99). The resulting ground-level flow rate is

$$\dot{m}_L = \rho \frac{\pi}{4} (aL^n) \left(8 \frac{K_r}{K_z} \right)^{2/3} \left(\frac{12}{\pi} V \right)^{1/3} \frac{n^{1/3}}{(n+1)^{2/3}} \tag{10.102}$$

with the observation that (aL^n) is a constant, cf. Eq. (10.100).

In conclusion, $\dot{m}_L$ depends on root shape (n) according to the function $n^{1/3}/(n+1)^{2/3}$. This function is maximum when

$$n = 1 \tag{10.103}$$

Working back, we find that the constructal root design must have this length and aspect ratio:

$$L = \left(\frac{3V K_z}{\pi K_r} \right)^{1/3} \tag{10.104}$$

$$\frac{L}{D_L} = \frac{1}{2} \left(\frac{K_z}{K_r} \right)^{1/2} \tag{10.105}$$

The constructal root shape is conical. The slenderness of this cone is dictated by the anisotropy of the porous structure, $(K_z/K_r)^{1/2}$. The root is more slender when the vascular structure is more permeable longitudinally.

Another important feature of the discovered root geometry is that the longitudinal volume averaged fluid velocity (u) is independent of longitudinal position (z), because $n = 1$ means that $dP/dz =$ constant, and

$$u = \frac{K_z}{\mu}\frac{P_g - P_L}{L} \qquad (0 < z < L) \tag{10.106}$$

$$\begin{aligned} \dot{m}_L &= \rho\frac{\pi}{4}(bL)^2 u \\ &= \frac{K_z}{\nu}(P_g - P_L)(3V)^{1/3}\left(\pi\frac{K_r}{K_z}\right)^{2/3} \end{aligned} \tag{10.107}$$

The morphological implications of this theoretical feature are important. If the porous structure is a bundle of tiny capillary channels, then the fluid velocity through each tube must be constant, and must not depend on the size of the root cross-section $(\pi/4)D^2(z)$ that the channel pierces. On the other hand, earlier work with constructal design (e.g., Problem 1.1) has shown the following: flow strangulation is not good for flow performance, the constructal configuration of a long capillary with specified flow rate and volume is the one where the cross-section does not vary with longitudinal position, and the cross-section is round (section 3.1.2). Combining this with the new conclusion that u must not depend on z, we *discover* the internal structure of the constructal root body. The longitudinal tubes must be round, with diameters that do not vary with z, even though some tubes are longer than other tubes. The external shape and internal structure of the root body discovered in this section are sketched on the right side of Fig. 10.28.

Another feature of the constructal root design is visible in Eq. (10.90). Because both $(P_g - P)$ and D are proportional to z, we conclude that v must also be z-independent. One can show that

$$\frac{v}{u} = \left(\frac{K_r}{K_z}\right)^{1/2} \tag{10.108}$$

The anisotropy of the vascular porous structure dictates the ratio between constant-v and constant-u, in the same way that it dictates the root slenderness ratio D_L/L, cf. Eq. (10.105).

10.5.2 Trunk and Canopy Shapes

The water stream guided by the root from underground to ground level continues to flow upward through the trunk and canopy of the plant. To continue with the same analytical ease as in the analysis of root geometry, for the trunk and canopy of the plant we make the simplifying assumptions hinted at in Fig. 10.29, which is based

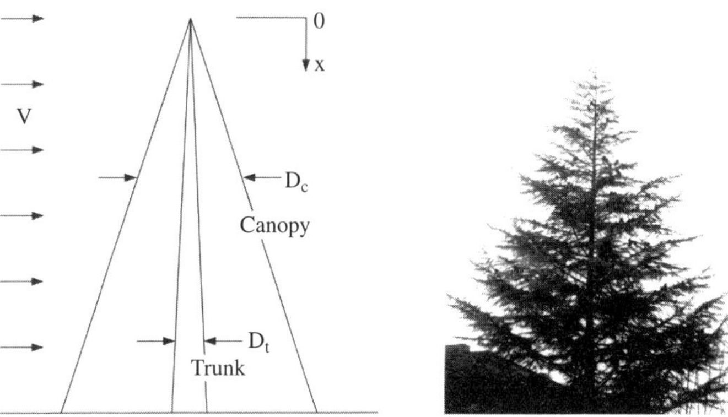

Figure 10.29 Tree with slender canopy and trunk, exposed to a horizontal wind with uniform velocity.

on a problem proposed in Ref. [15], pp. 831–832. We assume that both the canopy and the trunk are sufficiently *slender*. This allows us to analyze the forces exerted by the wind on the canopy as a problem of two-dimensional flow, in a horizontal plane that cuts the trunk and the canopy.

The trunk and the canopy are modeled as two bodies of revolution, with unknown diameters $D_t(x)$ and $D_c(x)$, where x is measured downward from the top of the tree. The drag force per unit length (x) experienced by the tree canopy is

$$F' = C_D D_c \frac{1}{2} \rho V^2 \tag{10.109}$$

where V is the horizontal wind speed, and $D_c(x)$ is the radius of the canopy at the distance x from the tree top. We assume that the Reynolds number VD_c/ν is greater than 10^3 so that the drag coefficient C_D is a constant approximately equal to 1 (cf. Fig. 1.1).

To give our search for geometry sufficient generality, assume that the canopy has a shape that belongs in the family of power-law functions

$$D_c = ax^n \tag{10.110}$$

where a is a constant, and the shape exponent n is not known. The bending moment experienced by the trunk at the distance x from the tree top is

$$M(x) = \int_0^x F'(x - \xi) d\xi = \frac{a' x^{n+2}}{(n+1)(n+2)} \tag{10.111}$$

where a' is another constant

$$a' = \frac{a}{2} \rho V^2 C_D \tag{10.112}$$

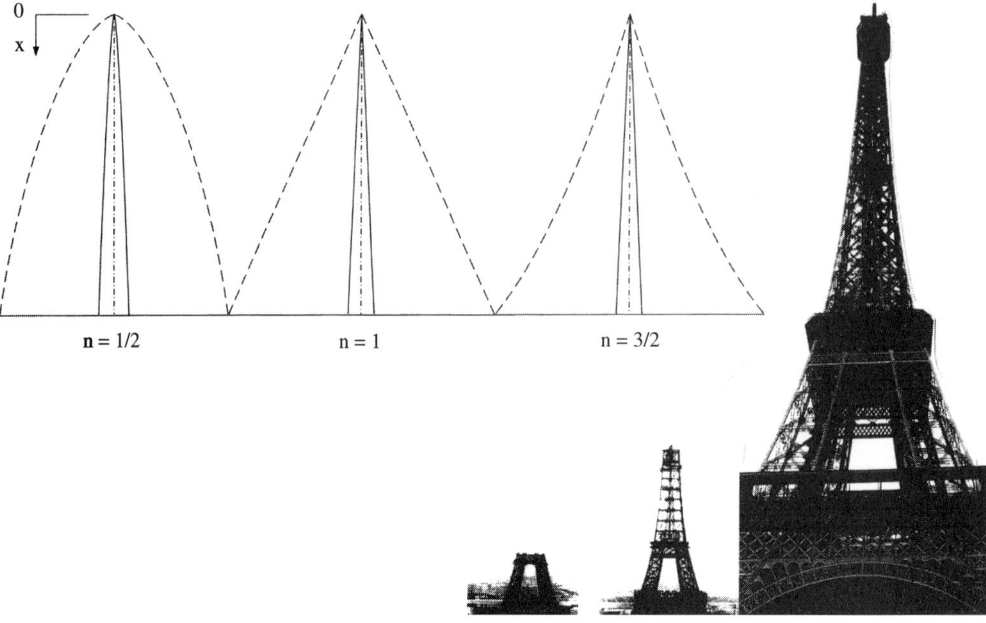

Figure 10.30 Three canopy shapes, showing that the trunk shape is near-conical in all cases. See also Color Plate 17.

We now turn our attention to the maximum bending stresses in the cross-section of the trunk of diameter $D_t(x)$, cf. Eq. (10.3),

$$s_m = \frac{M(x)}{I_t(x)} \frac{D_t(x)}{2} \tag{10.113}$$

where $I_t = \pi D_t^4/64$. The stress s_m occurs in the dorsal and ventral fibers of the trunk, as the trunk bends in the wind that pushes the canopy. Optimal distribution of imperfection means that s_m must be the same over the entire height of the tree. According to Eq. (10.112), the trunk diameter must vary as

$$D_t(x) = \left[\frac{32a'/\pi}{s_m(n+1)(n+2)}\right]^{1/3} x^{(n+2)/3} \tag{10.114}$$

This is an important result, but it is not the end of the story. It says that if we know the canopy shape (n), then we can predict the trunk shape, and vice versa (Fig. 10.30). To determine the trunk and canopy shapes uniquely we need an additional idea (see section 10.5.3).

If the canopy is shaped as a cone ($n = 1$), then the trunk is also shaped as a cone, $D_t/x =$ constant. Figure 10.30 shows that if the canopy has a round top (e.g. $n = 1/2$), then the trunk diameter must vary as $D_t/x^{5/6} =$ constant, which is not much different than $D_t/x =$ constant. If the canopy has a very sharp tip (e.g., $n = 3/2$), then $D_t(x)$ must vary as $D_t/x^{7/6} =$ constant, which again is not far off the conical trunk shape. In sum, we have discovered that the shape of the trunk that is uniformly stressed is relatively insensitive to how the canopy is shaped. A conical trunk is essentially a uniform-stress body in bending for a wide variety of canopy shapes that deviate (concave vs. convex) from the conical canopy shape sketched in Fig. 10.29.

A simpler version of the problem solved in this section is to search for the optimal shape of the trunk $D_t(x)$ when there is no canopy. The trunk alone is the obstacle in the wind, and its bending is due to the distributed drag force F' of Eq. (10.109), in which D_c is replaced by D_t. The analysis leads to Eq. (10.114) where $M(x)$ varies as x^{n+2}, and s_m(constant) is proportional to $M(x)/D_t^3$. The conclusion is that the trunk (or solitary pole) is the strongest to bending when it is conical, $n = 1$. The same result follows from the subsequent discussion of Eq. (10.114), if we assume $D_c = D_t$.

A famous structure that only now reveals its bending resistance design is the Eiffel Tower [19]. The shape of the structure is not conical (Fig. 10.30) because, in addition to bending in the wind, the structure must be strong in vertical compression. The optimization of tower shape for uniform distribution of compressive stress (Problem 10.3) leads to a tower profile that becomes exponentially narrower with altitude. The shape of a tower that is uniformly resistant to lateral bending and axial compression is between the conical and the exponential. This apparent "imperfection" (deviation from the exponential) of the Eiffel Tower has been a puzzle until recently [19].

10.5.3 Conical Trunks, Branches and Canopies

The preceding section unveiled the architecture of a tree that is polished (shaped, torn, shaved, trimmed, cut to size) so that its maximum allowable stress is distributed uniformly. Like the beams of section 10.2, this tree supports the largest load (i.e., it resists the strongest wind) when the tree volume is specified. Conversely, the same architecture withstands a specified load (wind) by using minimum tree volume. In summary, the multitude of near-conical designs discovered in Eq. (10.114) and Fig. 10.30 refer to the mechanical design of the structure, that is, to the flow of stresses (section 10.1), not to the flow of fluid that seeps from thick to thin, along the trunk and its branches.

There is no question that the maximization of access for fluid flow plays a major role in the configuring of the tree. This is why the tree is "tree-shaped," dendritic, one trunk with branches, and branches with many more smaller branches. How do the designs of Eq. (10.114) facilitate the maximization of access for fluid flow?

The answer is provided by the constructal root discovered in section 10.5.1 and Eqs. (10.104) through (10.108). The constructal shape for a body permeated by Darcy flow with two permeabilities (K_z, K_r) is conical. The longitudinal and lateral seepage velocities (u, v) are uniform, independent of the longitudinal position z. For a root, the lateral seepage is provided by direct (contact) diffusion from the soil, and indirect seepage from root branches, strands and moustaches. The same holds for the tree trunk above the ground, except that the lateral flow that accounts for v is facilitated (ducted) almost entirely by lateral branches. Above the ground, the lateral v is contributed *discretely* by branches that are distributed appropriately along and around the trunk.

The theoretical leap that we make here is this: the constructal flow design of the root is the same as the flow design of the trunk and canopy. From this, we deduce that out of the multitude of near-conical trunk shapes for wind resistance, Eq. (10.114), the constructal law selects the conical shape, $n = 1$. The conical shape is also the constructal choice for the large and progressively smaller lateral branches, provided that their mechanical design is dominated by wind resistance considerations, not by the resistance to their own body weight.

Recognition of the conical trunk and canopy shapes means that the analysis in this section begins with Eqs. (10.105) and (10.110), which for the tree trunk and canopy reduce to

$$\frac{D_t(x)}{x} = 2\left(\frac{K_r}{K_x}\right)^{1/2} = b \tag{10.115}$$

$$\frac{D_c(x)}{x} = a \tag{10.116}$$

Here, it should be noted that for the tree trunk the axial coordinate (x) is measured downward (from the tree top, Fig. 10.29), whereas the axial coordinate of the root (z) is measured upward (from the root tip, Fig. 10.28). The proportionality between $D_t(x)$ and $D_c(x)$ is provided by Eq. (10.114) with $n = 1$, in combination with Eqs. (10.112), (10.115) and (10.116):

$$\frac{D_c(x)}{D_t(x)} = \frac{a}{b} = \frac{3\pi s_m}{2C_D\rho V^2}\frac{K_r}{K_x} \tag{10.117}$$

Equation (10.117) recommends a large D_c/D_t ratio for trees with hard wood in moderate winds, and a small D_c/D_t ratio for trees with soft wood in windy climates. A hardwood example is the walnut tree (*Juglans regia*) with $s_m \simeq 1.2 \times 10^8\,\text{N/m}^2$, in a mild wind climate represented by $V \sim 50\,\text{km/hour}\,(14\,\text{m/s})$. Equation (10.117) with $C_D \sim 1$ yields $D_c/D_t\ 2.42 \times 10^6\ (K_r/K_x)_{\text{walnut}}$ and, after additional algebra, $D_c/L_{\text{trunk}} \sim 4.8\ \times 10^6 (K_r/K_x)^{3/2}_{\text{walnut}}$. The corresponding estimates for a pine tree (*Pinus silvestris*) with $s_m \simeq 6.6 \times 10^7\,\text{N/m}^2$ in a windy climate of $V \sim 100\,\text{km/hour}\,(28\,\text{m/s})$ are $D_c/D_t \sim 3.4 \times 10^5 (K_r/K_x)_{\text{pine}}$ and $D_c/L_{\text{trunk}} \sim 6.8 \times 10^5 (K_r/K_x)^{3/2}_{\text{pine}}$. If L_{trunk} and K_r/K_x are the same for the walnut and the pine, then, in this numerical example $D_{c,\text{walnut}}/D_{c,\text{pine}} \simeq 7$.

How many branches should be placed in the canopy, and at what level x? We answer this question with reference to Fig. 10.31, where the aspect ratios of the trunk ($D_t/x = b$) and canopy ($D_c/x = a$) also hold for the branch $L_B(x)$ located at level x:

$$\frac{D_{t,B}}{L_B} = b \qquad \frac{D_{c,B}}{L_B} = a \tag{10.118}$$

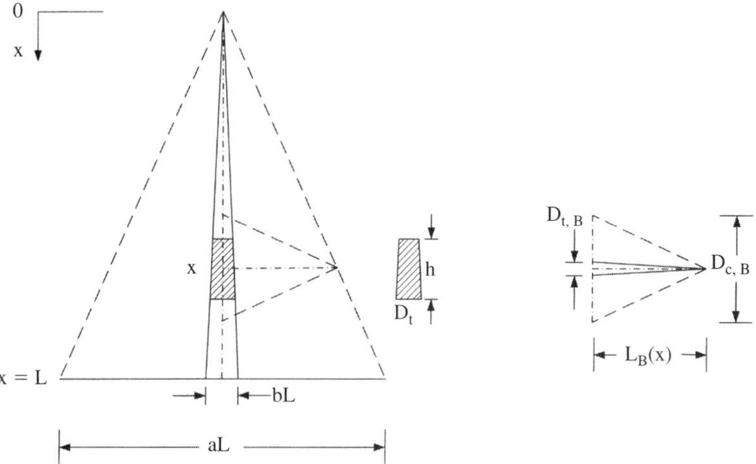

Figure 10.31 Conical canopy with conical branches and branch canopies.

Furthermore, in accordance with Eq. (10.116) for the canopy, $D_c(x)$ is the same as $2\,L_B(x)$, which means that

$$L_B(x) = \frac{1}{2}ax \tag{10.119}$$

$$D_{t,B}(x) = \frac{1}{2}abx \tag{10.120}$$

$$D_{c,B}(x) = \frac{1}{2}a^2x \tag{10.121}$$

A single branch $L_B(x)$ resides in a frustum of the conical canopy: the frustum height is $h(x)$ and the base radius is $L_B(x)$. In the center of this frustum there is a trunk segment (another conical frustum) of height $h(x)$ and diameter $D_t(x)$. The trunk frustum can be approximated as a cylinder of diameter $D_t(x)$. The total flow rate of fluid that seeps laterally from this trunk segment is

$$\dot{m}_B = \rho v \pi D_t h \tag{10.122}$$

If u_B is the longitudinal seepage velocity along the branch L_B, then the same fluid mass flow rate can be written as

$$\dot{m}_B = \rho u_B \frac{\pi}{4} D_{t,B}^2 \tag{10.123}$$

where $D_{t,B}$ is the diameter of branch L_B at the junction with the trunk. Eliminating $\dot{m}_B$ between Eqs. (10.122) and (10.123), and using Eqs. (10.115) and (10.119) through (10.120), we find that h is proportional to x,

$$\frac{h}{x} = \frac{u_B}{u}\frac{a^2}{8} \tag{10.124}$$

The ratio u_B/u is a constant determined as follows. Let $P(x)$ be the pressure at level x inside the trunk, and P_0 the pressure at the tip of the trunk ($x = 0$). The pressure at the tip of the branch L_B is also P_0. In accordance with Eq. (10.106) we write

$$u = \frac{K_x}{\mu} \frac{P(x) - P_0}{x} \tag{10.125}$$

$$u_B = \frac{K_{x,B}}{\mu} \frac{P(x) - P_0}{L_B} \tag{10.126}$$

which yield

$$\frac{u}{u_B} = \frac{K_x}{K_{x,B}} \frac{L_B}{x} \tag{10.127}$$

It is reasonable to approximate the longitudinal permeability of the wood to be the same in the trunk and the branch, $K_x \cong K_{x,B}$, such that Eq. (10.124) reduces to

$$h = \frac{1}{4}ax = \frac{1}{2}L_B \tag{10.128}$$

In conclusion, the vertical segment of trunk (h) that is responsible for the flow rate into one lateral branch is proportional to the length of the branch. Another dimension that is proportional to $L_B(x)$ is the diameter of the conical "branch canopy" circumscribed to the horizontal L_B, namely, $D_{c,B} = aL_B$, cf. Eq. (10.121). Comparing h with $D_{c,B}$, we find that

$$\frac{h(x)}{D_{c,B}(x)} = \frac{1}{2a} \tag{10.129}$$

which is a constant of order 1. In other words, there is room in the global canopy (L, D_c) to install one L_B-long branch on every h-tall segment of tree trunk. The geometrical features discovered in this section have been sketched in Fig. 10.31.

Now we return to the discussion of the Eiffel Tower at the end of the preceding section, where we noted that strength for compression (under the weight) near the base was combined with strength in bending (subject to lateral wind) in the upper body of the tower. This discussion is relevant in the modeling of the horizontal branches, which in this section was based on the assumption that the loading is due to lateral wind. The branch is also loaded in the vertical direction, because of its own weight. If we assume that the distributed weight of the branch is the only load, then the branch shape of constant strength (i.e., with x-independent s_m) is $D = ax^2$, cf. Problem 10.10. Such a branch, however, would have zero thickness near the tip ($dD/dx = 0$ at $x = 0^+$), which is not a shape found in nature. This result alone indicates that the tips of branches are not shaped by weight loading alone, and that wind loading (which prescribes $D = ax$, i.e., finite D at small x) is the more appropriate model there. For the thick end of the branch, it can be argued that $D = ax^2$ is a realistic shape, and that near the trunk the weight loading of the beam is the dominant shaping mechanism, just like in the Eiffel Tower.

10.5.4 Forest

The fluid flow rate ducted by the entire tree from the ground to the tips of the trunk and branches is

$$\begin{aligned}\dot{m} &= \rho u \frac{\pi}{4} D_t^2(x = L) \\ &= \frac{\pi}{4} \frac{b^2}{a\nu} K_x [P(x = L) - P_0] D_c(x = L)\end{aligned} \tag{10.130}$$

where $x = L$ indicates ground level, and $D_c(x = L)$ is the diameter of the canopy projected as a disc on the ground. The important feature of the tree design is the proportionality between $\dot{m}$ and $D_c(x = L)$. This proportionality will be modified somewhat if we take into account the additional flow resistance encountered by $\dot{m}$ as it flows from the smallest branches (P_0) through the leaves and into the atmosphere (P_a).

Seen from above, an area covered with trees of many sizes ($D_{c,i}$) is an area covered with fluid mass sources ($\dot{m}_i$), where each $\dot{m}_i$ is proportional to the diameter of the circular area allocated to it. From the constructal law of maximization of ground-to-air fluid flow access follows the design of the forest.

The principle is to morph the area design into a design with mass sources (or disc-shaped canopy projections) such that the total fluid-flow rate lifted from the area is the largest. From this invocation of the constructal law follows, first, the prediction that the forest must have trees of many sizes, few large trees interspaced with more and more numerous smaller trees. This is illustrated in Fig. 10.32a with a triangular area covered by canopy projections arranged according to the algorithm that a single disc is inserted in the curvilinear triangle that emerges where three discs touch. If the side of the large triangle is X_t, then the diameter of the largest canopy disc is $D_0 = X_t$, and the number of D_0-size canopies present on one X_t triangle is $n_0 = 1/2$. For the next smaller canopy the parameters are $D_1 = (3^{-1/2} - 1/2)\, X_t$ and $n_1 = 1$. At the next smaller size, the number of canopies is $n_2 = 3$, and the disc size is $D_2 = 0.0613X_t$. The construction continues in an infinite number of steps ($n_3 = 3$, $n_4 = 6, \ldots$) until the X_t triangle is covered completely. The total fluid-flow rate vehicled by the design from the triangular area of Fig. 10.32a is proportional to

$$\begin{aligned} m_a &= \sum_{i=0}^{\infty} n_i D_i \\ &= \frac{1}{2} D_0 + D_1 + 3D_2 + \ldots \\ &= (0.761 + \ldots) X_t \end{aligned} \tag{10.131}$$

Because a canopy disc D contributes more to the global production (m) when D is large and when the number of D-size discs is large, a better forest architecture is the one where the larger discs are more numerous. This observation leads to Fig. 10.32b, where the X_t triangle is covered more uniformly by larger discs, in this

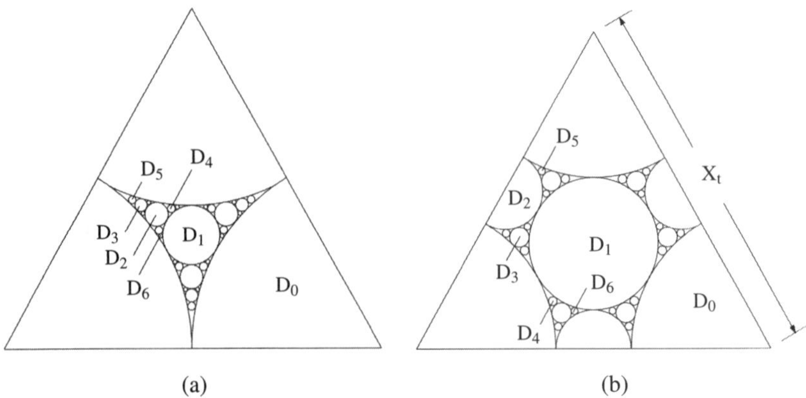

Figure 10.32 Multiscale canopies projected on the forest floor: (a) triangular pattern with algorithm-based generation of smaller scales; (b) triangular pattern with more large-size canopies.

sequence: $D_0 = [(3^{-1/2} + 1)/2]\, X_t$ and $n_0 = 1/2$, $D_1 = [(3^{1/2} - 1)/2]\, X_t$ and $n_1 = 1$, $D_2 = [(1 - 3^{-1/2})/2]\, X_t$ and $n_2 = 3/2$, etc. The total mass flow rate is

$$\begin{aligned} m_b &= \sum_{i=0}^{n} n_i D_i \\ &= \frac{1}{2} D_0 + D_1 + \frac{3}{2} D_2 + \ldots \\ &= (1.077 + \ldots) X_t \end{aligned} \tag{10.131}$$

This flow rate is significantly greater than that of Fig. 10.32a. The numbers of canopies of smaller scales that would complete the construction of Fig. 10.32b are $n_3 = 6$, $n_4 = 6$, $n_5 = 6$, $n_6 = 6, \ldots$, but their contributions to the global flow rate (m_b) are minor.

The important aspect of the comparison between Figs. 10.32a and 10.32b is that *there is a choice* [(*b*) is better than (*a*)], because each tree contributes to the global flow rate in proportion to its length scale.

One may ask, why should (*b*) look different than (*a*), and why should (*b*) have three large scales (D_0, D_1, D_2) instead of just one? There is nothing strange about the *evolution* of the drawing (in time) from (a) to (b). This is the time arrow of the constructal law. It may be possible to find triangular designs that are (marginally) better than (b), but that should not be necessary in view of the global picture that will be discussed in relation to Figs. 10.34 through 10.36.

Discs arranged in a square pattern also cover an area completely. One can draw and evaluate the square equivalent of Fig. 10.32a, by replacing the X_t triangle with a square of side X_s. The result is Fig. 10.33a. The numbers of discs of decreasing scales ($D_0 \gg D_1, D_2, \ldots$) present on this square will be $n_0 = 1$, $n_1 = 1$, $n_2 = 4$, etc.

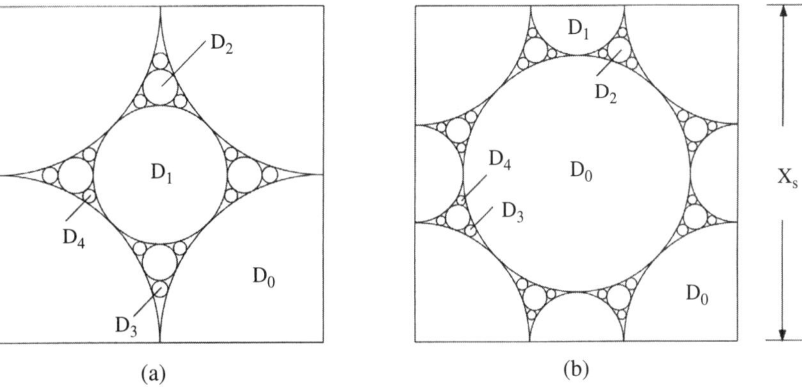

Figure 10.33 Square pattern of canopy assemblies: (a) algorithm-based generation of smaller scales; (b) more numerous large-size canopies, for greater ground-air flow conductance.

The performance of this regular design will be significantly inferior to that of the square pattern shown in Fig. 10.33b, which is the square equivalent of Fig. 10.32b. The canopy sizes and numbers in the square design are $D_0 = 2^{-1/2}X_s$ and $n_0 = 2$, $D_1 = (1 - 2^{-1/2})X_s$ and $n_1 = 2$, etc. The total mass flow rate extracted from the X_s-square is

$$\begin{aligned} m_s &= \sum_{i=0}^{\infty} n_i D_i \\ &= 2D_0 + 2D_1 + 8D_2 ... \\ &= 2.608 X_s \end{aligned} \tag{10.132}$$

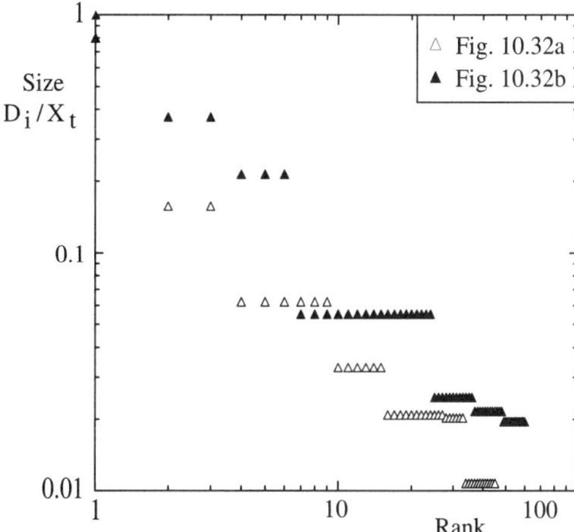

Figure 10.34 Hierarchies of canopy scales in the triangular patterns of Fig. 10.32.

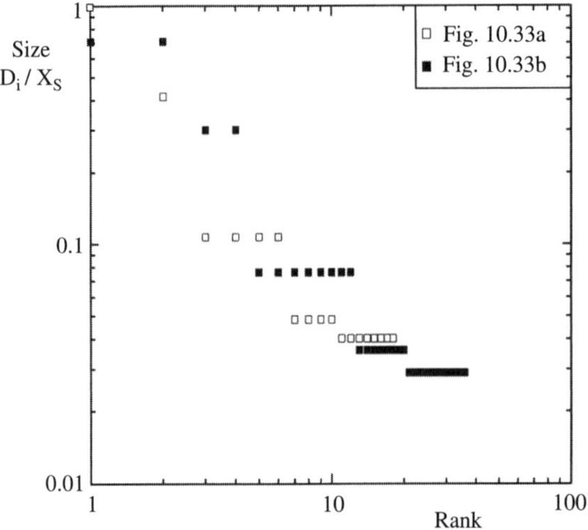

Figure 10.35 Hierarchies of canopy scales in the square patterns of Fig. 10.33.

One can show that the m values of Figs. 10.33a and 10.33b form the same ratio (namely, 0.71) as the m values of Figs. 10.32b and 10.32b.

Finally, we compare Eq. (10.132) with Eq. (10.131) to decide whether the square design (Fig. 10.33b) is better than the triangular design (b). The area is the same in both designs, therefore $X_t/X_s = 2/3^{1/4}$ and Eqs. (10.131) and (10.132) yield

$$\frac{m_b}{m_s} = 0.826 \tag{10.133}$$

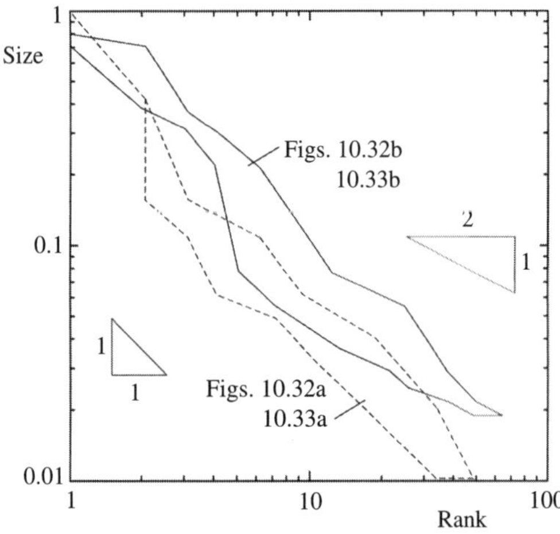

Figure 10.36 Distribution of canopy sizes versus rank, as a summary of Figs. 10.34 and 10.35. See also Color Plate 18.

Table 10.1 Sizes, numbers, and ranks for the multiscale canopies populating the forest designs of Fig. 10.32

	Size, D_i/X_t		$2n_i$		Rank	
i	(a)	(b)	(a)	(b)	(a)	(b)
0	1	0.789	1	1	1	1
1	0.155	0.366	2	2	2, 3	2, 3
2	0.0613	0.211	6	3	4–9	4–6
3	0.0325	0.054	6	12	10–15	7–24
4	0.0206	0.024	12	12	16–27	25–36
5	0.02	0.021	6	12	28–33	37–48
6	0.0106	0.019	12	12	34–45	49–60

The square design is better, but not by much. Random effects (geology, climate) will make the distribution of multiscale trees switch back and forth between triangle and square and maybe hexagon, creating in this way multiscale patterns that appear even more random than the triangle alone, the square alone, and the hexagon alone. The key feature, however, is that the design is with multiple scales arranged hierarchically, and that this sort of design is demanded by the constructal law of maximization of ground–air flow access.

The hierarchical character of the large and small trees of the forest is revealed in Fig. 10.34, where we plotted the size (D_i) and rank of the canopies shown in Figs. 10.32a and 10.32b. The calculation of the rank is explained in Table 10.1. The largest canopy has the rank 1, and after that, the canopies are ordered according to size and counted sequentially. For example, the canopies of size D_2 in Fig. 10.32b are tied for places 4 through 6. The sizes indicated with "$\sim$" are not exact: they were estimated graphically by inscribing a circle in the respective curvilinear triangle in which the projected canopy would fit.

The data collected for designs (a) and (b) in Table 10.1 are displayed as canopy size versus canopy rank in Fig. 10.34. To one very large canopy belongs an entire "organization," namely, two canopies of next (smaller) size, followed by increasingly larger numbers of progressively smaller scales. This conclusion is reinforced by Fig. 10.35, which in combination with Table 10.2 summarizes the ranking of scales visible in the square arrangements of canopies drawn in Fig. 10.33a and b. There are no significant differences between Figs. 10.34 and 10.35.

The most amazing feature is the alignment of these data as approximately straight lines on the log-log field of Figs. 10.34 and 10.35. A bird's-eye view of this hierarchy is presented in Fig. 10.36. This type of alignment is associated empirically with the Zipf distribution, and it was discovered theoretically in the constructal theory of the distribution of multiscale human settlements (cities) on a large territory (Ref. [15], pp. 774–779). It is discovered theoretically again here.

Table 10.2 Sizes, numbers, and ranks for the multiscale canopies populating the square forest design of Fig. 10.33

	Size, D_i/X_s		n_i		Rank	
i	(a)	(b)	(a)	(b)	(a)	(b)
0	1	0.707	1	2	1	1, 2
1	0.414	0.3	1	2	2	3, 4
2	0.107	0.076	4	8	3–6	5–12
3	0.048	0.036	4	8	7–10	13–20
4	0.040	0.029	8	16	11–18	21–36

Additional discoveries made based on this theory are Leonardo da Vinci's rule, Huber's rule and the the Fibonacci sequence. These and other comparisions with the biology literature are detailed in Ref. [20].

REFERENCES

1. A. Bejan, *Shape and Structure, from Engineering to Nature*. Cambridge, UK: Cambridge University Press, 2000.
2. G. A. Ledezma, A. Bejan and M. R. Errera, Constructal tree networks for heat transfer. *J Appl Phys*, Vol. 82, 1997, pp. 89–100.
3. S. Lorente and A. Bejan, Combined "flow and strength" geometric optimization: Internal structure in a vertical insulating wall with air cavities and prescribed strength. *Int J Heat Mass Transfer*, Vol. 45, 2002, pp. 3313–3320.
4. A. Bejan, *Convection Heat Transfer*, 3rd ed. Hoboken, NJ: Wiley, 2004.
5. S. Lorente, Heat losses through building walls with closed, open and deformable cavities. *Int J Energy Research*, Vol. 26, 2002, pp. 611–632.
6. B. Lartigue, S. Lorente, and B. Bourret, Multicellular natural convection in a high aspect ratio cavity: Experimental and numerical results. *Int J Heat Mass Transfer*, Vol. 43, 2000, pp. 3159–3170.
7. A. Bejan, Note on Gill's solution for free convection in a vertical enclosure. *J Fluid Mech*, Vol. 90, 1979, pp. 561–568.
8. S. W. Churchill and R. Usagi, A standardized procedure for the production of correlations in the form of a common empirical equation. *Indust Eng Chem Fund*, Vol. 13, 1974, pp. 39–46.
9. L. Gosselin, A. Bejan, and S. Lorente, Combined "heat flow and strength" optimization of geometry: Mechanical structures most resistant to thermal attack. *Int J Heat Mass Transfer*, Vol. 47, 2004, pp. 3477–3489.
10. R. M. Lawson, Fire engineering design of steel and composite buildings. *J Construct Steel Res*, Vol. 57, 2001, pp. 1233–1247.
11. V. S. Arpaci, *Conduction Heat Transfer*. Reading, MA: Addison-Wesley, 1966.
12. P. M. Ferguson, *Reinforced Concrete Fundamentals*, 4th ed. New York: Wiley, 1979.

13. M. Saafi, Effect of fire on FRP reinforced concrete members. *Compos Struct*, Vol. 58, 2002, pp. 11–20.
14. C. Avram, I. Facaoaru, I. Filimon, O. Mirsu, and I. Tertea, *Concrete Strength and Strains*. Amsterdam: Elsevier, 1981.
15. A. Bejan, *Advanced Engineering Thermodynamics*, 3rd ed. p. 770. Hoboken NJ: Wiley, 2006.
16. A. Bejan, *Advanced Engineering Thermodynamics*, 2nd ed. pp. 798–804. New York: Wiley, 1997.
17. A. Ciobanas, A. Bejan, and Y. Fautrelle, Dendritic solidification morphology viewed from the perspective of constructal theory. *J Phys D: Appl Phys*, Vol. 39, 2006, pp. 5252–5266.
18. A. F. Miguel, Constructal pattern formation in stony corals, bacterial colonies and plant roots under different hydrodynamics conditions. *J Theoretical Biology*, Vol. 242, 2006, pp. 954–961.
19. La tour Eifel a livré son équation (The Eiffel Tower delivered its equation) *Science & Vie*, No. 1050, March 2005, pp. 18–19.
20. A. Bejan, S. Lorente, and J. Lee, Unifying constructal theory of tree roots, canopies and forests. *J. Theoretical Biology*, 2008, doi: 10.1016/j.jtbi.2008.06.026
21. C. Zener, *Elasticity and Anelasticity of Metals*. Chicago: University of Chicago Press, 1948.
22. V. K. Kinra and K. B. Milligan, A second-law analysis of thermoelastic damping. *J Appl Mech*, Vol. 61, 1994, pp. 71–76.

PROBLEMS

10.1. An elastic bar is placed in tension by the end forces F shown in the figure. The bar has a nonuniform cross-section: A_1 over the length L_1, and A_2 over the remaining length $(L - L_1)$. The total length L is fixed, and A_1 is smaller than A_2. The maximal allowable stress s_{ma} is specified. Show that the bar that performs this function while using the least material is the one with uniform cross-section $(A_1 = A_2)$.

10.2. The Y-shaped linkage of bars shown in Fig. 10.2 consists of one vertical bar of length L_1 and cross-sectional area A_1, which is connected to two identical bars of length L_2 and cross-sectional area A_2. All the joints are frictionless: the bars can rotate freely around the joints, and can be loaded only axially (tension, or compression). The distance (z) between the upper and lower joints is fixed. Also fixed is the horizontal spacing between the two lower joints $(2x)$. The Y-shaped structure supports the vertical compressive load F_1. The maximal allowable compressive stress for this bar material (s_{ma}) is specified. The position of the bifurcation joint (y) can vary. Determine the optimal position y such that the total volume of bar material is minimum. What is the shape of the linkage in the constructal design? Does this configuration dominate the design of multibar trusses in engineering?

10.3. A vertical slender structure of variable cross-sectional area $A(z)$ must be shaped such that the compressive stress at every level is the same. The base area A_0 and the height H are specified. Assume that the structure is solid with the density ρ. Determine the shape $A(z)$, the top cross-sectional area A_H, the volume of the structure, and the weight of the superstructure that is supported by A_H. Does your optimized structure look like the Eiffel Tower?

10.4. The leaning tower of Pisa may be approximated as a slender cylindrical beam implanted into the ground. The tower aspect ratio is $H_0/D_0 \cong 3.4$, where H_0 and D_0 are the tower height and diameter. In the Middle Ages and in the modern era until the invention of Portland cement (1830), tall structures were built as piles of large stones held together by gravity (dry-stone construction). Such structures were not able to withstand tension. The tower of Pisa is one structure of this kind. It is now in danger because the stones on the dorsal (high) side of the leaning tower may be pulled apart by tensile bending stresses. This would happen in places where the bending stresses are greater than the compressive stresses that are due to the weight of the stones of the tower. Determine the maximum angle of inclination that the stone structure of the tower can withstand before its dorsal stones separate. Compare this angle with the angle calculated by treating the tower as a solid block that rests on a table (in this case, the block threatens to tip over when the diagonal of its $H_0 \times D_0$ longitudinal cross-section becomes aligned with the vertical). Which *disaster scenario* is more threatening, the bending of a pile of stones or the tipping of a solid cylinder?

10.5. Reexamine Problem 10.4 and think whether by shaping the tower one could have enhanced its survivability. The answer to Problem 10.4 is that the critical angle for incipient stone separation in a cylindrical tower is $\sin\alpha_0 = D_0/4H_0$, and that the first stones to be pulled apart are at the base. The way to improve the tower design is to distribute its imperfection (the separation of stones) optimally, that is, along the entire height. Assume a new tower shape, which is conical, $D = c(H - y)$, where c is a constant, D and H are the diameter and the height, and y is the altitude. The conical tower contains the same volume of stones as the cylindrical tower. Show that the condition for stone separation is $\sin\alpha = c/6$, where α is the small angle of inclination of the cone axis. This condition holds everywhere, from $y = 0$ to $y = H$. Show further that if $\alpha = \alpha_0$, the conical tower can be made taller than the cylindrical tower. Conversely, show that if the two towers have the same height, the conical one can lean more before its dorsal stones come apart.

10.6. Assume that the hollow cylinder of Fig. 10.3(2) is such that its thickness is always proportional to its radius, $t_2/R_2 = \varepsilon \ll 1$. Derive Eqs. (10.9) through (10.11) by using the method shown for the solid rod in Eqs. (10.4), (10.6), and (10.7).

10.7. Consider the tapered hollow beam shown in Fig. 10.3(3), and assume that $t_3/R_3 = \varepsilon \ll 1$, constant. Derive Eqs. (10.12) through (10.14), and show that (relative to Problem 10.6) the external shaping of the beam reduces the volume of beam material by one third, $V_3/V_2 = 1/3$.

10.8. The length scales of the optimally shaped root sketched in Fig. 10.28 can be deduced more directly by scale analysis of Darcy flow in a porous medium with two permeabilities (K_r, K_z). For the volume-averaged fluid velocity we write $u \sim (K_z/\mu)\Delta P_z/L$, where ΔP_z is the scale of the longitudinal pressure drop. For the transversal velocity, we write $v \sim (K_r/\mu)\ \Delta P_r/D$. Mass conservation means that the water flow rate $(\dot{m})$ entering through the lateral surface equals the flow rate (the "water uptake") through the ground-level end of the root. The volume (V) of the root is fixed, but its shape (L/D) may vary. The overall pressure difference that drives the flow is $\Delta P \sim \Delta P_r + \Delta P_z$. Minimize the overall flow resistance $\Delta P/\dot{m}$ with respect to L (or D), and show that the optimal aspect ratio of the root is $L/D \sim (K_z\,/\,K_r)^{1/2}$.

10.9. The unidirectional flow through a porous medium can be modeled as the flow through a bundle of capillary tubes of diameter $2r_0$, as shown in Fig. P10.9. Assume that the density of such tubes per unit frontal area is N/A (tubes/m^2). Assume further that the flow through each tube can be modeled as Poiseuille flow. Demonstrate that based on these assumptions, the Darcy law (10.89) can be derived analytically and that the effective permeability K of the capillary tube bundle porous medium is

$$K = \frac{\pi r_0^4}{8}\frac{N}{A}$$

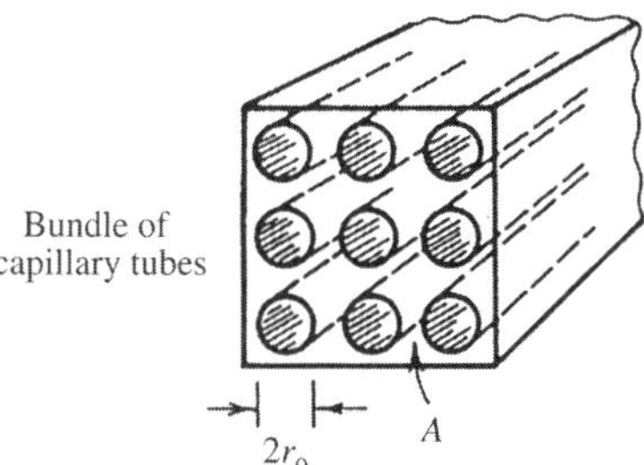

Figure P10.9

10.10. The tree branch model used in section 10.5.3 is the same as the trunk model: the branch is a constant-strength beam loaded laterally by wind forces. This model led to conical shapes for the branches and the trunk. Consider the alternative view that the branch is horizontal and its constant-strength shaping is intended to make it the strongest beam that supports its own weight. Assume that the branch diameter has the form $D = ax^n$, where a and n are two constants and x is measured from the tip of the branch toward the trunk. The branch weight per unit length is $F' = \rho \frac{\pi}{4} D^2 g$. Continue with the analysis following Eq. (10.109), and show that the constant-strength shape of the weight loaded branch is $D = ax^2$. Note that $dD/dx = 0$ at $x = 0$, and comment on the realism of such a branch design. In other words, where is the weight loading model more likely to be applicable, near the tip or near the trunk? By the same token, where is the wind loading model more likely to be realistic, near the tip or near the trunk?

10.11. The phenomenon of *volumetric cracking* can be explained as a manifestation of the natural tendency expressed by the constructal law. Consider an isolated elastic material that is initially in a state of uniform tension throughout its volume. When internal cracks appear, the solid chunks between cracks vibrate, as their stored spring energy is converted periodically into kinetic energy. These vibrations do not go on forever because of internal damping. The most intrinsic damping mechanism, which is present even in the most perfectly elastic bodies, is the *thermoelastic damping* [21, 22]. During vibrations, regions that are momentarily in compression become warmer than regions in tension. Heat currents flow from the warmer to the colder, and this "imperfection" accounts for the gradual destruction of the useful spring-in-tension energy (or exergy), which was stored initially in the stressed material.

Thermoelastic damping is most intense when the time scale of thermal diffusion (t_t) along the solid chunks of length L (the distance between successive cracks) matches the period of longitudinal mechanical vibrations (t_m) of the elastic chunk. Because the constructal-law tendency is for the isolated body to select a configuration for a faster approach to equilibrium (Chapter 1, p. 2), the body should break up into chunks sized such that each chunk vibrates with maximum damping. Estimate the time scales t_t and t_m, and from the balance $t_t \sim t_m$ show that the length scale of the pattern of cracks should be, within a factor of order one,

$$L \sim \alpha \left(\frac{\rho}{E} \right)^{1/2}$$

where α, ρ and E are the thermal diffusivity, density and modulus of elasticity of the material. Show that this length scale formula does not change if the cracking occurs in two or three dimensions, that is, if the chunks between cracks have the size $L \times L$ (with the width W perpendicular to $L \times L$), or the size $L \times L \times L$. Estimate the L scales for steel and aluminum, and report the ratio $L_{\text{steel}}/L_{\text{aluminum}}$.

10.12. One elastic column of diameter D_1 and height H is placed in end to end compression with the force P, Fig. P10.12. Assume that P is large enough so that the column just begins to buckle. In this critical condition, the force is $P = n\pi^2 EI_1/L^2$, where n is a constant factor that depends on how the two ends of the column are attached (hinged) to the upper and lower bodies that compress it with P. Note that E is the modulus of elasticity, and $I_1 = \pi D_1^4/64$. Determine the required volume of this column (V_1) as a function of P, L and other known constants.

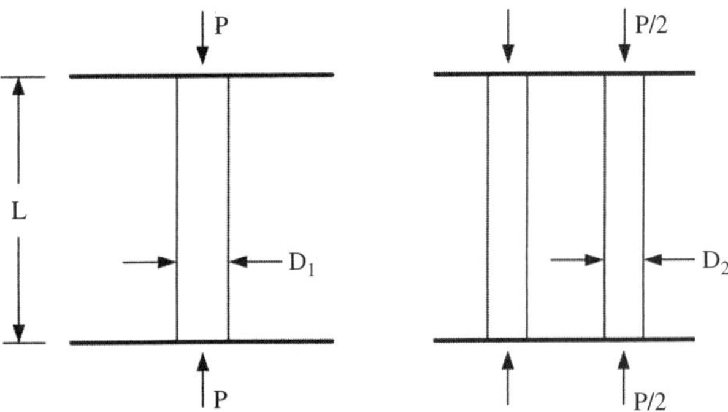

Figure P10.12

It is proposed to replace the single-column design with two parallel columns, each having the diameter D_2, height L, and compressive force $P/2$. Assume that each column just begins to buckle, therefore $P/2 = n\pi^2 EI_2/L^2$. Determine the total volume required by the two columns, and decide which of the two designs requires less material.

10.13. The critical axial buckling force (P) for a round column of length L and diameter D is $P = n\pi^2 EI/L^2$, where $I = \pi D^4/64$ and n is a factor accounting for the type of connection between the end of the column and the larger body that pushes the column. Assume that n, P and L are fixed. Consider the proposal sketched in Fig. P10.13, where the support structure consists of one round column of diameter D_1 and length xL, and two round columns of diameter D_2 and length $(1 - x)\,L$. The D_2 columns are almost collinear with the D_1 column, so that the compressive force in each of them (P_2) is half of the compressive force (P_1) experienced by the D_1 column. Assume that all the columns are sized such that their compressive loads (P_1, P_2) are critical buckling loads. In other words, if the structure collapses, all three columns collapse together, at the same time. Determine the total amount of column material required, and show that it is minimum when $x = 0.59$, which corresponds to $D_1 = 2^{1/2} D_2$. This optimal configuration is sketched in Fig. P10.13.

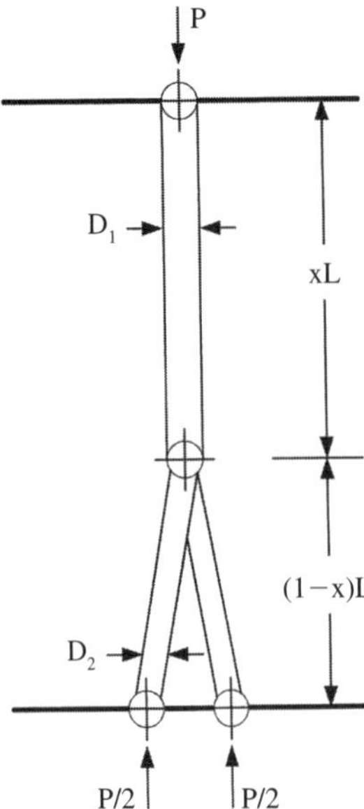

Figure P10.13

10.14. Consider the alternate design in which the columns shown in Fig. P10.13 are pipes, i.e. thin-walled cylinders of diameters D_1 and D_2 and wall thicknesses t_1 and t_2. Assume that the cross-sectional images are geometrically similar,

$$\frac{t_1}{D_1} = \frac{t_2}{D_2} = \varepsilon \ll 1$$

where ε is a known constant. Redo the analysis and show that the results do not change: the configuration for distribution of critical buckling in all three columns is represented by $x = 0.559$ and $D_1/D_2 = 2^{1/2}$.

10.15. Much needed for aerospace structures are adhesive joints with minimum mass. In this chapter we showed that architectures with minimum mass can be discovered by morphing them so that their stresses are distributed uniformly and at the highest allowable level throughout the material. Use this philosophy in the shapping of the two-dimensional adhesive joint sketched in Fig. P10.15. The plate of thickness $H(x)$ makes a sandwich of width L with an adhesive layer of constant thickness δ. The substrate under the adhesive is rigid. The plate profile is sufficiently thin (slender) so that it is in pure longitudinal tension or compression. The specified end load F_{end} is transmitted through the plate to the adhesive layer, which is modeled as a layer in pure shear. The plate axial stress is uniform (x-independent) and at the highest allowable level, σ_{ma}. The nonuniform distribution of shear stress in the adhesive layer is to be determined, however, the maximum allowable shear stress level in the adhesive (τ_{ma}) is specified. Also known are the modulus of elasticity of the plate material (E), and the shear modulus of elasticity of the adhesive (G).

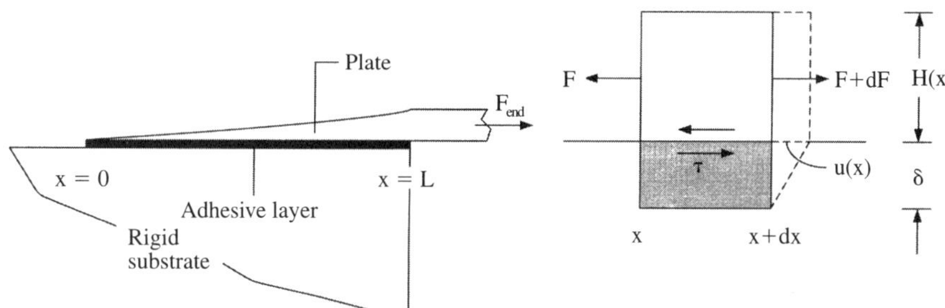

Figure P10.15

Determine analytically the shape of the plate profile, and show that H must be proportional to x^2. Show that the required contract width and adhesive thickness are

$$L = 2\frac{F_{end}}{\tau_{ma}} \qquad \delta = 2\,F_{end}\frac{G\sigma_{ma}}{E\tau_{ma}^2}$$

Finally, imagine that all the values (or narrow ranges) of the properties τ_{ma}, G and E are known, but σ_{ma} varies over a wide range because of many materials and many application. Which σ_{ma} value promises the joint with minimum mass of plate and adhesive? Show that if ρ_1 and ρ_2 are the densities of plate and adhesive materials, the σ_{ma} value for the joint with least total mass is

$$\sigma_{ma} = \tau_{ma}\left(\frac{\rho_1 E}{6\rho_2 G}\right)^{1/2}$$

10.16. The adhesive joint can be improved further by distributing the shear stress uniformly along the adhesive layer as well. Repeat the steps of the preceding analysis by setting $\tau = \tau_{ma}$ (constant) throughout the plate. Show that in this new design the plate and the adhesive layer are wedge shaped, Fig. P10.16, with

$$L = \frac{F_{end}}{\tau_{ma}} \qquad \delta = F_{end}\frac{G\sigma_{ma}}{E\tau_{ma}^2}$$

Show that relative to the design of Problem 10.15, the volume of plate material decreased by the factor 3/4, and the amount of adhesive material decreased by the factor 1/8.

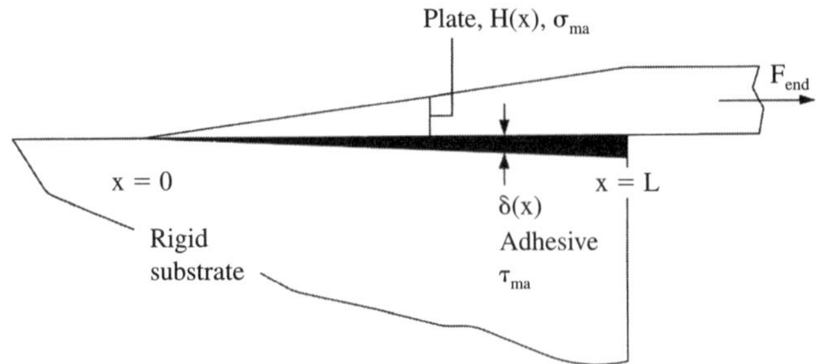

Figure P10.16

10.17. Two plates are joined by overlapping over a strip of width L, Fig. P10.17. The tensile force transmitted from plate to plate is F_{end}. A layer of adhesive material is sandwiched between the two plates. The design must have uniform longitudinal tensile stresses in both plates, $\sigma = \sigma_{\text{ma}}$ (constant) and uniform shear stresses (τ_{ma}) throughout the adhesive material. Show that this new design is obtained by sandwiching two of the designs developed in the preceding problem. Note that the two blades are wedge shaped, and that their common adhesive layer has constant thickness, $\delta = F_{\text{end}}G\sigma_{\text{ma}}/(E\tau^2{}_{\text{ma}})$. Review again the path from Problems 10.15 and 10.16, which led to this uniform stresses (minimal mass) architecture.

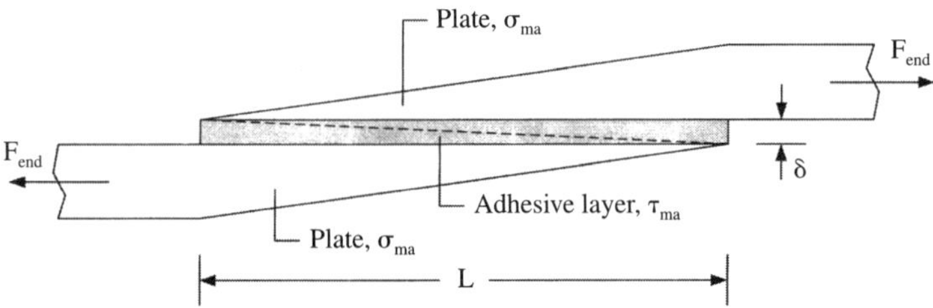

Figure P10.17

11

QUO VADIS CONSTRUCTAL THEORY?

The route traced in this course was long but clear: from principle to better configurations for engineering. At no time did we stray off the path to wander in the woods. We did not have to because the constructal law pointed the way, and because we invoked it and applied it at every turn. The route covered a huge diversity of domains, fluid flow and solid structures, distribution of fluid, heat and mass, multiscale design (macro, micro, nano), multiobjective design, and the vascularization of every thing.

Where is this route heading? In this chapter we paint a bird's-eye view of the entire tree of constructal theory thought, from its trunk (the constructal law in thermodynamics) and the growing volume of physical evidence in support of the constructal law, to fields of application that have not been dug and planted until now. We believe that *design with constructal theory* can be practiced with benefit for everybody—the larger the scales of the flow system the greater the benefit and the group of beneficiaries. Large-scale systems that occupy the agenda today are globalization, security, energy, climate, and sustainability. They survive (with us inside of them) because of the evolution of science and technology. Design with constructal theory covers all these agenda items, the global-scale flows and their evolution.

11.1 THE THERMODYNAMICS OF SYSTEMS WITH CONFIGURATION

The generation of configuration (shape, structure, design) is a phenomenon of physics. It occurs everywhere, in animate and inanimate systems. Pick any configuration (the tree, for example), and you see how the drawing tears down the presumed walls between biology (lungs, vascularized tissues) and geophysics (river basins, deltas, lightning). The physics phenomenon leads to the physics question: Why is "designedness" a characteristic of all nature? On what thought can we base the hierarchy, time evolution, complexity, and rhythm of natural structures? Is there a single physics principle from which form and rhythm can be *deduced*, without any use of empiricism? There is such a principle, and it is a summary of common observations that if a flow system is endowed with sufficient freedom to change its configuration, then the system exhibits configurations that provide progressively better access routes for the currents that flow. Observations of this kind come in huge numbers, and they mean one thing: a time arrow is associated with the sequence of flow configurations that constitutes the *existence* of the system. In this movie, existing drawings are replaced by easier flowing drawings.

This principle was formulated in 1996 as the *constructal law* of the generation of flow configuration (section 1.1), and it is worth repeating: "For a finite-size flow system to persist in time (to live) it must evolve in such a way that it provides easier and easier access to the currents that flow through it" [1]. This law is the basis for the constructal theory of the generation of flow configuration in nature (Fig. 11.1). Today, this entire body of work represents a new extension of thermodynamics: the thermodynamics of nonequilibrium systems with configuration [3,4].

To see why the constructal law is a law of thermodynamics (physics), we ask why the constructal law is different than (i.e., distinct from or complementary to) the other laws of thermodynamics (Chapter 2). Consider an isolated thermodynamic system that is initially in a state of internal nonuniformity (e.g., regions of higher and lower pressures or temperatures, separated by internal partitions that suddenly break). The first law proclaims that the system energy does not change. The second law accounts for observations that describe a tendency in time: if enough time passes, the isolated system settles into a state of equilibrium (no internal flows, maximum entropy at constant energy). The first law and second law speak of a black box. They say nothing about the configurations (the drawings) of the things that flow. Classical thermodynamics is not concerned with the configurations of its nonequilibrium (flow) systems.

The tendency to go with the flow (i.e., the time sequence of drawings that the flow system exhibits as it evolves) is the phenomenon covered by the constructal law. Not the drawings *per se*, but the time direction in which they morph if given freedom. No configuration in nature is "predetermined" or "destined" to be or to become a particular image. The actual evolution or lack of evolution (rigidity) of the drawing depends on many factors, which are mostly random.

The same can be said about the second law. No isolated system in nature is predetermined or destined to end up in a state of uniform intensive properties so that all future flows are ruled out. One cannot count on the removal of all the internal constraints.

The second law proclaims the existence of the state of equilibrium in an isolated system, at sufficiently long times when all internal constraints have been removed. The constructal law proclaims the existence of the concept of equilibrium flow

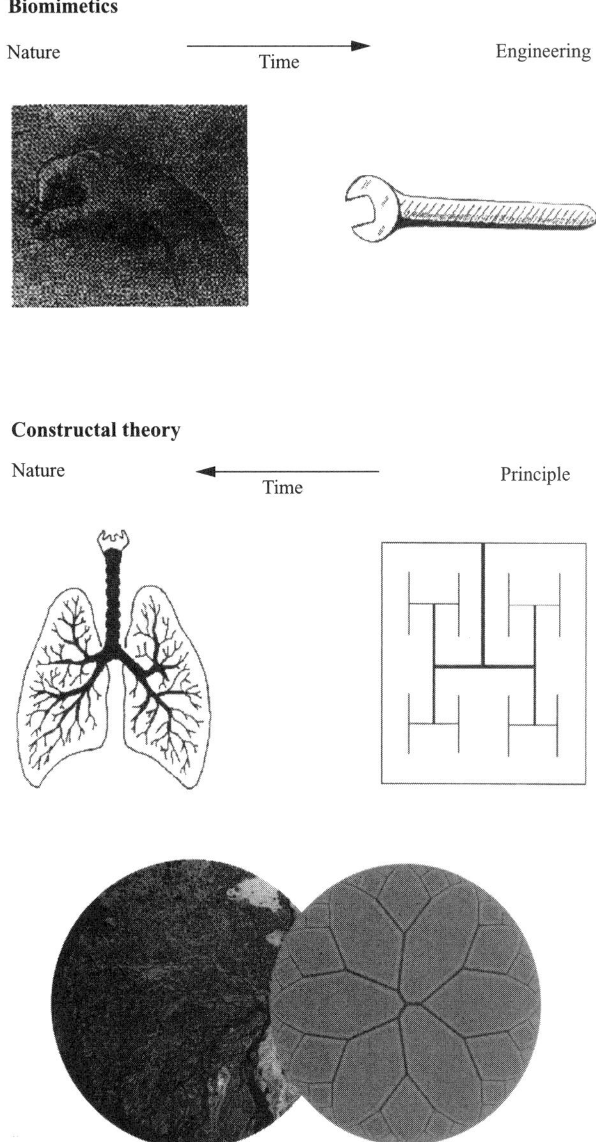

Figure 11.1 Constructal theory proceeds in time against empiricism or copying from nature. Bottom: The Lena delta and dendritic architecture derived from the constructal law. [2]. See also Color Plate 1.

architecture, when all possibilities of increasing morphing freedom have been exhausted [3,4].

Constructal theory is now a fast growing field with contributions from many sources, and with leads in many directions. This body of work has two sides. One is the use of the constructal law to predict and explain the occurrence of *natural* flow configurations, inanimate and animate: this was reviewed recently in Refs. [2], [5], and [6], and a concise view is provided in the essay of section 11.2. The other side is the application of the constructal law as a physics principle in *engineering*. This philosophy of *design as science* was the main object of Chapters 1 through 10.

11.2 TWO WAYS TO FLOW ARE BETTER THAN ONE

When we travel through the Atlanta airport (Fig. 11.2) we move in two ways, by walking along a concourse and riding on the train [7]. The two ways are *very different*. The walk is individual, disorganized, and impossible to draw. The ride is organized along a well-defined channel. The time to walk is the same as the time to ride: five minutes. Coincidence?

Respiration, river basins, and other natural structures combine the same two mechanisms [1]. The movement of oxygen and carbon dioxide through the lung occurs in two dissimilar ways: convection along the bronchial tubes (the air stream is the vehicle), and diffusion across the vascularized tissues that surround the alveoli. The time scale of convection (inhaling, exhaling) is the same five seconds as the time of O_2 and CO_2 diffusion across the alveolar tissue.

Recent measurements in hydrology reveal a similar coincidence. In all the river basins on earth, the time scale of seepage down the hill slope is the same as the residence time in the entire basin [8]. This balance between dissimilar flow mechanisms makes the river basin flow the same way as the lung. Both maximize the flow access between one point and an infinity of points (area, volume). The visible imprint of the two-mechanism balance (the allocation of channels to diffusion volume elements, i.e. the tree architecture) is shared by river basins, lungs, and many other animate and inanimate flow systems (e.g., blood vascularization, lightning, botanical trees).

The turbulent eddy is born out of the same diffusion-convection balance [1,2, 8]. Consider the flow of momentum by shear, from a semi-infinite fluid of uniform speed V to the rest of the fluid, which is stationary. The shear plane becomes a layer of thickness D that increases in time. If momentum transport across D is by viscous diffusion (laminar flow), then D grows as $(\nu t)^{1/2}$, where ν is the kinematic viscosity. If momentum is transported by streams (rolls), then D acquires the scale Vt. Maximum flow access means that laminar shear is preferred at short times, and eddies at longer times. The transition between the two is the first eddy. The intersection of the $D(t)$ curves for the laminar and eddy shear leads to the prediction [2] that transition occurs when the *local* Reynolds number VD/ν is 10^2, where V and D are the peripheral speed and size of the first eddy. This agrees with all observations of transition in jets, wakes, plumes, shear layers, and boundary layers in forced and natural convection [2, 8].

Dendritic solidification has the same origin. Consider a motionless fluid medium, which is at a temperature slightly below the solidification temperature. Latent heat is

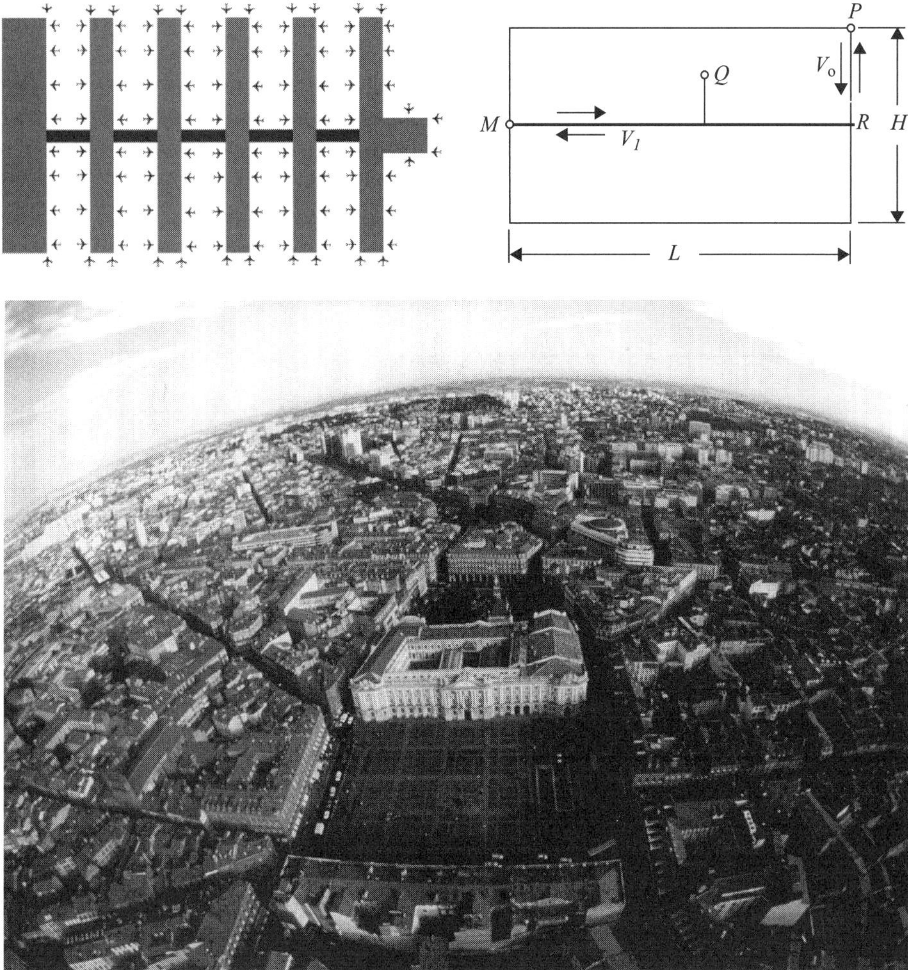

Figure 11.2 Two ways to flow are better than one. A large airport without a train, or without walking, cannot compete on the same area with the design that combines walking with riding on a vehicle. The Atlanta airport is a modern illustration of the seed from which all forms of urban and natural flow networks have grown. On a fixed area ($A = HL$) with variable shape (H/L) and two speeds (walking V_0, train V_1), the time of travel from P to M (or from all points Q to M, averaged over A) is minimum when the shape is $H/L = 2V_0/V_1$. In such a design, the walking time (PR) is equal to the riding time (RM). See also Color Plates 4 and 5.

released at the solidification site and flows into the subcooled medium [1, 8, 9]. How does nature facilitate the flow of heat? It does by placing thermal diffusion at short times and needles at longer times. Thermal diffusion spreads as a spherical wave of radius $(\alpha\ t)^{1/2}$, where α is the thermal diffusivity of the medium. Needles grow at constant speed parallel to themselves. We have the two-mechanism balance again, and from it a theoretical basis on which to expect "rapid solidification"—needles growing on a background swept by thermal diffusion.

The same scenario rules the formation of needles and dendritic aggregates of dust particles in air and filter surfaces [10]. Stony corals, bacterial colonies and plant roots favor a small-size ball shape in the beginning of their growth [11]. Later, corals opt for larger structures with branches that invade the water.

Electrodiffusion is another manifestation of the competition between two mass-flow mechanisms: classical diffusion driven by the imposed concentration difference, and ion convection facilitated by an electric current (Chapter 9). Nature selects diffusion at short times, and ion convection at longer times. This combination of two very different mechanisms is the reason why measurements of electrodiffusion could not be explained and correlated until recently [12].

River basins of all sizes are united by several scaling laws, such as an average of four tributaries to a larger stream. These rules of flow architecture could not be predicted until recently [8, 13] when river basins were analyzed as patchworks of areas with diffusion (seepage areas) allocated optimally to streams.

The puzzle of animal and geophysical design is unraveling because of the evolutionary flow-access principle that the two flow mechanisms illustrate. All the examples discussed above are sumarized by Fig. 11.3. For a flow system to persist in time, it must evolve such that its configuration provides greater and greater

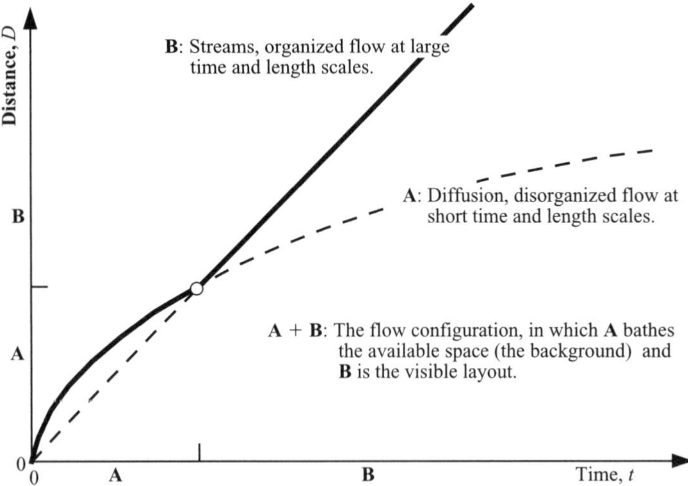

Figure 11.3 Turbulence, dendritic solidification, electrodiffusion, mud cracks, dust aggregates, stony corals and bacterial colonies owe their configurations to the natural tendency to flow more easily. The distance traveled by diffusion increases in time as $t^{1/2}$, and the speed of the diffusion front decreases as $t^{-1/2}$. Greater flow access calls for a mechanism that is more effective at long times: this second mechanism is "organized" flow, that is convection (streams). Together, the two mechanisms provide greater flow access than one mechanism alone, provided that they are arranged in this order: diffusion at short time and length scales, optimally woven with convection at larger scales. Nature always chooses this arrangement, not the opposite.

access to the currents that flow through it (cf. section 11.1). To endow a flow system with progressively greater flow access means to endow it with configuration; that is, with channels drawn on a background. From this universal tendency results "evolution" and the "designedness" of nature.

11.3 DISTRIBUTED ENERGY SYSTEMS

One lesson that nature teaches us is that in highly complex systems the generation and use of motive power is *distributed* throughout the body. It is not centered in a single spot, nodule or organ. The animal muscle is a "tapestry" of patches served by two kinds of flow systems: (1) tissues that generate movement (contraction), and (2) vascularization that feeds, cleanses, and endows the tissue with the ability to sense and act.

So perfect is the allocation of power generation to the networks for supply and distribution that the untrained eye sees the tissue as one or, at the most, as a complicated (multiscale) porous flow structure. The "allocating" of one flow system to the other flow system, in the same confined space, is the secret of the design. How are such designs made? How do they function?

These questions are interesting and highly promising when placed in the context of a sustainable energy future for our planet. The inhabited surface of the earth is covered by the same two classes of flow systems: (1) nodules, large channels of power generation, embedded in (2) networks of supply and distribution. Systems (1) and (2) are allocated to elemental areas, forming a patchwork that covers countries and continents. One example is shown in Figs. 11.4 and 11.5. The air mass-transit map has history and memory. In time, new channels appear and old ones become thicker.

Like the animal muscle, the patchwork of power generation, distribution, and use happened *naturally*. Unlike the animal muscle, which has spent millions of years in the factory of evolution, our energy systems evolve in front of our eyes. They morph while they grow. They produce more power, and they produce the power more efficiently (Fig. 11.6). Why do humans need power? For the same reason that animals need muscle power: to move mass on the earth's surface. Recent theoretical work on the origins of animal locomotion [16, 17] has shown that for all types of locomotion (running, flying, swimming), animal force is roughly equal to the body weight, and the minimum work that the body performs is proportional to the body weight times the distance traveled. The consumed food or fuel is "converted" into mass moved. Our cars, construction sites, and everything else we do (our legacy) are the product of this. All the animals and all of us consume food and fuel, and the result is the shaping and reshaping (the mixing) of the earth's surface.

These are the questions that underpin the fundamental direction proposed in this book. In addition to the importance of these fundamentals to the human design of

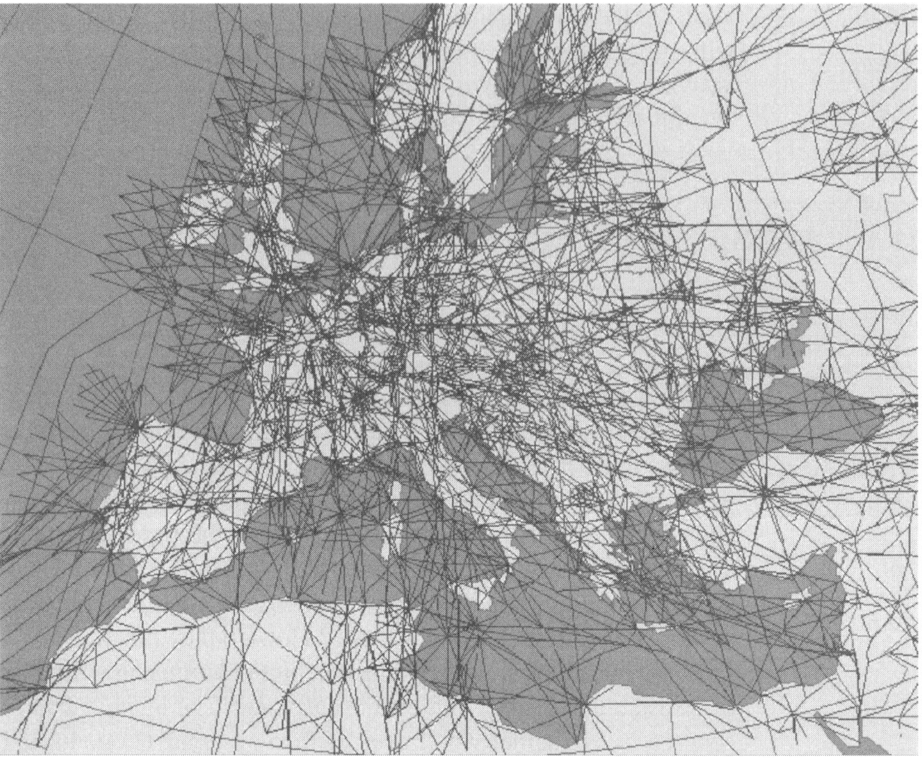

Figure 11.4 Tapestry of air traffic over Europe [14]. The burning of jet fuel is for moving people and goods: this flow is hierarchical and nonuniformly distributed. Large centers and thick channels are allocated to numerous smaller channels. The fine channels are allocated to area elements (yellow) that are covered by ground movement of people, and all the animate and the inanimate flows of the environment. See also Color Plate 6.

the best possible sustainable and safe future, we see three more reasons for our confidence in this direction of inquiry:

First, the need to consider the whole is universal. Yet, most of the research writing on energy science and engineering is devoted to the "energy" side, that is, to more efficient and cleaner production (systems of type 1 above). The remainder of the effort is devoted to the "environment", that is, to our flow networks that *interact* with the environment (systems of type 2 above). In constructal theory, energy (1) and the environment (2) are contemplated together, from the beginning, and from the smallest elemental area on which 1 and 2 come together to form a self-sustaining and long lasting tissue of flow systems.

Second, the widespread occurrence of *distributed* energy systems in nature is a very loud hint that the future of human energy design belongs to distributed systems. *The future belongs to the vascularized*. In nature, distributed energy systems occur not only in animal design but also in inanimate flow systems such as river basins.

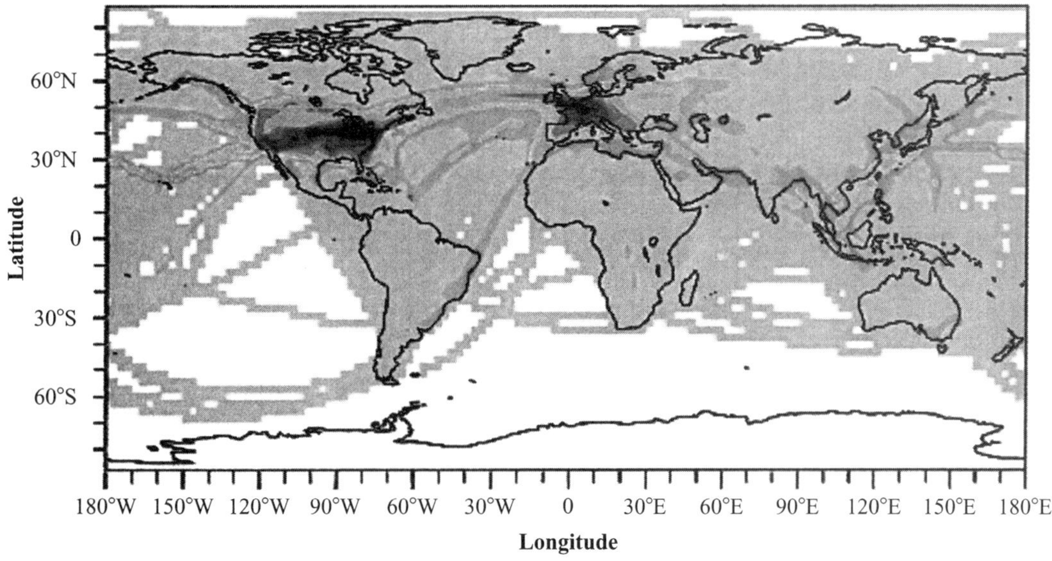

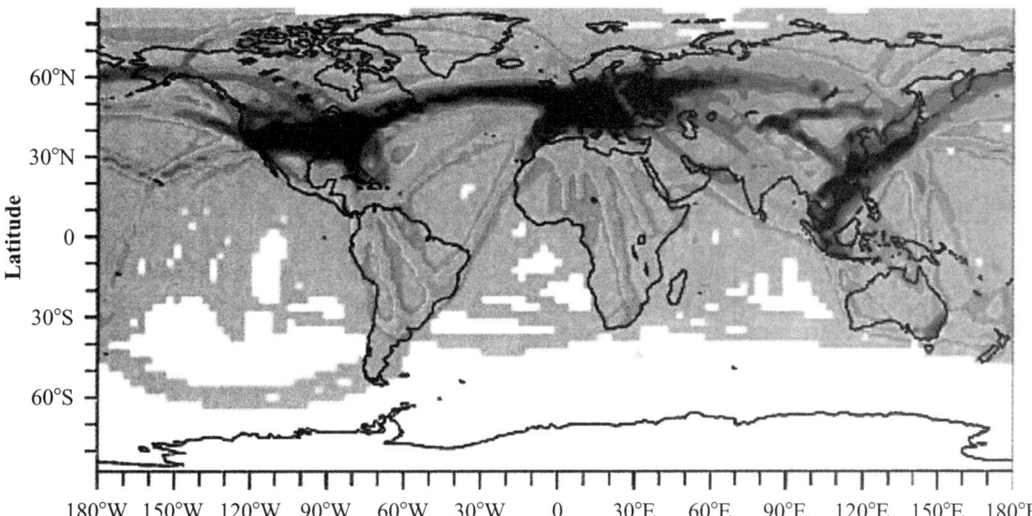

Figure 11.5 Where aircraft flew in 1992 and where aircraft will fly in 2050: the persistent contrail coverage (in % area cover) for the 1992 aviation fleet (from Gierens et al. [15], with permission from Springer). See also Color Plate 7.

Each sloped channel in a river basin is an optimal combination of (1) motive power (the slope, i.e., the driving gravitational potential); (2) distribution, use, dissipation (friction along the channel); and the allocation of 1 and 2 to the elemental territory bathed by the channel. The time arrow of evolution in natural flow systems points toward distributed energy systems.

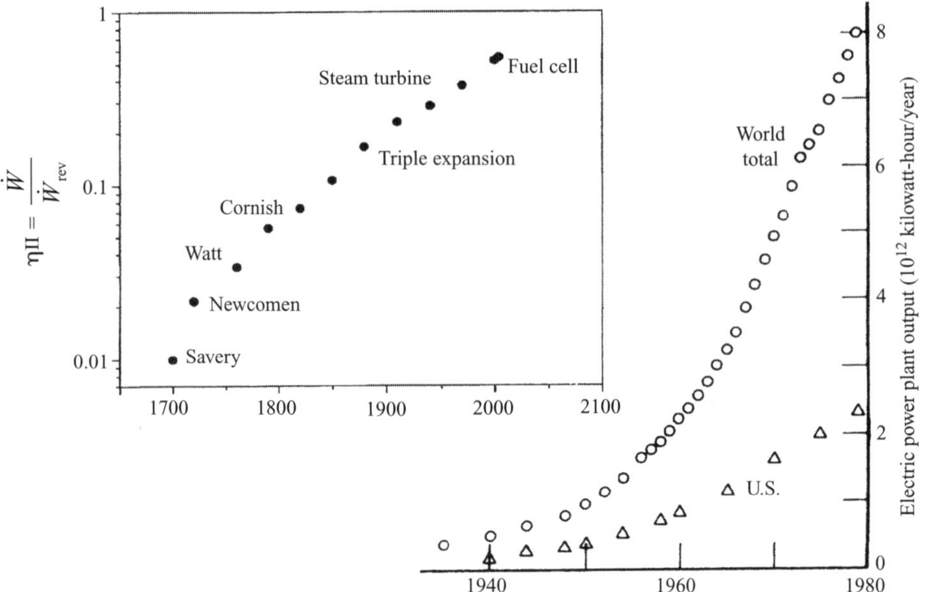

Figure 11.6 Time evolution of the second law efficiency of power plants and the growth of power generation worldwide [8].

Third, natural flow architectures have survived to this day because they have evolved into flow configurations that are the most efficient under the constraints, for example, tree-shaped flows oriented point-to-area or area-to-point. They survived because they are *free to adapt*; that is, their configurations are malleable in spite of the constraints. *Freedom is good for design*. We have and use freedom in the design of our own energy flow systems. Technology evolves by morphing, improving and spreading (cf. sections 3.7 and 5.4). There is a complete analogy between natural flow architectures and the flow architectures constructed by humans and powered by the sources designed and positioned by humans on the landscape.

Here, we show how to uncover the most fundamental principles of distributed energy systems—what makes them more efficient, more resilient, and more adaptable than other architectures. We search for the principles that govern the generation of patterns, clusters of energy systems, centralization vs. decentralization, and transitions (in time) from one configuration to another. We start from the simplest setting for pursuing the above questions, and look toward more complex, more realistic, and more interdisciplinary manifestations of the *phenomenon* of distributed energy design.

Consider the design of energy systems for heating. Humanity needs heating all over the globe, and for this reason the burning of fuel occurs all over the globe. Key is the observation that all the generated heat (the used and the unused) is eventually discharged into the environment. The challenge is to channel most of this heat

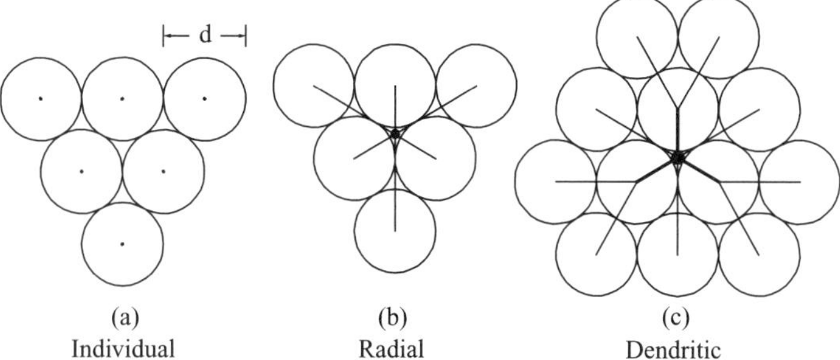

Figure 11.7 Distributed energy systems for heating: (a) individual heaters; (b) central heater and radial distribution lines; (c) central heater and dendritic distribution network.

through our homes and enterprises before discharging it into the environment. The challenge is to place humans and enterprises in the right places on the landscape, as optimally positioned interceptors. When this tapestry of interceptors of heat is designed from principle, two major objectives are achieved simultaneously:

1. The heating needs of humanity are met by burning minimum fuel.
2. The total heating dumped into the environment is the smallest that it can be.

To illustrate the approach, assume that our heating needs are served by streams of hot water of temperature $T_\infty + \Delta T$, where T_∞ is the environment temperature, and ΔT is specified. These streams are heated in imperfect installations that burn fuel, heat water, and leak a portion of the heat of combustion to the environment.

The hot water is used by individuals and their enterprises at discrete sites (Fig. 11.7a–c). Assume that all the sites are identical in size and need. Size is indicated by the length scale of one site, d, which is fixed. Need is indicated by the hot water mass m_1 used per-unit-time at one site. The following scale analysis refers to hot water generation and use on a per-unit-time basis. According to the rules of scale analysis (Ref. [18] and Appendix A), factors of order 1 are neglected, and the results are correct and accurate within a factor of order 1.

Here, we consider two designs for distributing water heating and use. First, every user produces its hot water on site in a heater-tank installation. An individual water-heating tank is modeled as a sphere of diameter D_1. This tank is filled with hot water (mass m_1), and leaks heat to the ambient in proportion to the tank surface,

$$q_1 \sim U D_1^2 \Delta T \tag{11.1}$$

where U is the overall heat-transfer coefficient (between water mass and ambient) multiplied by the time unit. In view of the tank size, $m_1 \sim \rho D_1^3$, where ρ is the water density, the heat loss from the water tank is

$$q_1 \sim U \left(\frac{m_1}{\rho}\right)^{2/3} \Delta T \tag{11.2}$$

There are N user sites on the territory of size A. The total heat loss from the N sites is

$$q = Nq_1 \sim a\,N\,m_1^{2/3} \tag{11.3}$$

where $a = U\,\rho^{-2/3}\,\Delta T$. We see that the loss of fuel burned (q) is proportional to the size of the population (N) and the individual water consumption raised to the power 2/3. Can this penalty be made smaller?

The answer is yes, and the solution consists of *organizing* the users on the landscape. One design with organization is where N users are arranged around a single water-heating site. The central tank has the size

$$m_c = Nm_1 \tag{11.4}$$

and the heat loss [cf. Eq. (11.3)]

$$q_c \sim a\,m_c^{2/3} \tag{11.5}$$

The N users are positioned on a circle of radius L; therefore,

$$L \sim Nd \tag{11.6}$$

Each user receives its hot water allocation (m_1) through an L-long radial pipe of standard diameter D_p. Each pipe loses heat in amount

$$q_{1p} \sim U_p D_p\,L\,\Delta T \tag{11.7}$$

where U_p is the pipe-ambient heat transfer coefficient multiplied by the time unit. The loss from the N radial pipes is

$$q_p = Nq_{1p} \sim b\,L\,N \tag{11.8}$$

where $b = U_p\,D_p\,\Delta T$. The global loss from the entire construct (central heater + radial pipes) is

$$q \sim q_c + q_p \tag{11.9}$$

The global loss per user is

$$\frac{q}{N} \sim \frac{a\,m_1^{2/3}}{N^{1/3}} + b\,N\,d \tag{11.10}$$

This expression shows that the density of heat loss is minimum when the number of users grouped around a single central heater is

$$N_{\text{opt}} \sim \left(\frac{a}{bd}\right)^{3/4} m_1^{1/2} \tag{11.11}$$

The minimum heat loss density is

$$\left(\frac{q}{N}\right)_{\text{min}} \sim a^{3/4}(bd)^{1/4} m_1^{1/2} \tag{11.12}$$

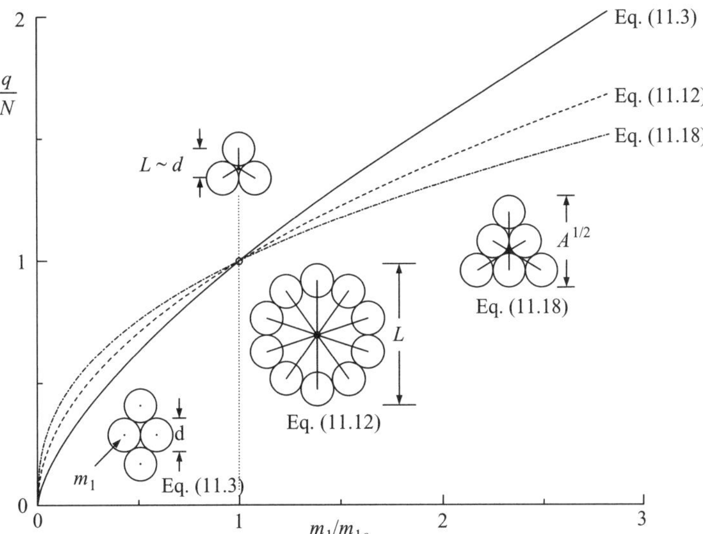

Figure 11.8 The fuel burned per user in three configurations: individual heaters, central heater with radial lines to users on a circle, and central heater with radial lines to users clustered near the heater.

The significance of this result becomes evident when we compare it with Eq. (11.3), where $q/N \sim a\, m_1^{2/3}$. In both designs, q/N increases with the individual water use, but the rate of increase depends on m_1. This is shown in Fig. 11.8, where the abscissa can be interpreted as the direction of time, or the direction of increase in living standard (namely, amount of hot water use per unit time). When m_1 is small, preferable is the decentralized design: one heater on every site. On the other hand, when m_1 is sufficiently large, preferable is the centralized water heating and distribution design. The *transition* between the two designs occurs when m_1 reaches the critical size

$$m_{1c} \sim \left(\frac{bd}{a}\right)^{3/2} \sim \rho \left(\frac{U_p}{U} D_p d\right)^{3/2} \tag{11.13}$$

for which the critical number of organized sites is:

$$N_c \sim 1 \tag{11.14}$$

The geometric meaning of $N_c \sim 1$ is derived from Eq. (11.6), which now reads

$$L_c \sim d \tag{11.15}$$

In conclusion, at transition the users should cluster tightly at a distance of order d around a central source. Arrangements of this type are triangular ($N = 3$), square ($N = 4$), hexagonal ($N = 6$) (e.g., Fig. 11.7b). Larger clusters ($N > N_c$) are more attractive when m_1 becomes greater than $m_{1,c}$.

When the m_1 sites are tightly packed, the minimization of q/N is synonymous with the minimization of the loss per unit territory, q/A. When $m_1 > m_{1c}$, the radial scale L exceeds d, and the area (of order L^2) is covered only partially by user sites (area of order Ld).

If land is at such a premium that it is covered continuously by users, then Eqs. (11.4) through (11.15) can be repeated for an area A covered completely by N sites of size d, namely, $A \sim Nd^2$. The length scale of area A is then $L \sim A^{1/2} \sim N^{1/2}d$. This is the length scale of any of the pipes connecting one user to the single hot water generation center located on A. Substituting $N^{1/2}d$ in place of L in Eq. (11.8) we obtain $q_p \sim b\,N^{3/2}d$. In place of Eq. (11.10) we write

$$\frac{q}{A} \sim \frac{a\,m_1^{2/3}}{d^2\,N^{1/3}} + \frac{b}{d}N^{1/2} \tag{11.16}$$

The optimal number N for minimum q/A is

$$N_{\text{opt}} \sim \left(\frac{a}{bd}\right)^{6/5} m_1^{4/5} \tag{11.17}$$

for which Eq. (11.16) yields

$$\left(\frac{q}{A}\right)_{\text{min}} \sim b^{2/5}\,a^{3/5}d^{-8/5}m_1^{2/5} \tag{11.18}$$

Written in terms of q/A, Eq. (11.3) becomes

$$\frac{q}{A} \sim \frac{a}{d^2}m_1^{2/3} \tag{11.19}$$

The transition between the disorganized design (11.19) and the organized design (11.18) occurs at the intersection of the two, which is the same as in Eq. (11.13). This is to be expected, because when $m_1 \le m_{1c}$ the area A is proportional to N.

The next question is what should happen when m_1 exceeds m_{1c}. In this domain, we compare Eq. (11.18) with Eq. (11.12) on a q/N basis. The two equations become

$$\frac{q_{11.12}}{N} \sim bd \sim m_1^{1/2} \tag{11.20}$$

$$\frac{q_{11.18}}{N} \sim bd \sim m_1^{2/5} \tag{11.21}$$

where, in view of Eq. (11.13), $\tilde{m}_1 = m_1/m_{1c}$. When $\tilde{m}_1 > 1$, the heat loss minimized in Eq. (11.18) is smaller than in the design of Eq. (11.12). This is indicated by the curve drawn for Eq. (11.18) in Fig. 11.8.

The chief message of this very simple example is that the emergence of configuration in time can be derived from principle. The drawings and their times of emergence can be *predicted*. Organization is good, provided that the level of advancement (m_1) calls for it. For example, Fig. 11.8 indicates that when the individual need (m_1) is less than m_{1c} the better configuration is the individual pattern (Fig. 11.7a). When m_1 increases above the critical level, the preferred pattern is radial and tightly packed (Fig. 11.7b).

Chapters 4 through 10 suggest that farther in time (i.e., to the right in Fig. 11.8) we will discover even better configurations, such as the dendritic design sketched in Fig. 11.7c. This is the direction of the constructal paradigm taught in this book: thinking ahead (in time), and searching *without bias* for the best organization that benefits every member of the organization.

The illustration of the generation of distributed-energy configuration (Figs. 11.7 and 11.8), can be thought of as a trade off between losses concentrated in the nodes of production and losses spread along the lines of the distribution network. When these two kinds of losses are balanced (i.e., summed up and minimized on the available territory), and when the flow paths are free to morph, the configuration takes shape.

This trade-off generates other important designs. The generation of electricity in power plants is an important application of this approach. Economies of scale are well established in power generation. Larger stationary power plants are more efficient than miniature power plants. The attractiveness of a large power plant makes the clustering of more and more users attractive. At the same time, the territory served by the power plant increases, and so does the length scale of the power distribution lines. The losses due to distributing the power grow as the power plant grows. There is a trade-off between the savings associated with using a central (efficient) power plant and the losses of a dissipative distribution network. This trade-off establishes the length scale of the pattern in which power plants must be allocated to users, and how each such cluster must be allocated to its area on the landscape. This is the conceptual route to discovering the architecture of the distributed power system.

Along this route, we take full advantage of the dynamics inherent in this phenomenon of distributed-energy generation. How does the pattern selection evolve in time? As societies become more advanced and the power per capita increases, will the clusters of users increase or decrease in size? In other words, will the generation of power become so distributed that each user site will have its own power plant? If so, when? These are important and very basic questions, especially in view of analogies that are being made nowadays between the decentralization of computing (the "PC" phenomenon) and the presumed decentralization of power generation. We question the validity of this analogy.

Economies of scale are also a characteristic of refrigeration plants: larger machines operate closer to the ideal limit than smaller machines [1, 8]. The argument made above for distributed power generation can be made identically for distributed refrigeration or distributed liquefaction of refrigerant. How large should the refrigeration plant be, and for what territory? Furthermore, how will this rule of organization change as the demand of refrigeration increases?

Figure 11.9 Round patches of vegetation in muddy terrain at low tide. See also Color Plate 16. The pattern is reminiscent of Fig. 11.5. Why the similarities? What is the global flow access that the configuring of vegetation is facilitating?

Constructal theory flows both ways, from the mind to engineering (principles for improving the man and machine species), and from principles to predicting and explaining nature, that is back to the mind, for new ideas, to restart the thinking cycle. For example, the discovery of efficient and sustainable design as "distributed energy systems" had its inspiration in natural design (section 11.2). The discovery is so fundamental that its message resonates in biology and geophysics, where there are many distributed systems that are waiting to be explained. An example is pictured in Fig. 11.9, for which the caption question is answered in section 10.5. This book is an invitation to follow our curiosity back to nature, and to show on a case-by-case basis how the principles developed here can be used to tear down the presumed walls between "science" and "engineering."

11.4 SCALING UP

The work line sketched in this book is a promising direction to solving the toughest of all problems in engineering design: scaling, that is, how to use the results from a desk-size model in order to predict the behavior and performance of "the same" system but at much larger scales. The difficulty stems from the nature of all flow systems: the larger is not "the same" as the laboratory model. The larger is not a magnified replica of the model.

What happens during the magnification exercise is suggested by the abscissa of Fig. 11.8. The configuration *changes*, because the flow system must be the best that it can be *at any size*. One cannot predict the performance of a large-use heating system ($m_1 > m_{1c}$, Fig. 11.8) by extrapolating from the tested performance of a small-use system ($m_1 < m_{1c}$).

The only way to crack the scaling nut is to have a firm grip on the hammer of principles, that is, to know how the system configuration changes as its size increases. Once we know the drawing, large or small, we can analyze (or test) the flow system and describe its performance with confidence. This means that if we know that Fig. 11.7c will be the configuration at large scales, then we must test in the laboratory a miniature of Fig. 11.7c, not of Fig. 11.7a. This knowledge is a powerful new tool, and a very timely one for placing the subject of "design" on a scientific basis.

11.5 SURVIVAL VIA GREATER PERFORMANCE, SVELTENESS AND TERRITORY

All the designs discussed in this book, the large and the small, and the concrete and the abstract are flow systems. The evolution of each such system in the constructal-law direction can be viewed on a performance-freedom field such as Figs. 3.2 and 4.12. In closing, it is useful to view this image in the most general terms [3,4] that apply to all flow systems.

A flow system has *properties* that distinguish it from a static (nonflow) system. The properties of a flow system are (1) global external size, for example, the length scale of the body bathed by the tree flow L; (2) global internal size, for example, the total volume of the flow channels V; (3) at least one global objective, or performance, for example, the global flow resistance of the tree R; (4) configuration, drawing, architecture; and (5) freedom to morph, that is, freedom to change the configuration. The global external and internal sizes (L,V) mean that a flow system has two length scales L and $V^{1/3}$. These form a dimensionless ratio—the *svelteness* Sv—which is a global property of the flow configuration, $\text{Sv} = L/V^{1/3}$ (cf. section 1.1).

The flow structures generated in accordance with the constructal law populate and move in the V = constant plane shown in Fig. 11.10. This plane is home to a performance versus freedom diagram: in time, and if the architecture is free to change, R decreases (i.e., performance increases) at constant L and V. The configuration with the smallest R value represents the equilibrium flow structure. The configurations that preceded it are nonequilibrium flow structures.

The evolution of configurations in the constant-V cut (and at constant L, Fig. 11.10) represents survival through increasing performance—survival of the fittest. This is the physics law that rules not only the animate flow systems but also the natural inanimate flow systems and all the human and machine species. The constructal law defines the meaning of "the survivor" or of the equivalent concept of "the more fit." The constructal-law idea that freedom to morph is good for

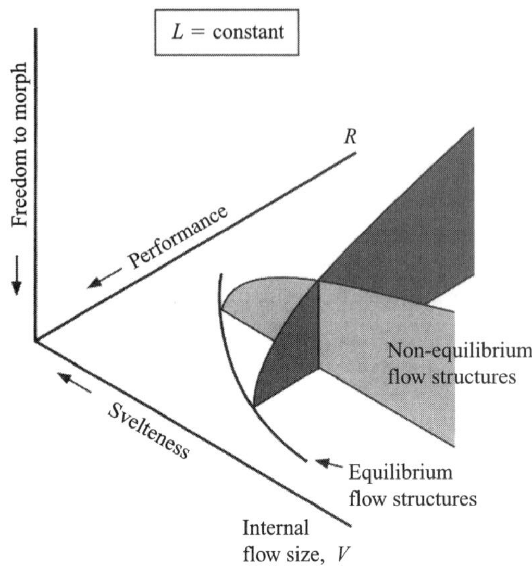

Figure 11.10 Performance versus freedom to change configuration, at fixed global external size [2, 3]. See also Color Plate 8.

performance (Fig. 11.10) also accounts for the Darwinian argument that the survivor is the one most capable to adapt.

In the bottom plane of Fig. 11.10 the locus of equilibrium structures is a curve with negative slope, $(\partial R/\partial V)_L < 0$, because of flow physics: the resistance decreases when the size of the internal space inhabited by the flow increases. This slope means that the nonequilibrium flow structures occupy the design space suggested by the three-dimensional surface sketched in Fig. 11.10. The time evolution of nonequilibrium flow structures toward the bottom edge of the surface (the equilibrium structures) is the action of the constructal law.

The same time arrow can be described alternatively with reference to the constant-R cut through the three-dimensional space of Fig. 11.10. Flow architectures with the same global performance (R) and global size (L) evolve toward compactness and svelteness—smaller volumes dedicated to internal ducts, that is, larger volumes reserved for the working tissue (the interstices). Paraphrasing the original statement of the constructal law (pp. 2, 468), we may describe the evolution at constants L and R as follows:

> For a system with fixed global size and global performance to persist in time (to live), it must evolve in such a way that its flow structure occupies a smaller fraction of the available space.

This is survival based on the maximization of the use of the available space. *Survival by increasing svelteness (compactness)* is equivalent to survival by increasing performance. Both statements are the constructal law.

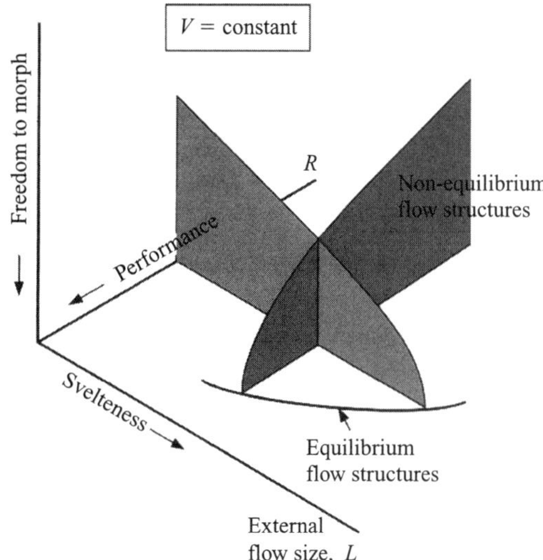

Figure 11.11 Performance vs. freedom to change configuration, at fixed global internal size [2, 3]. See also Color Plate 8.

A third statement of the constructal law becomes evident if we recast the constant-L design world of Fig. 11.10 in the constant-V design space of Fig. 11.11. In this new figure, the constant-L cut is the same performance versus freedom diagram as in Fig. 11.10, and the constructal law means survival by increasing performance. The contribution of Fig. 11.11 is the shape and orientation of the hypersurface of nonequilibrium flow structures: the slope of the curve in the bottom plane $(\partial R/\partial L)_V$ is positive because of flow physics, that is, because the flow resistance increases when the distance traveled by the stream increases.

The world of possible designs can be viewed in the constant-R cut made in Fig. 11.11, to see that flow structures of a certain performance level (R) and internal flow volume (V) morph into new flow structures that cover progressively larger territories. Again, flow configurations evolve toward greater svelteness Sv. The constructal law statement becomes:

> In order for a flow system with fixed global resistance (R) and internal size (V) to persist in time, the flow architecture must evolve in such a way that it covers a progressively larger territory.

There is a limit to the spreading of a flow structure, and it is set by global properties such as performance (technology) and internal flow volumes R and V.

River deltas in the desert, animal species on the plain, and the Roman Empire spread to their limits. Such is the constructal law of survival by spreading, by increasing territory for flow and movement, up to a limit. Now we know why it has always been this way.

11.6 SCIENCE AS A CONSRUCTAL FLOW ARCHITECTURE

In the thermal sciences that emerge, the readjustment of fossil and renewable fuel streams (i.e., new equilibria of *how* to flow) are being predicted and optimized based on principles [19]. In this new field, the shrinking of the environment (i.e., new equilibria between our flows and the external ones) is predicted and optimized based on principles. Systems have new properties such as configuration, objective, svelteness and freedom to morph. The new field is by its very nature transdisciplinary.

No flow system is an island. No river exists without its wet plain. No city thrives without its farmland and open spaces. Everything that flowed and lived to this day to "survive" is in an optimal balance with the flows that surround it and sustain it. The air flow to the alveolus is optimally matched to the blood that permeates through the vascularized tissue, and vice versa. The "system" and its "environment" live and morph together [20].

And so, we arrive at the essence of constructal theory, or the thermodynamics of systems with configuration—the union that it forges between physics, engineering science and life sciences. We see this union in Fig. 11.12. Earth, with its solar heat input, heat rejection, and wheels of atmospheric and oceanic circulation, is a heat engine without shaft. Its maximized (but not ideal) mechanical power output cannot be delivered to an extraterrestrial system. Instead, the earth engine is destined to dissipate through air and water friction and other irreversibilities (e.g., heat leaks) all the mechanical power that it produces. It does so by "spinning in its brake" the fastest that it can (hence the winds and the ocean currents, which proceed along easiest routes). Because the flowing earth is a constructal heat engine, its flow configuration has evolved in such a way that it is the least imperfect that it can be. It produces maximum power, which it then dissipates at maximum rate. A principle of maximum dissipation is now being invoked ad hoc in geophysics: all such writings refer only to what goes on in the brake and are already covered by the constructal law.

The heat engines of engineering and biology (power plants, animal motors) have shafts, rods, legs, and wings that deliver the mechanical power to external entities that use the power (e.g., vehicles and animal bodies needing propulsion). Because the engines of engineering and biology are constructal, they morph in time toward flow configurations that make them the least imperfect that they can be. Therefore, they evolve toward producing maximum mechanical power (under finiteness constraints), which, for them, means a time evolution toward minimum dissipation (minimum entropy generation rate).

If we look outside an engineering or biology engine, we see that all the mechanical power that the engine delivers is destroyed through friction and other irreversibility mechanisms (e.g., transportation and manufacturing for man, animal locomotion and body heat loss to ambient). The engine and its immediate environment (the brake), as one thermodynamic system, are analogous to the whole earth

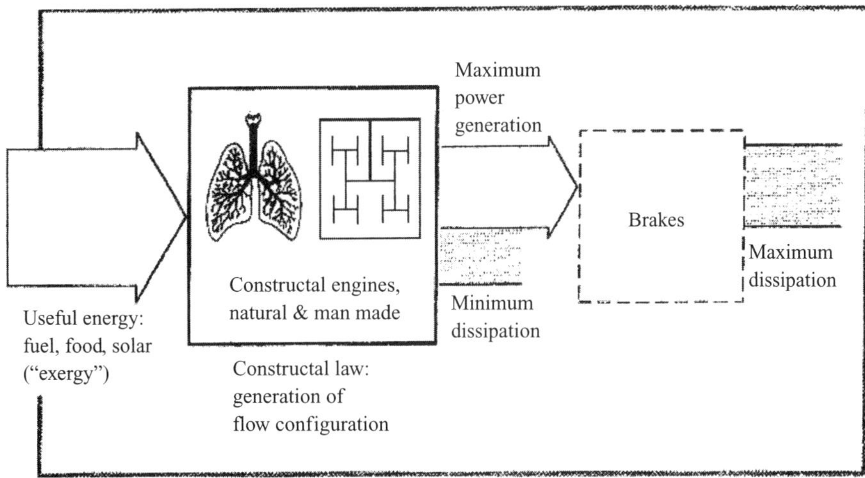

Figure 11.12 Every nonequilibrium (flow) component of the Earth functions as an engine that drives a brake [21]. The constructal law governs "how" the system functions: by generating a flow architecture that distributes imperfections through the flow space and endows it with configuration. The "engine" evolves in time toward generating maximum power (or minimum dissipation), and as a consequence, the "brake" exhibits maximum dissipation. Evolution means that each flow system assures its persistence in time by freely morphing into easier and easier flow structures under finiteness constraints. The arrows proceed from left to right because this is the general drawing for a flow (nonequilibrium) system, in steady state or unsteady state. When equilibrium is reached, all the flows cease, and the arrows disappear. See also Fig. 2.3.

(Fig. 11.12). After everything is said and done, the flowing earth (with all its engine and brake components, rivers, fish, turbulent eddies, etc.) accomplishes as much as any other flow architecture, animate or inanimate: it mixes the earth's lithosphere, atmosphere, hydrosphere, and biosphere most effectively—more effectively than in the absence of constructal phenomena of generation of flow configuration.

Irrefutable evidence of this accomplishment is how all the large eddies of biological matter have morphed and spread over larger areas and altitudes, in this sequence in time: fish in water, walking fish and other animals on land, flying animals in the atmosphere, flying man and machine species, and man and machine species in the outer space. The balanced and intertwined flows that generate our engineering, economics, and social organization are no different than the natural flow architectures of biology (animal design) and geophysics (river basins, global circulation).

The evolution of thermodynamics from Carnot to entropy generation minimization and now constructal theory is an illustration of the universal phenomenon of evolution of flow configuration in time. Science, ideas, news, and education flow and cover the globe like the streams of river basins [22]. They cover a multidimensional territory better known as history, geography, and civilization. The flow architecture of science continues to change, to improve, and to grow.

Physics is our knowledge of how nature works. Physics (or nature) is everything, including engineering: the biology and medicine of human and machine species. Our knowledge is condensed in simple statements (thoughts, connections), which evolve in time by being replaced by simpler statements. We know more because of this evolution in time, not because brains become bigger and neurons smaller and more numerous. Our finite-size brains keep up with the steady inflow of new information through a process of simplification by replacement: in time, and stepwise, bulky catalogs of empirical information (e.g., measurements, data, complex empirical models) are replaced by much simpler summarizing statements (e.g., concepts, formulas, constitutive relations, principles, laws). A hierarchy of statements emerges along the way: it emerges naturally, because it is better.

The simplest and most universal are the laws. The bulky and the laborious are being replaced by the compact and the fast. In time, science optimizes and organizes itself in the same way that a river basin evolves: toward configurations (links, connections) that provide faster access or easier flowing. The bulky measurements of pressure drop versus flow rate through round pipes and saturated porous media were rendered unnecessary by the formulas of Poiseuille and Darcy. The measurements of how things fall (faster and faster, and always from high to low) were rendered unnecessary by Galilei's principle and the second law of thermodynamics.

The hierarchy (specialization) that science exhibited at every stage in the history of its development is an expression of its never-ending struggle to optimize and redesign itself. Hierarchy means that measurements, ad hoc assumptions, and empirical models come in huge numbers, a "continuum" above which the compact statements (the laws) rise as needle-shaped peaks. Both are needed, the numerous and the singular [23]. One class of flows (information links) sustains the other. The many and unrelated heat engine builders of Britain fed the imagination of one Sadi Carnot. In turn, Sadi Carnot's mental viewing (thermodynamics today) feeds the minds of contemporary and future builders of all sorts of machines throughout the world.

Civilization with all its constructs (science, religion, language, and writing, etc.) is this never-ending physics of generation of new configurations, from the flow of mass, energy, and knowledge to the world migration of the special persons to whom ideas occur (the creative). Good ideas travel. Better-flowing configurations replace existing configurations. Empirical facts (observations) are extremely numerous, like the hill slopes of a river basin. The laws are the extremely few big rivers, *La Têt* and the Danube.

REFERENCES

1. A. Bejan, *Advanced Engineering Thermodynamics*, 2nd ed., New York: Wiley, 1997.
2. A. Bejan and S. Lorente, Constructal theory of generation of configuration in nature and engineering. *J Appl Phys* 100, 2006, 041301.

3. A. Bejan and S. Lorente, The constructal law and the thermodynamics of flow systems with configuration. *Int J Heat Mass Transfer*, Vol. 47, 2004, pp. 3203–3214.
4. A. Bejan and S. Lorente, *La loi constructale*. Paris: L'Harmattan, 2005.
5. A. H. Reis, Constructal theory: From engineering to physics, and how systems flow develop shape and structure. *Applied Mechanics Reviews*, Vol. 59, 2006, pp. 269–281.
6. A. Bejan and G. W. Merkx, eds., *Constructal Theory of Social Dynamics*. New York: Springer, 2007.
7. A. Bejan and S. Lorente, Thermodynamic optimization of flow geometry in mechanical and civil engineering. *Journal of Non-Equilibrium Thermodynamics*, Vol. 26, 2001, pp. 305–354.
8. A. Bejan, *Advanced Engineering Thermodynamics*, 3rd ed. Hoboken NJ: Wiley, 2006.
9. A. Ciobanas, A. Bejan, and Y. Fautrelle, Dendritic solidification morphology viewed from the perspective of constructal theory. *J Phys D: Appl Phys*, Vol. 39, 2006, pp. 5252–5266.
10. A. H. Reis, A. F. Miguel, and A. Bejan, Constructal theory of particle agglomeration. *J Phys D: Appl Phys*, Vol. 39, 2006, pp. 2311–2318.
11. A. F. Miguel, Constructal pattern formation in stony corals, bacterial colonies and plant roots under different hydrodynamic conditions. *J Theoretical Biology*, Vol. 242, 2006, pp. 954–961.
12. P. Bégué and S. Lorente, Migration versus diffusion through porous media: Time-dependent scale analysis. *J Porous Media*, Vol. 9, 2006, pp. 637–650.
13. A. H. Reis, Constructal view of scaling laws of river basins. *Geomorphology*, Vol. 78, 2006, pp. 201–206.
14. S. Périn, The constructal nature of the air traffic system. In *Constructal Theory of Social Dynamics*, A. Bejan and G. W. Merkx, eds., New York: Springer, 2007.
15. K. Gierens, R. Sausen, and U. Schumann, A diagnostic study of the global distribution of contrails, Part 2: Future air traffic scenarios. *Theoretical and Applied Climatology*, Vol. 63, 1999, pp. 1–9.
16. A. Bejan and J. H. Marden, Unifying constructal theory for scale effects in running, swimming and flying. *J Exp Biol*, Vol. 209, 2006, pp. 238–248.
17. A. Bejan and J. H. Marden, Constructing animal locomotion from new thermodynamics theory. *American Scientist*, July–August, 2006, pp. 343–349.
18. A. Bejan, *Convection Heat Transfer*, 3rd ed. Hoboken, NJ: Wiley, 2004.
19. A. Bejan and S. Lorente, Constructal theory of energy-system and environment flow configurations. *Int J Exergy*, Vol. 2, 2004, pp. 335–347.
20. A. Bejan, Design in nature: tinkering and the constructal law. *The Quarterly Review of Biology*, Vol. 83, No. 1, March 2008, pp. 91–94.
21. A. H. Reis and A. Bejan, Constructal theory of global circulation and climate. *Int J Heat Mass Transfer*, Vol. 49, 2006, pp. 1857–1875.
22. A. Bejan, Why university rankings do not change: Education as a natural hierarchical flow architecture. *Int J Design & Nature*, Vol. 2, No. 4, 2008, pp. 319–327.
23. A. Bejan, The many and the few. *Mechanical Engineering*, No. 7, July 2007, pp. 42–43.

PROBLEMS

11.1. Larger animals travel faster, are stronger and wave their bodies (legs, wings, tail) less frequently. These characteristics of animal design can be predicted based on the constructal law: animal locomotion seen as the time evolution toward greater flow access for the movement of animal mass on the globe [16, 17]. Greater flow access means less consumption of useful energy (exergy) for the movement of mass — less potential energy for the water flow down the river basin, less food (chemical exergy) for the flow of animal mass, and less fuel for our vehicles, i.e. for the movement of human mass.

Running is a periodic movement composed of two acts. The animal performs work during both acts, and this work is eventually dissipated into heat and rejected to the ambient.

Model the runner as a one-scale body of mass $M_b \sim \rho_b L_b^3$, or body length scale $L_b \sim (M_b/\rho_b)^{1/3}$. During the first part of the cycle, the animal lifts itself off the ground to a height of order L_b The gravitational acceleration is g. The scale of the work required by this lifting process is $W_y \sim M_b g L_b$.

The second part of cycle is the horizontal movement at speed V, during the time t, over the horizontal distance Vt. The scale of the horizontal work is $W_x \sim F_D V t$, where F_D is the drag force. For analytical simplicity, model the drag as follows: the ground is covered with L_b-deep snow of density ρ_s, the drag coefficient is constant and of order 1.

Obtain the two-term expression for the total work spent per distance traveled, $W' = \left(W_y + W_x\right)/Vt$, and use the fact that t is the time scale of free fall from a height of order L_b. Minimize W' with respect to V, and report the formulas for the optimal speed V, frequency t^{-1}, body force and minimum W'. Do these formulas confirm the observations of animal design stated at the top of this problem?

11.2. The speed of animal locomotion on the ground and on the water can be estimated based on the following simple model. The animal body (mass M_b) has a single length scale (L_b) and the density ρ_b. It tips and falls forward to the distance L_b during the time of free fall from the height L_b subject to the gravitational acceleration g. Show that the horizontal speed $V \sim L_b/t$ estimated in this manner is $V \sim g^{1/2}\rho^{-1/6}M_b^{1/6}$.

Appendix

A. THE METHOD OF SCALE ANALYSIS

The object of scale analysis [1] is to use the basic principles of heat and fluid flow to produce order-of-magnitude estimates for the quantities of interest. This means that if one of the quantities of interest is the thickness of the boundary layer in forced convection, the object of scale analysis is to determine whether the boundry layer thickness is measured in millimeters or centimeters. Note that scale analysis goes beyond dimensional analysis (whose objective is to determine the dimension of boundary layer thickness, namely, length). When done properly, scale analysis anticipates within a factor of order one (or within percentage points) the expensive results produced by "exact" analyses. The value of scale analysis is remarkable.

As an example, consider a plate plunged at $t = 0$ into a highly conducting fluid (Fig. A1), such that the surfaces of the plate instantaneously assume the fluid temperature $T_\infty = T_0 + \Delta T$. Suppose that we are interested in estimating the time needed by the thermal front to penetrate the plate, that is, the time until the center plane of the plate "feels" the heating imposed on the outer surfaces.

To answer this question, we focus on a half-plate of thickness $D/2$ and the energy equation for pure conduction in one direction:

$$\rho c_P \frac{\partial T}{\partial t} = K \frac{\partial^2 T}{\partial x^2} \tag{1}$$

Next, we estimate the order of magnitude of each of the terms appearing in Eq. (1). On the left-hand side, we have

$$\rho c_P \frac{\partial T}{\partial t} \sim \rho c_P \frac{\Delta T}{t} \tag{2}$$

In other words, the scale of the temperature change (in the chosen space and in a time of order t) is ΔT. On the right-hand side, we obtain

$$k \frac{\partial^2 T}{\partial x^2} = k \frac{\partial}{\partial x}\left(\frac{\partial T}{\partial x}\right) \sim \frac{k}{D/2}\frac{\Delta T}{D/2} = \frac{k \Delta T}{(D/2)^2} \tag{3}$$

Equating the two orders of magnitude (2) and (3), as required by the energy equation (1), we find the answer to the problem

$$t \sim \frac{(D/2)^2}{\alpha} \tag{4}$$

where α is the thermal diffusivity of the medium, $k/\rho c_P$. The penetration time (4) compares well with any order-of-magnitude interpretation of the exact solution to this classical problem. However, the time and effort associated with deriving Eq. (4) do not compare with the labor required by Fourier analysis and the graphical presentation of Fourier series.

Based on this example, the rules of scale analysis have been stated in Ref. [1] as follows:

- *Rule 1*. Always define the spatial extent of the region in which you perform the scale analysis. In the example of Fig. A1, the size of the region of interest is $D/2$. In other problems, such as boundary layer flow, the size of the region of interest is unknown: the scale analysis begins by selecting the region and by labeling the unknown thickness of this region δ. Any scale analysis of a flow or a flow region that is not uniquely defined is nonsense.
- *Rule 2*. One equation constitutes an equivalence between the scales of two dominant terms appearing in the equation. In the transient conduction example, the left-hand side of Eq. (1) could only be of the same order of magnitude as the right-hand side. The two terms appearing in Eq. (1) are the dominant terms (considering that the discussion referred to pure conduction). In general, the energy equation contains more terms, but not all of them important. The reasoning for selecting the dominant scales from many scales is condensed in rules 3–5.
- *Rule 3*. If in the sum of two terms,

$$c = a + b \tag{5}$$

the order of magnitude of one term is greater than the order of magnitude of the other term,

$$O(a) > O(b) \tag{6}$$

then the order of magnitude of the sum is dictated by the dominant term:

$$O(c) = O(a) \tag{7}$$

The same conclusion holds if instead of Eq. (5), we have the difference $c = a - b$ or $c = -a + b$.
- *Rule 4*. If in the sum of two terms, Eq. (5), the two terms are of the same order of magnitude,

$$O(a) = O(b) \tag{8}$$

the sum is of the same order of magnitude:

$$O(c) \sim O(a) \sim O(b) \tag{9}$$

- *Rule 5*. In any product

$$p = ab \tag{10}$$

the order of magnitude of the product is equal to the product of th
magnitude of the two factors

$$O(p) = O(a)O(b)$$

If, instead of Eq. (10), we have the ratio

$$r = \frac{a}{b} \tag{12}$$

then

$$O(r) = \frac{O(a)}{O(b)} \tag{13}$$

In addition to having its own set of rules, scale analysis requires special care with regard to notation. In rules 1 through 5, we used the following symbols:

$\sim$ is of the same order of magnitude as

O(a) the order of magnitude of a

$>$ greater than, in an order-of-magnitude sense

For brevity, the scale analyses included in this book do not employ the repetitive potentially confusing notation $O(\cdot)$ for "order of magnitude." Scale analysis [1] has been adopted by many researchers and is now employed widely in fluid mechanics, heat transfer, and design.

B. METHOD OF UNDETERMINED COEFFICIENTS (LAGRANGE MULTIPLIERS)

In analytical design, the method of Lagrange multipliers is used to determine the extremum (maximum or minimum) of a function

$$Z = F(x, y) \tag{1}$$

in which the arguments (x, y) vary while satisfying a constraint,

$$G(x, y) = \text{constant} \tag{2}$$

The location (x_0, y_0) of the extremum of F is determined by constructing the aggregate function

$$H(x, y) = F(x, y) + \lambda G(x, y) \tag{3}$$

where λ is an undetermined coefficient called Lagrange multiplier, and solving

$$\frac{\partial H}{\partial x} = 0 \qquad \frac{\partial H}{\partial y} = 0 \tag{4,5}$$

The system formed by Eqs. (2), (4) and (5) delivers the values of x_0, y_0 and λ. To decide whether at this extremum F is maximum or minimum, one must calculate

$$D = AC - B^2 \tag{6}$$

where

$$A = \frac{\partial^2 F}{\partial x^2}(x_0, y_0) \quad B = \frac{\partial^2 F}{\partial x \partial y}(x_0, y_0) \quad C = \frac{\partial^2 F}{\partial y^2}(x_0, y_0) \tag{7}$$

If D is positive (negative) then $F\ (x_0, y_0)$ is a minimum (maximum).

C. VARIATIONAL CALCULUS

The basic problem in variational calculus consists of determining, from among functions possessing certain properties, that function for which a given integral (functional) assumes its maximum or minimum value. The integrand of the integral in question depends on the function and its derivatives. Consider the many values of the integral

$$I = \int_a^b F(x, y, y')dx$$

where $y(x)$ is unknown, and $y' = dy/dx$. The special function y for which I reaches an extremum satisfies the Euler equation:

$$\frac{\partial F}{\partial y} - \frac{d}{dx}\left(\frac{\partial F}{\partial y'}\right) = 0$$

If, in addition to minimizing or maximizing I, the wanted function $y(x)$ satisfies an *integral constraint* of the type

$$C = \int_a^b G(x, y, y')\,dx, \text{ constant}$$

then $y(x, \lambda)$ satisfies the new Euler equation:

$$\frac{\partial H}{\partial y} - \frac{d}{dx}\left(\frac{\partial H}{\partial y'}\right) = 0$$

where

$$H = F + \lambda G$$

Constant λ is a Lagrange multiplier; its value is determined by substituting the $y(x, \lambda)$ solution into the integral constraint C.

In general, the integral may depend on the first n derivatives of the unknown function $y(x)$:

$$I = \int_a^b F(x, y, y', y'', \ldots, y^{(n)})\, dx$$

In this case, the condition for I to reach an extremum is

$$\frac{\partial F}{\partial y} - \frac{d}{dx}\left(\frac{\partial F}{\partial y'}\right) + \frac{d^2}{dx^2}\left(\frac{\partial F}{\partial y''}\right) - \cdots + (-1)^n \frac{d^n}{dx^n}\left[\frac{dF}{\partial y^{(n)}}\right] = 0$$

The differential equation for $y(x)$ above is called the *Euler-Poisson equation*. If the integral depends not on one but on k unknown functions, y_i $(i = 1, \ldots, k)$ and their derivatives, y_i',

$$I = \int_a^b F(x,\ y_1, \ldots, y_k,\ y_1', \ldots, y_k')\, dx$$

the k functions $y_i(x)$ that minimize or maximize I satisfy the system of Euler equations:

$$\frac{\partial F}{\partial y_i} - \frac{d}{dx}\left(\frac{\partial F}{\partial y_1'}\right) = 0 \quad (i = 1, \ldots, k)$$

D. CONSTANTS

Universal ideal gas constant	$\overline{R} = 8.314$ kJ/kmol · K
	$= 1.9872$ cal/mol · K
	$= 1.9872$ Btu/lbmol · °R
	$= 1545.33$ ft · lbf/lbmol · °R
Boltzmann's constant	$k = 1.38054 \times 10^{-23}$ J/K
Planck's constant	$h = 6.626 \times 10^{-34}$ J · s
Speed of light in vacuum	$c = 2.998 \times 10^{8}$ m/s
Avogadro's number	$N = 6.022 \times 10^{23}$ molecules/mol
Stefan–Boltzmann constant	$\sigma = 5.669 \times 10^{-8}$ W/m^2· K^4
	$= 0.1714 \times 10^{-8}$ Btu/h · ft^2 · °R^4
Atmospheric pressure	$P_{\text{atm}} = 0.101325$ MPa
	$= 1.01325$ bar
	$= 1.01325 \times 10^{5}$ N/m^2
Ice point at 1 atm	$T_{\text{ice}} = 0$ °C $= 273.15$ K
Gravitational acceleration	$g = 9.807$ m/s^2
	$= 32.17$ ft/s^2

Calorie	1 cal	= 4.187 J
Mole	1 mol	= sample containing 6.022 $\times 10^{23}$ elementary entities (e.g., molecules); also abbreviated as 1 gmol, or
	1 mol	$= 10^{-3}$ kmol
		$= \frac{1}{453.6}$ lbmol
Important numbers	$\ln x$	$= 2.30258 \log_{10} x$
	$\log_{10} x$	$= 0.4343 \ln x$
	e	= 2.71828
	π	= 3.14159
	1°	= 0.01745 rad

E. CONVERSION FACTORS

Acceleration	1 m/s^2	$= 4.252 \times 10^7$ ft/h^2
Area	1 in^2	= 6.452 cm^2
	1 ft^2	= 0.0929 m^2
	1 yd^2	= 0.8361 m^2
	1 mi^2	= 2.59 km^2
	1 hectare	= $(100\ m)^2$
	1 acre	= 4047 m^2
Density	1 kg/m^3	= 0.06243 lbm/ft^3
	1 lbm/ft^3	= 16.018 kg/m^3
Energy	1 kJ	= 737.56 ft · lbf
		= 0.9478 Btu
		$= 3.725 \times 10^{-4}$ hp · h
		$= 2.778 \times 10^{-4}$ kW · h
	1 Btu	= 1055 J
		= 778.16 ft · lbf
		= 3412.14 kW · h
		= 2544.5 hp · h
	1 cal	= 4.187 J
	1 erg	$= 10^{-7}$ J
Force	1 lbf	= 4.448 N
		= 0.4536 kgf
	1 dyne	$= 10^{-5}$ N
Heat flux	1 W/m^2	= 0.317 Btu/h · ft^2
	1 Btu/h · ft^2	= 3.154 W/m^2

Quantity	Unit	Equivalent
Heat transfer coefficient	$1\ W/m^2 \cdot K$	$= 0.1761\ Btu/h \cdot ft^2 \cdot °F$
		$= 0.8598\ kcal/h \cdot m^2 \cdot °C$
	$1\ Btu/h \cdot ft^2 \cdot °F$	$= 5.6786\ W/m^2 \cdot K$
Heat transfer rate	1 Btu/s	= 1055 W
	1 Btu/h	= 0.2931 W
	1 hp	= 745.7 W
	$1\ ft \cdot lbf/s$	= 1.3558 W
Kinematic viscosity (ν), thermal diffusivity (α), mass diffusivity (D)	$1\ m^2/s$	$= 10^4\ cm^2/s$
		$= 10^4$ stokes
		$= 3.875 \times 10^4\ ft^2/h$
		$= 10.764\ ft^2/s$
Latent heat, specific energy, specific enthalpy	1 kJ/kg	= 0.4299 Btu/lbm
		= 0.2388 cal/g
	1 Btu/lbm	= 2.326 kJ/kg
Length	1 in	= 2.54 cm
	1 ft	= 0.3048 m
	1 yd	= 0.9144 m
	1 mile	= 1.609 km
Mass	1 lbm	= 0.4536 kg
	1 kg	= 2.2046 lbm
		$= 1.1023 \times 10^{-3}$ U.S. ton
		$= 10^{-3}$ tonne
	1 oz	= 28.35 g
Mass transfer coefficient	1 m/s	$= 1.181 \times 10^4$ ft/h
	1 ft/h	$= 8.467 \times 10^{-5}$ m/s
Power	1 Btu/s	= 1055 W = 1.055 kW
	1 Btu/h	= 0.293 W
	1 W	= 3.412 Btu/h
		$= 9.48 \times 10^{-4}$ Btu/s
	1 HP	= 0.746 kW
		= 0.707 Btu/s
Pressure, stress	1 Pa	$= 1\ N/m^2$
	1 psi	$= 6895\ N/m^2$
	1 atm	= 14.69 psi
		$= 1.013 \times 10^5\ N/m^2$
	1 bar	$= 10^5\ N/m^2$
	1 torr	= 1 mmHg
		$= 133.32\ N/m^2$
	1 psi	= 27.68 in H_2O
	1 ft H_2O	= 0.4335 psi

Specific heat, specific entropy	1 kJ/kg · K	= 0.2388 Btu/lbm · °F
		= 0.2389 cal/g · °C
	1 Btu/lbm · °F	= 4.187 kJ/kg · K
Speed	1 mi/h	= 0.447 m/s
		= 1.609 km/h
	1 km/h	= 0.278 m/s
		= 0.622 mi/h
	1 m/s	= 3.6 km/h
		= 2.237 mi/h
Temperature	1 K	= 1°C
	1 K	= (9/5) °F
	T (K)	= T (°C) + 273.15
	T (°C)	= 5/9 [T (°F) – 32]
	T (°F)	= T (°R) – 459.67
Temperature difference	ΔT (K)	= ΔT (°C)
		= 5/9 ΔT (°F)
		= 5/9 ΔT (°R)
Thermal conductivity	1 W/m · K	= 0.5782 Btu/h · ft · °F
		= 0.01 W/cm · K
		= 2.39×10^{-3} cal/cm · s · °C
	1 Btu/h · ft · °F	= 1.7307 W/m · K
Thermal resistance	1 K/W	= 0.5275°F/Btu · h
	1°F/Btu · h	= 1.896 K/W
Viscosity (μ)	1 N · s/m^2	= 1 kg/s · m
		= 2419.1 lbm/ft · h
		= 5.802×10^{-6} lbf · h/ft^2
	1 poise	= 1 g/s · cm
Volume	1 L	= 10^{-3} m^3 = 1 dm^3
	1 in^3	= 16.39 cm^3
	1 ft^3	= 0.02832 m^3
	1 yd^3	= 0.7646 m^3
	1 gal (U.S.)	= 3.785 L
	1 gal (impcrial)	= 4.546 L
	1 pint	= 0.5683 L
Volumetric heat generation rate	1 Btu/h · ft^3	= 10.35 W/m^3
	1 W/m^3	= 0.0966 Btu/h · ft^3

F. DIMENSIONLESS GROUPS*

Bejan number	$\mathrm{Be} = (\Delta P L^2)/\mu\alpha$
Biot number	$\mathrm{Bi} = hL/k_s$
Fourier number	$\mathrm{Fo} = \alpha t/L^2$
Nusselt number	$\mathrm{Nu} = hL/k_f$
Péclet number	$\mathrm{Pe} = UL/\alpha = \mathrm{Re\,Pr}$
Prandtl number	$\mathrm{Pr} = \nu/\alpha = \mathrm{Sc/Le}$
Rayleigh number	$\mathrm{Ra} = g\beta\Delta TH^3/\alpha\nu$
Rayleigh number based on heat flux	$\mathrm{Ra}_* = g\beta q''H^4/\alpha\nu k$
Reynolds number	$\mathrm{Re} = UL/\nu$
Svelteness number	Sv = (external length scale) / (internal length scale)

*Subscripts: $(\cdot)_s$ = solid, $(\cdot)_f$ = fluid.

G. NONMETALLIC SOLIDS

Material	T (°C)	ρ (kg/m^3)	c_P (kJ/kg · K)	k (W/m · K)	α (cm^2/s)
Asbestos					
Cement board	20			0.6	
Felt (16 laminations/cm)	40			0.057	
Fiber	50	470	0.82	0.11	0.0029
Sheet	20			0.74	
	50			0.17	
Asphalt	20	2120	0.92	0.70	0.0036
Bakelite	20	1270	1.59	0.230	0.0011
Bark	25	340	1.26	0.074	0.0017
Beef (*see* Meat)					
Brick					
Carborundum	1400			11.1	
Cement	10	720		0.34	
Chrome	100			1.9	
Common	20	1800	0.84	0.38–0.52	0.0028–0.0034
Facing	20			1.3	
Firebrick	300	2000	0.96	0.1	0.00054
Magnesite (50% MgO)	20	2000		2.68	
Masonry	20	1700	0.84	0.66	0.0046

(*Continued*)

NONMETALLIC SOLIDS (*Continued*)

Material	T (°C)	ρ (kg/m^3)	c_P (kJ/kg · K)	k (W/m · K)	α (cm^2/s)
Silica (95% SiO_2)	20	1900		1.07	
Zircon (62% ZrO_2)	20	3600		2.44	
Brickwork, dried in air	20	1400–1800	0.84	0.58–0.81	0.0049–0.0054
Carbon Diamond (type IIb)	20	3250	0.51	1350	8.1
Graphite (firm, natural)	20	2000–2500	0.61	155	1.02–1.27
Carborundum (SiC)	100	1500	0.62	58	0.62
Cardboard	0–20	~ 790		~ 0.14	
Celluloid	20	1380	1.67	0.23	0.001
Cement (Portland, fresh, dry)	20	3100	0.75	0.3	0.0013
Chalk ($CaCO_3$)	20	2000–3000	0.74	2.2	0.01–0.015
Clay	20	1450	0.88	1.28	0.01
Fireclay	100	1700–2000	0.84	0.5–1.2	0.35–0.71
Sandy clay	20	1780		0.9	
Coal	20	1200–1500	1.26	0.26	0.0014–0.0017
Anthracite	900	1500		0.2	
Brown coal	900			0.1	
Bituminous in situ		1300		0.5–0.7	0.003–0.004
Dust	30	730	1.3	0.12	0.0013
Concrete, made with gravel, dry	20	2200	0.88	1.28	0.0066
cinder	24			0.76	
Cork					
Board	20	150	1.88	0.042	0.0015
Expanded	20	120		0.036	
Cotton	30	81	1.15	0.059	0.0063
Earth					
Coarse-grained	20	2040	1.84	0.59	0.0016
Clayey (28% moisture)	20	1500		1.51	
Sandy (8% moisture)	20	1500		1.05	
Diatomaceous	20	466	0.88	0.126	0.0031
Fat	20	910	1.93	0.17	0.001
Felt, hair	–7	130–200		0.032–0.04	
	94	130–200		0.054–0.051	
Fiber insulating board	20	240		0.048	

NONMETALLIC SOLIDS (*Continued*)

Material	T (°C)	ρ (kg/m^3)	c_P (kJ/kg · K)	k (W/m · K)	α (cm^2/s)
Glass					
Borosilicate	30	2230		1.09	
Fiber	20	220		0.035	
Lead	20	2890	0.68	0.7–0.93	0.0036–0.0047
Mirror	20	2700	0.80	0.76	0.0035
Pyrex	60–100	2210	0.75	1.3	0.0078
Quartz	20	2210	0.73	1.4	0.0087
Window	20	2800	0.80	0.81	0.0034
Wool	0	200	0.66	0.037	0.0028
Granite	20	2750	0.89	2.9	0.012
Gypsum	20	1000	1.09	0.51	0.0047
Ice	0	917	2.04	2.25	0.012
Ivory	80			0.5	
Kapok	30			0.035	
Leather, dry	20	860	1.5	0.12–0.15	~ 0.001
Limestone (Indiana)	100	2300	0.9	1.1	~ 0.005
Linoleum	20	535		0.081	
Lunar surface dust, in high vacuum	250	1500 ± 300	~ 0.6	~ 0.0006	
Magnezia (85%)	38–204			0.067–0.08	
Marble	20	2600	0.81	2.8	0.013
Meat	25	~ 1060		0.3–0.6	~ 0.0014
Mica	20	2900		0.52	
Mortar	20	1900	0.8	0.93	0.0061
Paper	20	700	1.2	0.12	0.0014
Paraffin	30	870–925	2.9	0.24–0.27	~ 0.001
Plaster	20	1690	0.8	0.79	0.0058
Plexiglas (acrylic glass)	20	1180	1.44	0.184	0.0011
Polyethylene	20	920	2.30	0.35	0.0017
Polystyrene	20	1050		0.157	
Polyurethane	20	1200	2.09	0.32	0.0013
Polyvinyl chloride (PVC)	20	1380	0.96	0.15	0.0011
Porcelain	95	2400	1.08	1.03	0.004
Quartz	20	2100–2500	0.78	1.40	~ 0.008
Rubber					
Foam	20	500	1.67	0.09	0.0011
Hard (ebonite)	20	1150	2.01	0.16	0.0006
Soft	20	1100	1.67	~ 0.2	~ 0.001
Synthetic	20	1150	1.97	0.23	0.001

(*Continued*)

NONMETALLIC SOLIDS (*Continued*)

Material	T (°C)	ρ (kg/m^3)	c_P (kJ/kg · K)	k (W/m · K)	α (cm^2/s)
Salt (rock salt)	0	2100–2500	0.92	7	0.03–0.036
Sand					
Dry	20			0.58	
Moist	20	1640		1.13	
Sandstone	20	2150–2300	0.71	1.6–2.1	0.01–0.013
Sawdust, dry	20	215		0.07	
Silica aerogel	0	140		0.024	
Silica stone (85% SiC)	700	2720	1.05	1.56	0.055
Silicon	20	2330	0.703	153	0.94
Silk (artificial)	35	100	1.33	0.049	0.0037
Slag	20	2500–3000	0.84	0.57	0.0023–0.0027
Slate					
Parallel to lamination	20	2700	0.75	2.9	0.014
Perpendicular to lamination	20	2700	0.75	1.83	0.009
Snow, firm	0	560	2.1	0.46	0.0039
Soil (*see also* Earth)					
Dry	15	1500	1.84	1	0.004
Wet	15	1930		2	
Strawberries, dry	−18			0.59	
Sugar (fine)	0	1600	1.25	0.58	0.0029
Sulfur	20	2070	0.72	0.27	0.0018
Teflon (polytetrafluoroethylene)	20	2200	1.04	0.23	0.001
Wood, perpendicular to grain					
Ash	15	740		0.15–0.3	
Balsa	15	100		0.05	
Cedar	15	480		0.11	
Mahogany	20	700		0.16	
Oak	20	600–800	2.4	0.17–0.25	~ 0.0012
Pine, fir, spruce	20	416–421	2.72	0.15	0.0012
Plywood	20	590		0.11	
Wool					
Sheep	20	100	1.72	0.036	0.0021
Mineral	50	200	0.92	0.046	0.0025
Slag	25	200	0.8	0.05	0.0031

Source: Constructed based on data compiled in Refs. 2–10.

H. METALLIC SOLIDS

Metals, Alloys	Properties at 20°C (293 K)				Thermal Conductivity k (W/m · K)						
	ρ (kg/m^3)	c_P (kJ/kg · K)	k (W/m · K)	α (cm^2/s)	–100°C (173 K)	0°C (273 K)	100°C (373 K)	200°C (473 K)	400°C (673 K)	600°C (873 K)	1000°C (1273 K)
Aluminum											
Pure	2,707	0.896	204	0.842	215	202	206	215	249		
Duralumin (94–96% Al, 3–5% Cu, trace Mg)	2,787	0.883	164	0.667	126	159	182	194			
Silumin (87% Al, 13% Si)	2,659	0.871	164	0.710	149	163	175	185			
Antimony	6,690	0.208	17.4	0.125	19.2	17.7	16.3	16.0	17.2		
Beryllium	1,850	1.750	167	0.516	126	160	191	215			
Bismuth, polycrystalline	9,780	0.124	7.9	0.065	12.1	8.4	7.2	7.2			
Cadmium, polycrystalline	8,650	0.231	92.8	0.464	97	93	92	91			
Cesium	1,873	0.230	36	0.836							
Chromium	7,190	0.453	90	0.276	120	95	88	85	77		
Cobalt (97.1% Co), polycrystalline	8,900	0.389	70	0.202							
Copper											
Pure	8,954	0.384	398	1.16	420	401	391	389	378	366	336
Commercial	8,300	0.419	372	1.07							
Aluminum bronze (95% Cu, 5% Al)	8,666	0.410	83	0.233							
Brass (70% Cu, 30% Zn)	8,522	0.385	111	0.341	88		128	144	147		

(Continued)

H. METALLIC SOLIDS (*Continued*)

	Properties at 20°C (293 K)				Thermal Conductivity k (W/m · K)						
Metals, Alloys	ρ (kg/m^3)	c_P (kJ/kg · K)	k (W/m · K)	α (cm^2/s)	−100°C (173 K)	0°C (273 K)	100°C (373 K)	200°C (473 K)	400°C (673 K)	600°C (873 K)	1000°C (1273 K)
Brass (60% Cu, 40% Zn)	8, 400	0.376	113	0.358							
Bronze (75% Cu, 25% Sn)	8, 666	0.343	26	0.086							
Bronze (85% Cu, 6% Sn, 9% Zn, 1% Pb)	8, 800	0.377	61.7	0.186							
Constantan (60% Cu, 40% Ni)	8, 922	0.410	22.7	0.061	21		22.2	26			
German silver (62% Cu, 15% Ni, 22% Zn)	8, 618	0.394	24.9	0.073	19.2		31	40	48		
Gold	19, 300	0.129	315	1.27		318		309			
Iron											
Pure	7, 897	0.452	73	0.205	87	73	67	62	48	40	35
Cast (5% C)	7, 272	0.420	52	0.170							
Carbon steel, 0.5% C	7, 833	0.465	54	0.148		55	52	48	42	35	29
1.0% C	7, 801	0.473	43	0.117		43	43	42	36	33	28
1.5% C	7, 753	0.486	36	0.097		36	36	36	33	31	28
Chrome steel, 1% Cr	7, 865	0.460	61	0.167		62	55	52	42	36	33
5% Cr	7, 833	0.460	40	0.111		40	38	36	33	29	29
20% Cr	7, 689	0.460	22	0.064		22	22	22	24	24	29
Chrome–nickel steel, 15% Cr, 10% Ni	7, 865	0.460	19	0.053							
20% Cr, 15% Ni	7, 833	0.460	15.1	0.042							

Invar (36% Ni)	8,137	0.460	10.7	0.029							
Manganese steel, 1% Mn	7,865	0.460	50	0.139							
5% Mn	7,849	0.460	22	0.064							
Nickel–chrome steel											
80% Ni, 15% Cr	8,522	0.460	17	0.045							
20% Ni, 15% Cr	7,865	0.460	14	0.039		14	15.1	15.1	17	19	
Silicon steel, 1% Si	7,769	0.460	42	0.116							
5% Si	7,417	0.460	19	0.056							
Stainless steel, type 304	7,817	0.460	13.8	0.040			15	17	21	25	
type 347	7,817	0.420	15	0.044	13		16	18	20	23	28
Tungsten steel, 2% W	7,961	0.444	62	0.176		62	59	54	48	45	36
10% W	8,314	0.419	48	0.139							
Wrought (0.5% CH)	7,849	0.460	59	0.163		59	57	52	45	36	33
Lead	11,340	0.130	34.8	0.236	36.9	35.1	33.4	31.6	23.3		
Lithium	530	3.392	61	0.340		61	61				
Magnesium Pure	1,746	1.013	171	0.970	178	171	168	163			
6–8% Al, 1–2% Zn, electrolytic	1,810	1.000	66	0.360		52	62	74			
2% Mn	1,778	1.000	114	0.640	93	111	125	130			
Manganese											
Pure	7,300	0.486	7.8	0.022							
Manganin (85% Cu, 4% Ni, 12% Mn)	8,400	0.406	21.9	0.064							
Molybdenum	10,220	0.251	123	0.480	138	125	118	114	109	106	99
Monel 505 (at 60°C)	8,360	0.544	19.7	0.043							
Nickel											
Pure	8,906	0.445	91	0.230	114	94	83	74	64	69	78
Nichrome (24% Fe, 16% Cr)	8,250	0.448	12.6	0.034							
90% Ni, 10% Cr	8,666	0.444	17	0.044		17.1	18.9	20.9	24.6		

(Continued)

H. METALLIC SOLIDS (*Continued*)

	Properties at 20°C (293 K)				Thermal Conductivity k (W/m · K)						
Metals, Alloys	ρ (kg/m^3)	c_P (kJ/kg · K)	k (W/m · K)	α (cm^2/s)	–100°C (173 K)	0°C (273 K)	100°C (373 K)	200°C (473 K)	400°C (673 K)	600°C (873 K)	1000°C (1273 K)
Niobium	8, 570	0.270	53	0.230							
Palladium	12, 020	0.247	75.5	0.254		75.5	75.5	75.5	75.5		
Platinum	21, 450	0.133	71.4	0.250	73	72	72	72	74	77	84
Potassium	860	0.741	103	1.62							
Rhenium	21, 100	0.137	48.1	0.166							
Rhodium	12, 450	0.248	150	0.486							
Rubidium	1, 530	0.348	58.2	1.09							
Silver, 99.99% Ag	10, 524	0.236	427	1.72	431	428	422	417	401	386	
99.90% Ag	10, 524	0.236	411	1.66	422	405		373	364		
Sodium	971	1.206	133	1.14							
Tantalum	16, 600	0.138	57.5	0.251		57.4					
Tin, polycrystalline	7, 304	0.220	67	0.417	76	68	63				
Titanium, polycrystalline	4, 540	0.523	22	0.093	26	22	21	20	19	21	22
Tungsten, polycrystalline	19, 300	0.134	179	0.692		182					
Uranium	18, 700	0.116	28	0.129	24	27	29	31	36	41	
Vanadium	6, 100	0.502	31.4	0.103		31.3					
Wood's metal (50% Bi, 25% Pb, 12.4% VCd, 12.5% Sn)	1, 056	0.147	12.8	0.825							
Zinc	7, 144	0.388	121	0.437	122	122	117	110	100		
Zirconium, polycrystalline	6, 570	0.272	22.8	0.128		23.2					

Source: Constructed based on data compiled in Refs. 2–8.

I. POROUS MATERIALS

Material	Porosity ϕ	Permeability K (cm^2)	Contact Surface per Volume (cm^{-1})
Agar-agar		2×10^{-10}–4.4×10^{-9}	
Black slate powder	0.57–0.66	4.9×10^{-10}–1.2×10^{-9}	7×10^3–8.9×10^3
Brick	0.12–0.34	4.8×10^{-10}–2.2×10^{-9}	
Catalyst (Fischer-Tropsch, granules only)	0.45		5.6×10^5
Cigarette		1.1×10^{-5}	
Cigarette filters	0.17–0.49		
Coal	0.02–0.12		
Concrete (ordinary mixes)	~0.10		
Concrete (bituminous)		1×10^{-9}–2.3×10^{-7}	
Copper powder (hotcompacted)	0.09–0.34	3.3×10^{-6}–1.5×10^{-5}	
Cork board		2.4×10^{-7}–5.1×10^{-7}	
Fiberglass	0.88–0.93		560–770
Granular crushed rock	0.45		
Hair (on mammals)	0.95–0.99		
Hair felt		8.3×10^{-6}–1.2×10^{-5}	
Leather	0.56–0.59	9.5×10^{-10}–1.2×10^{-9}	1.2×10^4–1.6×10^4
Limestone (dolomite)	0.04–0.10	2×10^{-11}–4.5×10^{-10}	
Sand	0.37–0.50	2×10^{-7}–1.8×10^{-6}	150–220
Sandstone ("oil sand")	0.08–0.38	5×10^{-12}–3×10^{-8}	
Silica grains	0.65		
Silica powder	0.37–0.49	1.3×10^{-10}–5.1×10^{-10}	6.8×10^3–8.9×10^3
Soil	0.43–0.54	2.9×10^{-9}–1.4×10^{-7}	
Spherical packings (well shaken)	0.36–0.43		
Wire crimps	0.68–0.76	3.8×10^{-5}–1×10^{-4}	29–40

Source: Refs. 11–14.

J. LIQUIDS

Water at Atmospheric Pressure

T (°C)	ρ (g/cm^3)	c_P (kJ/kg · K)	c_v (kJ/kg · K)	h_{fg} (kJ/kg)	β (K^{-1})
0	0.9999	4.217	4.215	2501	-0.6×10^{-4}
5	1	4.202	4.202	2489	$+0.1 \times 10^{-4}$
10	0.9997	4.192	4.187	2477	0.9×10^{-4}
15	0.9991	4.186	4.173	2465	1.5×10^{-4}
20	0.9982	4.182	4.158	2454	2.1×10^{-4}
25	0.9971	4.179	4.138	2442	2.6×10^{-4}
30	0.9957	4.178	4.118	2430	3.0×10^{-4}
35	0.9941	4.178	4.108	2418	3.4×10^{-4}
40	0.9923	4.178	4.088	2406	3.8×10^{-4}
50	0.9881	4.180	4.050	2382	4.5×10^{-4}
60	0.9832	4.184	4.004	2357	5.1×10^{-4}
70	0.9778	4.189	3.959	2333	5.7×10^{-4}
80	0.9718	4.196	3.906	2308	6.2×10^{-4}
90	0.9653	4.205	3.865	2283	6.7×10^{-4}
100[a]	0.9584	4.216	3.816	2257	7.1×10^{-4}

T (°C)	μ (g/cm · s)	ν (cm^2/s)	k (W/m · K)	α (cm^2/s)	Pr	$\frac{g\beta}{\alpha\nu} = \frac{\mathrm{Ra}_H}{H^3\Delta T}$ (K^{-1} · cm^{-3})
0	0.01787	0.01787	0.56	0.00133	13.44	-2.48×10^3
5	0.01514	0.01514	0.57	0.00136	11.13	$+0.47 \times 10^3$
10	0.01304	0.01304	0.58	0.00138	9.45	4.91×10^3
15	0.01137	0.01138	0.59	0.00140	8.13	9.24 × 103
20	0.01002	0.01004	0.59	0.00142	7.07	14.45×10^3
25	0.00891	0.00894	0.60	0.00144	6.21	19.81×10^3
30	0.00798	0.00802	0.61	0.00146	5.49	25.13×10^3
35	0.00720	0.00725	0.62	0.00149	4.87	30.88×10^3
40	0.00654	0.00659	0.63	0.00152	4.34	37.21×10^3
50	0.00548	0.00554	0.64	0.00155	3.57	51.41×10^3
60	0.00467	0.00475	0.65	0.00158	3.01	66.66×10^3
70	0.00405	0.00414	0.66	0.00161	2.57	83.89×10^3
80	0.00355	0.00366	0.67	0.00164	2.23	101.3×10^3
90	0.00316	0.00327	0.67	0.00165	1.98	121.8×10^3
100	0.00283	0.00295	0.68	0.00166	1.78	142.2×10^3

Source: Data collected from Refs. 4, 15 and 16.

[a] Saturated.

Water at Saturation Pressure

T (°C)	ρ (g/cm^3)	c_P (kJ/kg · K)	μ (g/cm · s)	ν (cm^2/s)	k (W/m · K)	α (cm^2/s)	Pr
0	0.9999	4.226	0.0179	0.0179	0.56	0.0013	13.7
10	0.9997	4.195	0.0130	0.0130	0.58	0.0014	9.5
20	0.9982	4.182	0.0099	0.0101	0.60	0.0014	7
40	0.9922	4.175	0.0066	0.0066	0.63	0.0015	4.3
60	0.9832	4.181	0.0047	0.0048	0.66	0.0016	3
80	0.9718	4.194	0.0035	0.0036	0.67	0.0017	2.25
100	0.9584	4.211	0.0028	0.0029	0.68	0.0017	1.75
150	0.9169	4.270	0.00185	0.0020	0.68	0.0017	1.17
200	0.8628	4.501	0.00139	0.0016	0.66	0.0017	0.95
250	0.7992	4.857	0.00110	0.00137	0.62	0.0016	0.86
300	0.7125	5.694	0.00092	0.00128	0.56	0.0013	0.98
340	0.6094	8.160	0.00077	0.00127	0.44	0.0009	1.45
370	0.4480	11.690	0.00057	0.00127	0.29	0.00058	2.18

Source: Data collected from Refs. 4 and 15.

Ammonia, Saturated Liquid

T (°C)	ρ (g/cm^3)	c_P (kJ/kg · K)	μ (kg/s · m)	ν (cm^2/s)	k (W/m · K)	α (cm^2/s)	Pr
−50	0.704	4.46	3.06×10^{-4}	4.35×10^{-3}	0.547	1.74×10^{-3}	2.50
−25	0.673	4.49	2.58×10^{-4}	3.84×10^{-3}	0.548	1.81×10^{-3}	2.12
0	0.640	4.64	2.39×10^{-4}	3.73×10^{-3}	0.540	1.82×10^{-3}	2.05
25	0.604	4.84	2.14×10^{-4}	3.54×10^{-3}	0.514	1.76×10^{-3}	2.01
50	0.564	5.12	1.86×10^{-4}	3.30×10^{-3}	0.476	1.65×10^{-3}	2.00

Source: Constructed based on data compiled in Ref. 2.

Carbon Dioxide, Saturated Liquid

T (K)	P (bar)	ρ (g/cm^3)	c_P (kJ/kg · K)	μ (kg/s · m)	ν (cm^2/s)	k (W/m · K)	α (cm^2/s)	Pr
216.6	5.18	1.179	1.707	2.10×10^{-4}	1.78×10^{-3}	0.182	9.09×10^{-4}	1.96
220	6.00	1.167	1.761	1.86×10^{-4}	1.59×10^{-3}	0.178	8.26×10^{-4}	1.93
240	12.83	1.089	1.933	1.45×10^{-4}	1.33×10^{-3}	0.156	7.40×10^{-4}	1.80
260	24.19	1.000	2.125	1.14×10^{-4}	1.14×10^{-3}	0.128	6.03×10^{-4}	1.89
280	41.60	0.885	2.887	0.91×10^{-4}	1.03×10^{-3}	0.102	4.00×10^{-4}	2.57
300	67.10	0.680		0.71×10^{-4}	1.04×10^{-3}	0.081		
304.2	73.83	0.466		0.60×10^{-4}	1.29×10^{-3}	0.074		

Source: Constructed based on data compiled in Refs. 17 and 18.

Fuels, Liquids at $P \cong 1$ atm

T (°C)	ρ (g/cm^3)	c_P (kJ/kg · K)	μ (kg/s · m)	ν (cm^2/s)	k (W/m · K)	α (cm^2/s)	Pr
			Gasoline				
20	0.751	2.06	5.29×10^{-4}	7.04×10^{-3}	0.1164	7.52×10^{-4}	9.4
50	0.721	2.20	3.70×10^{-4}	5.13×10^{-3}	0.1105	6.97×10^{-4}	7.4
100	0.681	2.46	2.25×10^{-4}	3.30×10^{-3}	0.1005	6.00×10^{-4}	5.5
150	0.628	2.74	1.56×10^{-4}	2.48×10^{-3}	0.0919	5.34×10^{-4}	4.6
200	0.570	3.04	1.11×10^{-4}	1.95×10^{-3}	0.0800	4.62×10^{-4}	4.2
			Kerosene				
20	0.819	2.00	1.49×10^{-3}	1.82×10^{-2}	0.1161	7.09×10^{-4}	25.7
50	0.801	2.14	9.56×10^{-4}	1.19×10^{-2}	0.1114	6.50×10^{-4}	18.3
100	0.766	2.38	5.45×10^{-4}	7.11×10^{-3}	0.1042	5.72×10^{-4}	12.4
150	0.728	2.63	3.64×10^{-4}	5.00×10^{-3}	0.0965	5.04×10^{-4}	9.9
200	0.685	2.89	2.62×10^{-4}	3.82×10^{-3}	0.0891	4.50×10^{-4}	8.5
250	0.638	3.16	2.01×10^{-4}	3.15×10^{-3}	0.0816	4.05×10^{-4}	7.8

Source: Constructed based on data compiled in Ref. 19.

Helium, Liquid at $P = 1$ atm

T (K)	ρ (g/cm^3)	c_P (kJ/kg · K)	μ (Kg/s · m)	ν (cm^2/s)	k (W/m · K)	α (cm^2/s)	Pr
2.5	0.147	2.05	3.94×10^{-6}	2.68×10^{-4}	0.0167	5.58×10^{-4}	0.48
3	0.143	2.36	3.86×10^{-6}	2.69×10^{-4}	0.0182	5.38×10^{-4}	0.50
3.5	0.138	3.00	3.64×10^{-6}	2.64×10^{-4}	0.0191	4.62×10^{-4}	0.57
4	0.130	4.07	3.34×10^{-6}	2.57×10^{-4}	0.0196	3.71×10^{-4}	0.69
4.22[a]	0.125	4.98	3.17×10^{-6}	2.53×10^{-4}	0.0196	3.15×10^{-4}	0.80

Source: Data collected from Ref. 20.
[a]Saturated.

Lithium, Saturated Liquid

T (K)	P (bar)	ρ (g/cm^3)	c_P (kJ/kg · K)	μ (kg/s · m)	ν (cm^2/s)	k (W/m · K)	α (cm^2/s)	Pr
600	4.2×10^{-9}	0.503	4.23	4.26×10^{-4}	0.0085	47.6	0.223	0.038
800	9.6×10^{-6}	0.483	4.16	3.10×10^{-4}	0.0064	54.1	0.270	0.024
1000	9.6×10^{-4}	0.463	4.16	2.47×10^{-4}	0.0053	60.0	0.312	0.017
1200	0.0204	0.442	4.14	2.07×10^{-4}	0.0047	64.7	0.355	0.013
1400	0.1794	0.422	4.19	1.80×10^{-4}	0.0043	68.0	0.384	0.011

Source: Constructed based on data compiled in Refs. 17 and 18.

Mercury, Saturated Liquid

T (K)	P (bar)	ρ (g/cm^3)	c_P (kJ/kg · K)	μ (kg/s · m)	ν (cm^2/s)	k (W/m · K)	α (cm^2/s)	Pr	β (K^{-1})
260	6.9×10^{-8}	13.63	0.141	1.79×10^{-3}	1.31×10^{-3}	8.00	0.042	0.0316	1.8×10^{-4}
300	3.1×10^{-6}	13.53	0.139	1.52×10^{-3}	1.12×10^{-3}	8.54	0.045	0.0248	1.8×10^{-4}
340	5.5×10^{-5}	13.43	0.138	1.34×10^{-3}	1.00×10^{-3}	9.06	0.049	0.0205	1.8×10^{-4}
400	1.4×10^{-3}	13.29	0.137	1.17×10^{-3}	8.83×10^{-4}	9.80	0.054	0.0163	1.8×10^{-4}
500	0.053	13.05	0.135	1.01×10^{-3}	7.72×10^{-4}	10.93	0.062	0.0125	1.8×10^{-4}
600	0.578	12.81	0.136	9.10×10^{-4}	7.10×10^{-4}	11.94	0.071	0.0100	1.9×10^{-4}
800	11.18	12.32	0.140	8.08×10^{-4}	6.56×10^{-4}	13.57	0.079	0.008	1.9×10^{-4}
1000	65.74	11.79	0.149	7.54×10^{-4}	6.40×10^{-4}	14.69	0.084	0.0076	1.9×10^{-4}

Source: Constructed based on data compiled in Refs. 17 and 18.

Nitrogen, Liquid at $P = 1$ atm

T (K)	ρ (g/cm^3)	c_P (kJ/kg · K)	μ (kg/s · m)	ν (cm^2/s)	k (W/m · K)	α (cm^2/s)	Pr
65	0.861	1.988	2.77×10^{-5}	3.21×10^{-3}	0.161	9.39×10^{-4}	3.42
70	0.840	2.042	2.12×10^{-5}	2.53×10^{-3}	0.151	8.77×10^{-4}	2.88
75	0.819	2.059	1.77×10^{-5}	2.17×10^{-3}	0.141	8.36×10^{-4}	2.59
77.3[a]	0.809	2.065	1.64×10^{-5}	2.03×10^{-3}	0.136	8.15×10^{-4}	2.49

Source: Interpolated from data in Ref. 21.
[a]Saturated.

Potassium, Saturated Liquid

T (K)	P (bar)	ρ (g/cm^3)	c_P (kJ/kg · K)	μ (kg/s · m)	ν (cm^2/s)	k (W/m · K)	α (cm^2/s)	Pr
400	1.84×10^{-7}	0.814	0.805	4.13×10^{-4}	0.0051	52.0	0.794	0.0064
600	9.26×10^{-4}	0.767	0.771	2.38×10^{-4}	0.0031	43.9	0.742	0.0042
800	0.0612	0.720	0.761	1.71×10^{-4}	0.0024	37.1	0.677	0.0035
1000	0.7322	0.672	0.792	1.35×10^{-4}	0.0020	31.3	0.589	0.0034
1200	3.963	0.623	0.846	1.14×10^{-4}	0.0018	26.3	0.499	0.0037
1400	12.44	0.574	0.899	0.98×10^{-4}	0.0017	21.5	0.416	0.0041

Source: Constructed based on data compiled in Refs. 17 and 18.

Sodium, Saturated Liquid

T (K)	P (bar)	ρ (g/cm^3)	c_P (kJ/kg · K)	μ (kg/s · m)	v (cm^2/s)	k (W/m · K)	α (cm^2/s)	Pr
500	7.64×10^{-7}	0.898	1.330	4.24×10^{-4}	0.0047	80.0	0.67	0.0070
600	5.05×10^{-5}	0.873	1.299	3.28×10^{-4}	0.0038	75.4	0.66	0.0057
700	9.78×10^{-4}	0.850	1.278	2.69×10^{-4}	0.0032	70.7	0.65	0.0049
800	0.00904	0.826	1.264	2.30×10^{-4}	0.0028	65.9	0.63	0.0044
900	0.0501	0.802	1.258	2.02×10^{-4}	0.0025	61.4	0.61	0.0041
1000	0.1955	0.776	1.259	1.81×10^{-4}	0.0023	56.7	0.58	0.0040
1200	1.482	0.729	1.281	1.51×10^{-4}	0.0021	54.5	0.58	0.0036
1400	6.203	0.681	1.330	1.32×10^{-4}	0.0019	52.2	0.58	0.0034
1600	17.98	0.633	1.406	1.18×10^{-4}	0.0019	49.9	0.56	0.0033

Source: Constructed based on data compiled in Refs. 17 and 18.

Unused Engine Oil

T (K)	ρ (g/cm^3)	c_P (kJ/kg · K)	μ (kg/s · m)	v (cm^2/s)	k (W/m · K)	α (cm^2/s)	Pr	β (K^{-1})
260	0.908	1.76	12.23	135	0.149	9.32×10^{-4}	144500	7×10^{-4}
280	0.896	1.83	2.17	24.2	0.146	8.90×10^{-4}	27200	7×10^{-4}
300	0.884	1.91	0.486	5.50	0.144	8.53×10^{-4}	6450	7×10^{-4}
320	0.872	1.99	0.141	1.62	0.141	8.13×10^{-4}	1990	7×10^{-4}
340	0.860	2.08	0.053	0.62	0.139	7.77×10^{-4}	795	7×10^{-4}
360	0.848	2.16	0.025	0.30	0.137	7.48×10^{-4}	395	7×10^{-4}
380	0.836	2.25	0.014	0.17	0.136	7.23×10^{-4}	230	7×10^{-4}
400	0.824	2.34	0.009	0.11	0.134	6.95×10^{-4}	155	7×10^{-4}

Source: Constructed based on data compiled in Refs. 17 and 18.

Critical Point Data

	Critical Temperature		Critical Pressure		Critical Specific
Liquid	K	°C	MPa	atm	Volume (cm^3/g)
Air	133.2	−140	3.77	37.2	2.9
Alcohol (ethyl)	516.5	243.3	6.39	63.1	3.6
Alcohol (methyl)	513.2	240	7.98	78.7	3.7
Ammonia	405.4	132.2	11.3	111.6	4.25
Argon	150.9	−122.2	4.86	48	1.88
Butane	425.9	152.8	3.65	36	4.4
Carbon dioxide	304.3	31.1	7.4	73	2.2
Carbon monoxide	134.3	−138.9	3.54	35	3.2

Liquid	Critical Temperature K	°C	Critical Pressure MPa	atm	Critical Specific Volume (cm^3/g)
Carbon tetrachloride	555.9	282.8	4.56	45	1.81
Chlorine	417	143.9	7.72	76.14	1.75
Ethane	305.4	32.2	4.94	48.8	4.75
Ethylene	282.6	9.4	5.85	57.7	4.6
Helium	5.2	−268	0.228	2.25	14.4
Hexane	508.2	235	2.99	29.5	4.25
Hydrogen	33.2	−240	1.30	12.79	32.3
Methane	190.9	−82.2	4.64	45.8	6.2
Methyl chloride	416.5	143.3	6.67	65.8	2.7
Neon	44.2	−288.9	2.7	26.6	2.1
Nitric oxide	179.3	−93.9	6.58	65	1.94
Nitrogen	125.9	−147.2	3.39	33.5	3.25
Octane	569.3	296.1	2.5	24.63	4.25
Oxygen	154.3	−118.9	5.03	49.7	2.3
Propane	368.7	95.6	4.36	43	4.4
Sulfur dioxide	430.4	157.2	7.87	77.7	1.94
Water	647	373.9	22.1	218.2	3.1

Source: Based on a compilation from Ref. 22.

K. GASES

Dry Air at Atmospheric Pressure

T (°C)	ρ (kg/m^3)	c_P (kJ/kg · K)	μ (kg/s · m)	ν (cm^2/s)	k (W/m · K)	α (cm^2/s)	Pr	$\frac{g\beta}{\alpha\nu} = \frac{\mathrm{Ra}_H}{H^3\Delta T}$ (cm^{-3} · K^{-1})
−180	3.72	1.035	6.50×10^{-6}	0.0175	0.0076	0.019	0.92	3.2×10^4
−100	2.04	1.010	1.16×10^{-5}	0.057	0.016	0.076	0.75	1.3×10^3
−50	1.582	1.006	1.45×10^{-5}	0.092	0.020	0.130	0.72	367
0	1.293	1.006	1.71×10^{-5}	0.132	0.024	0.184	0.72	148
10	1.247	1.006	1.76×10^{-5}	0.141	0.025	0.196	0.72	125
20	1.205	1.006	1.81×10^{-5}	0.150	0.025	0.208	0.72	107
30	1.165	1.006	1.86×10^{-5}	0.160	0.026	0.223	0.72	90.7
60	1.060	1.008	2.00×10^{-5}	0.188	0.028	0.274	0.70	57.1
100	0.946	1.011	2.18×10^{-5}	0.230	0.032	0.328	0.70	34.8
200	0.746	1.025	2.58×10^{-5}	0.346	0.039	0.519	0.68	9.53
300	0.616	1.045	2.95×10^{-5}	0.481	0.045	0.717	0.68	4.96
500	0.456	1.093	3.58×10^{-5}	0.785	0.056	1.140	0.70	1.42
1000	0.277	1.185	4.82×10^{-5}	1.745	0.076	2.424	0.72	0.18

Source: Data collected from Refs. 4, 15, and 16.

Ammonia, Gas at $P = 1$ atm

T (°C)	ρ (kg/m^3)	c_P (kJ/kg · K)	μ (kg/s · m)	ν (cm^2/s)	k (W/m · K)	α (cm^2/s)	Pr
0	0.793	2.18	9.35×10^{-6}	0.118	0.0220	0.131	0.90
50	0.649	2.18	1.10×10^{-5}	0.170	0.0270	0.192	0.88
100	0.559	2.24	1.29×10^{-5}	0.230	0.0327	0.262	0.87
150	0.493	2.32	1.47×10^{-5}	0.297	0.0391	0.343	0.87
200	0.441	2.40	1.65×10^{-5}	0.374	0.0467	0.442	0.84

Source: Constructed based on data compiled in Ref. 2.

Carbon Dioxide, Gas at $P = 1$ bar

T(K)	ρ (kg/m^3)	c_P (kJ/kg · K)	μ (kg/s · m)	ν (cm^2/s)	k (W/m · K)	α (cm^2/s)	Pr
300	1.773	0.852	1.51×10^{-5}	0.085	0.0166	0.109	0.78
350	1.516	0.898	1.75×10^{-5}	0.115	0.0204	0.150	0.77
400	1.326	0.941	1.98×10^{-5}	0.149	0.0243	0.195	0.77
500	1.059	1.014	2.42×10^{-5}	0.229	0.0325	0.303	0.76
600	0.883	1.075	2.81×10^{-5}	0.318	0.0407	0.429	0.74
700	0.751	1.126	3.17×10^{-5}	0.422	0.0481	0.569	0.74
800	0.661	1.168	3.50×10^{-5}	0.530	0.0551	0.714	0.74
900	0.588	1.205	3.81×10^{-5}	0.648	0.0618	0.873	0.74
1000	0.529	1.234	4.10×10^{-5}	0.775	0.0682	1.043	0.74

Source: Constructed based on data compiled in Refs. 17 and 18.

Helium, Gas at $P = 1$ atm

T(K)	ρ (kg/m^3)	c_P (kJ/kg · K)	μ (kg/s · m)	ν (cm^2/s)	k (W/m · K)	α (cm^2/s)	Pr
4.22	16.9	9.78	1.25×10^{-6}	7.39×10^{-4}	0.011	6.43×10^{-4}	1.15
7	7.53	5.71	1.76×10^{-6}	2.34×10^{-3}	0.014	3.21×10^{-3}	0.73
10	5.02	5.41	2.26×10^{-6}	4.49×10^{-3}	0.018	6.42×10^{-3}	0.70
20	2.44	5.25	3.58×10^{-6}	0.0147	0.027	0.0209	0.70
30	1.62	5.22	4.63×10^{-6}	0.0286	0.034	0.0403	0.71
60	0.811	5.20	7.12×10^{-6}	0.088	0.053	0.125	0.70
100	0.487	5.20	9.78×10^{-6}	0.201	0.074	0.291	0.69
200	0.244	5.19	1.51×10^{-5}	0.622	0.118	0.932	0.67
300	0.162	5.19	1.99×10^{-5}	1.22	0.155	1.83	0.67
600	0.0818	5.19	3.22×10^{-5}	3.96	0.251	5.94	0.67
1000	0.0487	5.19	4.63×10^{-5}	9.46	0.360	14.2	0.67

Source: Data collected from Ref. 20.

n-Hydrogen, Gas at $P = 1$ atm

T(K)	ρ (kg/m^3)	c_P (kJ/kg · K)	μ (kg/s · m)	ν (cm^2/s)	k (W/m · K)	α (cm^2/s)	Pr
250	0.0982	14.04	7.9×10^{-6}	0.804	0.162	1.17	0.69
300	0.0818	14.31	8.9×10^{-6}	1.09	0.187	1.59	0.69
350	0.0702	14.43	9.9×10^{-6}	1.41	0.210	2.06	0.69
400	0.0614	14.48	1.09×10^{-5}	1.78	0.230	2.60	0.68
500	0.0491	14.51	1.27×10^{-5}	2.59	0.269	3.78	0.68
600	0.0408	14.55	1.43×10^{-5}	3.50	0.305	5.12	0.68
700	0.0351	14.60	1.59×10^{-5}	4.53	0.340	6.62	0.68

Source: Constructed based on the data compiled in Refs. 17 and 18.

Nitrogen, Gas at $P = 1$ atm

T(K)	ρ (kg/m^3)	c_P (kJ/kg · K)	μ (kg/s · m)	ν (cm^2/s)	k (W/m · K)	α (cm^2/s)	Pr
77.33	4.612	1.123	5.39×10^{-6}	0.0117	0.0076	0.0147	0.80
100	3.483	1.073	6.83×10^{-6}	0.0197	0.0097	0.0261	0.76
200	1.711	1.044	1.29×10^{-5}	0.0754	0.0185	0.103	0.73
300	1.138	1.041	1.78×10^{-5}	0.156	0.0259	0.218	0.72
400	0.854	1.045	2.20×10^{-5}	0.258	0.0324	0.363	0.71
500	0.683	1.056	2.58×10^{-5}	0.378	0.0386	0.535	0.71
600	0.569	1.075	2.91×10^{-5}	0.511	0.0442	0.722	0.71
700	0.488	1.098	3.21×10^{-5}	0.658	0.0496	0.925	0.71

Source: Data collected from Refs. 21.

Oxygen, Gas at $P = 1$ atm

T(K)	ρ (kg/m^3)	c_P (kJ/kg · K)	μ (kg/s · m)	ν (cm^2/s)	k (W/m · K)	α (cm^2/s)	Pr
250	1.562	0.915	1.79×10^{-5}	0.115	0.0226	0.158	0.73
300	1.301	0.920	2.07×10^{-5}	0.159	0.0266	0.222	0.72
350	1.021	0.929	2.34×10^{-5}	0.229	0.0305	0.321	0.71
400	0.976	0.942	2.58×10^{-5}	0.264	0.0343	0.372	0.71
500	0.780	0.972	3.03×10^{-5}	0.388	0.0416	0.549	0.71
600	0.650	1.003	3.44×10^{-5}	0.529	0.0487	0.748	0.71
700	0.557	1.031	3.81×10^{-5}	0.684	0.0554	0.963	0.71

Source: Constructed based on the data compiled in Refs. 17 and 18.

Steam at $P = 1$ bar

T(K)	ρ (kg/m^3)	c_P (kJ/kg · K)	μ (kg/s · m)	ν (cm^2/s)	k (W/m · K)	α (cm^2/s)	Pr
373.15	0.596	2.029	1.20×10^{-5}	0.201	0.0248	0.205	0.98
400	0.547	1.996	1.32×10^{-5}	0.241	0.0268	0.246	0.98
450	0.485	1.981	1.52×10^{-5}	0.313	0.0311	0.324	0.97
500	0.435	1.983	1.73×10^{-5}	0.398	0.0358	0.415	0.96
600	0.362	2.024	2.15×10^{-5}	0.594	0.0464	0.633	0.94
700	0.310	2.085	2.57×10^{-5}	0.829	0.0581	0.899	0.92
800	0.271	2.151	2.98×10^{-5}	1.10	0.0710	1.22	0.90
900	0.241	2.219	3.39×10^{-5}	1.41	0.0843	1.58	0.89
1000	0.217	2.286	3.78×10^{-5}	1.74	0.0981	1.98	0.88
1200	0.181	2.43	4.48×10^{-5}	2.48	0.130	2.96	0.84
1400	0.155	2.58	5.06×10^{-5}	3.27	0.160	4.00	0.82
1600	0.135	2.73	5.65×10^{-5}	4.19	0.210	5.69	0.74
1800	0.120	3.02	6.19×10^{-5}	5.16	0.330	9.10	0.57
2000	0.108	3.79	6.70×10^{-5}	6.20	0.570	13.94	0.45

Source: Constructed based on data compiled in Refs. 17 and 18.

Ideal Gas Constants and Specific Heats[a]

Gas		M (kg/kmol)	R (kJ/kg · K)	c_P (kJ/kg · K)	c_v (kJ/kg · K)
Air, dry	–	28.97	0.287	1.005	0.718
Argon	Ar	39.944	0.208	0.525	0.317
Carbon dioxide	CO_2	44.01	0.189	0.846	0.657
Carbon monoxide	CO	28.01	0.297	1.040	0.744
Helium	He	4.003	2.077	5.23	3.15
Hydrogen	H_2	2.016	4.124	14.31	10.18
Methane	CH_4	16.04	0.518	2.23	1.69
Nitrogen	N_2	28.016	0.297	1.039	0.743
Oxygen	O_2	32.000	0.260	0.918	0.658
Water vapor	H_2O	18.016	0.461	1.87	1.40

Source: After Ref. 22.

[a]The c_P and c_v values correspond to the temperature 300 K. This ideal gas model is valid at low and moderate pressures ($P \lesssim 1$ atm).

REFERENCES

1. A. Bejan, *Convection Heat Transfer*. New York: Wiley, 1984.
2. E. R. G. Eckert and R. M. Drake, *Analysis of Heat and Mass Transfer*. New York: McGraw-Hill, 1972.
3. Y. S. Touloukian and C. Y. Ho, eds., *Thermophysical Properties of Matter*. New York: Plenum, 1972.

4. K. Raznjevic, *Handbook of Thermodynamic Tables and Charts*. Washington, DC: Hemisphere, 1976.
5. U. Grigull and H. Sandner, *Heat Conduction* (Appendix E). Translated by J. Kestin, Washington, DC: Hemisphere, 1984.
6. F. Kreith and W. Z. Black, *Basic Heat Transfer*, (Appendix E), New York: Harper & Row, 1980.
7. J. H. Lienhard, *A Heat Transfer Textbook*, 2nd ed. (Appendix A). Englewood Cliffs, NJ: Prentice-Hall, 1987.
8. L. C. Witte, P. S. Schmidt and D. R. Brown, *Industrial Energy Management and Utilization*, New York: Hemisphere, 1988.
9. M. S. Qashou, R. I. Vachon, and Y. S. Touloukian. Thermal conductivity of foods, *ASHRAE Trans.*, Vol. 78, Pt. 1, 1972, pp. 165–183.
10. R. Dickerson Jr. and R. B. Reed Jr., Thermal diffusivity of meats, *ASHRAE Trans.*, Vol. 81, 1975, pp. 356–364.
11. A. Bejan, I. Dincer, S. Lorente, A. F. Miguel, and A. H. Reis, *Porous and Complex Flow Structures in Modern Technologies*. New York: Springer-Verlag, 2004.
12. D. A. Nield and A. Bejan, *Convection in Porous Media*, 2nd ed. New York: Springer-Verlag, 1999.
13. A. E. Scheidegger, *The Physics of Flow through Porous Media*. Toronto: University of Toronto Press, 1974.
14. A. Bejan and J. L. Lage, Heat transfer from a surface covered with hair. In S. Kakac, B. Kilkis, and F. A. Kulacki, eds., *Convective Heat and Mass Transfer in Porous Media*, Dordrecht, The Netherlands: Kluwer Academic, 1991.
15. A. Bejan, *Convection Heat Transfer*, 3rd ed. Hoboken, NJ: Wiley, 2004.
16. G. K. Batchelor, *An Introduction to Fluid Dynamics*. Cambridge: Cambridge University Press, 1967.
17. D. W. Green and J. O. Maloney, eds., *Perry's Chemical Engineers' Handbook*, 6th ed. (pp. 3–1 to 3.263). New York: McGraw-Hill, 1984.
18. P. E. Liley, Thermophysical properties. In S. Kakac, R. K. Shah, and W. Aung, eds., *Handbook of Single-Phase Convective Heat Transfer* (Ch. 22). New York: Wiley, 1987.
19. N. B. Vargaftik, *Tables on the Thermophysical Properties of Liquids and Gases*, 2nd ed. Washington, DC: Hemisphere, 1975.
20. R. D. McCarty, Thermophysical properties of Helium-4 from 2 to 1500 K with pressures to 1000 atmospheres. NBS TN 631, National Bureau of Standards, Washington, DC, November 1972.
21. R. T. Jacobsen, R. B. Stewart, R. D. McCarty, and H. J. M. Hanley, Thermophysical properties of nitrogen from the fusion line to 3500 R (1944 K) for pressures to 150,000 psia (10342×10^5 N/m^2). NBS TN 648, National Bureau of Standards, Washington, DC, December 1973.
22. A. Bejan, *Advanced Engineering Thermodynamics*, 3rd ed. Hoboken, NJ: Wiley, 2006.

Author Index

Subject Index

p 141 fig 4.15 the "l" appears typeset as a "1"
fig 4.15 could use A indicated